Ylid Chemistry

ORGANIC CHEMISTRY

A SERIES OF MONOGRAPHS

Edited by

ALFRED T. BLOMQUIST

Department of Chemistry, Cornell University, Ithaca, New York

Volume 1. Wolfgang Kirmse. CARBENE CHEMISTRY. 1964

Volume 2. Brandes H. Smith. BRIDGED AROMATIC COMPOUNDS. 1964

Volume 3. Michael Hanack. CONFORMATION THEORY. 1965

Volume 4. Donald J. Cram. FUNDAMENTALS OF CARBANION CHEMISTRY. 1965

Volume 5. Kenneth B. Wiberg (Editor).
OXIDATION IN ORGANIC CHEMISTRY
PART A. 1965
PART B. *In preparation.*

Volume 6. R. F. Hudson. STRUCTURE AND MECHANISM IN ORGANO-PHOSPHORUS CHEMISTRY. 1965

Volume 7. A. William Johnson. YLID CHEMISTRY. 1966

IN PREPARATION

Jan Hamer (Editor). 1,4-CYCLOADDITION REACTIONS

M. P. Cava and M. J. Mitchell. CYCLOBUTADIENE AND RELATED COMPOUNDS

YLID CHEMISTRY

A. WILLIAM JOHNSON

Department of Chemistry
University of Saskatchewan
Regina, Saskatchewan, Canada

ACADEMIC PRESS New York and London 1966

ACADEMIC PRESS INC.
111 Fifth Avenue, New York, New York 10003

United Kingdom Edition published by
ACADEMIC PRESS INC. (LONDON) LTD.
Berkeley Square House, London W.1

LIBRARY OF CONGRESS CATALOG CARD NUMBER: 66-14471

PRINTED IN THE UNITED STATES OF AMERICA

This book is dedicated to an inspiring teacher, Reuben Benjamin Sandin, University of Alberta, Alberta, Canada

PREFACE

The first ylid was prepared and isolated by Michaelis and Gimborn in 1894. In the next sixty years approximately five hundred articles appeared in the chemical literature concerning the chemical and physical properties of ylids. Much of this activity occurred within the last decade, and probably was stimulated by the development of the Wittig synthesis of olefins which involves the reaction between phosphonium ylids and carbonyl compounds.

Several review articles have appeared which deal with the Wittig reaction, and in most of these the chemistry of phosphonium ylids is discussed in varying detail but only as incidental to the major topic. There have not been any reviews covering other types of ylids. Therefore, it now seemed appropriate, in view of the development of ylid chemistry, to gather together and analyze all aspects of ylid chemistry. This monograph represents such an effort. The foremost goals of this work have been to provide a complete survey of ylid chemistry, to set the field in order, and to make more apparent the significant problems yet to be tackled. Some of the problems yet to be solved when this manuscript was completed have been tackled in the interim. Appendixes elucidating this later work have been added to the end of several chapters. By this means literature coverage of significant work in ylid chemistry is complete through January 1966. An attempt has been made to analyze in a critical manner the data on ylids where appropriate. In many instances, however, too little data are available, and detailed mechanistic speculation at this point would be idle. A special effort has been made to indicate profitable areas of research.

Acknowledgment must be rendered to the University of North Dakota (where most of this volume was written) for the use of their facilities, and to Cornell University for the use of their library during the summer of 1964. Appreciation must be extended to my graduate and undergraduate research students whose questions and discussion helped clarify many points. An acknowledgment must also be made to my wife who proofread the entire manuscript, and to the rest of my family for their patience and understanding which permitted the completion of this task.

A. WILLIAM JOHNSON

Regina, Saskatchewan, Canada
May, 1966

CONTENTS

Preface vii

1. Introduction 1
References 4

Part 1
YLIDS OF PHOSPHORUS

2. Introduction to Phosphorus Ylids 7
References 14

3. Phosphonium Ylids 16
I. Preparation of Phosphonium Ylids 17
II. Structure and Physical Properties of Phosphonium Ylids 61
III. Reactions of Phosphonium Ylids 88
Appendix 123
References 125

4. The Wittig Reaction 132
I. Experimental Conditions of the Wittig Reaction 134
II. Scope and Limitations of the Wittig Reaction 138
III. Applications of the Wittig Reaction 145
IV. Mechanism of the Wittig Reaction 152
V. Stereochemistry of the Wittig Reaction 171
Appendix 187
References 189

5. Other Phosphorus Ylids 193
I. Phosphinoxy Carbanions 193
II. Phosphonate Carbanions 203
III. Miscellaneous Phosphorus Carbanions 212
References 215

6. Iminophosphoranes 217

I. Preparation of Iminophosphoranes 217
II. Reactions of Iminophosphoranes 222
III. Iminotrihalophosphoranes 230
IV. Oxygenated Iminophosphoranes 233
V. The Structure of Iminophosphoranes 236
VI. Phosphinazines 238
References 245

Part 2

YLIDS OF OTHER HETEROATOMS

7. Nitrogen Ylids 251

I. Ammonium Ylids 253
II. Pyridinium Ylids 260
III. Nitrogen Imines 266
IV. Diazo Compounds 270
V. Rearrangements via Nitrogen Ylids 273
VI. The Role of Nitrogen Ylids in the Hofmann Degradation 277
References 281

8. Arsenic and Antimony Ylids 284

I. Introduction 284
II. Arsonium Ylids 288
III. Iminoarsenanes 299
IV. Stibonium Ylids and Imines 301
References 303

9. Sulfur Ylids 304

I. Introduction 304
II. Sulfonium Ylids 310
III. Other Sulfur Ylids 344
IV. Iminosulfuranes 356
Appendix 360
References 362

Author Index 367

Subject Index 381

1

INTRODUCTION

An ylid can be defined as a substance in which a carbanion is attached directly to a heteroatom carrying a high degree of positive charge—represented by the general formula I. This definition is intended

$$>\overset{\ominus}{\ddot{C}}-\overset{\oplus}{X}$$

I

to include those resonance hybrid molecules in which there is an important contributing structure which meets the original definition. Therefore, ylids may have an enolate structure (II). The definition also

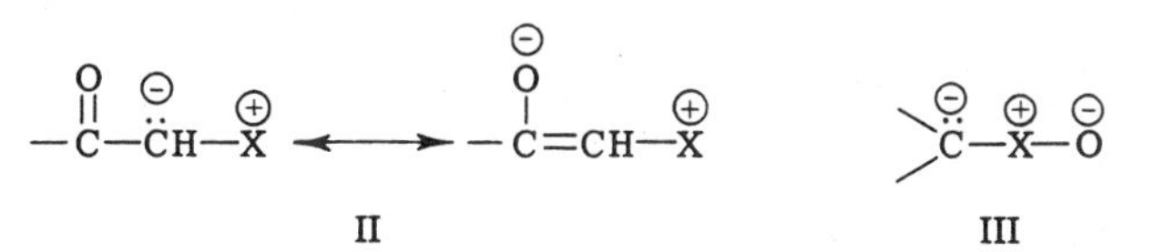

II III

includes those molecular systems whose heteroatoms carry less than a formal full positive charge—structures such as III. According to this definition an ylid is but a special type of zwitterion or betaine, and they have been called such upon occasion. However, these terms should be reserved for those doubly but oppositely charged species, such as the zwitterionic form of amino acids, in which the formal charges are not on adjacent atoms.

The special characteristics of ylids that make them worthy of study in their own right is the unique stabilization afforded the carbanions by the presence of the adjacent 'onium atom group. Thus, many ylids have been isolated as crystalline, stable substances whereas normal carbanions are seldom isolable and are very reactive toward atmospheric components.

This book is not concerned just with ylids as defined above but is concerned, in addition, with those substances that resemble ylids in their chemical and physical properties and/or by virtue of their molecular structure. Therefore, an important part of this book consists of discussion of the chemistry of 'onium imines of general structure IV. These sub-

$$\overset{\ominus}{\ddot{\underset{\cdot\cdot}{N}}}\!\!-\!\overset{\oplus}{X}$$

$$-\overset{\ominus}{\underset{\cdot\cdot}{\ddot{N}}}-\overset{\oplus}{X}$$

IV

stances are isoelectronic with ylids, and the two classes of compounds exhibit a remarkable similarity in their properties.

The term "ylid" was first coined in the German language by George Wittig in 1944 (*1*). It was derived by use of the ending -yl to imply an open valence (i.e., meth*yl*) and the ending -id to imply anionicity (i.e., acetyl*id*) both on a carbon atom. Later, Wittig and his students also used the term "ylene," especially with reference to phosphonium ylids to imply the formation of a double bond between the phosphorus atom and the carbanionic carbon (*2*). The use of the latter term has not grown and it is recommended that it be discarded completely since it implies a structural characteristic that varies from ylid to ylid and has not been explicitly evaluated for any ylid system to date. Thus, its implication is too arbitrary for meaningful scientific use.

The term "ylid" most often has been translated into the English form "ylide." This author does not agree with the addition of the "e" ending and it will not be used in this book. He recommends that it not be used in the literature. The use of such a spelling would necessitate a pronunciation that is not used in fact (i.e., ylīde as in hīde). The spoken pronunciation usually is "ylĭd" as in "hĭd." Furthermore, dropping the "e" saves space.

A general term, "stabilized ylid," has been used throughout the literature and is taken to imply an ylid which can be isolated, purified, usually stored in the atmosphere and used in a subsequent experiment. Such ylids usually are those with powerful electron-withdrawing groups attached to the ylid carbanion.

Ylids undergo two basic types of reactions, those in which only the carbanion is involved mechanistically and those in which both the carbanion and the heteroatom portion are involved. The former group consists basically of those reactions which any carbanion, regardless of structure, would undergo. The presence of the heteroatom portion of the ylid usually is reflected only in its effect on the nucleophilicity exhibited by the carbanion. The usefulness of ylids in this type of reaction is due mainly to their availability in a wide variety of structural environments. Since carbanion reactions inevitably are those which permit the formation of new carbon-carbon bonds the availability of almost any carbanion without worry of isomeric possibilities has been a boon to synthetic organic chemistry. The most interesting reactions of ylids are the second group which involve both the carbanion and the heteroatom portion. The Wittig reaction falls into this category and the discovery of this and related reactions incited the burst of activity in the field of ylid chemistry during the last decade. Studies of the mechanisms of these reactions and of the physical properties of ylids have evoked interest in and provided a substrate for the study of valence shell expansion by elements of the second and lower periods.

As mentioned in the preface, the first phosphonium ylid seems to have been prepared in 1894 by Michaelis and Gimborn (*3*). This appeared to be an isolated event, however, and the first flurry of activity in the field of ylids and the related imines occurred in the early 1920's in Staudinger's laboratory (*4*). His work almost exclusively was on phosphonium systems. In the mid-1930's Krohnke (*5*) began work on a series of pyridinium ylids, efforts which have continued to the present day. Wittig and his students (*6*) began by working with ammonium ylids in the 1940's in the course of attempts to obtain pentavalent nitrogen compounds. He followed the same procedure with phosphonium compounds in the early 1950's (*7*) and developed the Wittig reaction of phosphonium ylids with carbonyl compounds in 1953. This event signaled the beginning of tremendous activity in organophosphorus chemistry in general which has not ceased. Sulfur ylids and arsenic ylids have been studied only in recent years, mainly as an extrapolation of phosphorus chemistry (*8*). The preceding is the outline of the general advance of the ylid field but it must be pointed out that there were individuals, usually far ahead of their time, who reported isolated work on ylids. C. K. Ingold (*9*) and C. S. Marvel (*10*) are but two examples.

This book is divided into two parts. Part I deals with the chemistry of all types of phosphorus ylids and includes discussion of the Wittig reaction and iminophosphoranes. Part II includes discussions of nitrogen, arsenic, antimony and sulfur ylids and imines.

Before commencing the discussion of ylid chemistry it would be appropriate to comment on the nomenclature problems involved. Phosphonium ylids have been named as phosphoniumalkylides, phosphinemethylenes and, more recently, as phosphoranes. For example, V has

$$\overset{\ominus}{C}H_2-\overset{\oplus}{P}(C_6H_5)_3 \longleftrightarrow CH_2{=}P(C_6H_5)_3$$

V

been named triphenylphosphoniummethylide, triphenylphosphinemethylene and methylenetriphenylphosphorane. The latter name, a derivative of PH_5, phosphorane, is the recommended form (*11*) and will be used throughout this book. However, the use of an alternate form in some cases is suggested at this time because to use the phosphorane name one assumes that it always will be rather simple to identify the carbanionic portion of an ylid. This is not necesarily the case as examination of the recent literature will indicate. Therefore, it is proposed that the $R^1R^2R^3P{=}$ group be named a phosphoranyl group and considered capable of substitution on any carbon skeleton. It seems much easier and more consistent to name the simple heteroatom group as a substituent on a complex carbon skeleton for which a nomenclature system already has been well developed than to attempt the reverse. Thus, VI could

$$\begin{matrix} CH_3CH_2CH_2 \diagdown \\ \quad C{=}P(C_6H_5)_3 \\ CH_3CH_2 \diagup \end{matrix}$$

VI

better be named 3-triphenylphosphoranylhexane than as triphenyl(ethylpropyl)methylenephosphorane.

It is also proposed that analogous nomenclature rules be adopted for sulfur and arsenic ylids. Therefore VII would be named triphenylethylidenearsenane or triphenylarsenanylethane and VIII would be named phenacylidenediphenylsulfurane or ω-diphenylsulfuranylacetophenone.

$$CH_3CH{=}As(C_6H_5)_3 \qquad C_6H_5COCH{=}S(C_6H_5)_2$$

VII VIII

These systems of nomenclature will be used throughout this book and it is hoped they will be adopted and used in the chemical literature.

REFERENCES

1. G. Wittig and G. Felletschin, *Ann.* **555,** 133 (1944).
2. G. Wittig, *Angew. Chem.* **68,** 505 (1956).
3. A. Michaelis and H. V. Gimborn, *Ber. deut. chem. Ges.* **27,** 272 (1894).
4. H. Staudinger and J. Meyer, *Helv. Chim. Acta* **2,** 619 (1919).
5. F. Krohnke, *Ber. deut. chem. Ges.* **68,** 1177 (1935).
6. G. Wittig and M. H. Wetterling, *Ann.* **557,** 193 (1947).
7. G. Wittig and M. Rieber, *Ann.* **562,** 177 (1949).
8. A. Wm. Johnson and R. B. LaCount, *Chem. & Ind.* (*London*) 1440 (1958); G. Wittig and H. Laib, *Ann.* **580,** 57 (1953).
9. C. K. Ingold and J. A. Jessop, *J. Chem. Soc.* 2357 (1929), 713 (1930).
10. D. D. Coffmann and C. S. Marvel, *J. Am. Chem. Soc.* **51,** 3496 (1929).
11. Editorial Report on Phosphorus Nomenclature, *J. Chem. Soc.* 5122 (1952).

PART I

YLIDS OF PHOSPHORUS

2

INTRODUCTION TO PHOSPHORUS YLIDS

Phosphorus ylids have a general structure often written as a resonance hybrid [2.1]. The minimum structural requirement for a phos-

$$\overset{\oplus}{\underset{/}{\overset{\diagdown}{-P}}}-\overset{\ominus}{\ddot{C}}\!\!\begin{smallmatrix}\diagup\\ \diagdown\end{smallmatrix} \longleftrightarrow \underset{/}{\overset{\diagdown}{-P}}=C\!\!\begin{smallmatrix}\diagup\\ \diagdown\end{smallmatrix} \qquad [2.1]$$

phorus ylid is that it contains an anionic carbon attached to a phosphorus atom which carries a high degree of positive charge. There is a variety of phosphorus structures which meet this minimum requirement but the most common is the phosphonium ylid, $R_3P{=}C\langle$. Many, but not all, of the conceivably applicable phosphorus groups have been incorporated into ylid structures and will be discussed in this book.

As would be expected on the basis of the above general formula for phosphorus ylids these substances are reactive and, unless special structural features have been incorporated, are usually not capable of isolation. However, the number of isolable phosphorus ylids has risen sharply in recent years and permitted a detailed study of the ylids as pure substances.

Phosphorus ylids have a history reaching back to the 1890's but virtually nothing was known of their chemistry until the 1950's with the exception of the results of a brief study in the 1920's. Michaelis and Gimborn (*1*) appear to have prepared the first phosphonium ylid, $(C_6H_5)_3P{=}CH{-}COOC_2H_5$, in 1894 although they proposed a different structure for the substance. They obtained the ylid by treating an aqueous solution of triphenyl(carbethoxymethyl)phosphonium chloride with cold potassium hydroxide solution. Aksnes (*2*) subsequently has confirmed the ylid structure of the product. In 1899 Michaelis and Kohler (*3*) reported the preparation of two additional "phosphonium betaines" which subsequently have been shown to be ylids (*4*, Ramirez and Dershowitz). It apparently was fortuitous that Michaelis chose to use phosphonium salts that produced the stable and isolable ylids, $(C_6H_5)_3P{=}CH{-}COCH_3$ and $(C_6H_5)_3P{=}CH{-}COC_6H_5$.

Staudinger and his students studied the chemistry of a variety of substances closely related to ylids but actually obtained only one ylid, benzhydrylidenetriphenylphosphorane, and that in 1919 (5) by pyrolysis of the corresponding phosphinazine. Staudinger was the first to examine the reactions of ylids, and his pioneering experiments laid the groundwork for the important synthetic applications of phosphonium ylids developed much later by Wittig and his students.

During the next thirty years there only was sporadic activity in phosphorus ylid chemistry. Worrall (*6*) and Schonberg and Ismail (*7*) unknowingly appear to have prepared ylids. Pinck and Hilbert (*8*), however, were the first methodically to undertake and complete the synthesis of an isolable phosphonium ylid, fluorenylidenetriphenylphosphorane (I), using the now well-established "salt method." The awakening of the latent field of phosphorus ylid chemistry, however, was triggered by the observation of Wittig and Rieber in 1949 (*9*) that treatment of tetramethylphosphonium salts with phenyllithium led to the formation of a yellow solution which contained methylenetrimethylphosphorane, $(CH_3)_3P{=}CH_2$. The importance of the chemistry of phosphorus ylids later was assured by Wittig and Geissler (*10*) through their observation that methylenetriphenylphosphorane reacted with benzophenone to form 1,1-diphenylethene and triphenylphosphine oxide [2.2]. This experiment

$$(C_6H_5)_3P{=}CH_2 + (C_6H_5)_2C{=}O \longrightarrow (C_6H_5)_2C{=}CH_2 + (C_6H_5)_3PO \qquad [2.2]$$

signaled the birth of the Wittig reaction, a novel method for the conversion of carbonyl groups into olefinic functions. It also altered the role of ylids, moving them from the realm of chemical curiosities into the arsenal of important synthetic tools.

The fact that phosphorus ylids exist at all and that some have sufficient stability to be capable of isolation has been attributed to the structural and electronic factors which contribute to stabilization of the ylidic carbanion. This stabilization has been thought to result from delocalization of the non-bonded electrons of the carbanion. In a given ylid, $X^+{-}C^-R_2$, stabilization for the carbanion could be afforded by both the heteroatom portion (X) and the two carbanion substituents (R). From the discussion that will follow in later chapters it becomes apparent that the ability of the groups R to delocalize the carbanionic electrons does affect the stability of the ylid. However, it is equally apparent that this stabilization is not sufficient in itself to account for the unique stability of phosphorus ylids. The phosphorus atom itself must play an important role in the stabilization of the carbanion.

These relationships perhaps can be clarified by an examination of several compounds in the fluorene series. It is well known that the fluorenyl anion (III) can be formed from fluorene by treatment with a

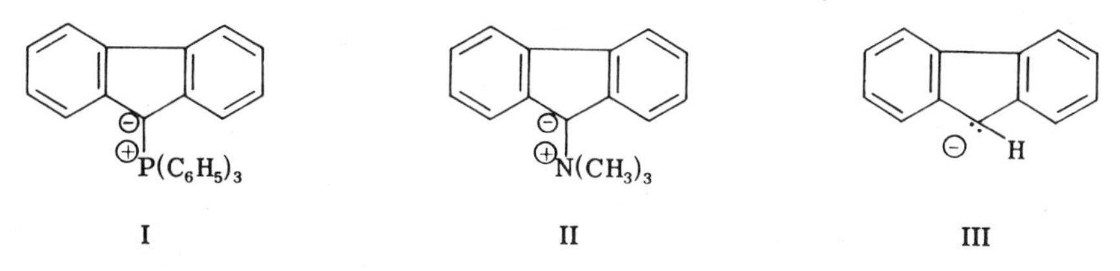

variety of bases such as butyllithium. The conjugate acid, fluorene, has a pK_a of about 25 (*11*, McEwen). By comparison, the conjugate acid of the nitrogen ylid (II) appears to have a pK_a only a little less than that of fluorene (*12*, Wittig and Felletschin). In other words, the inductive effect of the trimethylammonium group does not alter the stability of the fluorenyl anion very much. On the other hand, the pK_a of the conjugate acid of the phosphonium ylid (I) was less than 10 (*13*, Johnson and LaCount), the proton being removed by dilute aqueous ammonia. It is apparent that the phosphorus group exerts an effect above and beyond any inductive effect which may be stabilizing the carbanion. This stabilization effect has been attributed to the use of the vacant 3*d*-orbitals of the phosphorus atom, the carbanion taking advantage of the ability of the phosphorus atom to expand its outer shell to accommodate more than eight electrons.

There is considerable experimental evidence indicating that phosphorus can use its 3*d*-orbitals in σ bonding and that a phosphorus atom can be pentavalent. Pentavalent phosphorus atoms were proposed at an early date by Staudinger and Meyer (*5*) to account for the formation of pentaphenylmethylenephosphorane, $(C_6H_5)_2C{=}P(C_6H_5)_3$, and by Meyer (*14*) to account for the properties of a phosphonium betaine (IV).

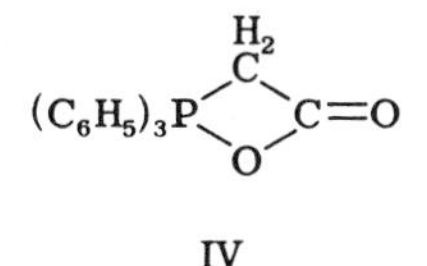

IV

The ability of phosphorus to expand its octet and become pentavalent is more clearly recognizable in a compound such as phosphorus pentachloride. This substance has been shown to exist as a trigonal bipyramid structure in the liquid (*15*, Moureu *et al.*) and in the vapor phase (*16*, Rouault), consistent with the proposal of sp^3d hybridization for the phosphorus atom. In the solid phase, PCl_5 exists as a salt of the tetrahedral PCl_4 cation (sp^3) and the octahedral PCl_6 anion (sp^3d^2) (*17*, Clark *et al.*). It is obvious that the penta- and hexavalent phosphorus atoms must use the 3*d*-orbitals in σ bonding. A more recent and most interesting example is pentaphenylphosphorane, $(C_6H_5)_5P$, prepared by Wittig and his students (*18*). X-ray crystallographic examination of this substance recently has shown that it too possesses a trigonal bipyramid structure (*19*, Wheatley) although infrared evidence is claimed to favor a square pyramid structure (*20*, Degani *et al.*). In the above examples valence shell expansion is certainly involved in the construction of the σ bond framework about phosphorus.

A much more difficult problem is met in attempting to determine whether or not the 3*d*-orbitals of phosphorus can be used in multiple bonding between phosphorus and another atom. It should be mentioned at the outset of this discussion that it appears virtually impossible to *prove* that *d*-orbitals are involved in π bonding. Nonetheless, a large body of experimental data is accounted for by such a proposal and until it can be proven to be inadequate its use is certainly justified. Several examples of the application of this proposal will be presented.

It has been shown that the basicity of a series of trivalent phosphorus compounds correlates well with the σ bonding ability of these phosphines in systems where π bonding is not possible (*21*, Meriwether and Fiene). The order of this bonding ability was $PR_3 > P(OR)_3 > PCl_3 > PF_3$. On the other hand, the bonding ability of these same phosphorus derivatives in complexes of metals with filled *d*-orbitals was in exactly the opposite order. This was shown by Chatt and Williams (*22*) in studies of the "*trans* effect" in $(PF_3)_2PtCl_2$ and by Meriwether and Fiene (*21*) in studies of the carbonyl stretching frequencies of the nickel complexes, $Ni(CO)_2(PX_3)_2$. These observations have led to the proposal that the bonding in such metal complexes involves the ligand (PX_3) sharing its non-bonded electrons with the metal to form the σ bond and the filled *d*-orbitals of the metal atom overlapping with the vacant *d*-orbitals of the ligand to form a π bond. A ligand such as PF_3 would be expected to form a weak σ bond but a fairly strong π bond. On the other hand, a ligand such as $P(CH_3)_3$ would be expected to form a strong σ bond but a weaker π bond. The proposed involvement of the 3*d*-orbitals of the phosphorus atom seems to account reasonably well for the bonding characteristics in such metal complexes.

It is well known that the addition of nucleophiles to unactivated olefins fails unless the adding nucleophile is very strong. In a particularly pertinent experiment Doering and Schreiber (*23*) found that under no circumstances could bases (oxygen, sulfur or carbon bases) be forced to participate in a Michael-type addition to vinyltrimethylammonium bromide. In contrast there have been several reports of the addition of nucleophiles to vinylphosphorus systems. Pudovik and Imaev (*24*) reported the addition of thiolate anions to vinyldibutylphosphonate, $CH_2=CH—P(O)(OC_4H_9)_2$, and Kabachnik and coworkers (*25*) reported the addition of a variety of nucleophiles to vinyldiphenylphosphine oxide. More recently, Keough and Grayson (*26*) have reported analogous additions to vinyltri-*n*-butylphosphonium bromide while Schweizer and Bach (*27*) have reported analogous experiments with vinyltriphenylphosphonium bromide.

The difference between the vinylammonium ion and the various vinylphosphorus systems mentioned above can be attributed to the ability of the phosphorus systems to stabilize the carbanionic intermediate of the Michael addition by overlap of the filled 2*p*-orbital of the carbon atom with the vacant 3*d*-orbital of the phosphorus atom (i.e., V). Such

$$\overset{\ominus}{B:} + CH_2{=}CH{-}\overset{\oplus}{P}R_3 \longrightarrow \left[B{-}CH_2{-}\overset{\ominus}{\ddot{C}}H{-}\overset{\oplus}{P}R_3 \longleftrightarrow B{-}CH_2{-}CH{=}PR_3 \right] \xrightarrow{H^{\oplus}} B{-}CH_2{-}CH_2{-}\overset{\oplus}{P}R_3 \qquad [2.3]$$

V

valence shell expansion appears impossible for the nitrogen analog since the next highest vacant orbital is the 3*s*-orbital, and this is of too high energy for effective bonding. Clearly, the relative coulombic effects of the ammonium and phosphonium groups are such that the ammonium group should be able to stabilize better an adjacent carbanion were electrostatic interactions dominant. The carbon-nitrogen bond (1.47 Å) is considerably shorter than the carbon-phosphorus bond (1.87 Å) and should exert about 30% *more* coulombic attraction (*28*, Huggins). Other polarizability effects may favor phosphorus stabilization of the anion but it is doubtful whether they could outweigh the coulombic effect. It is apparent that yet another effect must afford considerable stabilization for the phosphonium carbanion, and that effect is proposed to be the ability of the phosphorus atom to expand its valence shell using the vacant, low-energy 3*d*-orbitals for overlap with the filled 2*p*-orbitals of the carbanion in a form of π bonding (V).

Doering and Hoffmann (*29*) have provided a quantitative estimate of the importance of valence shell expansion of phosphorus in providing stabilization for an adjacent carbanion. They found that tetramethylphosphonium iodide incorporated 73.9 atom percent of deuterium when treated with a solution of deuteroxide anion in deuterium oxide for three hours at 62°. Under the same conditions, except for 504 hours reaction time, tetramethylammonium iodide underwent no observable exchange and raising the temperature to 100° for 358 hours led to the incorporation of only 1.13 atom percent of deuterium. The heats of activation for the ammonium and phosphonium cases were 32.2 and 25.6 kcal/mole, respectively. The authors predicted that in the absence of any resonance effect in the stabilization of the ylidic intermediates and assuming that coulombic effects were the only energy factor operating, the heat of activation for the phosphonium system would have been approximately 41 kcal/mole since the phosphorus–carbon bond is longer than the nitrogen–carbon bond. Thus, the effective stabilization of the phosphonium carbanion was actually in the vicinity of 15 kcal/mole, rather than the 6.6 kcal/mole difference in the observed heats. This stabilization was attributed to valence shell expansion of the phosphorus atom with concomitant overlap of the vacant 3*d*-orbitals with the filled 2*p*-orbitals of the carbanion intermediate (VI).

$$(CH_3)_3\overset{\oplus}{P}-CH_3 \xrightarrow{\overset{\ominus}{O}D} \left[\begin{array}{c} (CH_3)_3\overset{\oplus}{P}-\overset{\ominus}{C}H_2 \\ \text{VI} \updownarrow \\ (CH_3)_3P=CH_2 \end{array} \right] \xrightarrow{D_2O} (CH_3)_3\overset{\oplus}{P}-CH_2D \qquad [2.4]$$

VII

The difference in the entropies of activation for the ammonium and phosphonium cases also were significant, the values being —15 and +4 cal/degree, respectively. This difference certainly would account for a fair measure of the observed difference in the rates of deuterium incorporation. The difference also is in the direction expected if valence shell expansion of the phosphorus atom were an important factor in the stabilization of the carbanionic intermediate. If the resonance form VII was an important contributor to the structure of the intermediate a release of solvent would be expected to accompany conversion of the ionic salt into the less polar ylid intermediate (VI $\leftrightarrow$ VII). This should lead to a much lower entropy of activation than for the ammonium case as was observed.

In all of the above examples it is apparent that the proposed ability of the phosphorus atom to expand its outer valence shell through use of the low energy, vacant 3*d*-orbitals permits a rational explanation of the experimental data. It should be emphasized again that these examples do not *prove* the existence of such a phenomenon but they are certainly consistent with the proposal. Accordingly, we propose to account for the properties of phosphorus ylids on the basis that the ylid carbanion is stabilized to an appreciable extent by overlap of the filled $2p$-orbital of carbon with the vacant $3d$-orbital of phosphorus. The majority of the people working in the ylid field operate on the assumption that such a proposal is valid.

Several groups have treated the $p\pi$–$d\pi$ bonding problem theoretically. They seem to agree, for the most part, that π bonding between $2p$- and $3d$-orbitals is feasible but they tend to differ in their explanation of how the atom goes about using its vacant d-orbitals. Jaffe (*30*) and Craig and coworkers (*31*) agree that multiple bonding involving overlap between d- and p-orbitals can provide effective stabilization for a system. Furthermore, they have proposed that such bonding will be more effective if the atom carrying the vacant d-orbitals carries either a formal positive charge or very electronegative ligands so that the d-orbitals, which are normally diffuse in the free atom, are contracted in the bonding state. This would permit an appreciable overlap integral and effective bonding. Craig (*32*) has quoted unpublished calculations by Buckingham and Carter on the hypothetical molecule, PH_5, which indicated that the $3d$-orbitals of the free phosphorus atom would be contracted by a factor of from two to three when put in the ligand field of the protons. The chemistry of phosphorus ylids, as it will be discussed in the following chapters, fits into the pattern described above in that one can obtain carbanions rather easily when they are adjacent to positively charged groups capable of valence shell expansion but they are difficult to obtain when they are adjacent to the same heteroatom if it does not carry a positive charge.

In a recent note Cruickshank and Webster (*33*) criticized the Craig approach in the case of the closely related sulfur systems. Doing their calculations using the Hartree-Fock SCF method they claimed it was unnecessary to propose contraction of d-orbitals for effective bonding since the probability distribution function showed a maximum of the $3d$-orbital nearer the nucleus even in the free atom than proposed by Craig (*31*) after orbital contraction. The related calculations for the phosphorus system and especially those involving $p\pi$–$d\pi$ bonding will be awaited with interest.

On the basis of the theoretical conclusions of Craig and coworkers (*31*) and the experimental evidence available it would be predicted that ylids ought to be obtainable from any phosphorus system which carries a hydrogen atom on a carbon atom adjacent to a phosphorus atom carrying a reasonable degree of positive charge. As will be seen in later chapters this prediction has been successfully tested in fair measure using a variety of phosphorus systems ranging from phosphonium groups through phosphine oxides to phosphinates and phosphonates (VIII a–d).

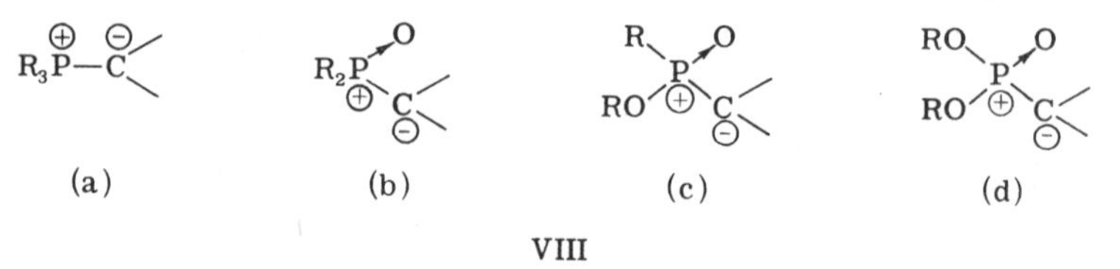

VIII

The initial investigations of phosphorus ylid chemistry concentrated on phosphonium ylids and most of the physical and chemical data to be discussed derives from their chemistry. However, and especially in recent years, more effort has been placed on other phosphorus systems. The chemistry of the various phosphorus ylids will be discussed in detail in the following chapters of Part I.

REFERENCES

1. A. Michaelis and H. V. Gimborn, *Ber. deut. chem. Ges.* **27**, 272 (1894).
2. G. Aksnes, *Acta Chem. Scand.* **15**, 438 (1961).
3. A. Michaelis and E. Kohler, *Ber. deut. chem. Ges.* **32**, 1566 (1899).
4. F. Ramirez and S. Dershowitz, *J. Org. Chem.* **22**, 41 (1957).
5. H. Staudinger and J. Meyer, *Helv. Chim. Acta* **2**, 635 (1919).
6. D. E. Worrall, *J. Am. Chem. Soc.* **52**, 2933 (1930).
7. A. Schonberg and A. F. A. Ismail, *J. Chem. Soc.* 1374 (1940).
8. L. Pinck and G. E. Hilbert, *J. Am. Chem. Soc.* **69**, 723 (1947).
9. G. Wittig and M. Rieber, *Ann.* **562**, 177 (1949).
10. G. Wittig and G. Geissler, *Ann.* **580**, 44 (1953).
11. W. K. McEwen, *J. Am. Chem. Soc.* **58**, 1124 (1936).
12. G. Wittig and G. Felletschin, *Ann.* **555**, 133 (1944).
13. A. Wm. Johnson and R. B. LaCount, *Tetrahedron* **9**, 130 (1960).
14. A. H. Meyer, *Ber. deut. chem. Ges.* **4**, 734 (1871).
15. H. Moureu, M. Mogat, and G. Wetroff, *Compt. rend.* **203**, 257 (1936); **205**, 276, 545 (1937).
16. M. Rouault, *Compt. rend.* **207**, 620 (1938); *Ann. Phys. (Lpz.)* **14**, 78 (1940).
17. D. Clark, H. M. Powell and A. F. Wells, *J. Chem. Soc.* 642 (1942).
18. G. Wittig and M. Rieber, *Ann.* **562**, 187 (1949); G. Wittig and D. Hellwinkel, *Chem. Ber.* **97**, 741 (1964).
19. P. J. Wheatley, *J. Chem. Soc.* 2206 (1964).
20. C. Degani, M. Halmann, I. Laulicht and S. Pinchas, *Spect. Acta* **20**, 1289 (1964).
21. L. S. Meriwether and M. L. Fiene, *J. Am. Chem. Soc.* **81**, 4200 (1959).
22. J. Chatt and A. A. Williams, *J. Chem. Soc.* 3061 (1951).
23. W. von E. Doering and K. C. Schreiber, *J. Am. Chem. Soc.* **77**, 514 (1955).
24. A. N. Pudovik and M. C. Imaev, *Izvest. Akad. Nauk S.S.S.R.* 916 (1952); *Chem. Abstr.* **47**, 10463 (1953).
25. M. I. Kabachnik, T. Y. Medved, Y. M. Polikarpov and K. S. Yudina, *Izvest. Akad. Nauk S.S.S.R.* 1548 (1962).

26. P. T. Keough and M. Grayson, *J. Org. Chem.* **29,** 631 (1964).
27. E. E. Schweizer and R. D. Bach, *J. Org. Chem.* **29,** 1746 (1964).
28. M. L. Huggins, *J. Am. Chem. Soc.* **75,** 4126 (1953).
29. W. von E. Doering and A. K. Hoffmann, *J. Am. Chem. Soc.* **77,** 521 (1955).
30. H. H. Jaffe, *J. Phys. Chem.* **58,** 185 (1954).
31. D. P. Craig and E. A. Magnusson, *J. Chem. Soc.* 4895 (1956); D. P. Craig, A. Maccoll, R. S. Nyholm, L. E. Orgel and L. E. Sutton, *J. Chem. Soc.* 332 (1954).
32. Buckingham and Carter as quoted by D. P. Craig, *Chem. Soc. Special Pub. No.* **12,** 350 (1958).
33. D. W. J. Cruickshank and B. C. Webster, *J. Chem. Phys.* **40,** 3733 (1964).

3

PHOSPHONIUM YLIDS

In the narrowest sense a phosphonium ylid is one which is derived, at least formally so, from a quaternary phosphonium salt, that is, from a phosphorus atom having four carbon atoms attached to it. However, in a broader sense, and in the manner in which we shall use the term, a phosphonium ylid is derived from a quaternary phosphorus group having at least one carbon atom attached to the phosphorus and three other substituents of any type covalently bound to the phosphorus atom. The real distinction between the two definitions is that the latter operational definition includes ylids whose phosphorus atom is attached to *any* three other covalently bound groups, whether the atoms immediately attached to the phosphorus are carbon, oxygen, nitrogen or any other atom. Accordingly, phosphonium ylids have the general structure I with virtually no limitation on the nature of the X groups on phosphorus. The definition is meant to exclude a group that is coordinately bound to the phosphorus atom such as the oxygen atom of phosphine oxides and phosphonates.

Interest in the chemistry of phosphonium ylids resulted primarily from the development by Wittig and his coworkers (*1*) of their broad application in the synthesis of olefins from carbonyl compounds [3.1].

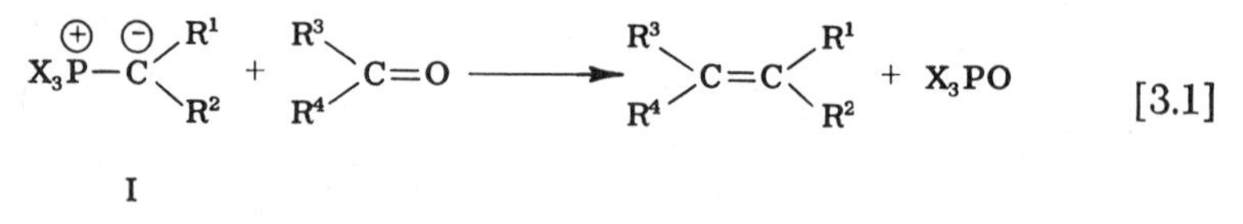

Since most of the earlier pioneering work on phosphorus ylids (see Chapter 2) also had been done using phosphonium ylids and since they are the only type of phosphorus ylid isolated to date, their chemistry will be discussed first.

Since the first phosphonium ylid was reported by Michaelis and Gimborn in 1894 (*2*) many different phosphonium ylids have been prepared. Considerable information has been obtained on the means of preparing ylids of varying structure, their electronic and physical properties and on the reactions undergone by them. In this chapter we will discuss these topics in that order.

I. Preparation of Phosphonium Ylids

All of the phosphonium ylids reported in the literature before 1949 were triphenylphosphonium ylids. Presumably this was because of the availability of the triphenylphosphine needed in their preparation. However, and in spite of the meager number of ylids reported before that date, there was considerable variety in the nature of the substituents attached to the ylid carbanion. With the development of the olefin synthesis by Wittig and Schollkopf in 1954 (*1*), the interest in ylids grew and lead to the preparation of a wide variety of ylids. The nature of the substituents on both the phosphorus and the carbanion portions has been varied considerably.

A. Nature of the Phosphorus Substituents

As indicated above, the early stages of the development of phosphonium ylid chemistry saw the nearly exclusive use of *triphenyl*phosphonium ylids. The earliest example of the preparation of a phosphonium ylid with other than three phenyl groups on phosphorus was reported by Wittig and Rieber in 1949 (*3*). Reaction of tetramethylphosphonium iodide with phenyllithium gave an etheral solution containing methylene*trimethyl*phosphorane (II). As expected, the ylid acted as a typical

$$(CH_3)_4\overset{\oplus}{P}\ \overset{\ominus}{I} + C_6H_5Li \longrightarrow \underset{\text{II}}{(CH_3)_3P{=}CH_2} \qquad [3.2]$$

nucleophile in that it could be alkylated with methyl iodide but the ylid would not undergo the Wittig reaction when treated with benzophenone. By comparison it was noted four years later that methylene*triphenyl*phosphorane (III) would convert benzophenone to 1,1-diphenylethene

$$\underset{\text{II}}{(CH_3)_3P{=}CH_2} + (C_6H_5)_2C{=}O \not\longrightarrow (C_6H_5)_2C{=}CH_2$$
$$\underset{\text{III}}{(C_6H_5)_3P{=}CH_2} + (C_6H_5)_2C{=}O \longrightarrow (C_6H_5)_2C{=}CH_2 \qquad [3.3]$$

(*4*, Wittig and Geissler). As a result of this one experiment the literature became replete with statements to the effect that the Wittig olefin synthesis would not occur with trialkylphosphonium ylids (5, Levisalles).

In 1959 Johnson and LaCount (*6*) prepared ethereal solutions of fluorenylidene*trimethyl*phosphorane and showed that it would undergo the Wittig reaction with aldehydes. They subsequently isolated fluorenylidene*tri-n-butyl*phosphorane and showed it to be a more basic and more nucleophilic ylid than the triphenyl analog (*7*). At a later date Trippett and Walker (*8*) demonstrated that under forcing conditions the trimethyl ylid (II) would react with benzophenone. Shortly thereafter several groups examined the relative reactivity of trialkyl- and triphenylphosphonium ylids and found that the former generally were the more nucleophilic (*8–10*). However, it should be noted that these examples all involved ylids that were stabilized by electron-withdrawing groups on the carbanion (i.e., R_1 and R_2 of I), groups such as benzoyl, acetyl and carbomethoxy. Wittig *et al.* (*11*) and later Bestmann and Kratzer (*10*) confirmed that all trialkylphosphonium ylids were more basic and nucleophilic than their triphenyl counterparts but noted that the former appeared to be more reactive in the Wittig olefin synthesis only if they carried electron-withdrawing groups on the carbanion. The rationale for this observation will be discussed in Chapter 4.

The nature of the alkyl groups on phosphorus has varied widely. Many have used primary alkyl groups (*6–9*). Bestmann and Kratzer (*10*) used cyclohexyl groups while Wittig *et al.* (*11*) used *p*-tolyl, *p*-anisyl, piperidyl and *N*-morpholinyl groups. Trippett and Walker (*12*) used diphenyl(*p*-dimethylaminophenyl)phosphonium ylids in the Wittig reaction but only in order that the phosphine oxide produced should contain a basic group to permit acid extraction from the reaction mixture.

One feature of triphenylalkylphosphonium salts which makes them the most widely used is that in their conversion to ylids by removal of a proton from a carbon alpha to the phosphorus atom, there is only one carbon—the alkyl carbon—which can give up a proton. Thus, there is no question that the ylid is a triphenylphosphoniumalkylide. In a tetraalkylphosphonium salt such as trimethylethylphosphonium bromide, however, treatment with a base such as *n*-butyllithium would form a mixture of two ylids, ethylidenetrimethylphosphorane (IV) and methylenedimethylethylphosphorane (V), since both the methyl and ethyl group carry hydrogen. Therefore, often in spite of reactivity considerations,

$$(CH_3)_3\overset{\oplus}{P}{-}CH_2CH_3\ \ Br^{\ominus} \xrightarrow{C_6H_5Li} \underset{\text{IV}}{(CH_3)_3P{=}CHCH_3} + \underset{\text{V}}{(CH_3)_2\underset{\displaystyle C_2H_5}{\underset{|}{P}}{=}CH_2} \qquad [3.4]$$

triphenylphosphoniumalkylides will be used in order to avoid structural ambiguities in the products.

As far as the Wittig olefin synthesis is concerned the phosphonium group makes no difference in the nature of the olefinic products since the group is eliminated as a phosphine oxide. Accordingly, the only justification for varying the nature of the groups on phosphorus in an ylid is an attempt to alter the nucleophilicity of that ylid. As will be explained in Section II of this chapter, any group on phosphorus which tends to decrease the degree of positive charge on the phosphorus atom will increase the nucleophilicity of the ylid.

All of the known phosphonium ylids are listed in Tables 3.1 through 3.4 (see pages 24–32). The nature of the phosphorus substituents has been noted in each instance.

B. Nature of the Carbanion Substituents

An amazingly wide variety of substituents has been attached to the ylid carbanion in phosphonium ylids. In principle the groups R_1 and R_2 in a phosphonium ylid, $R_1R_2C{=}PX_3$, can be any molecular group whatsoever. In practice the choice is somewhat limited. In our discussion we shall concern ourselves not with the question of whether the ylid or its precursor can be formed but instead with the question of whether the ylid can exist once it is formed.

One of the key limitations on the existence of a given ylid is the requirement that the ylid not contain any active hydrogen which is acidic enough to be removed by the basic ylid carbanion. In other words, the most acidic hydrogen in the ylid precursor must be on the α-carbon. For example, Denney and Smith (*13*) could not prepare an ylid from triphenyl(2-carboxyethyl)phosphonium chloride (VI) because the carboxyl proton was the most acidic. They did obtain the phosphonium betaine. Interestingly, Corey *et al.* (*14*) recently have noted that treat-

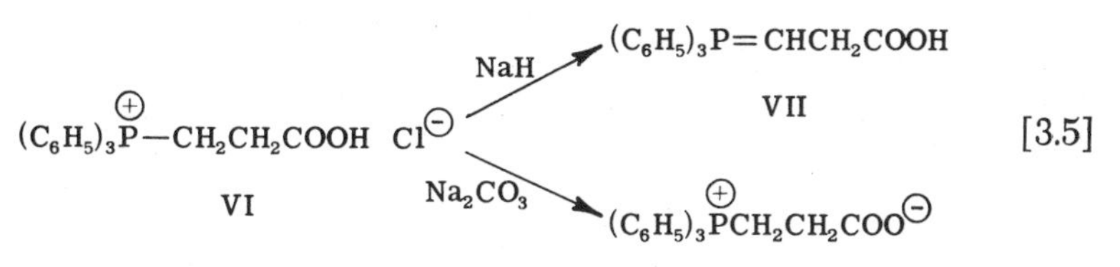

ment of VI with sodium hydride in the presence of dimethylsulfoxide did afford the desired ylid (VII) in about 60% yield since it could be trapped with a ketone in a Wittig reaction. This may be the result of the solvation characteristics of the sulfoxide solvent.

Another important limitation on the nature of the groups R_1 and R_2 which can be attached to the ylid carbanion (I) requires that the groups themselves are stable and not subject to spontaneous fragmentation when next to a carbanion. Several such fragmentation reactions have been observed. Wittig and coworkers (*15*) found that attempts to convert 1,2-bis(triphenylphosphonio)ethane dibromide (VIII) into a bis-ylid (IX, $n = 0$) with phenyllithium instead led to the isolation of

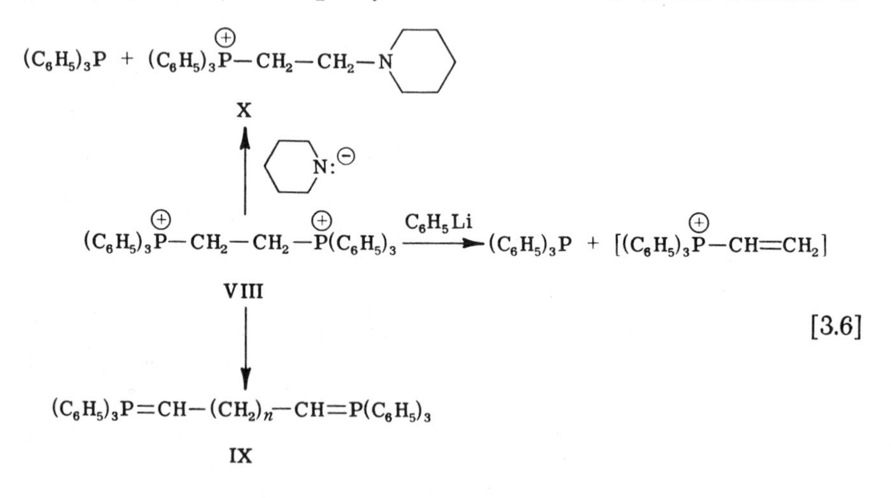

$$(C_6H_5)_3\overset{\oplus}{P}-\overset{\ominus}{C}H-CH_2-\overset{\oplus}{P}(C_6H_5)_3$$

XI

triphenylphosphine. The suspicion of an elimination reaction was confirmed when the attempt was repeated and lithium piperidide was used as the base. The isolated products were triphenylphosphine and the piperidyl derivative (X), indicating that an elimination probably formed vinyltriphenylphosphonium bromide which underwent a Michael addition of the piperidide anion. The observations can be rationalized equally well by supposing that the monoylid (XI) was formed but rapidly underwent a unimolecular elimination (fragmentation) to triphenylphosphine and the vinylphosphonium salt.

When two phosphonium groups are separated by three or more methylene groups the corresponding bis-ylids appear to be stable. Wittig *et al.* (*15*) prepared and used the ylid (IX, $n = 1$) in a reaction with a dicarbonyl compound to form a ring system. Two other groups (*16, 17*) used the bis-butylide (IX, $n = 2$) in syntheses of squalene without any sign of ylid fragmentation. Fragmentation of 1,4-bis(triphenylphosphonio)-2-butene dibromide into triphenylphosphine and triphenyl(1-butadienyl)phosphonium bromide upon treatment with phenyllithium was reported by Burger (*18*). This appears to be simply a vinylogous

$$
\begin{array}{c}
(C_6H_5)_3P \;+\; [(C_6H_5)_3\overset{\oplus}{P}-CH=CH-CH=CH_2] \\
\uparrow C_6H_5Li \\
(C_6H_5)_3\overset{\oplus}{P}-CH_2CH=CHCH_2-\overset{\oplus}{P}(C_6H_5)_3 \quad 2\,Br^{\ominus} \\
\downarrow NaOCH_3 \\
(C_6H_5)_3P=CH-CH=CH-CH=P(C_6H_5)_3 \\
\textbf{XII}
\end{array} \qquad [3.7]
$$

case of the fragmentation of VIII via the monoylid (XI). Heitman and coworkers (*9, 19*) later found that use of a weaker base, sodium methoxide, led to the successful formation of the bis-ylid (XII). Ford and Wilson (*20*) claimed that attempts to form XII using aqueous sodium carbonate as the base still afforded the fragmentation products.

An interesting fragmentation reaction of another type was reported by Trippett and Walker (*21*). Treatment of nitromethyltriphenylphosphonium bromide with hydroxide ion led to the formation of triphenylphosphine oxide and the fulminate ion. These products were proposed to have resulted from the spontaneous decomposition of the ylid (XIII).

$$
\underset{\textbf{XIII}}{(C_6H_5)_3\overset{\oplus}{P}-CH=\overset{\oplus}{N}(-O^{\ominus})-O^{\ominus}} \longrightarrow (C_6H_5)_3PO \;+\; [HCNO] \qquad [3.8]
$$

Horner and Oediger (*22*) reported the same observations when the ylid was prepared by an alternate route.

There have been two reports of phosphonium ylids spontaneously fragmenting into triphenylphosphine and a carbene. *n*-Butoxymethylenetriphenylphosphorane (XIV) was prepared in solution from the bromide salt and *n*-butyllithium. Warming the solution to 45° for 20 hours afforded a mixture of products, the major ones being 1,2-dibutoxyethene,

$$
\underset{\textbf{XIV}}{(C_6H_5)_3\overset{\oplus}{P}-\overset{\ominus}{C}H-OC_4H_9} \longrightarrow (C_6H_5)_3P \;+\; C_4H_9OCH:
$$

$$
C_4H_9OCH: \;+\; \textbf{XIV} \longrightarrow \left[C_4H_9O\overset{\ominus}{C}H-\underset{OC_4H_9}{\overset{\overset{\oplus}{P}(C_6H_5)_3}{CH}} \right] \longrightarrow \begin{array}{c} (C_6H_5)_3P \\ + \\ C_4H_9OCH=CHOC_4H_9 \end{array} \qquad [3.9]
$$

dibutoxymethane, 1-butoxypentene and triphenylphosphine (*23*, Wittig and Boll). The products were accounted for by proposing an initial spontaneous cleavage of the ylid into *n*-butoxycarbene. The carbene was thought to react with a variety of nucleophiles present in the medium, for example, the original ylid (XIV), to form the observed products. The same group found that bis(phenoxy)methylenetriphenylphosphorane behaved in a similar manner to afford triphenylphosphine and tetraphenoxyethylene (*23*).

In a similar vein Trippett (*24*) accounted for the formation of stilbene and triphenylphosphine upon treatment of triphenylbenzylphosphonium bromide with sodium methoxide by proposing the benzylide underwent a spontaneous cleavage to phenylcarbene which was trapped by the ylid. It must be noted that in neither of the above examples has

$$(C_6H_5)_3\overset{\oplus}{P}-\overset{\ominus}{C}HC_6H_5 \longrightarrow C_6H_5CH: + (C_6H_5)_3P$$

$$C_6H_5CH: + (C_6H_5)_3\overset{\oplus}{P}-\overset{\ominus}{C}HC_6H_5 \longrightarrow \left[C_6H_5\overset{\ominus}{C}H-\underset{\overset{|}{\oplus P(C_6H_5)_3}}{CH}-C_6H_5 \right] \longrightarrow C_6H_5CH{=}CHC_6H_5 + (C_6H_5)_3P \quad [3.10]$$

the presence of a carbene been demonstrated experimentally. These proposals warrant skeptical consideration when it is noted that phosphines tend to *combine* with carbenes to form relatively stable ylids (see Section I,C of this chapter). Clearly, experimental studies tuned to these proposals are needed.

Phosphonium ylids are not prone to undergo spontaneous rearrangements as are similar nitrogen and sulfur ylids. Thus, whereas dimethylsulfoniumfluorenylide underwent a Sommelet rearrangement to 1-methylthiomethylfluorene (*25*, Pinck and Hilbert) and diethylbenzylammoniumfluorenylide underwent a Sommelet rearrangement to 9-(diethylamino)-9-benzylfluorene, diethylbenzylphosphoniumfluorenylide did not rearrange (*26*, Wittig and Laib), even upon pyrolysis or strenuous hydrolysis.

A final limitation to the formation and/or existence of phosphonium ylids with a variety of substituents on the ylid carbanion is that there be no other group in the molecule with which the ylid function can react. This problem has been exemplified by Mondon's report that a solution containing the bromoylid (XV) slowly lost the characteristic ylid coloration and afforded triphenylcyclobutylphosphonium bromide. The ylid apparently underwent an intramolecular alkylation (*27*).

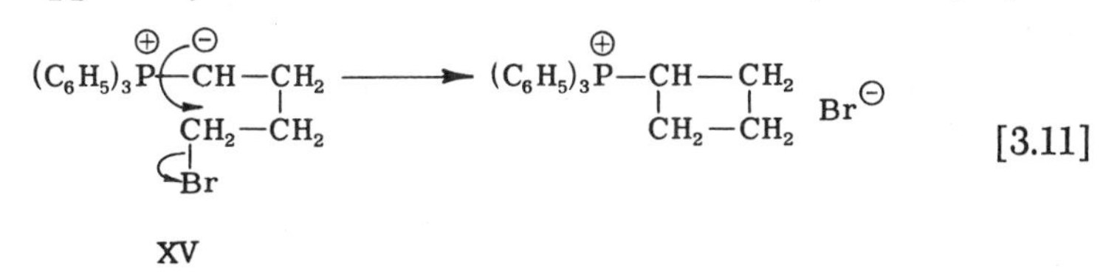

In a similar reaction Bergelson and coworkers (*28*) found that the carbethoxy ylid (XVI) underwent a spontaneous *intra*molecular acylation to form a new keto-ylid (XVII). Griffin and Witschard (*29*) found that the keto-ylid (XVIII) underwent a spontaneous *inter*molecular Wittig reaction. These examples are sufficient to make the point that ylids with

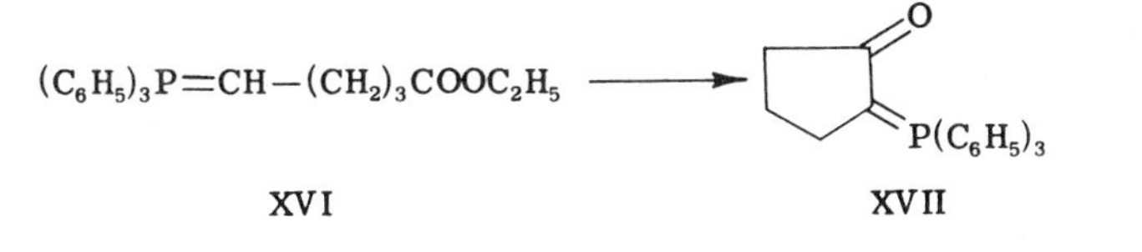

[3.12]

$$2\ (C_6H_5)_3P{=}CHCH_2C({=}O){-}C_6H_5 \longrightarrow \text{1,4-}(C_6H_5)_2\text{-cyclohexa-1,4-diene} + 2\ (C_6H_5)_3PO$$

XVIII

certain substituents can and will undergo intra- and intermolecular reactions spontaneously and lead to the destruction of the original ylid.

Apart from the limitations mentioned above which, in total perspective are relatively minor hindrances to their use, ylids with almost any substitution pattern can be prepared and used successfully in typical ylid reactions. The nature of the substituents obviously will affect the properties of the ylid and its ease of formation, topics to be discussed later in this chapter.

Tables 3.1–3.4 list virtually every phosphonium ylid reported in the literature with the exception of those few in which the ylid bond is incorporated in a ring system, i.e., cyclic ylids. The tables indicate the nature of all substituents, the method(s) used in the preparation, a reference to the preparation of each and a notation indicating whether the ylid has been isolated. The tables are divided into several categories. In Table 3.1 are listed those ylids which carry one hydrogen atom on the ylid carbon. Table 3.2 lists those ylids with two substituents on the ylid carbanion. Table 3.3 lists bis-ylids and [*text continues on p. 29*]

TABLE 3. 1

Monosubstituted Phosphonium Ylids
($X_3P{=}CHR$)

R[a]	X	Preparation[b]	Ref.
H—	Trimethyl	11	*3*
H—	Tricyclohexyl	11	*10*
H—	Tri-*N*-piperidyl	11	*11*
H—	Tri-*N*-morpholino	11	*11*
H—	Triphenyl	2, 4, 11	*1,4,48, 54,127*
H—	Tri-*p*-tolyl	11	*11*
H—	Tri-*p*-anisyl	11	*11*
CH_3—	Triphenyl	11	*117*
C_2H_5—	Triphenyl	11	*134*
C_3H_7—	Triphenyl	11	*191*
Br—$(CH_2)_3$—	Triphenyl	11	*27*
CH_3—$(CH_2)_3$—	Triphenyl	11	*253*
$(CH_3)_2CH$—CH_2—	Triphenyl	11	*246*
Cyclo-C_6H_{11}—	Triphenyl	11	*245*
4-Bornyl—	Triphenyl	11	*244*
C_6H_5—CH_2—	Triphenyl	3	*50*
N-(2-Formylpyrrolo)methylene	Triphenyl	3	*52*
HC≡C—	Triphenyl	11	*250*
CH_2=CH—	Triphenyl	11	*1*
CH_3CH=CH—	Triphenyl	11	*106*
C_4H_9—CH=CH—	Triphenyl	11	*243*
C_6H_{13}—CH=CH—	Triphenyl	11	*243*
C_6H_5—CH=CH—	Triphenyl	11	*147, 181*
C_6H_5—CH=CH—	Tricyclohexyl	11	*10*
Cyclohexylidenemethylene	Triphenyl	11	*107*
Cyclohexenyl —	Triphenyl	11	*107*
CH_3—$(CH{=}CH)_2$—	Triphenyl	11	*106*
C_3H_7—$(C{\equiv}C)_2$—CH_2CH_2—	Triphenyl	11	*231*
$(CH_3)_2C$=CH—$(CH_2)_2$—$C(CH_3)$=CH—	Triphenyl	11	*238*
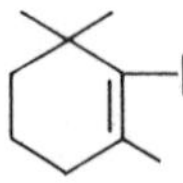 —[CH=CH—$C(CH_3)$=CH]$_2$—	Triphenyl	11	*184*
C_6H_5CO—CH_2—	Triphenyl	11	*29, 236*

TABLE 3. 1 (Continued)

R[a]	X	Preparation[b]	Ref.
$C_6H_5CO-(CH_2)_2-$	Triphenyl	11	*236*
$C_6H_5CO-(CH_2)_3-$	Triphenyl	11	*230*
$C_6H_5CO-(CH_2)_4-$	Triphenyl	11	*236*
$HOOCCH_2-$	Triphenyl	11	*14*
$C_2H_5O_2CCH_2-$	Triphenyl	5	*66*
$C_2H_5O_2C(CH_2)_n-$	Triphenyl	11	*88, 123*
$NC-CH_2-$	Triphenyl	5	*66*
CHO; $-O-(CH_2)_n-$	Triphenyl	11	*51, 240*
$ROOC-CH=CH-$	Triphenyl	11	*247*
OH; $=CH-$	Triphenyl	11	*150*
* $(C_6H_5)_2NCO-$	Triphenyl	11	*94*
* CH_3OOC-	Triphenyl	11	*10*
* CH_3OOC-	Tricyclohexyl	11	*10*
* CH_3OOC-	Tris(dimethylamino)	11	*251*
* C_2H_5OOC-	Triphenyl	11	*2*
* C_2H_5OOC-	Tris(*p*-biphenylyl)	11	*229*
* C_2H_5OOC-	Diphenyl(*p*-dimethyl-aminophenyl)	11	*12*
* C_2H_5OOC-	Tri-*n*-butyl	–	*210*
* C_2H_5OOC-	Trimethyl	11	*8*
* C_2H_5OOC-	Phenylbis(dimethyl-amino)	11	*249*
* C_2H_5OOC-	Tricyclohexyl	11	*254*
* C_2H_5OOC-	Tri-*n*-hexyl	11	*254*
* C_2H_5OOC-	Tri-*n*-octyl	11	*254*
* C_2H_5OOC-	Tri-*n*-decyl	11	*254*
* C_2H_5OOC-	Tri-*p*-anisyl	11	*254*
* $-CHO$	Triphenyl	7	*86, 87*
* $-CHO$	Diphenyl-*p*-dimethyl-aminophenyl	11	*12*
* $-COCH_3$	Triphenyl	11	*110, 129*
* $-COCH_3$	Diphenyl-*p*-dimethyl-aminophenyl	11	*12*

TABLE 3. 1 (Continued)

R[a]	X	Preparation[b]	Ref.
* $-COCH_2Cl$	Triphenyl	11	*139, 237*
* $-CO-C(CH_3)_2COCH(CH_3)_2$	Triphenyl	–	*177*
* $-COC_6H_5$	Triphenyl	11	*110, 129*
* $-COC_6H_5$	$(CH_3)_n(C_6H_5)_{3-n}$	11	*8*
* $-COC_6H_4Br(p)$	Triphenyl	–	*209*
* $-COC_6H_4OCH_3(p)$	Triphenyl	11	*139*
* $-COC_6H_4NO_2(p)$	Triphenyl	11	*139*
* $-COCH{=}CHC_6H_5$	Triphenyl	11	*237*
* $-C_6H_5$	Triphenyl	11	*1*
* $-C_6H_5$	Tricyclohexyl	11	*10*
* $-C_6H_5$	Methylethylphenyl	11	*211*
* $-C_6H_5$	$(CH_3)_n(C_6H_5)_{3-n}$	11	*8*
* $-C_6H_5$	Diphenyl-*p*-dimethylaminophenyl	11	*12*
* $-C_6H_5$	Tris(dimethylamino)	11	*251*
* $-C_6H_5$	Phenylbis(dimethylamino)	11	*249*
* $-C_6H_4NO_2(p)$	Triphenyl	11	*119*
$-C_6H_4OCH_3(p)$	Triphenyl	11	*239*
$-C_6H_4COOCH_3(p)$	Triphenyl	11	*232*
* $-C_6H_4SO_2C_6H_5(p)$	Triphenyl	11	*241*
* $-C_6H_4SO_2C_6H_4Br(p,p)$	Triphenyl	11	*241*
$-C_6H_3$(2-OCH_3, 5-CHO)	Triphenyl	11	*97*
$-OCH_3$	Triphenyl	11	*33, 225*
$-OC_4H_9$	Triphenyl	11	*23, 242*
$-OC_6H_4-CH_3(p)$	Triphenyl	11	*242*
$-SCH_3$	Triphenyl	11	*33*
$-Si(CH_3)_3$	Triphenyl	11	*234*
$-Si(CH_3)_3$	Trimethyl	11	*136*
* $-\overset{+}{P}(C_6H_5)_3$	Triphenyl	11	*91, 112*
* $-P(O)(C_6H_5)_2$	Triphenyl	–	*91*
* $-CN$	Triphenyl	11	*140, 248*
$-Cl$	Triphenyl	11	*32, 33*
$-Br$	Triphenyl	11	*107*
$-I$	Triphenyl	11	*124*

[a] Those ylids marked with an asterisk have been isolated and characterized.

[b] The methods listed by number correspond to those numerically discussed in Section I, C of this chapter (following section).

TABLE 3. 2

Disubstituted Phosphonium Ylids
($X_3P{=}CR^1R^2$)

R^1 [a]	R^2	X	Preparation [b]	Ref.
* $-C_6H_5$	$-COOCH_3$	$(C_6H_5)_3$	7	*83*
* $-CH_3$	$-COOCH_3$	$(C_6H_5)_3$	6	*74*
* $-C_2H_5$	$-COOCH_3$	$(C_6H_5)_3$	6	*74*
* $-C_3H_7$	$-COOCH_3$	$(C_6H_5)_3$	7	*83*
$-C_6H_{11}$(cyclo)	$-COOCH_3$	$(C_6H_5)_3$	7	*83*
* $-CH_2C_6H_5$	$-COOCH_3$	$(C_6H_5)_3$	6	*72*
* $-CH_2CH{=}CH_2$	$-COOCH_3$	$(C_6H_5)_3$	6	*74*
* $-CN$	$-COOCH_3$	$(C_6H_5)_3$	9	*22*
* $-CHO$	$-COOCH_3$	$(C_6H_5)_3$	–	*75*
* $-N{=}N-C_6H_5$	$-COOCH_3$	$(C_6H_5)_3$	6	*208*
* $-SO_3Na$	$-COOCH_3$	$(C_6H_5)_3$	11	*201*
* $-Cl$	$-COOCH_3$	$(C_6H_5)_3$	11	*94*
* $-Br$	$-COOCH_3$	$(C_6H_5)_3$	8, 11	*92, 94*
* $-I$	$-COOCH_3$	$(C_6H_5)_3$	8, 11	*92, 94*
* $-CH(COOCH_3)CH_2COC_6H_5$	$-COOCH_3$	$(C_6H_5)_3$	5	*67*
* $-C(COOCH_3){=}CHCH_3$	$-COOCH_3$	$(C_6H_5)_3$	5	*71*
* $-C(COOCH_3){=}CHC_3H_7$	$-COOCH_3$	$(C_6H_5)_3$	5	*71*
* $-C(COOCH_3){=}CHC_6H_5$	$-COOCH_3$	$(C_6H_5)_3$	5	*71*
* $-C(COOCH_3){=}CHCH{=}CHC_6H_5$	$-COOCH_3$	$(C_6H_5)_3$	5	*71*
* $-C(COOCH_3){=}C(CH_3)C_6H_5$	$-COOCH_3$	$(C_6H_5)_3$	5	*71*
* $-C(COOCH_3){=}C(C_6H_5)_2$	$-COOCH_3$	$(C_6H_5)_3$	5	*70*
* $-C(COOCH_3){=}C(o-C_6H_4)_2$	$-COOCH_3$	$(C_6H_5)_3$	5	*70*
* $-COCH_3$	$-COOC_2H_5$	$(C_6H_5)_3$	9	*82*
* $-CH_2COOC_2H_5$	$-COOC_2H_5$	$(C_6H_5)_3$	5	*60*
* $-COOC_2H_5$	$-COOC_2H_5$	$(C_6H_5)_3$	9	*22*
* $-CONHC_6H_5$	$-COOC_2H_5$	$(C_6H_5)_3$	7	*21*
* $-Cl$	$-COOC_2H_5$	$(C_6H_5)_3$	8	*92, 93*
* $-Br$	$-COOC_2H_5$	$(C_6H_5)_3$	11	*94*
* $-SeC_6H_5$	$-COOC_2H_5$	$(C_6H_5)_3$	6	*76*
* $-SO_2C_6H_5$	$-COOC_2H_5$	$(C_6H_5)_3$	7	*76*
$-CH_3$	$-COO-$menthyl	$(C_6H_5)_3$	–	*198*
$-CH_3$	$-COO-$(2-octyl)	$(C_6H_5)_3$	–	*198*
* $-CONHC_6H_5$	$-CONHC_6H_5$	$(C_6H_5)_3$	7	*21*
* $-Cl$	$-CONHC_6H_5$	$(C_6H_5)_3$	11	*94*
* $-Br$	$-CONHC_6H_5$	$(C_6H_5)_3$	11	*94*

TABLE 3. 2 (Continued)

R^1 [a]	R^2	X	Preparation[b]	Ref.
*—C_6H_5	—COC_6H_5	$(C_6H_5)_3$	7, 11	*8, 21*
*—CH_3	—COC_6H_5	$(C_6H_5)_3$	7	*85*
*—$C(COOCH_3)=CHCOOCH_3$	—COC_6H_5	$(C_6H_5)_3$	5	*222*
*—Cl	—COC_6H_5	$(C_6H_5)_3$	8	*93*
*—Br	—COC_6H_5	$(C_6H_5)_3$	–	*140*
*—Br	—COC_6H_5	$(nC_4H_9)_3$	–	*140*
*—I	—COC_6H_5	$(C_6H_5)_3$	11	*94*
*—HgCl	—COC_6H_5	$(C_6H_5)_3$	11	*205*
*—$COCH_3$	—COC_6H_5	$(C_6H_5)_3$	7	*90*
*—COC_6H_5	—COC_6H_5	$(C_6H_5)_3$	7	*90*
*—$CH_2COC_6H_5$	—COC_6H_5	$(C_6H_5)_3$	5	*62*
*—$CH_2COC_6H_5$	—COC_6H_5	$(C_6H_5)_2OC_2H_5$	5	*63*
*—$CH_2COC_6H_5$	—COC_6H_5	$(OCH_3)_3$	5	*61*
*—$CH_2COC_6H_5$	—COC_6H_5	$(nC_4H_9)_3$	5	*62*
*—C_6H_5	—$COCH_3$	$(C_6H_5)_3$	7, 11	*8, 82*
*—Cl	—$COCH_3$	$(C_6H_5)_3$	8	*93*
*—$COCH_3$	—$COCH_3$	$(C_6H_5)_3$	7	*90*
*—$COC_6H_4OCH_3(p)$	—$COCH_3$	$(C_6H_5)_3$	7	*90*
*—$COC_6H_4NO_2(p)$	—$COCH_3$	$(C_6H_5)_3$	7	*90*
—CH_3	—C_6H_5	$(C_6H_5)_3$	11	*71*
*—C_6H_5	—C_6H_5	$(C_6H_5)_3$	2, 11	*42, 117*
*—$C_6H_4NO_2(p)$	—C_6H_5	$(C_6H_5)_3$	–	*226*
*—CHO	—C_6H_5	$(C_6H_5)_3$	7	*75*
*—CHO	—CH_3	$(C_6H_5)_3$	7	*86*
—CH_3	—CH_3	$(C_6H_5)_3$	11	*117*
—$CH=CHCOOCH_3$	—CH_3	$(C_6H_5)_3$	11	*252*
—OC_6H_5	—OC_6H_5	$(C_6H_5)_3$	11	*23*
*—$SO_2C_6H_5$	—$SO_2C_6H_5$	$(C_6H_5)_3$	9, 11	*22, 111*
*—$N=N-C_6H_5$	—$N=N-C_6H_5$	$(C_6H_5)_3$	6	*228*
*—CN	—CN	$(C_6H_5)_3$	9	*22*
*—Br	—CN	$(C_6H_5)_3$	11	*140*
*—$CH=C(COOC_2H_5)_2$	—CN	$(C_6H_5)_3$	5	*68*
*—$C(CN)=C(CN)_2$	—CN	$(C_6H_5)_3$	5	*68*
*—$C(COOCH_3)=CHCOOCH_3$	—CN	$(C_6H_5)_3$	5	*68*
—Cl	—Cl	$(C_6H_5)_3$	1, 11	*31, 114*
—Cl	—Cl	$(nC_4H_9)_3$	1	*31*
—F	—Cl	$(C_6H_5)_3$	1	*31*

TABLE 3. 2 (Continued)

R^1 [a]	R^2	X	Preparation [b]	Ref.
—Br	—Br	$(C_6H_5)_3$	1, 11	*31, 115*
* $-\overset{+}{P}(C_6H_5)_3$	—Br	$(C_6H_5)_3$	8	*91*
—F	—F	$(C_6H_5)_3$	1	*34*
—F	—F	$(N(CH_3)_2)_3$	1	*37*
$-SCH_3$	$-SCH_3$	$(C_6H_5)_3$	1	*38*
$-SC_2H_5$	$-SC_2H_5$	$(C_6H_5)_3$	1	*38*

[a] Those ylids marked with an asterisk have been isolated and characterized.
[b] The methods listed by number correspond to those numerically discussed in Section I, C of this chapter (following section).

Table 3.4 records those compounds in which the phosphoranyl group is exocyclic to a ring system. A perusal of these tables will indicate the tremendous variety of ylids that have been prepared to date.

C. Methods of Preparation of Phosphonium Ylids

Numerous methods have been explored as possible routes to phosphonium ylids. The most flexible of these is the so-called "salt method," but there are a variety of others available. The latter usually are used for unique cases but some have wider application than has been used to advantage. The methods to be discussed are listed below in the order in which they will be discussed. The numerical indication of preparative methods found in the preceding tables refers to the following numerical order.

1. Synthesis of Ylids via Carbenes
2. Synthesis of Ylids via Phosphinazines
3. Synthesis of Ylids via Addition to Vinylphosphonium Salts
4. Synthesis of Ylids via Addition to Benzyne
5. Synthesis of Ylids via Addition to Olefins and Alkynes
6. Synthesis of Complex Ylids via Alkylation of Simple Ylids
7. Synthesis of Complex Ylids via Acylation of Simple Ylids
8. Synthesis of Haloylids via Halogenation of Simple Ylids
9. Ylids from Active Methylene Compounds and Tertiary Phosphine Dihalides
10. Synthesis of Quino-Ylids
11. Synthesis of Ylids from Phosphonium Salts (Salt Method)

1. Synthesis of Ylids via Carbenes

If a metathetic coupling reaction were designed to form phosphonium ylids the two fragments would have to be a tertiary phosphine and a

TABLE 3.3

bis-Phosphonium Ylids
($X_3P{=}A{=}PX_3$)

A[a]	X	Preparation[b]	Ref.
$=CH-C(CH_3)=CH-CH=CH-CH=C(CH_3)-CH=$	$(C_6H_5)_3$	11	*220*
$=CH-COOCH_2CH_2OOC-CH=$	$(C_6H_5)_3$	11	*20*
$=CH-O-CH=$	$(C_6H_5)_3$	11	*233*
$=CH-CH=CH-CH=$	$(C_6H_5)_3$	11	*9,19*
$=CH-CH_2-CH=$	$(C_6H_5)_3$	11	*15*
$=CH-(CH_2)_2-CH=$	$(C_6H_5)_3$	11	*15,27,185*
$=CH-(CH_2)_3-CH=$	$(C_6H_5)_3$	11	*185*
$=CH-(CH_2)_4-CH=$	$(C_6H_5)_3$	11	*185*
$=CH-(CH_2)_5-CH=$	$(C_6H_5)_3$	11	*185*
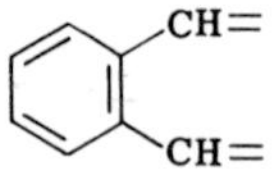 $CH=$ $CH=$	$(C_6H_5)_3$	11	*235*
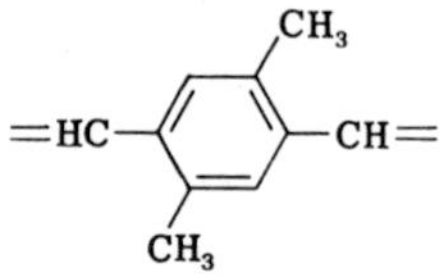 CH_3 $=HC-$ $-CH=$ CH_3	$(C_6H_5)_3$	11	*232*
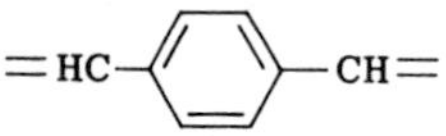 $=HC-$ $-CH=$	$(C_6H_5)_3$	11	*97*
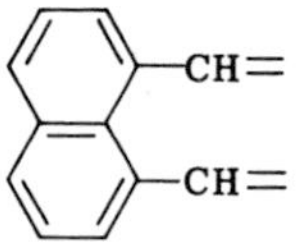 * $-CH=$ $-CH=$	$(C_6H_5)_3$	–	*182*
*$=C=$	$(C_6H_5)_3$	11	*91,112*
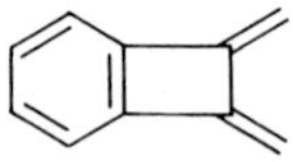	$(C_6H_5)_3$	11	*227*
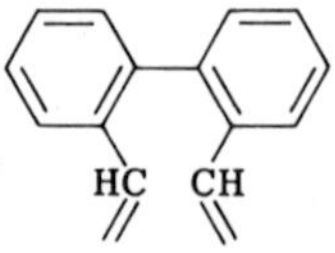 * HC CH	$(C_6H_5)_3$	–	*185*

[a] Those ylids marked with an asterisk have been isolated and characterized.

[b] The methods listed by number correspond to those numerically discussed in Section I, C of this chapter (following section).

TABLE 3. 4

Exocyclic Phosphonium Ylids

$(X_3P{=}R)$

R[a]	X	Preparation[b]	Ref.
$=C=C(C_6H_5)$	$(C_6H_5)_3$	5	*234*
* 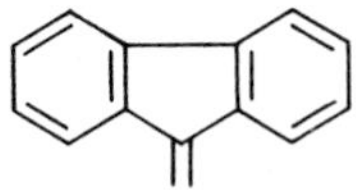	$(C_6H_5)_3$	11	*116*
* 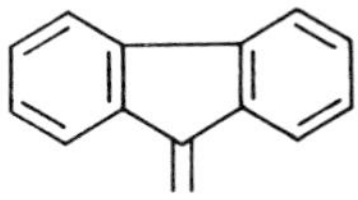	$(nC_4H_9)_3$	11	*7*
* 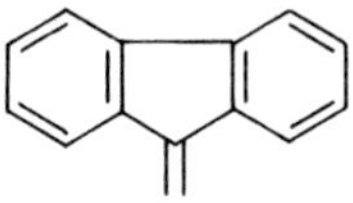	$(CH_3)_3$	11	*6,26*
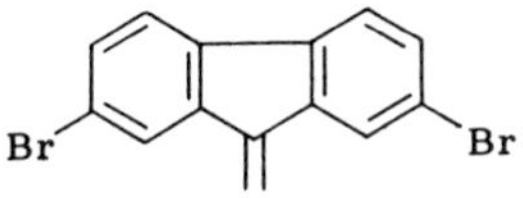 * Br Br	$(C_6H_5)_3$	11	*206*
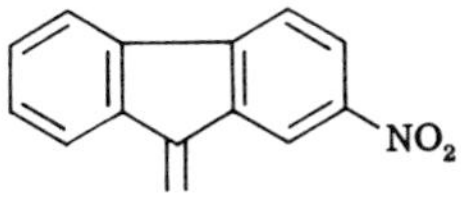 * NO_2	$(C_6H_5)_3$	11	*255*
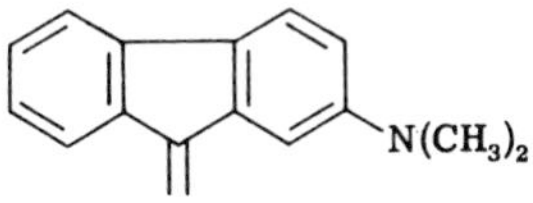 * $N(CH_3)_2$	$(C_6H_5)_3$	11	*255*
* 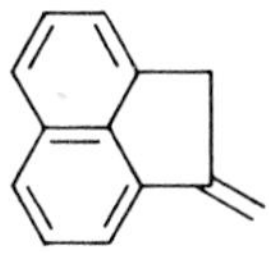	$(C_6H_5)_3$	6	*77*

TABLE 3. 4 (Continued)

R[a]	X	Preparation[b]	Ref.
* 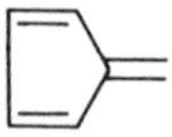	$(C_6H_5)_3$	11	*130*
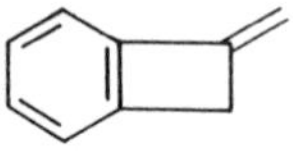	$(C_6H_5)_3$	11	*227*
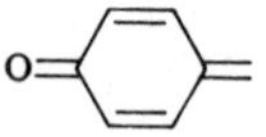 O= (?)	$(C_6H_5)_3$	11	*98*
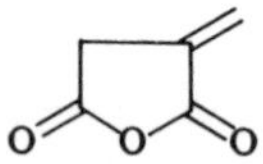 * O O O	$(C_6H_5)_3$	5	*58,59*
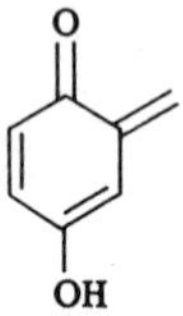 * O OH	$(C_6H_5)_3$	5	*56,96*
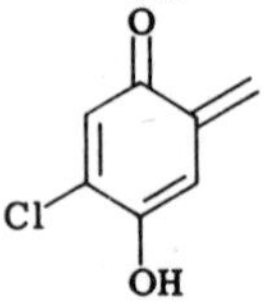 * O Cl OH	$(C_6H_5)_3$	11	*56*
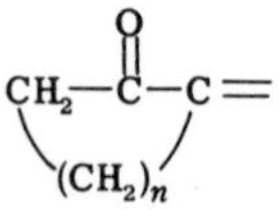 * CH_2—C(=O)—C= $(CH_2)_n$	$(C_6H_5)_3$	7	*28,88*

[a] Those ylids marked with an asterisk have been isolated and characterized.

[b] The methods listed by number correspond to those numerically discussed in Section I, C of this chapter (following section).

carbene. The fact that phosphines are known to be nucleophilic and carbenes had been shown to possess electrophilic character (*30*, Doering and Hoffmann) led to the initial test of the coupling reaction.

Speziale and his coworkers (*31*) were the first to illustrate this method of preparation of phosphonium ylids. They found that addition of chloroform to a cold mixture of triphenylphosphine and potassium *tert*-butoxide in heptane afforded a yellow suspension of the ylid (XIX) whose presence was confirmed by the isolation of 1,1-dichloro-2,2-diphenylethene and triphenylphosphine oxide upon addition of benzophenone to the solution. Treatment of chloroform with potassium *tert*-butoxide

$$(C_6H_5)_3P + CHCl_3 \xrightarrow{KOBu^t} \underset{\text{XIX}}{(C_6H_5)_3P{=}CCl_2} \xrightarrow{(C_6H_5)_2CO} (C_6H_5)_2C{=}CCl_2 + (C_6H_5)_3PO \quad [3.13]$$

$$(C_6H_5)_3P\!: + :CCl_2 \longrightarrow (C_6H_5)_3\overset{\oplus}{P}{-}\overset{\ominus}{\ddot{C}}Cl_2$$

previously had been shown to afford dichlorocarbene (*30*), and Speziale proposed that it underwent a direct combination with triphenylphosphine. Shortly after this initial report Seyferth *et al.* (*32*) and Wittig and Schlosser (*33*) both reported an analogous preparation of chloromethylenetriphenylphosphorane from methylene chloride, *n*-butyllithium and triphenylphosphine in about 65% yield. In their full paper Speziale and Ratts (*31*) obtained the corresponding ylids from dibromocarbene and chlorofluorocarbene. However, they were unable to obtain the ylid (XX) from the difluorocarbene(chlorodifluoromethane and *tert*-butoxide) and were unable to repeat Franzen's report (*34*) of the preparation of XX from difluorodibromomethane and *n*-butyllithium. More recently, Fuqua and coworkers (*35*) reported the preparation of XX by heating a mixture of sodium difluorochloroacetate and triphenylphosphine. The ylid was reacted *in situ* with a variety of aldehydes to produce the corresponding 1,1-difluoroolefins. The authors proposed that the acetate ion

$$ClF_2C{-}COO^{\ominus} \xrightarrow{\Delta} [:CF_2] + CO_2 + Cl^{\ominus} \xrightarrow{(C_6H_5)_3P} \underset{\text{XX}}{(C_6H_5)_3P{=}CF_2} \quad [3.14]$$

fragmented into carbon dioxide, chloride and difluorocarbene, the latter being trapped by triphenylphosphine. This proposal is consistent with the studies of Hine and Duffey on the difluorochloroacetate ion (*36*). Mark (*37*) proposed a similar reaction path when tris(dimethylamino)-phosphine was treated with trifluoroacetophenone to afford difluoromethylenetris(dimethylamino)phosphorane.

Recently Lemal and Banitt (*38*) proposed that the tosylhydrazone (XXI) decomposed in the presence of sodium hydride via carbene formation. They were unable to trap the supposed carbene with cyclohexene but found that treatment of the solution with triphenylphosphine afforded a yellow suspension of the ylid (XXII) which reacted with *p*-nitrobenzaldehyde in 71% yield. Apparently the cyclohexene was not,

$$\underset{\text{XXI}}{Ts-NH-N=C(SC_2H_5)_2} \xrightarrow{NaH} [:C(SC_2H_5)_2] \xrightarrow{\text{cyclohexene}} N.R.$$

$$[:C(SC_2H_5)_2] \xrightarrow{(C_6H_5)_3P} \underset{\text{XXII}}{(C_6H_5)_3P=C(SC_2H_5)_2} \xrightarrow{ArCHO} ArCH=C(SC_2H_5)_2 \quad [3.15]$$

although the phosphine was, sufficiently nucleophilic to attack the "carbenoid" species, whatever its structure. Earlier Franzen and Wittig (*39*) had found trimethylammoniummethylide to dissociate to trimethylamine and methylene since the latter could be trapped with cyclohexene to afford norcarane. They also found that addition of triphenylphosphine and benzophenone in place of cyclohexene afforded diphenylethylene, presumably via the intermediate formation of methylenetriphenylphosphorane (III).

$$(CH_3)_3\overset{\oplus}{N}-\overset{\ominus}{C}H_2 \longrightarrow [:CH_2] \xrightarrow{(C_6H_5)_3P} \underset{\text{III}}{(C_6H_5)_3P=CH_2} \xrightarrow{(C_6H_5)_2C=O} (C_6H_5)_2C=CH_2 \quad [3.16]$$

It must be mentioned at this point that no attempt has been made to differentiate between free carbenes, if they exist in solution, and potential carbenes (carbenoids). That is a controversy unto itself and readers are referred to recent books by Hine (*40*) and Kirmse (*41*) for discussion. As far as ylid formation is concerned, the experimental data reported to date are consistent with either interpretation. The tertiary phosphine can be viewed as being capable of attacking either one of three conceivable species to produce, in the case of chloroform as illustrated in [3.17], dichloromethylenetriphenylphosphorane (XIX).

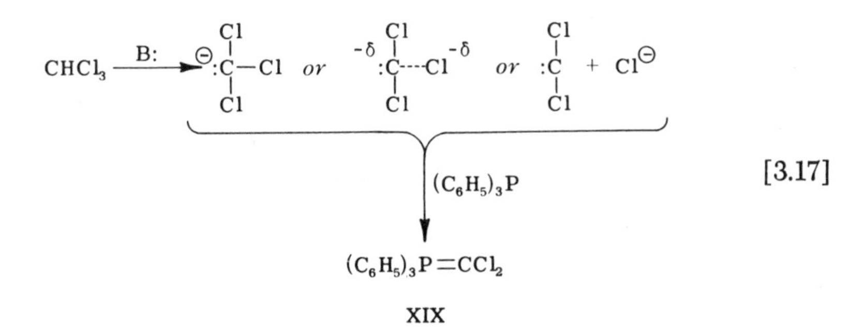

[3.17]

2. *Synthesis of Ylids via Phosphinazines*

The pyrolysis of phosphinazines to force the elimination of molecular nitrogen was one of the earliest methods developed for the synthesis of phosphonium ylids but it has turned out to be probably the least general. Staudinger and Meyer (*42*, *43*) first reported this reaction with the azine obtained by coupling triphenylphosphine with diphenyldiazomethane. The azine eliminated nitrogen upon heating to 195° for fifteen minutes and left a residue of benzhydrylidenetriphenylphosphorane

$$(C_6H_5)_2CN_2 + (C_6H_5)_3P \longrightarrow (C_6H_5)_3P{=}N{-}N{=}C(C_6H_5)_2 \xrightarrow[\Delta]{-N_2} (C_6H_5)_3P{=}C(C_6H_5)_2 \qquad [3.18]$$

XXIII

(XXIII). In spite of the fact that azines from a variety of diazo compounds and tertiary phosphines were prepared by these workers they were unable to isolate any other ylids by an analogous pyrolysis. Occasionally they obtained some triphenylphosphine, an indication of cleavage of the azine.

Seyferth *et al.* (*32*) proposed that the elimination of nitrogen in the single case reported above may have resulted from a reversal of the azine-formation reaction followed by decomposition of the diphenyldiazomethane into diphenylcarbene, the latter then being trapped by triphenylphosphine. If such were the case, it would be difficult to explain why the azine from diazofluorene and triphenylphosphine would not follow a similar course since the resulting ylid certainly would have survived the reaction conditions (*44*, Johnson). The same might be said for the diazocyclopentadiene case which formed an azine but could not be pyrolyzed into the well-known and perfectly stable ylid (*45*, Ramirez and Levy).

Wittig and Haag (*46*) were unable to thermally decompose methylenetriphenylphosphoranyl azine (XXIV) into methylenetriphenylphosphorane. Later (*47*) they repeated the decomposition in the presence of cyclohexene but were unable to trap any norcarane. It must be concluded that there is little reason to suspect the intermediacy of a carbene in the azine pyrolysis reaction.

$$\underset{\text{XXIV}}{(C_6H_5)_3P{=}N{-}N{=}CH_2} + \text{cyclohexene} \not\longrightarrow \text{norcarane } (C_6H_{10}{>}CH_2) \quad [3.19]$$

Wittig and Schlosser (*48*) recently have discovered a method for carrying out the pyrolysis of azines to produce phosphonium ylids. Treatment of triphenylphosphine with a diazo compound in the presence of cuprous chloride led to the formation of low yields of ylids contaminated with a variety of by-products. Use of phenyldiazomethane under these conditions followed by acidification of the solution led to the trapping of the ylid as its hydrochloride. Only 10% of the latter was obtained along with 45% of triphenylphosphine oxide and 41% of benzaldazine. The latter probably was formed by reaction of the ylid with phenyldiazomethane

$$(C_6H_5)_3P + C_6H_5CHN_2 \xrightarrow{CuCl} (C_6H_5)_3P{=}CHC_6H_5$$

$$(C_6H_5)_3P{=}CHC_6H_5 \xrightarrow{C_6H_5CHN_2} \left[(C_6H_5)_3\overset{\oplus}{P}{-}\underset{\overset{|}{\ominus N}{-}N{=}CHC_6H_5}{CH}{-}C_6H_5\right] \longrightarrow C_6H_5CH{=}N{-}N{=}CHC_6H_5 + (C_6H_5)_3P \quad [3.20]$$

$$(C_6H_5)_3P{=}CHC_6H_5 \xrightarrow{HCl} (C_6H_5)_3\overset{\oplus}{P}CH_2C_6H_5 \; Cl^{\ominus}$$

(*49*, Markl). Using diazomethane under these conditions in the presence of benzophenone led to a 23% yield of the Wittig reaction product, 1,1-diphenylethene. Treatment of the preformed azine (XXIV) with benzophenone and cuprous chloride gave the same olefin but in 35% yield.

$$(C_6H_5)_3P + CH_2N_2 \xrightarrow[(C_6H_5)_2CO]{CuCl} (C_6H_5)_2C{=}CH_2 \xleftarrow[(C_6H_5)_2CO]{CuCl} \underset{\text{XXIV}}{(C_6H_5)_3P{=}N{-}N{=}CH_2} \quad [3.21]$$

It seems reasonable to conclude that in spite of the ease of formation of phosphinazines their pyrolysis as a route to phosphonium ylids leaves much to be desired as a synthetic method.

3. *Synthesis of Ylids via Addition to Vinylphosphonium Salts*

The addition of nucleophiles to vinylphosphonium salts has been known for some time and, as mentioned in Chapter 2, has been used as evidence for the ability of phosphorus to expand its valence shell. Two groups recently discovered means of using the intermediates of such additions, which are in fact ylids, in typical ylid reactions. In the past the normal operation after such an addition had occurred was to quench the system with a proton source which did nothing more than form the ylid conjugate acid.

Seyferth *et al.* (*50*) observed that treatment of a solution of vinyltriphenylphosphonium bromide with phenyllithium and then with acetone afforded a 33% yield of 1-phenyl-3-methyl-2-butene. The product

$$(C_6H_5)_3\overset{\oplus}{P}-CH=CH_2 \xrightarrow{C_6H_5Li} [(C_6H_5)_3\overset{\oplus}{P}-\overset{\ominus}{C}H-CH_2C_6H_5] \quad \text{XXV} \xrightarrow{(CH_3)_2CO} (CH_3)_2C=CHCH_2C_6H_5 \qquad [3.22]$$

undoubtedly was formed by nucleophilic addition of the phenyl carbanion to the vinyl group followed by reaction of the resulting ylid (XXV) with acetone in a typical Wittig reaction. In a similar fashion treatment of 2-bromoethyltriphenylphosphonium bromide with two equivalents of methyllithium and then cyclohexanone afforded a 13% yield of *n*-propylidenecyclohexane and a 6% yield of methylene cyclohexane.

Schweizer and coworkers (*51, 52*) recently have applied this general method of ylid preparation to an elegant synthesis of heterocyclic ring systems. They have combined the ylid formation with an intramolecular Wittig reaction, making no attempt to isolate the ylids. The oxyanion (XXVI) was added to vinyltriphenylphosphonium bromide to form a new ylid (XXVII) which rapidly cyclized to 3,4-chromene. Similarly, reaction of the anion of 2-formylpyrrole with the vinyl salt afforded pyrrolizine, presumably via the ylid (XXVIII).

This general method holds some potential for the preparation of ylids of the general type $(C_6H_5)_3P=CH-CH_2-R$. The advantage is that the R group does not have to be attached until the very last step in the reaction sequence. The major limitation of the method is that ligand exchange is liable to occur, the carbanion group R displacing either the vinyl or a phenyl group from phosphorus, ultimately leading to the generation of another ylid. This alternate reaction accounted for the

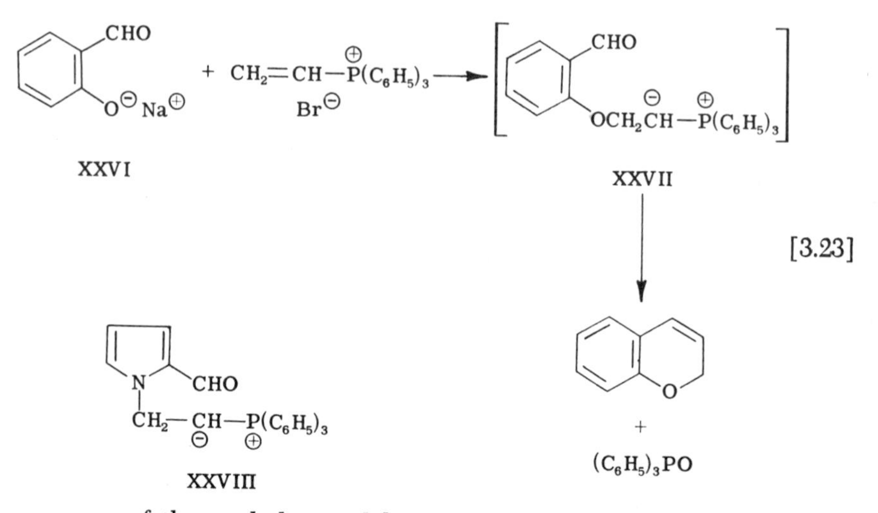

presence of the methylenecyclohexane along with the propylidenecyclohexane reported by Seyferth and outlined above (*50*). This limitation is likely to render the cyclization method of Schweizer applicable only to oxygen or nitrogen bases since most carbon bases would be nucleophilic enough to carry out ligand exchange.

4. Synthesis of Ylids via Addition to Benzyne

With the publication by Franzen and coworkers (*53*) of the synthesis of methylenemethylphenylsulfurane via the addition of methylsulfide to benzyne the reaction potentially became useful for an analogous synthesis of phosphonium ylids. To date there has been but one attempt to utilize the method, and not much promise can be held out for its general application.

Seyferth and Burlitch (*54*) reported that treatment of *o*-bromofluorobenzene with either lithium amalgam or magnesium in tetrahydrofuran and in the presence of methyldiphenylphosphine led to the formation of a red solution. Addition of cyclohexanone afforded 21% and 14% yields, respectively, of methylenecyclohexane. When the tertiary phosphine was used in elevenfold excess the yield of olefin increased to 52%. The reaction was thought to proceed via the addition of the phosphine to benzyne initially to afford a zwitterion (XXIX) which then underwent a prototropic shift to form the more stable ylid, methylenetriphenylphosphorane (III). The latter was then trapped in a Wittig reaction with cyclohexanone.

This synthesis should be a general one permitting the use of any alkyldiphenylphosphine, $RR'CH—P(C_6H_5)_2$. The reaction involves the

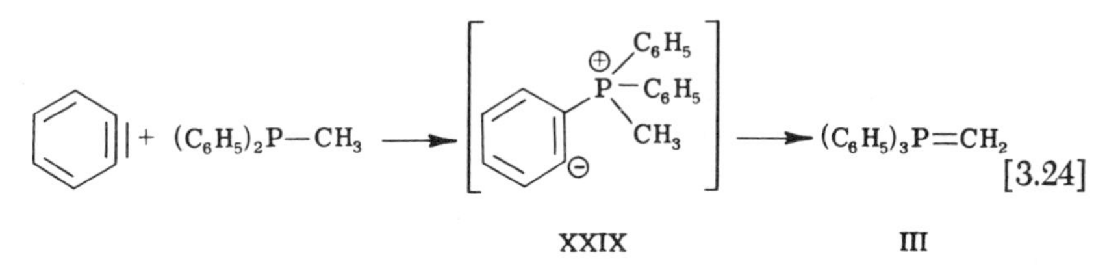

[3.24]

other basic approach to the formation of phosphonium ylids, that of putting one of the phenyl groups on phosphorus in the last operation rather than the more usual procedure of starting with triphenylphosphine and putting the potential ylid carbon on in the last step. However, the usefulness of the benzyne synthesis is limited by the availability of the appropriate phosphine. The latter could be prepared from an alkyl halide, RR′CHBr, by displacement with the diphenylphosphide anion but this introduces an extra step into the whole operation. Displacement on the same alkyl halide with triphenylphosphine instead of the phosphide anion would afford the alkyltriphenylphosphonium salt normally used as an ylid precursor (method no. 11). The latter would involve the use of much simpler experimental conditions and a weaker nucleophile, thereby lessening the chance for elimination reactions to compete.

5. *Synthesis of Ylids via Addition to Olefins and Alkynes*

Ylids have been prepared by the addition of triphenylphosphine to α,β-unsaturated carbonyl compounds. The general mechanism behind most of these additions probably is as shown in [3.25]. The intermediate

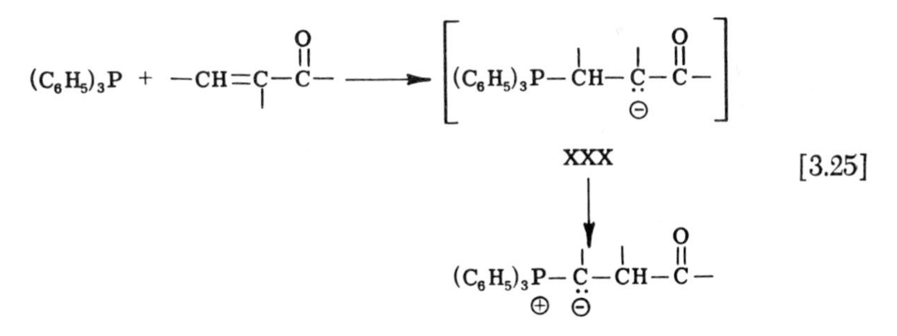

[3.25]

(XXX) from the conjugate addition to the olefinic group must have a proton on the α-carbon in order that a prototropic shift may take place to form the β-carbonyl ylid.

The earliest example of this type of reaction was reported by Schonberg and Michaelis (*55*) between triphenylphosphine and *p*-benzoquinone. They proposed the adduct to have structure XXXI but Ramirez and Dershowitz (*56*) more recently have shown the adduct to have the betaine structure (XXXII). The adduct must be viewed as a resonance hybrid which includes, as one of the important contributing structures,

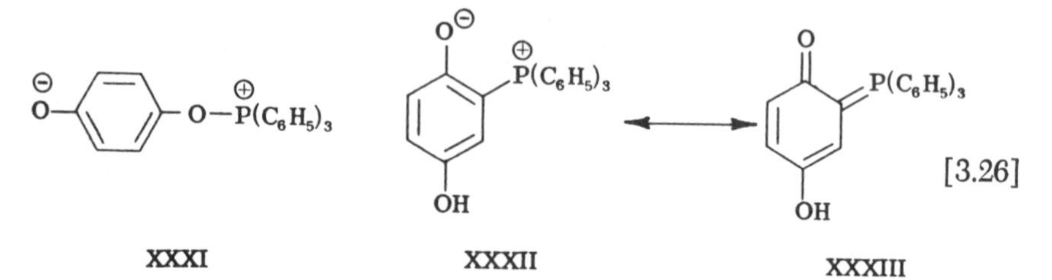

the ylid form (XXXIII). Most of the nucleophilic power of the adduct appears to be concentrated on the oxygen atom, however.

Schonberg and Ismail (*57*) discovered another example of this general reaction with their observation that triphenylphosphine reacted with maleic anhydride to afford an adduct for which they proposed structure XXXIV. More recently Aksnes (*58*) has proposed the ylid structure

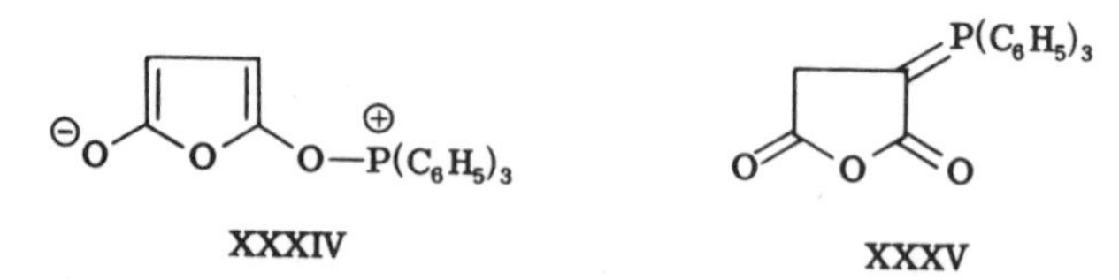

(XXXV) for the adduct but only on the basis of the anhydride-like infrared spectrum. Osuch *et al.* (*59*) have provided chemical confirmation of Aksnes' proposal, Hoffmann (*60*) has reported similar behavior for diethyl fumarate with triphenylphosphine, affording diethyl 2-triphenylphosphoranylsuccinate.

Ramirez and coworkers (*61–63*) recently have reported the addition of triphenylphosphine, tri-*n*-butylphosphine, trimethylphosphite and ethyl diphenylphosphinite all to *trans*-dibenzoylethylene. In each case the corresponding phenacylbenzoylmethylenephosphorane was obtained in accord with the general mechanism presented earlier. Horner and Klup-

$$R_3P + C_6H_5COCH{=}CHCOC_6H_5 \longrightarrow \left[C_6H_5{-}CO{-}\underset{R_3\overset{\oplus}{P}}{\underset{|}{C}H}{-}\underset{\ominus}{\ddot{C}H}{-}COC_6H_5 \right] \longrightarrow C_6H_5CO{-}\underset{R_3P}{\underset{\|}{C}}{-}CH_2COC_6H_5 \qquad [3.27]$$

fel (*64*) earlier had proposed zwitterionic phosphorus-oxygen adduct structures for these substances.

Takashina and Price (*65*) had found that triphenylphosphine catalyzed the conversion of acrylonitrile into its hexamer and proposed that the reaction proceeded by nucleophilic addition of triphenylphosphine to form an intermediate zwitterion which underwent a prototropic shift to afford an ylid (XXXVI). The ylid was expected to undergo

$$(C_6H_5)_3\overset{\oplus}{P}-\overset{\ominus}{C}H-CH_2CN \qquad C_6H_5CH{=}CH-CH_2CN$$

XXXVI XXXVII

subsequent reaction with additional monomer. Oda *et al.* (*66*) recently have demonstrated the validity of the Price proposal by repeating the reaction but in the presence of benzaldehyde. They isolated the unsaturated nitrile (XXXVII), presumably the result of a reaction between the aldehyde and the ylid intermediate (XXXVI). The same group subsequently demonstrated the generality of this approach to ylids by the preparation and trapping of ethyl 2-triphenylphosphoranylpropanoate from triphenylphosphine and ethyl acrylate.

Complex ylids also have been prepared by the addition of simple ylids to conjugated olefins. Bestmann and Seng (*67*) found that XXXVIII added to methyl 3-benzoylpropenoate afforded a new ylid (XXXIX). The

$$\underset{\text{XXXVIII}}{(C_6H_5)P{=}CHCO_2CH_3} + C_6H_5COCH{=}CHCO_2CH_3 \longrightarrow \underset{\text{XXXIX}}{C_6H_5COCH_2-\overset{\overset{CO_2CH_3}{|}}{C}H-\underset{\underset{CO_2CH_3}{|}}{C}{=}P(C_6H_5)_3} \quad [3.28]$$

formation of this product can be rationalized by a Michael-type addition of the ylid followed by a prototropic shift. In a somewhat related case, Trippett (*68*) noted that when the olefinic portion in such additions contained a group capable of being eliminated an unsaturated ylid resulted. Thus, cyanomethylenetriphenylphosphorane with tetracyano-

$$(C_6H_5)_3P{=}CHCN + (NC)_2C{=}C(CN)_2 \longrightarrow \left[(C_6H_5)_3\overset{\oplus}{P}-\underset{\underset{H}{|}}{\overset{\overset{CN}{|}}{C}}-\underset{\underset{CN}{|}}{\overset{\overset{CN}{|}}{C}}-\underset{\underset{CN}{|}}{\overset{\overset{CN}{|}}{C}}{}^{\ominus}\right] \longrightarrow \underset{\text{XL}}{(C_6H_5)_3P{=}\overset{\overset{CN}{|}}{C}-\overset{\overset{CN}{|}}{C}{=}\underset{\underset{CN}{|}}{\overset{\overset{CN}{|}}{C}}} \quad [3.29]$$

ethylene afforded the unsaturated tetracyanoylid (XL). He suggested that the intermediate carbanion ejected a cyanide ion and a proton.

Several groups only recently have investigated the reaction between phosphonium ylids and conjugated alkynes. Hendrickson (*69*) originally reacted benzoylmethylenetriphenylphosphorane with diethyl acetylenedicarboxylate (XLI) and obtained a 1:1 adduct for which he proposed a cyclic phosphoranyl structure (XLII). However, the experimental data also are in agreement with the more likely ylid structure (XLIII). The

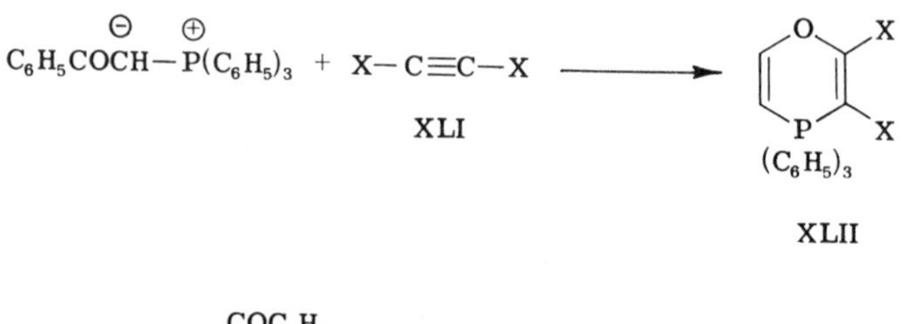

[3.30]

$$(C_6H_5)_3P{=}\underset{\displaystyle COC_6H_5}{C}{-}\underset{X}{C}{=}\underset{X}{C}H \qquad C_6H_5C{\equiv}C{-}\underset{X}{C}{=}\underset{X}{C}H$$

XLIII XLIV

$(X = COOCH_3)$

NMR (nuclear magnetic resonance) data are consistent with the latter formulation and such β-keto ylids are known to undergo pyrolytic elimination of triphenylphosphine oxide to afford alkynes (*21*, Trippett and Walker). Hendrickson's adduct did undergo such an elimination to afford a diester, probably XLIV, which could be hydrolyzed and reduced (3 moles of hydrogen) to β-phenylethylsuccinic acid. The ylid formulation (XLIII) is completely consistent with the other additions of ylids to conjugated carbonyl compounds, i.e., a normal Michael addition followed by a prototropic shift.

When the ylid carrying out a Michael addition to an acetylenic ester (XLI) does not have any α-hydrogen another course of reac-

$$(C_6H_5)_3P{=}C(C_6H_5)_2 + \underset{\text{XLI}}{X{-}C{\equiv}C{-}X} \longrightarrow \underset{\text{XLV}}{(C_6H_5)_3P{=}\underset{X}{C}{-}\underset{X}{C}{=}C(C_6H_5)_2}$$

[3.31]

$$\begin{array}{l}(C_6H_5)_2C{-}C{-}X \\ \quad\;\; | \quad\; \| \\ (C_6H_5)_3P{-}C{-}X\end{array}$$

XLVI

$(X = COOC_2H_5)$

tion prevails. The initial addition probably takes place as expected but the carbanionic intermediate must rearrange to form a new ylid. For example, Brown *et al.* (*70*) found that benzhydrylidenetriphenylphosphorane added to XLI to afford the new ylid (XLV). The structure of this ylid has not been proven unambiguously but Bestmann and Rothe (*71*) have discovered analogous reactions between XLI and a series of ylids $RCH{=}P(C_6H_5)_3$ where R was a variety of alkyl or aryl groups. These authors have proven the structure of the products to be analogous to XLV. A four-membered ring structure (XLVI) has been proposed by Brown (*70*) to be an intermediate in the reaction.

It would appear reasonable to assume that the initial step in all of these reactions with acetylenic esters (XLI) was the addition of the ylid to the alkyne in a typical Michael addition to form the intermediate

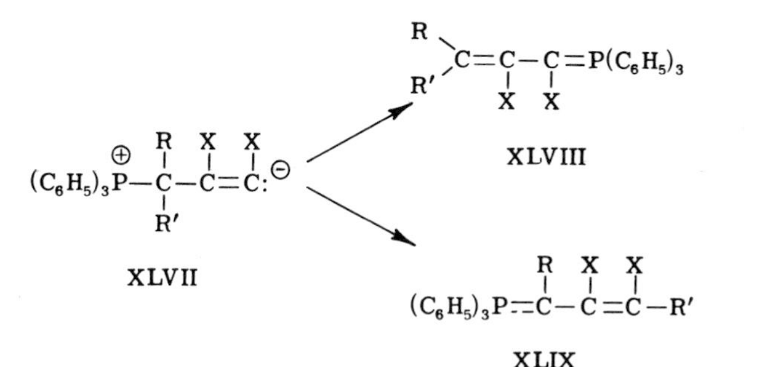

[3.32]

carbanion (XLVII, $X{=}COOC_2H_5$). If both R and R′ are other than hydrogen and are not prone to migrate the only course open to the intermediate would be the formation of a four-membered intermediate or transition state (i.e., XLVI) with the ultimate transfer of the phosphonium group and formation of a new ylid (XLVIII). However, when R or R′ are hydrogen and R or R′ is a group capable of providing considerable stabilization for a potential ylid, the intermediate could complete the normal Michael addition and form the ylid (XLIX) by prototropic shift. This proposal seems to account for all the reactions of phosphonium ylids with XLI.

It is apparent from the above discussions that the addition of triphenylphosphine or of simple phosphonium ylids to olefinic carbonyl compounds does provide a useful synthesis for complex ylids. It is equally apparent, however, that the addition of ylids to acetylenic carbonyl compounds is not an unambiguous route to complex ylids.

6. *Synthesis of Complex Ylids by Alkylation of Simple Ylids*

The synthesis of a dialkylmethylenephosphorane by the normal salt method would require the alkylation of a tertiary phosphine by a secondary alkyl halide in the first step. In some cases an elimination reaction may compete. An alternate route to the same disubstituted ylid is afforded by the preparation of a mono-substituted ylid by the normal "salt method" followed by its alkylation. The two alternatives are shown in [3.33].

$$
\begin{array}{c}
R'R\,CH{-}Br + (C_6H_5)_3P \longrightarrow R'R\,CH{-}\overset{\oplus}{P}(C_6H_5)_3\ Br^{\ominus} \xrightarrow{B:} R'R\,C{=}P(C_6H_5)_3 \\
\uparrow \text{RBr} \\
R'CH_2Br + (C_6H_5)_3P \xrightarrow{B:} R'CH{=}P(C_6H_5)_3
\end{array}
\qquad [3.33]
$$

Bestmann and Schulz (*72*) were the first to demonstrate the alkylation of ylids. They found, for example, that treatment of carbomethoxymethylenetriphenylphosphorane with benzyl bromide afforded carbomethoxybenzylmethylenetriphenylphosphorane (L) and carbomethoxymethyltriphenylphosphonium bromide each in 75% yield of an equimolar ratio. They proposed that the ylid carried out a nucleophilic substitution on the alkyl halide to afford the substituted phosphonium salt (LI). The latter was presumed converted into the corresponding ylid (L) by a second mole of the original ylid. This preparation of L certainly is much

$$
(C_6H_5)_3P{=}CHCOOCH_3 + C_6H_5CH_2Br \longrightarrow (C_6H_5)_3\overset{\oplus}{P}{-}CH(COOCH_3)(CH_2C_6H_5)\ Br^{\ominus} \quad \text{LI}
$$

$$
\text{LI} + (C_6H_5)_3P{=}CHCOOCH_3 \longrightarrow (C_6H_5)_3P{=}C(COOCH_3)(CH_2C_6H_5)\ (\text{L}) + (C_6H_5)_3\overset{\oplus}{P}{-}CH_2COOCH_3\ Br^{\ominus} \qquad [3.34]
$$

simpler than the salt method would be. The latter route would involve a preparation of methyl α-bromodihydrocinnamate followed by its alkylation with triphenylphosphine. The first step would be messy and the second might lead to elimination. In the Bestmann method all of the reagents are available and the reactions straightforward. Bestmann showed at a later date that the "*transylidation*" proposed in the last step does take place (*73*).

Bestmann and Schulz (*74*) provided many other examples of this method using a variety of alkyl halides. Markl (*75*) has used trialkyloxonium salts as the alkylating agent in similar reactions. Petragnini and Campos (*76*) used phenylbromoselenide as an alkylating agent and obtained the substituted ylid as in [3.35]. Cyclic alkylations also have

$$2\,(C_6H_5)_3P{=}CHCOOC_2H_5 + C_6H_5SeBr \longrightarrow (C_6H_5)_3P{=}C(SeC_6H_5)(COOC_2H_5) \quad [3.35]$$

been reported. Bestmann and Haberlein (*77*) noted that treatment of the phosphonium salt (LII) with sodium alkoxide resulted in the formation

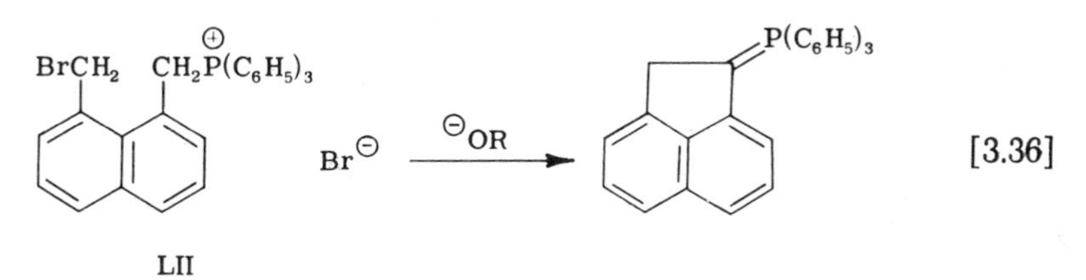

(but not the isolation) of acenaphthenylidenetriphenylphosphorane. Markl (*78, 79*) also has used the alkylation reaction in his synthesis of cyclic phosphonium ylids.

The above examples suffice to demonstrate that the alkylation of simple phosphonium ylids provides a useful route to complex ylids. The mechanism of the transformation seems obvious and uncomplicated.

7. *Synthesis of Complex Ylids by Acylation of Simple Ylids*

The acylation of ylids as a procedure for the synthesis of keto-ylids was first exploited by Bestmann (*80*). He found that one equivalent of benzoyl chloride reacted with two equivalents of methylenetriphenylphosphorane (III) to afford one equivalent of benzoylmethylenetriphenylphosphorane and one equivalent of methyltriphenylphosphonium chloride. The mechanistic interpretation in [3.37] was advanced. There

$$\underset{}{C_6H_5COCl} + \underset{\text{III}}{CH_2{=}P(C_6H_5)_3} \longrightarrow C_6H_5CO{-}CH_2{-}\overset{\oplus}{P}(C_6H_5)_3\ Cl^{\ominus}$$

$$\xrightarrow{\text{III}} C_6H_5COCH{=}P(C_6H_5)_3 + CH_3\overset{\oplus}{P}(C_6H_5)_3\ Cl^{\ominus} \quad [3.37]$$

seems to be little question regarding this mechanism since Bestmann (*73*) later demonstrated that III was a sufficiently strong base to convert phenacyltriphenylphosphonium salts into ylids. Bestmann and Arnason (*81*) and Trippett *et al.* (*8, 82*) later provided many additional examples of acylations with acid halides. Bestmann and Schulz (*83*) also used chlorocarbonates to prepare a variety of carboalkoxy ylids [3.38].

$$2\ RCH{=}P(C_6H_5)_3 + Cl{-}COOCH_3 \longrightarrow (C_6H_5)_3P{=}C\begin{matrix} R \\ COOCH_3 \end{matrix} \qquad [3.38]$$

Petragnini and Campos (*76*) found that sulfonyl halides with phosphonium ylids afforded sulfonylmethylenetriphenylphosphoranes.

The general acylation reaction is extremely useful for the preparation of any carbonyl-substituted ylid and provides a reliable alternative to the "salt method." The latter occasionally is complicated by side reactions such as the Perkow reaction (*84*, Trippett). The major disadvantage of the acylation method lies in the requirement of two moles of ylid per mole of acyl halide. One mole of ylid simply acts as a base to regenerate the acylated ylid from its conjugate acid. This disadvantage could be circumvented if the group displaced when the ylid originally attacked the acylating agent were itself a strong enough base to effect the ylid regeneration. This reasoning was behind the study by Bestmann and Arnason (*81, 85*) of the reaction of phosphonium ylids with thioesters. They found that methylenetriphenylphosphorane reacted with an equivalent of ethyl thiobenzoate to afford benzoylmethylenetriphenylphos-

$$(C_6H_5)_3P{=}CH_2 + C_6H_5{-}\overset{O}{\overset{\|}{C}}{-}SC_2H_5 \longrightarrow \left[C_6H_5{-}\overset{O}{\overset{\|}{C}}{-}CH_2{-}\overset{\oplus}{P}(C_6H_5)_3 \quad {}^{\ominus}SC_2H_5 \right]$$

$$\downarrow \qquad [3.39]$$

$$C_6H_5{-}\overset{O}{\overset{\|}{C}}{-}CH{=}P(C_6H_5)_3 + C_2H_5SH$$

phorane in 80% yield [3.39]. The thiolate ion ejected in the first step of the reaction apparently was sufficiently basic to convert the phosphonium salt into ylid. Therefore, acylation with thioesters is much more efficient than acylation with acyl halides.

Wittig and Schollkopf (*1*) had reported, at a much earlier date, the reaction between methylenetriphenylphosphorane and ethyl benzoate. They obtained phenacyltriphenylphosphonium bromide after quenching the reaction mixture with hydrobromic acid. This reaction presumably proceeded as for the thioester reaction. Somewhat later, Trippett and Walker (*8, 86*) found that methylenetriphenylphosphorane could be formylated in an analogous reaction with ethyl formate, and Staab and Sommer (*87*) found that *N*-formylimidazole also would function as a formylating agent. Since the imidazole anion was too weakly basic the latter reaction required two moles of ylid per mole of acylating agent.

The acylation of phosphonium ylids with esters to form keto-ylids also has been applied to the synthesis of cyclic ylids. House and Babad (*88*) found that treatment of a series of ω-carbethoxyalkyltriphenylphosphonium salts (LIII) with potassium *tert*-butoxide afforded the corresponding cyclic keto-ylids (LIV). The conversion probably proceeded by initial formation of the linear ylid followed by attack of the car-

$$(C_6H_5)_3\overset{\oplus}{P}{-}CH_2{-}(CH_2)_n{-}CH_2COOC_2H_5 \quad Br^{\ominus}$$

LIII

$$\downarrow KOBu^t$$

$$(C_6H_5)_3\overset{\oplus}{P}{-}\overset{\ominus}{C}H{-}(CH_2)_n{-}CH_2COOC_2H_5$$

$$\downarrow$$

$$\left[(C_6H_5)_3\overset{\oplus}{P}{-}CH{-}\overset{O}{\overset{\|}{C}}{-}CH_2 \;\; \overset{\ominus}{O}C_2H_5\right] \quad \text{(ring closed by } (CH_2)_n\text{)} \qquad [3.40]$$

$$\downarrow$$

$$(C_6H_5)_3P{=}C\diagup\overset{O}{\overset{\|}{C}}\diagdown CH_2 \quad \text{(ring closed by } (CH_2)_n\text{)}$$

LIV

banion on the carbethoxy group. The ethoxide ion ejected in the former step apparently acted as the base for the formation of the new ylid (LIV). This reaction afforded good yields when $n = 2$, 3, or 4. Bergelson *et al.* (*28*) also observed this reaction when $n = 2$ but claimed it did not take place with higher homologs.

Trippett and Walker (*21*) reported a novel acylation of ylids using phenylisocyanate. Reaction of the latter with carbethoxymethylenetriphenylphosphorane afforded the amido-ester (LV) and methylenetri-

$$(C_6H_5)_3P{=}CHCOOC_2H_5 + C_6H_5N{=}C{=}O \longrightarrow \left[\begin{array}{c} (C_6H_5)_3\overset{\oplus}{P}-\underset{\ominus O-\overset{|}{C}=N-C_6H_5}{CH}-COOC_2H_5 \\ \updownarrow \\ (C_6H_5)_3\overset{\oplus}{P}-\underset{O=\overset{|}{C}-\underset{\ominus}{N}-C_6H_5}{CH}-COOC_2H_5 \end{array}\right] \longrightarrow \underset{\textbf{LV}}{(C_6H_5)_3P{=}C\begin{matrix} COOC_2H_5 \\ CONHC_6H_5 \end{matrix}} \qquad [3.41]$$

phenylphosphorane could be doubly acylated with the same reagent. These reactions probably proceeded via initial addition of the ylid to the carbonyl group as for a typical Wittig reaction followed by a proton transfer directly to nitrogen or via the enol form. In this reaction, the intermediate betaine was a sufficiently strong base to carry out the regeneration of the acylated ylid. In those instances where there was no hydrogen on the carbon adjacent to phosphorus a normal Wittig reaction resulted (*89*, Staudinger and Meyer).

From the above examples it is clear that ylids can be acylated by a variety of acylating agents. Provided that there is at least one hydrogen on the initial ylid a new carbonyl-substituted ylid may result. This method, together with the alkylation of ylids as discussed in the previous section, provides a versatile alternative to the salt method for the preparation of a wide variety of phosphonium ylids. This flexibility is demonstrated in [3.42] for a general case. The only serious complication in

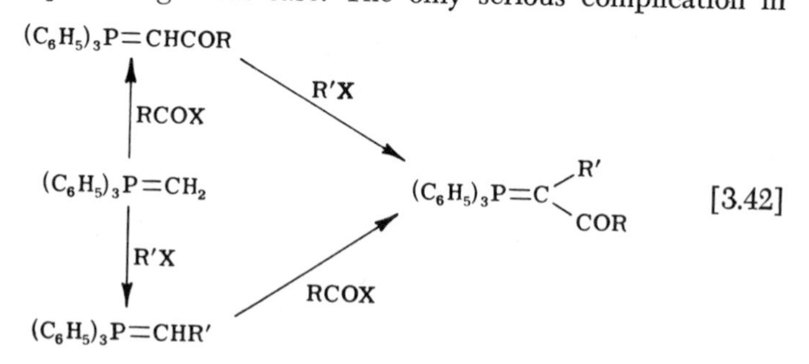

both the alkylation and acylation reactions is the possibility of C- and O-alkylation or acylation when using keto-ylids as starting materials (*90*, Chopard *et al.*). This aspect will be dealt with in Section III of this chapter under reactions of ylids.

8. *Synthesis of Haloylids by Halogenation of Simple Ylids*

The availability of α-haloylids is of some importance since their use in the Wittig reaction with carbonyl compounds is a valuable method for the preparation of vinyl halides. The formation of α-haloylids, $R—C(X)=P(C_6H_5)_3$, by the salt method is virtually unknown except in the case where R = H (*91*, Ramirez *et al.*). Markl (*92*) discovered a useful alternate route for the formation of such ylids, the direct halogenation of the simple ylid, $R—CH=P(C_6H_5)_3$. Carbethoxymethylenetriphenylphosphorane reacted with chlorine to afford the chloroylid (LVI) and the conjugate acid of the original ylid. Analogous reactions

$$(C_6H_5)_3P{=}CHCOOC_2H_5 + Cl_2 \longrightarrow (C_6H_5)_3\overset{\oplus}{P}{-}CH(Cl)(COOC_2H_5)\ Cl^{\ominus}$$

$$\xrightarrow{(C_6H_5)_3P{=}CHCOOC_2H_5} (C_6H_5)_3\overset{\oplus}{P}{-}CH_2COOC_2H_5\ Cl^{\ominus} + (C_6H_5)_3P{=}C(Cl)(COOC_2H_5) \quad [3.43]$$

LVI

occurred with bromine and iodine. Since the unhalogenated phosphonium salt was a by-product of the reaction it is evident that one equivalent of the original ylid was used as a base to reform the substituted ylid from its conjugate acid. Denney and Ross (*93*) later found that the same reaction, when run in pyridine solution, afforded the haloylid (LVI) in 80% yield. In this instance the pyridine presumably acted as the base in the last step of the reaction.

Several workers subsequently found that under certain conditions ylids would react with chlorine to afford only the chlorinated phosphonium salt (*93–95*). It appeared that the reaction proceeded as above with the initial formation of chloroylid and unsubstituted phosphonium salt. However, the phosphonium salt apparently underwent direct halogenation to form chlorophosphonium salt and hydrogen chloride, the latter reacting with chloroylid (LVI) also to form chlorophosphonium salt. Markl (*95*) has shown that phosphonium salts can be halogenated directly to chlorophosphonium salts. The latter, upon simple treatment with aqueous sodium carbonate, were converted to the chloroylids.

9. *Ylids from Active Methylene Compounds and Tertiary Phosphine Dihalides*

Horner and Oediger (*22*) discovered a simple method for the synthesis of functionally disubstituted phosphonium ylids. The key to the synthesis was the formation of a carbon-phosphorus bond by carrying out a nucleophilic substitution on triphenylphosphine dichloride. For example, the dihalide reacted with diethylmalonate in the presence of triethylamine to form dicarbethoxymethylenetriphenylphosphorane. The reaction appears to be general and has been applied to the cases where X and Y were cyano, sulfonyl, carbethoxy and acyl groups. The reaction probably proceeded as shown in [3.44]. This method is limited

$$X-CH_2-Y \xrightarrow{R_3N} X-\overset{\ominus}{\ddot{C}}H-Y \xrightarrow{(C_6H_5)_3PCl_2} (C_6H_5)_3\overset{\oplus}{P}-CH(X)(Y) \quad Cl^{\ominus}$$

$$(C_6H_5)_3\overset{\oplus}{P}-CH(X)(Y) \xrightarrow{R_3N} (C_6H_5)_3P{=}C(X)(Y) \qquad [3.44]$$

to the preparation of ylids carrying two electron-withdrawing groups but is a valuable alternative to what often might be the initial approach of acylation of a simple ylid. The method has not had any practical use since the ylids obtainable usually are too stable to undergo the Wittig reaction.

10. *Synthesis of Quino-Ylids*

Several groups have attempted to prepare ylids in which the ylid carbanion is part of a potential aromatic ring system. Such attempts have involved the formation of an aromatic carbon-phosphorus bond and this problem has been approached in interesting ways.

Schonberg and Michaelis (*55*) appear to be the first to have explored this problem in their reaction between *p*-benzoquinone and triphenylphosphine. They did obtain a 1:1 adduct for which they proposed the quasi-phosphonium salt structure (XXXI) but for which Ramirez and Dershowitz (*56*) proposed the tetraarylphosphonium salt structure (XXXII). The observation that the substance could be dialkylated with ethyl iodide clearly eliminated the Schonberg structure. Additional evidence for the Ramirez structure was provided by the observation that the same substance could be obtained from reaction of 2,5-dimethoxybromobenzene with triphenylphosphine in the presence of cobaltous chloride when it was followed by an ether cleavage (*96*, Hoffmann *et al.*).

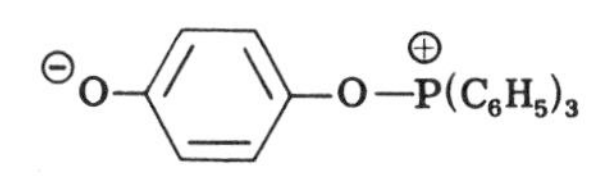

XXXI

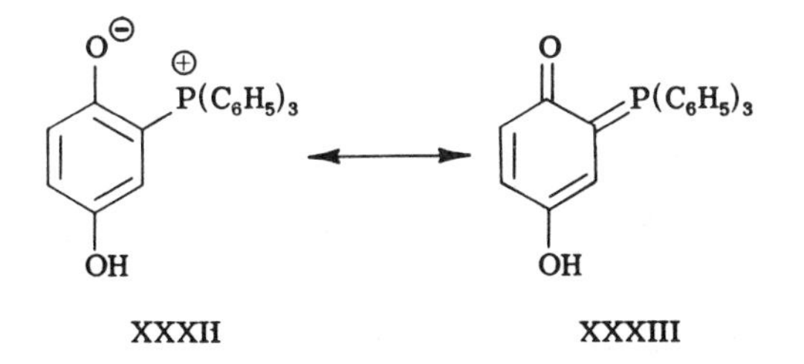

XXXII XXXIII

The method used by Schonberg (*55*) and Ramirez (*56*), the addition of a phosphine to an unsaturated carbonyl system, is simply that discussed previously in this section for open-chain structures. The application of this reaction for the preparation of potential aryl ylids is of little practical use, however, since there is no evidence that the carbon atom carrying the phosphonium group in the product has any carbanionic character. In other words, XXXIII contributes very little to the actual structure of the resonance hybrid. The only evidence bearing on this point, however, was the observation that alkylation took place on the oxygen atom rather than on carbon (*56*).

McDonald and Campbell (*97*) found that triphenylphosphine could be arylated directly by heating with *p*-bromophenol. The resulting *p*-hydroxyphenyltriphenylphosphonium bromide upon "treatment—in ethanol with lithium ethylate gave no evidence of reaction." However, Horner *et al.* (*98*) found that the same salt, upon treatment with aqueous hydroxide, afforded a halogen-free, high-melting substance whose anal-

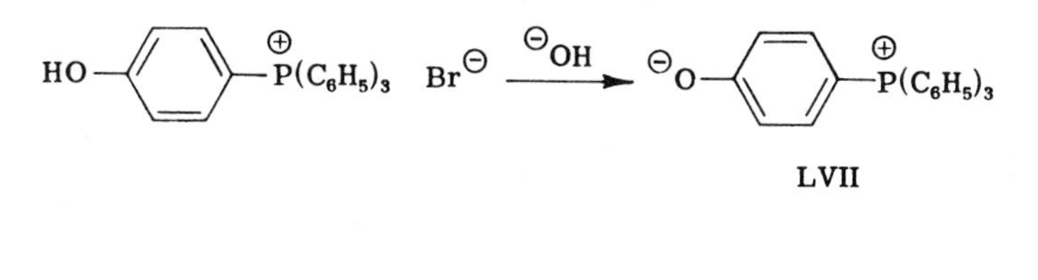

LVII

[3.45]

O=⟨ ⟩=P(C₆H₅)₃

LVIII

ysis corresponded to a dihydrate of the zwitterion (LVII). The substance reacted with methyl iodide in the absence of added base to afford *p*-anisyltriphenylphosphonium iodide.

As in the previous case, no evidence has been provided to indicate that the ylid structure (LVIII) contributes appreciably. In this instance, however, there is some reason to pursue evidence for LVIII since the "azolog" (LIX) has been prepared and is a stable substance (*99*, Horner and Schmelzer). This phosphinazine could not be alkylated with

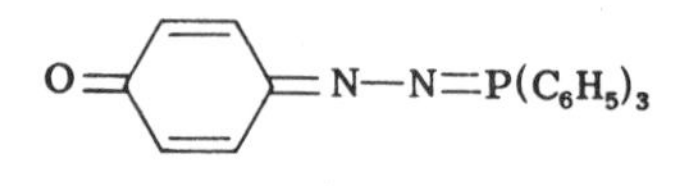

LIX

methyl iodide and was stable to hot aqueous base. However, it could be converted into the *p*-hydroxyphenylazotriphenylphosphonium bromide with dilute sulfuric acid.

The problem faced with the quino-ylids is analogous to that encountered with other ambident anions, and the same approaches, for example, studies of solvent effects (*100*, Kornblum *et al.*), could be applied in these cases.

11. Synthesis of Ylids from Phosphonium Salts (Salt Method)

The most widely applicable method for the formation of phosphonium ylids is treatment of the corresponding quaternary phosphonium salt with a suitable base [3.46]. The generality of the method depends on the availability of the appropriate phosphonium salt.

$$\begin{matrix} R^1 \\ \quad\diagdown \\ \end{matrix} CH-\overset{\oplus}{P}X_3 \xrightarrow{B:} \begin{matrix} R^1 \\ \quad\diagdown \\ \end{matrix} C{=}PX_3 \quad [3.46]$$

$$R^1R^2CH-\overset{\oplus}{P}X_3 \xrightarrow{B:} R^1R^2C{=}PX_3 \qquad [3.46]$$

a. Preparation of Phosphonium Salts. Most quaternary phosphonium salts are prepared by subjecting a tertiary phosphine to reaction with an electrophilic agent. If the salt ultimately is to be converted to an ylid it must carry at least one alkyl group which, in turn, carries at least one hydrogen. There are two ways in which the quaternary phosphonium salt could be prepared. The most common route is to subject a tertiary phosphine (PX_3) to an alkylation. The symmetrical tertiary phosphines are readily available. The alternative is to carry out a phenylation of the alkyl phosphine, $R_1R_2CH—PX_2$, perhaps using benzyne as per method no. 4. The phosphine most often used for alkylation is triphenylphosphine.

The preparation of the first phosphonium ylid by Michaelis and Gimborn in 1894 (*2*) involved, as the first step, the alkylation of tri-

$$(C_6H_5)_3P + ClCH_2COOC_2H_5 \longrightarrow (C_6H_5)_3\overset{\oplus}{P}CH_2COOC_2H_5 \quad Cl^{\ominus} \qquad [3.47]$$

phenylphosphine with ethyl chloroacetate [3.47]. In the intervening seventy years there have been innumerable analogous alkylations of triphenylphosphine. The alkylating agents, for the most part, have been alkyl halides. It has not been necessary to resort to the exotic alkylating agents such as oxonium salts. As expected for a nucleophilic substitution the reactions occur most readily with the alkyl iodides and least readily with the chlorides. Hellmann and Schumacher (*101*) did report the alkylation of triphenylphosphine with a quaternary ammonium salt but the reaction is not of wide application.

Many different alkyl halides have been employed in this method. The majority of the ylids listed in Tables 3.1 through 3.4 were prepared by this method, and a perusal of those tables will indicate the halides that must have been used in the initial alkylation. The R groups in the halides (R_1R_2CH—X) have been alkyl, aryl, alkoxyl, alkene, alkyne, cycloaliphatic, halogen, silyl, carboalkoxy, carbonyl and nitrile. In addition there have been many examples of bis-phosphonium salts prepared from dihalides. One of the earliest was the bis-salt from ethylene bromide and triphenylphosphine prepared by Wittig *et al.* (*15*) [3.48]. All of the ylids listed in Table 3.3 were prepared by this method.

$$BrCH_2CH_2Br + 2\,(C_6H_5)_3P \longrightarrow (C_6H_5)_3\overset{\oplus}{P}-CH_2-CH_2-\overset{\oplus}{P}(C_6H_5)_3 \quad 2\,Br^{\ominus} \qquad [3.48]$$

There have been some problems associated with the alkylation of triphenylphosphine. Elimination of the alkyl halide occasionally competes with its substitution. Trippett (*102*) reported that *sec*-butyl iodide was converted into 2-butene when refluxed with triphenylphosphine but that isopropyl iodide formed a normal salt. *tert*-Butyl chloride also was reported to eliminate but Horner and Mentrup (*103*) did obtain the expected *tert*-butyltriphenylphosphonium salt when the same reactants were heated under pressure. Seyferth *et al.* (*104*) obtained the bromide salt from *tert*-butyl bromide with no apparent complications.

Pommer (*105*) has shown that certain polyenic halides, particularly in the carotenoid field, tend to undergo elimination when treated with triphenylphosphine. However, the triphenylphosphonium halide has been shown, in a separate experiment, to add back to the elimination product to afford a quaternary phosphonium halide identical to that expected from a normal alkylation [3.49]. The use of allylic halides in the alkylation has led to some ambiguity in the structure of the resulting phosphonium salt. Bohlmann and Mannhardt (*106*) found that crotyl bromide was alkylated on the primary carbon to (2-butenyl)triphenylphos-

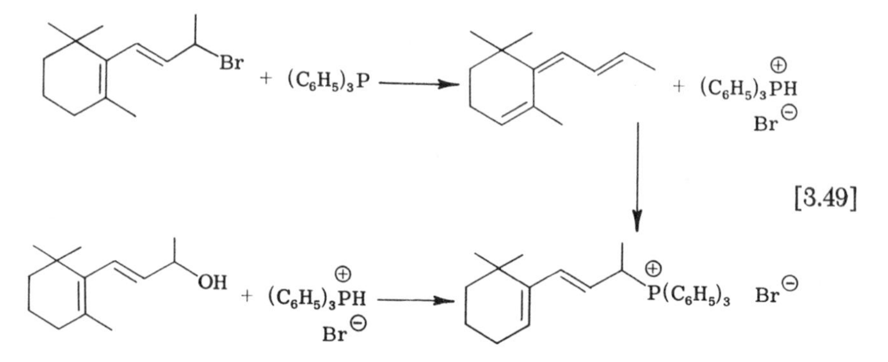

phonium bromide. However, Inhoffen *et al.* (*107*) found that 2-bromomethylenecyclohexane was converted to cyclohexenylmethyltriphenylphosphonium bromide by an allylic nucleophilic attack rather than to 2-methylenecyclohexyltriphenylphosphonium bromide [3.50]. While there

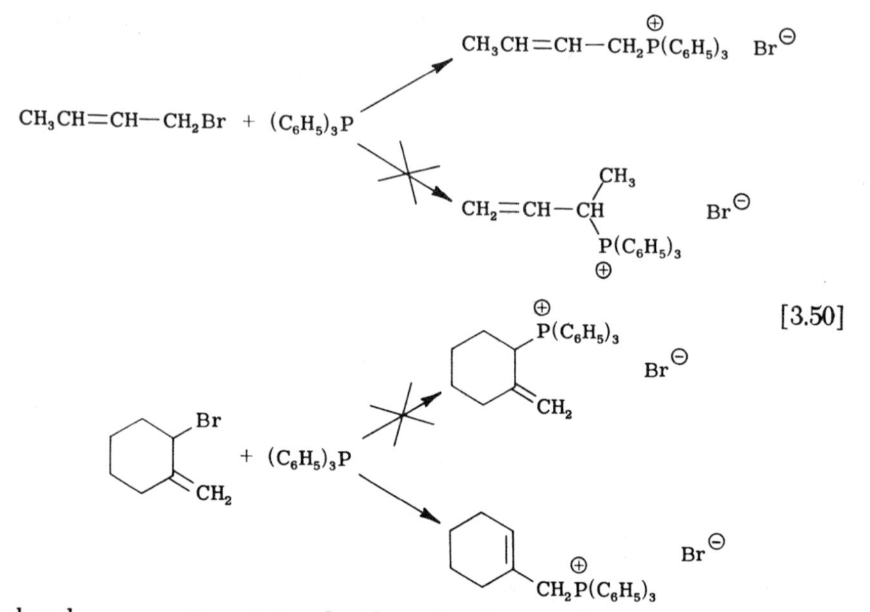

has been no attempt to elucidate the factors that control the site of nucleophilic attack in these alkylations it is apparent that the phosphine prefers to attack a primary carbon, whether it involves a S_N2 or a S_N2' reaction. This is consistent with normal behavior in the allylic halides.

Several compounds containing especially activated halogen, for example, α-haloketones, α-halonitro compounds and α-halosulfones, undergo dehalogenation rather than substitution on carbon when treated with triphenyl phosphine. The α-bromoketones occasionally undergo a reaction analogous to the Perkow reaction (*108*) to afford enolphosphonium salts (*84*, Trippett and Walker; *109*, Borowitz and Virkhaus). However, the simple systems such as α-chloroacetone and α-bromoacetophenone reacted normally with triphenylphosphine to afford the ketoalkyltriphenylphosphonium salts (*110*, Michaelis and Kohler). Hoffmann and Forster (*111*) found that α-bromo-α-phenylsulfonyl-*p*-nitrotoluene was dehalogenated with triphenylphosphine in methanolic solution. They proposed that the phosphine attacked the halogen and the resulting bromotriphenylphosphonium ion halogenated methanol to form methyl bromide. In a somewhat similar reaction Trippett and Walker (*84*) found that α-halo-α-nitroalkanes were dehalogenated and deoxygenated by triphenylphosphine affording triphenylphosphine oxide and nitriles.

Attempts to alkylate triphenylphosphine with polyhalomethanes have led to some interesting results. Ramirez *et al.* (*91*) found that methylene bromide and triphenylphosphine, when heated together at 150°, afforded a mixture of bromomethyltriphenylphosphonium bromide and the bis-salt (LX). Presumably these were straightforward alkylations. Recently, however, Matthews *et al.* (*112*) concluded that the course of this reaction was more complicated, perhaps involving ion pairs.

$$(C_6H_5)_3P + CH_2Br_2 \longrightarrow (C_6H_5)_3\overset{\oplus}{P}{-}CH_2Br\ \ Br^{\ominus} + \underset{\text{LX}}{(C_6H_5)_3\overset{\oplus}{P}{-}CH_2{-}\overset{\oplus}{P}(C_6H_5)_3\ \ 2\,Br^{\ominus}} \quad [3.51]$$

Triphenylphosphine was found not to react with bromoform at room temperature but reaction would take place under ultraviolet light or in the presence of benzoyl peroxide. On this basis Ramirez and McKelvie (*113*) proposed a radical chain process in which the initiation step involved the formation of a dibromomethyl radical. The latter was thought to attack the phosphine [3.52].

$$(C_6H_5)_3P + \cdot CHBr_2 \longrightarrow (C_6H_5)_3\dot{P}CHBr_2$$

$$(C_6H_5)_3\dot{P}{-}CHBr_2 + CHBr_3 \longrightarrow (C_6H_5)_3\overset{\oplus}{P}CHBr_2\ \ Br^{\ominus} + \cdot CHBr_2 \quad [3.52]$$

Carbon tetrachloride and triphenylphosphine were found to react at 65° to afford a solution which appeared to contain triphenylphosphine-dichloride and dichloromethylenetriphenylphosphorane (*114*, Rabinowitz and Marcus). The presence of these two substances was detected by reaction of the solution with benzaldehyde, the former producing benzal chloride and the latter affording β,β-dichlorostyrene. There was no indication that this reaction involved radicals so the authors proposed that the phosphine initially attacked halogen. A second molecule of the phosphine was thought to carry out a second attack on halogen [3.53].

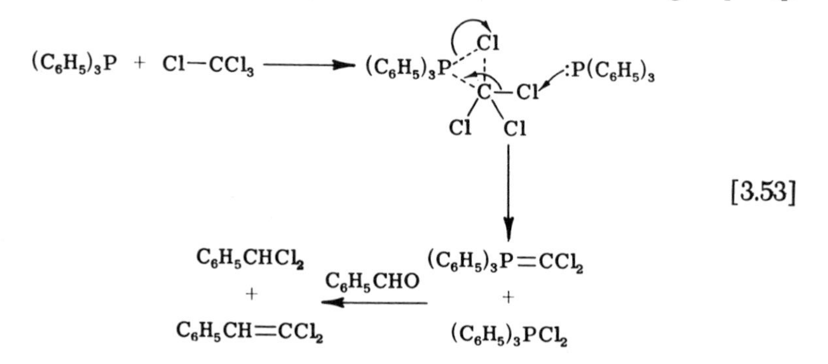

Ramirez *et al.* (*115*) subsequently examined the reaction between triphenylphosphine and carbon tetrabromide and also detected the presence of an ylid, dibromomethylenetriphenylphosphorane. They did not search for any phosphine dibromide but claimed the reaction possessed no radical characteristics. Hoffmann and Forster (*111*) found that di(phenylsulfonyl)dibromomethane and triphenylphosphine reacted in a similar fashion to afford the phosphine dibromide and di(phenylsulfonyl)methylenetriphenylphosphorane. These authors, however, proposed the initial attack of the phosphine to be on carbon followed by attack of a second phosphine on halogen [3.54].

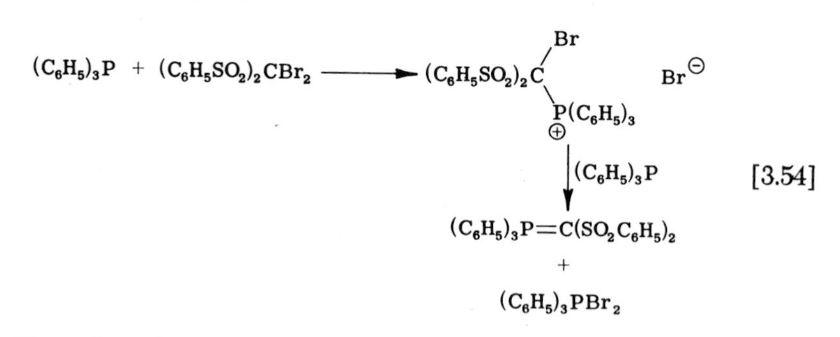

There have been no attempts to prove the mechanisms suggested in these reactions with polyhalomethanes. Nonetheless, it is apparent that simple alkylation by polyhalomethanes is not the normal result of reaction with triphenylphosphine and, therefore, not an unambiguous route to phosphonium salts and thence to phosphonium ylids.

b. Conversion of Phosphonium Salts into Phosphonium Ylids. Conversion of a phosphonium salt into a phosphonium ylid involves treating the salt with an appropriate base in order to remove an α-hydrogen [3.55]. The nature of the base depends on the structure of the phos-

$$\begin{matrix} R^1 \\ \\ R^2 \end{matrix}\!\!\!>\!CH-\overset{\oplus}{P}(C_6H_5)_3 \quad X^{\ominus} \xrightarrow{B:} \begin{matrix} R^1 \\ \\ R^2 \end{matrix}\!\!\!>\!\overset{\ominus}{C}-\overset{\oplus}{P}(C_6H_5)_3 + BH^{\oplus} \quad X^{\ominus} \qquad [3.55]$$

phonium salt, more specifically on the nature of the substituents R on the potential ylid carbanion. As would be expected any substituent which can delocalize the electron density on the carbanion will facilitate the formation of that ylid. In other words, the precursor phosphonium salt will be more acidic. These relationships will be discussed in detail in Section II of this chapter.

A qualitative picture of the relationship of phosphonium salt acidity to the nature of the substituents R can be obtained by an examination of the bases needed in the formation of various ylids. Methylenetriphenylphosphorane formation normally has required the treatment of methyltriphenylphosphonium salts with organolithium or other carbon bases. However, carbomethoxymethyltriphenylphosphonium salts were converted to their corresponding ylid in the presence of aqueous potassium hydroxide (*2*, Michaelis and Gimborn) and 9-fluorenyltriphenylphosphonium bromide only required aqueous ammonia for ylid formation (*116*, Pinck and Hilbert). In general, therefore, the more the ylid carbanion is stabilized by the substituents R or the heteroatom group the less powerful base is required for the formation of that ylid from its phosphonium salt precursor. Details on the base required for any given type of ylid is available by examination of Tables 3.1 through 3.4 for structural analogies.

A wide variety of different bases have been used in the formation of phosphonium ylids. Coffmann and Marvel (*117*) were the first to use a carbon base, using both trityl sodium and *n*-butyllithium to convert *n*-alkyltriphenylphosphonium salts into ylids. Other carbon bases that have been used successfully include phenyllithium (*3*), methylsulfinylcarbanion (*118*) and other, more basic, ylids (*73*). The latter is of little practical use.

Several nitrogen bases have been used for the formation of phosphonium ylids. As mentioned earlier, ammonia has been used in favorable cases (*116*). Others include sodium amide (*119*), lithium diethylamide (*120*) and lithium piperidide (*15, 121*). Oxygen bases used include sodium carbonate (*110*), sodium hydroxide (*2*), sodium ethoxide (*46*) and potassium *tert*-butoxide (*78*). Sodium hydride also has been used (*14*).

Welcher and Day (*122*) have reported an instance in which a base of insufficient strength was used in an attempted ylid synthesis. Treatment of the cyclic phosphonium salt (LXI) with aqueous sodium hydroxide led to an elimination reaction and cleavage of the ring system. That strength base would not be expected to convert LXI to the cyclic

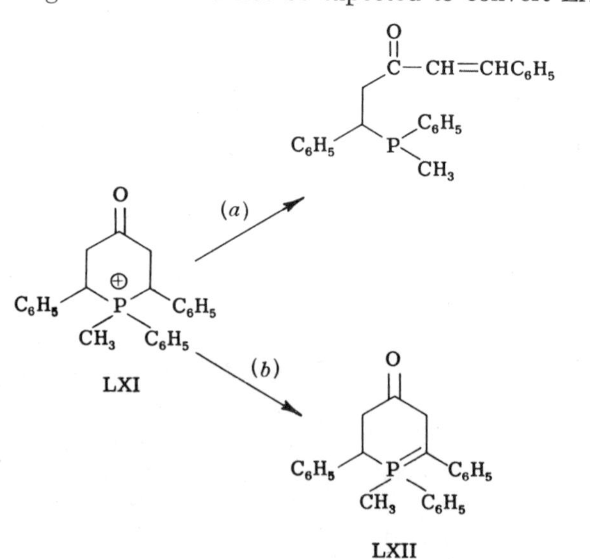

[3.56]

ylid (LXII) since the latter has little stabilization for the carbanion. Use of a stronger base could have taken advantage of a probable kinetic effect whereby reaction *a* likely would be slower than reaction *b*, thereby permitting the formation of some ylid.

A variety of solvents have been used in the formation of ylids from the phosphonium salts. These range from benzene and ethyl ether through dimethylsulfoxide and dimethylformamide. The solvent appears not to be crucial in the formation of an ylid but may be in the subsequent reaction to which the ylid is to be subjected. Often a slurry of salt is treated with the required base and the former gradually goes into solution as it is converted into ylid. Solvents which may react with the ylid obviously must be avoided. Therefore acetone, water and ethanol are seldom used except for stabilized ylids.

There are some circumstances in which the formation of a phosphonium ylid from a phosphonium salt is not as straightforward as may be implied by the preceding discussion. The choice of a base is especially crucial when forming an ylid carrying reactive substituents. Thus, Bergelson *et al.* (*123*) were able to prepare ω-carbethoxyalkylmethylenetriphenylphosphorane from the precursor salt using sodium ethoxide as the base whereas use of organolithium bases probably would have resulted in some attack on the ester group.

Attempts to form ylids from α-haloalkyltriphenylphosphonium salts using organolithium bases has led to ambiguous results. Wittig and Schlosser (*33*) found that treatment of chloromethyltriphenylphosphonium salts with phenyllithium led to proton abstraction and formation of chloromethylenetriphenylphosphorane, the latter being detected by its subsequent Wittig reaction with benzophenone or benzaldehyde. On the other hand, however, Seyferth *et al.* (*124*) found that the bromomethyl- and iodomethyl- analogs each afforded a mixture of two ylids. Bromomethyltriphenylphosphonium bromide underwent attack by phenyllithium both on hydrogen and on bromine to afford two ylids whose presence was detected by subsequent reaction with cyclohexanone [3.57]. On the basis of the ratio of bromobenzene to benzene

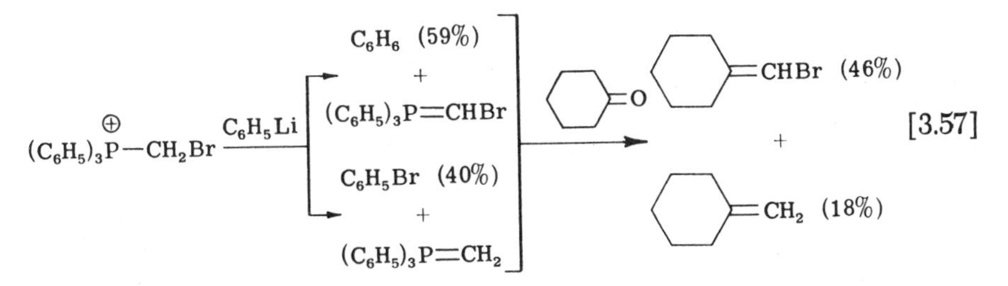

formed, the attack on hydrogen was favored by a ratio of 3:2. Kobrich (*121*) reported that using the same base but trapping the ylids with β-cyclocitral led to a ratio of attack on hydrogen to attack on bromine of just the opposite, 2:3. Changing to *n*-butyllithium in place of phenyllithium led to exclusive attack on halogen whereas use of the weaker lithium piperidide afforded exclusive attack on hydrogen. These trends are in accord with the general supposition that the stronger nucleophile should be better able to attack halogen.

In a similar vein Seyferth *et al.* (*124*) found that iodomethyltriphenylphosphonium iodide also was subject to attack both on hydrogen and halogen. In this instance, using cyclohexanone to trap the ylids, the ratio of attack on hydrogen to attack on halogen was 3:7. This trend also is consistent with that found in a normal halogen-metal interchange,

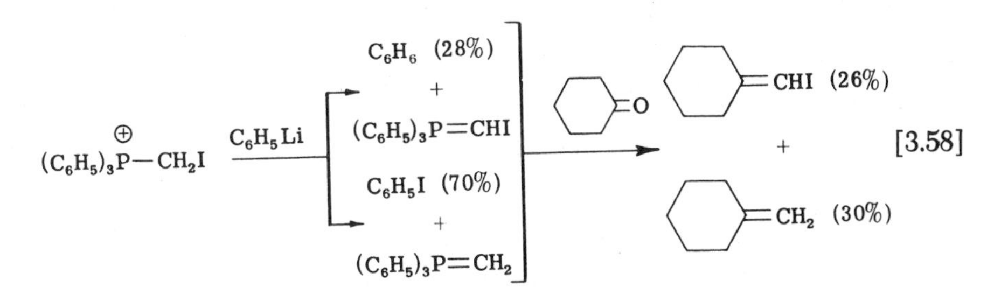

the order of susceptibility to attack by organolithium reagents being I > Br > Cl (*125*, Jones and Gilman).

Treatment of phosphonium salts with organolithium bases also can pose the problem of ligand exchange. Wittig and Rieber (*126*) previously had shown that tetra-substituted phosphonium salts were subject to nucleophilic attack on phosphorus since they prepared a pentaphenylphosphorane from tetraphenylphosphonium salts and phenyllithium. One might expect to find signs of similar behavior in attempts to prepare phosphonium ylids. Seyferth *et al.* (*127*) found that treatment of methyltriphenylphosphonium bromide with methyllithium gave less than a full equivalent of methane but did produce 21–26% of benzene. The ylid in the solution was methylenetriphenylphosphorane since it afforded methylenecyclohexane after reaction with cyclohexanone. They proposed that a pentavalent intermediate was formed by attack of methyllithium on phosphorus and that it subsequently ejected the phenyl anion rather than the methyl anion as shown in [3.59]. The

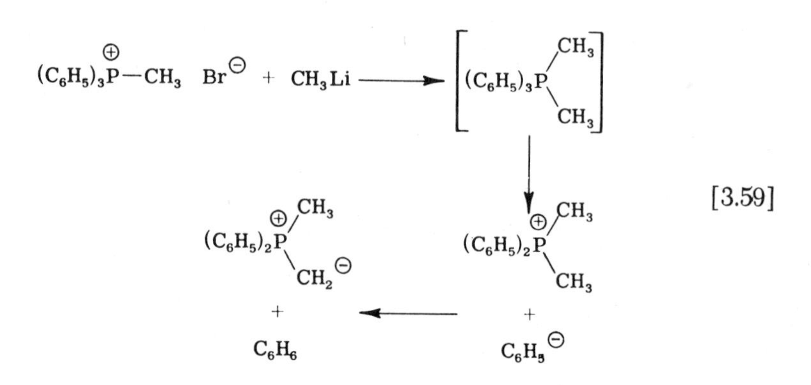

ejection of the phenyl anion from such an intermediate is consistent with the much earlier observations of Fenton and Ingold (*128*) with the hydrolyses of phosphonium salts. The benzene was thought to result from the phenyl anion converting the diphenyldimethylphosphonium salt into an ylid. Unfortunately, these workers did not attempt to isolate the phosphine oxide formed as a by-product of the Wittig reaction. It should have been diphenylmethylphosphine oxide rather than triphenylphosphine oxide.

A more striking example of ligand exchange was the reaction between tetraphenylphosphonium bromide and methyllithium which, in the presence of cyclohexanone, led to the formation of methylenecyclohexane in

$$(C_6H_5)_4\overset{\oplus}{P}\ \ Br^{\ominus} \xrightarrow{CH_3Li} (C_6H_5)_3P{=}CH_2 + C_6H_6 + LiBr \qquad [3.60]$$

58% yield (*127,* Seyferth *et al.*). Similarly, *tert*-butyltriphenylphosphonium bromide reacted with methyllithium to afford a solution which contained benzene and a methylenephosphorane (*104,* Seyferth *et al.*). In both of these examples ligand exchange followed by ylid formation must have taken place since the original phosphonium salts had no α-hydrogen and were incapable of forming ylids directly.

As far as the practical synthesis of ylids is concerned, the problem encountered is demonstrated by the observation that 2-bromoethyltriphenylphosphonium bromide reacted with methyllithium in the presence of cyclohexanone to afford a mixture of *n*-propylidenecyclohexane, the desired product, and methylenecyclohexane, the result of the ligand exchange reaction [3.61] (*50,* Seyferth *et al.*). Such ligand exchange reac-

$$(C_6H_5)_3\overset{\oplus}{P}{-}CH_2CH_2Br \xrightarrow{CH_3Li} (C_6H_5)_3P{=}CHCH_2CH_3 + (C_6H_5)_3P{=}CH_2$$

$$\downarrow \text{cyclohexanone } (C_6H_{10}{=}O) \qquad \downarrow \text{cyclohexanone } (C_6H_{10}{=}O) \qquad [3.61]$$

$$C_6H_{10}{=}CHCH_2CH_3 \qquad C_6H_{10}{=}CH_2$$

tions, the result of nucleophilic attack on phosphorus rather than on hydrogen, probably take place less readily when a less nucleophilic base is used. Therefore, it is advisable to use phenyllithium or the methylsulfinyl carbanion in place of an alkyllithium in the preparation of ylids. The ligand exchange reaction appears not to be a problem in the formation of stabilized ylids since benzyltriphenylphosphonium bromide did not undergo attack on phosphorus by methyllithium (*127*). However, this conclusion was based on the reported absence of benzene in the reaction mixture. The expected hydrocarbon from an exchange reaction is toluene, and from the communication as published it is not clear whether its presence would have been detected.

II. Structure and Physical Properties of Phosphonium Ylids

Considerably more information is available regarding the physical characteristics of phosphonium ylids than of any other type since many of these ylids have been isolated and purified. This has been especially the case since about 1957 at which time only five phosphonium ylids had been isolated and practically nothing was known of their physical or chemical properties. In this section are discussed the results of the physical studies that have been brought to bear on the question of the electronic and molecular structure of ylids.

A. Chemical Structure of Ylids

The first ylid about which much structural information was sought was phenacylidenetriphenylphosphorane. Michaelis and Kohler (*110*) found that treatment of triphenylphosphine with phenacyl bromide afforded the expected phosphonium salt ($C_{26}H_{22}POBr$). This salt, in the presence of aqueous hydroxide or carbonate, was converted into a compound of melting point 183–184° for which they assigned a molecular formula $C_{26}H_{23}PO_2$ and proposed a structural formula (LXIII) containing a pentavalent phosphorus.

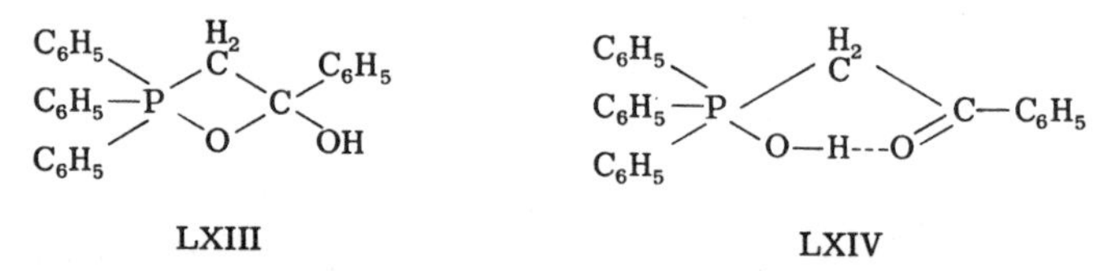

The structure proposed by Michaelis remained unchallenged in the literature until Wittig and Schollkopf (*1*) found that the same substance could be prepared by acylating methylenetriphenylphosphorane with ethyl benzoate, isolating the bromide salt by neutralization of the reaction mixture with hydrobromic acid and then treating the salt with aqueous sodium hydroxide. They accepted the molecular formula proposed by Michaelis but proposed that the substance was a "phosphonium betaine" of structure LXIV. This structure was proposed on the basis of the infrared spectrum of the analogous acetyl compound (methyl in place of phenyl on the carbonyl) which showed no hydroxyl absorption but had carbonyl absorption at 1529 cm^{-1}. Since the precursor phosphonium salt had carbonyl absorption at 1700 cm^{-1} the low frequency in the betaine was attributed to extensive hydrogen bonding.

Ramirez and Dershowitz (*129*) reexamined this whole problem in 1956. In their hands the supposed "betaine" analyzed for $C_{26}H_{21}PO$ after using special precautions for the removal of water. Such a formula indicates that the elements of hydrogen bromide were removed from the phosphonium salt; this led to the proposal of an ylid structure (LXV) for the substance. It exhibited the normal ylid characteristics—hydrolysis

$$(C_6H_5)_3\overset{\oplus}{P}-\overset{\ominus}{C}H-\underset{\underset{O}{\|}}{C}-C_6H_5 \longleftrightarrow (C_6H_5)_3\overset{\oplus}{P}-CH=\underset{\underset{O^{\ominus}}{|}}{C}-C_6H_5 \qquad [3.62]$$

LXV

to acetophenone and triphenylphosphine oxide, reconversion to the phosphonium bromide with hydrogen bromide and participation in a Wittig reaction with benzaldehyde to form benzalacetophenone. The ylid also underwent alkylation with ethyl iodide although on oxygen rather than on carbon. The ylid structure is in agreement with the high dipole moment, 5.45 D, and the low carbonyl frequency, 1518 cm^{-1}, which indicates extensive delocalization through the carbonyl group.

To date most of the structures of the various ylids have been proven on the basis of their chemical properties rather than by physical means. Even the application of NMR techniques to ylid structure proof has not progressed rapidly.

B. Effect of Carbanion Substituents

Phosphonium ylids are unique substances. As indicated by the dipole moments that have been reported, 7.0 D for cyclopentadienylidenetriphenylphosphorane (*130*), 5.5 D for acetylmethylenetriphenylphosphorane (*129*) and 7.1 D for fluorenylidenetriphenylphosphorane (*44*), they are certainly polar molecules. On the other hand, they exhibit solubility behavior characteristic of covalent compounds and certainly much different from other zwitterionic substances such as amino acids. Ylids normally are soluble in such solvents as ether, benzene and chloroform but they are insoluble in water and often in ethanol.

Phosphonium ylids usually are highly colored substances. Methylenetriphenylphosphorane is yellow in solution as are most non-isolable ylids. Phenacylidenetriphenylphosphorane (*129*) and most acyl ylids, cyclopentadienylidene- (*130*) and fluorenylidenetriphenylphosphorane (*44*) also are yellow substances but benzhydrylidenetriphenylphosphorane (*42*) is red and 2-nitrofluorenylidenetriphenylphosphorane (*131*) is the color of potassium permanganate. There has been one report of an ylid exhibiting a photochromic effect. Driscoll *et al.* (*132*) found that methylidebis(triphenylphosphonium)tetraphenylborate (LXVI), a colorless compound, became an orange-red upon exposure to ultraviolet or

$$(C_6H_5)_3\overset{\oplus}{P}-\overset{\ominus}{C}H-\overset{\oplus}{P}(C_6H_5)_3 \qquad \overset{\ominus}{B}(C_6H_5)_4$$

LXVI

incandescent radiation. The effect was reversible but occurred only slowly. The ultraviolet and infrared spectra of both forms were reported to be identical but the colored form gave an EPR (electron paramagnetic resonance) spectrum. Replacement of the tetraphenylborate anion of LXVI with bromide afforded a non-photochromic substance.

Little work has been done on the ultraviolet and visible spectra of phosphonium ylids. Fluorenylidenetriphenylphosphorane (*44*, Johnson) showed maxima, in chloroform solution, at 250 mμ (log ε 4.6), 258 (4.6), 284 (4.3) and 382 mμ (3.6). Below 300 mμ this spectrum was characteristic of a fluorene nucleus but the long wavelength absorption was unusual. It most likely is associated with the electrons involved in the ylid bond. Benzalfluorene shows absorption maxima at 248 mμ (4.3), 256 (4.4) and 325 mμ (4.1) with the lower region being similar to that of the ylid. The 325 mμ band probably is a π-π^* transition of the ethylenic bond, and it then may be surmised that the 382 mμ band of the ylid is also this type of transition but from a *p-d* π bond rather than from a *p-p* π bond as in the benzalfluorene. The spectrum of the fluorenylide showed no obvious relationship to that recently reported for the fluorenyl anion in cyclohexylamine (*133*, Streitwieser and Brauman). Bergelson *et al.* (*134, 135*) have reported the partial decolorization of solutions of benzylidenetriphenylphosphorane upon addition of lithium iodide, bromide or various amines. They have proposed that the decolorization was due to complexation of the halide ions with the phosphonium atom. This proposal has not been supported to date by convincing experimental evidence.

Most of the phosphonium ylids isolated and characterized to date have been solid, crystalline compounds. Some are low melting, such as the liquid trimethylsilylmethylenetriphenylphosphorane (*136*, Miller) but others, such as fluorenylidenetriphenylphosphorane (*44, 116*) and cyclopentadienylidenetriphenylphosphorane (*130*) melting at 253° and 229°, respectively, are very high melting and especially stable ylids.

The stability of ylids often is referred to and in most cases applies to the chemical stability of the substances. There is no evidence to indicate any inherent thermodynamic instability associated with phosphonium ylids in general. There is such evidence, however, in the case of sulfonium ylids since many decompose spontaneously to carbenoid intermediates and sulfides (see Chapter 9). The chemical instability associated with some phosphonium ylids appears to be due to their basicity and reaction with water. For example, methylenetriphenylphosphorane reacts with water, presumably by way of the methyltriphenylphosphonium hydroxide, to afford methyldiphenylphosphine oxide and benzene [3.63] (*4*, Wittig and Geissler). Most of the so-called "unstable ylids" react in

$$(C_6H_5)_3P{=}CH_2 + H_2O \longrightarrow [(C_6H_5)_3\overset{\oplus}{P}{-}CH_3 \quad \overset{\ominus}{O}H] \longrightarrow (C_6H_5)_2P(CH_3){=}O + C_6H_6 \qquad [3.63]$$

a similar fashion. The "stable ylids," generally meant to infer those which can be handled in the atmosphere, will not react with water because of their low basicity and inability to carry out the first step. Most of the ylids being discussed in this section have been isolated and fall into the latter category.

While phosphonium ylids usually are insoluble in aqueous solution they will dissolve in dilute aqueous acid solutions. In other words, they are basic and undergo the reverse reaction used in their formation from phosphonium salts. A study of the relative basicity of a series of ylids could shed light on the electronic characteristics of ylids in general. A more basic ylid generally infers that the electron density associated with the ylid carbanion is more localized. In other words there are structural features which do not permit the facile delocalization of the electrons on the carbanion.

The earliest indication of the relationship between the basicity of ylids and their molecular structure was obtained from the nature of the base required for the generation of an ylid from its conjugate acid. Thus a carbanion (organolithium, methylsulfinyl, etc.) was needed to remove a proton from methyltriphenylphosphonium salts but sodium ethoxide was a sufficiently strong base to remove a proton from benzyltriphenylphosphonium salts. Further, fluorenyltriphenylphosphonium bromide was acidic to ammonia (*116*). Therefore, the order of increasing ylid basicity was fluorenylide < benzylide < methylide. This is the order to be expected if the factor controlling the basicity is the ability of the hydrocarbon portion of the ylid to delocalize the electrons on the carbanion. This is the same order as that obtained for the acidity of the corresponding hydrocarbons; fluorene is more acidic than toluene and the latter is more acidic than methane (*137*, Streitwieser).

Bestmann (*73*) developed a different but still qualitative method for determining the effect of the carbanion substituents on the basicity of ylids. Reaction of a phosphonium salt with one equivalent of another ylid should result in an equilibrium between two salts and two ylids [3.64]. Bestmann has referred to this as an intermolecular "*umylid-*

$$RCH_2\overset{\oplus}{P}(C_6H_5)_3\ Br^{\ominus} + R'CH{=}P(C_6H_5)_3 \rightleftharpoons RCH{=}P(C_6H_5)_3 + R'CH_2\overset{\oplus}{P}(C_6H_5)_3\ Br^{\ominus} \qquad [3.64]$$

ierung"—perhaps translated as a "transylidation" reaction. Thus reaction of benzylidenetriphenylphosphorane (R′ = phenyl) with phenacyltriphenylphosphonium bromide (R = benzoyl) afforded a 76% yield of phenacylidenetriphenylphosphorane while reaction of carbomethoxymethylenetriphenylphosphorane (R′ = $COOCH_3$) with phenacyltriphenylphosphonium bromide (R = benzoyl) afforded a 78% yield of phenacylidenetriphenylphosphorane. By this procedure Bestmann found that the basicity of the ylids [$R'CH{=}P(C_6H_5)_3$] decreased in the following order:

$$R' = \text{alkyl} > \text{phenyl} > \text{carbomethoxy} > \text{benzoyl}$$

The electron-withdrawing power of these substituents increases from left to right as reflected by their (*para*) σ constants which are −0.17, −0.01, +0.31 and +0.46, respectively (*138*, Jaffe). It is apparent that the more electron withdrawing are the carbanion substituents, the less basic is the ylid. No quantitative conclusions may be drawn from Bestmann's work (*73*) since the yields he reported were not the result of freely established equilibria. The yields actually were obtained by isolating the products of the reaction after carrying out various evaporations, etc.

Some quantitative measurements on the basicity of phosphonium ylids have been reported but it is difficult to interrelate the data because of the different non-aqueous solvents used. Most workers report their data as the pK_a's of the conjugate acids of the ylids. Thus, a low pK_a implies a highly stabilized ylid. Johnson and LaCount (*7*) reported the pK_a of the hydrobromide of triphenylphosphoniumfluorenylide to be 7.5 in 31% water-dioxane. Methylenebis(triphenylphosphonium)dibromide was a dibasic acid with a pK_a of 5.4 for the first proton (*91*, Ramirez *et al.*)

Fliszar *et al.* (*139*) reported the pK_a's of the conjugate acids of a series of acylated methylenephosphoranes as determined in 80% alcohol-water. For the ylids $R{-}CO{-}CH{=}P(C_6H_5)_3$ the pK_a's of the salts increased in the order

$$R = p\text{-nitrophenyl } (4.2) < \text{phenyl } (6.0) < p\text{-anisyl } (6.7)$$

$$R = \text{chloromethyl } (4.5) < \text{methyl } (6.6) < \text{methoxy } (8.8) < \text{amino } (11)$$

This also is the order of increased basicity of the ylids. In these instances there can be no direct resonance of the R group with the ylid carbanion so it must be affecting the ability of the carbonyl group to delocalize the negative charge of the carbanion. Where $R = p$-nitrophenyl the

$$(C_6H_5)_3\overset{\oplus}{P}{-}\overset{\ominus}{C}H{-}\underset{\underset{O}{\|}}{C}{-}R \longleftrightarrow (C_6H_5)_3\overset{\oplus}{P}{-}CH{=}\underset{\underset{O^{\ominus}}{|}}{C}{-}R \qquad [3.65]$$

carbonyl group must become more electron deficient and therefore better able to accept electron density from the carbanion by resonance. The converse would be true for the *p*-anisyl case. Similarly, the carbonyl group of the ester should be more electron deficient than that of the amide and the former ylid is the less basic. All of these data are consistent with the hypothesis that the better able is the carbonyl group of the ylid to delocalize the negative charge of the carbanion, the less basic will be the ylid and the lower will be the pK_a of its conjugate acid.

Speziale and Ratts (*140*) carried out similar measurements but titrated the ylids potentiometrically in methanol solution with hydrochloric acid. Using similar ylids, $(C_6H_5)_3P{=}C(X){-}CO{-}R$, but varying the nature of both X and R, they found that the pK_a's of the salts increased in the following order:

$$X = H, R = \text{phenyl } (6.0) < \text{ethoxy } (9.2) < \text{diphenylamino } (9.7)$$

These observations are consistent with those of Fliszar *et al.* (*139*). In another series of salts the pK_a's increased in the following order:

$$R = \text{phenyl}, X = \text{chloro } (4.3) < \text{bromo } (5.0) < \text{iodo } (5.9) < \text{hydrogen } (6.0)$$

It is clear that substitution of halogen for hydrogen on the ylid carbanion to which a carbonyl group was attached provided increased stabilization for the ylid. The reason for this particular order of basicity for the halobenzoyl ylids ($I > Br > Cl$) must rest on speculation, however. The order of basicity is consistent with the known order of inductive electron withdrawal by the halogens—chlorine is the most electronegative in this series and the chloroylid is the least basic. The observed order of basicity also may be consistent with the relative abilities of the halogens to stabilize an adjacent carbanion by octet expansion through the use of *d*-orbitals. Doering and Hoffmann (*141*) felt that octet expansion varied little as a group in the periodic table was descended but Johnson and LaCount (*7*) felt that the ability decreased as the group was descended. This feeling was based on the observed properties of ylids involving various heteroatoms and on the complexing ability of various ligands. Such a decrease might be expected due to increasing atomic size and concomitant decrease in effective orbital overlap with the carbanion orbitals. If octet expanding ability does decrease as a group is descended the chlorobenzoyl ylid would be expected to be the least basic of the three halobenzoyl ylids as was observed. The fact that the iodobenzoyl ylid (pK_a 5.9) was almost as basic as the benzoyl ylid (pK_a 6.0) may be due to the size of the iodine atom preventing the carbonyl group from being coplanar with the carbanion, thereby decreasing the delocalizing influence of the carbonyl group (*140*).

In spite of the apparent consistency of the preceding rationale other considerations may nullify it. Hine *et al.* (*142*), on the basis of their studies of the acidity of trihalomethanes, concluded that halogen stabilization of an adjacent carbanion increased in the order $Cl < Br < I$. This is the exact opposite of the order found with the halobenzoyl ylids.

The inductive effect clearly was not dominant. The observed order may have been due to the steric effect (B-strain), a polarizability effect or the order of effective octet expansion being Cl < Br < I. The authors were unable to decide between these three possibilities (*142*). Speziale and Ratts (*140*) proposed that increasing steric inhibition by the halogen of the coplanarity, and therefore conjugation, of the carbonyl group and the carbanion was responsible for the basicity of the halobenzoyl ylids increasing in the order Cl < Br < I (i.e., the apparent effectiveness of the halogens in delocalizing the negative charge of the carbanion increasing in the order I < Br < Cl).

Miller (*136*) recently reported trimethylsilylmethylenetriphenylphosphorane to be less basic than methylenetriphenylphosphorane. The silyl group would be expected to stabilize the carbanion by overlap of the vacant *3d*-orbitals of silicon with the filled *2p*-orbitals of the carbanion [3.66].

$$(CH_3)_3Si-\overset{\ominus}{C}H-\overset{\oplus}{P}(C_6H_5)_3 \longleftrightarrow (CH_3)_3\overset{\ominus}{Si}=CH-\overset{\oplus}{P}(C_6H_5)_3 \longleftrightarrow (CH_3)_3Si-CH=P(C_6H_5)_3 \qquad [3.66]$$

From this discussion of the basicity of ylids it is clear that, in general, substitution of a more powerful electron-withdrawing group on the ylid carbanion decreases the electron density on the carbanion by inductive and/or resonance effects. The presence of such groups on a potential ylid carbon facilitates the formation of that ylid from its conjugate acid (phosphonium salt) using weaker bases.

Evidence for resonance interaction of the ylid carbanion with a carbonyl group can be obtained by an examination of the infrared spectra of acyl ylids and their conjugate acids. The carbonyl stretching frequency of phenacyltriphenylphosphonium salts occurs near 1670 cm^{-1} whereas that of the corresponding ylid, phenacylidenetriphenylphosphorane, occurs near 1500 cm^{-1} (*58, 129, 140, 143*), a shift of 170 cm^{-1}. Similar shifts have been observed for a wide variety of acyl ylids (X = H) and their salts, the shift being attributed to increased single bond character due to delocalization of the carbanion electrons through the carbonyl group [3.67]. Speziale and Ratts (*140*) studied the carbonyl frequencies in cases

$$(C_6H_5)_3\overset{\oplus}{P}-\underset{\ominus}{\overset{X}{\overset{|}{C}}}-\overset{O}{\overset{\|}{C}}-R \longleftrightarrow (C_6H_5)_3\overset{\oplus}{P}-\overset{X}{\overset{|}{C}}=\overset{O^{\ominus}}{\overset{|}{C}}-R \longleftrightarrow (C_6H_5)_3\overset{\oplus}{P}-\overset{X^{\ominus}}{\overset{\|}{C}}-\overset{O}{\overset{\|}{C}}-R \qquad [3.67]$$

where X = H and where X = halogen. In the ester ylids ($R{=}OCH_3$) the substitution of halogen for hydrogen led to an increase in the carbonyl frequency, probably due to competing halogen delocalization of the negative charge. However, when R = phenyl or R = diphenylamino the replacement of hydrogen by halogen led to a decrease in the carbonyl frequency. This was attributed to the influence of the halogen field on the carbonyl polarization, such influence depending on the steric repulsion between R and X. It is equally likely that in those two cases there is sufficient repulsion between the halogen and R groups so that the latter is forced out of coplanarity with the carbonyl group. This would lead to increased delocalization through the carbonyl group, and result in a lower stretching frequency. Credence is added to this proposal by the observation that in the ester ylids ($R{=}OCH_3$) replacement of X = H by X = CH_3 also led to a lowering of the carbonyl stretching frequency.

There is considerable chemical evidence which indicates delocalization of the charge on the ylid carbanion throughout its substituents. Evidence for contribution of the phosphonium enolate structure of phenacylidenetriphenylphosphorane was provided by the observation that it reacted with ethyl iodide in the absence of a catalyst to form an enol ether [3.68] (*129,* Ramirez and Dershowitz):

$$\left[(C_6H_5)_3\overset{\oplus}{P}-\overset{\ominus}{C}H-\overset{O}{\overset{\|}{C}}C_6H_5 \longleftrightarrow (C_6H_5)_3\overset{\oplus}{P}-CH{=}\overset{O^{\ominus}}{\overset{|}{C}}C_6H_5\right] \xrightarrow{C_2H_5I} (C_6H_5)_3\overset{\oplus}{P}-CH{=}\overset{OC_2H_5}{\overset{|}{C}}C_6H_5\ I^{\ominus} \xrightarrow{\overset{\ominus}{O}H} CH_2{=}C\begin{matrix}OC_2H_5\\ C_6H_5\end{matrix} \quad [3.68]$$

The fact that the same ylid underwent a Wittig reaction with benzaldehyde indicated that there was considerable electron density on the carbon atom next to phosphorus as well.

In some instances attempts to demonstrate the mesomeric nature of a carbonyl-substituted ylid have led to the conclusion that most of the electron density was located on the oxygen atom of an enolate form. Thus, methylation of the anion of *p*-hydroxyphenyltriphenylphosphonium bromide led to exclusive O-alkylation, indicating little or no contribution from the ylid structure (LVIII) (*98,* Horner *et al.*). Clearly, the potential stability of the ylid structure was not sufficient to compensate for the loss of the aromaticity of LVII. Likewise, alkylation of the adduct from triphenylphosphine and *p*-benzoquinone led to exclusive O-alkylation (*56,* Ramirez and Dershowitz).

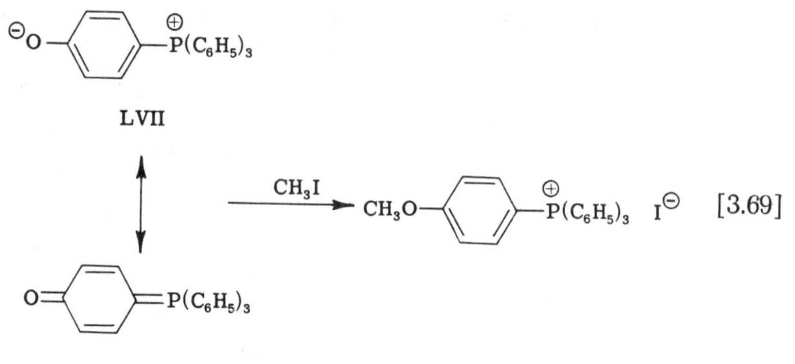

Several systems which could involve carbanion-carbanion mesomerism rather than carbanion-enolate mesomerism (as described in the above examples) have been examined. The most striking case was discovered by Ramirez and Levy in 1956 (*130*). They found that cyclopentadienylidenetriphenylphosphorane (LXVII) was an amazingly stable ylid.

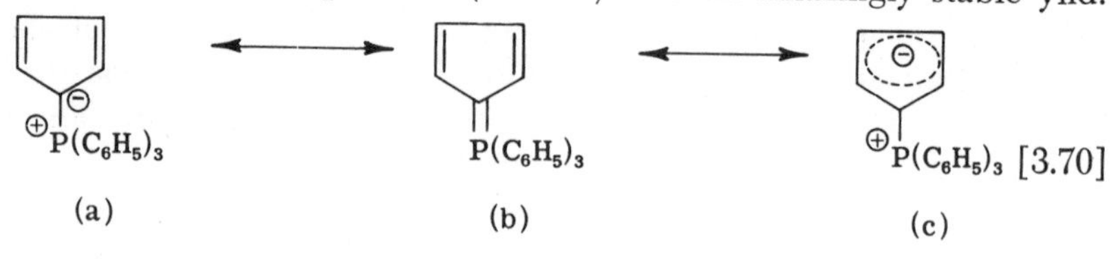

It was high melting (229–231°) and could be recovered unchanged after refluxing with alcoholic potassium hydroxide, a procedure that would convert most ylids to triphenylphosphine oxide and a hydrocarbon. In addition, the ylid would not undergo a Wittig reaction with carbonyl compounds. Its ylid character was verified by its solubility in dilute aqueous acid. The ylid could not be hydrogenated with hydrogen over platinum but its conjugate acid did absorb two moles of hydrogen to afford cyclopentyltriphenylphosphonium chloride. The unique stability of this ylid can be attributed to the importance of resonance structure LXVIIc, an aromatic cyclopentadienyl anion form. On the basis of its dipole moment of 6.99 D Ramirez estimated the ylid to have about 50% double bond character between the phosphorus atom and the five-membered ring.

The aromaticity of the cyclopentadienylide (LXVII) clearly was indicated by the observation that it underwent electrophilic aromatic substitution. The ylid reacted with benzenediazonium chloride to afford 2-(phenylazo)-1-triphenylphosphoniumcyclopentadienylide (*144*, Ramirez and Levy). Depoorter *et al.* (*145*) also reported electrophilic substitutions on LXVII. It appears that the delocalization of electron density from the carbanion in LXVIIa has gone so far that the substance can barely be considered ylidic and perhaps is best described as a phosphonium-substituted cyclopentadienyl anion.

Attempts to detect carbanion-carbanion mesomerism in non-cyclic conjugated ylid systems has met with mixed success. Wittig and Schollkopf (*1*) found that allylidenetriphenylphosphorane (LXVIII) reacted normally with benzaldehyde to afford 1-phenylbutadiene. There was no evidence of nucleophilicity at the terminal carbon of the ylid, indicating

$$\underset{\text{LXVIII}}{(C_6H_5)_3\overset{\oplus}{P}-\overset{\ominus}{C}H-CH=CH_2} \longleftrightarrow \underset{\text{LXIX}}{(C_6H_5)_3\overset{\oplus}{P}-CH=CH-\overset{\ominus}{C}H_2} \qquad [3.71]$$

no need to propose an important contribution from the "homoylid" structure (LXIX). More recently, however, Bestmann and Schulz (*83*) have found that LXVIII underwent acylation with methyl chlorocarbonate on the terminal carbon. The new ester could be hydrolyzed to crotonic acid or could be condensed with benzaldehyde to form methyl cinnamylideneacetate. The latter observation, but not the former, proves the site of acylation of the ylid (LXVIII ↔ LXIX) to have been the γ-carbon. It

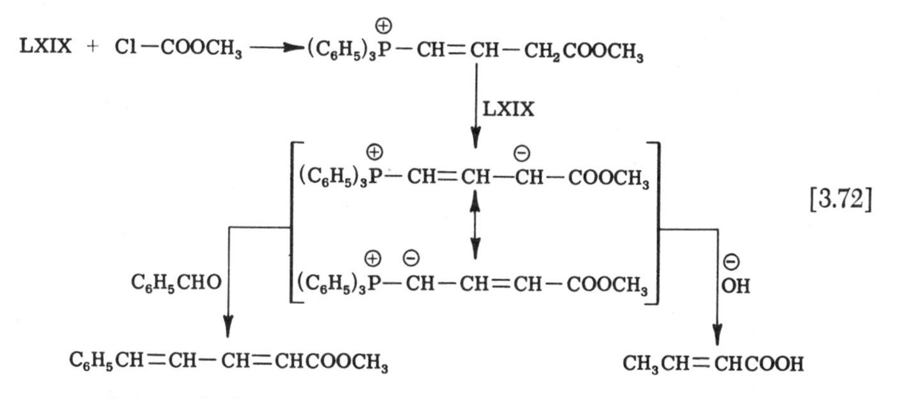

must be concluded, therefore, that both resonance structures are important in the actual structure of allylidenetriphenylphosphorane, the electron density being distributed between the α- and γ-carbon atoms.

Several groups have prepared and used cinnamylidenetriphenylphosphorane (LXX) for various reactions with carbonyl compounds. Campbell and McDonald (*146*) reported reaction of this ylid with benzaldehyde afforded *trans-trans*-1,4-diphenyl-1,3-butadiene. A tempting conclusion that this indicated the original *trans*-double bond of the ylid did not isomerize, and therefore was not involved in resonance, was nullified by the fact that the *trans-trans* isomer was the thermodynam-

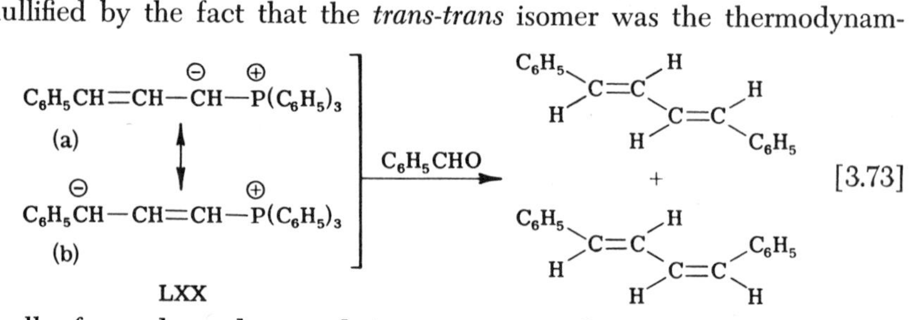

ically favored product and isomerism may have occurred after the reaction. However, the authors surmised that the ylid must be stabilized by resonance (i.e., the original double bond had some single bond character, LXXa ↔ LXXb) because of the red color of the ylid, similar to that of the cinnamyl anion. It should be pointed out that the original double bond would be expected to maintain its configuration whether or not the ylid existed as a resonance hybrid since in both cases (LXXa and LXXb) a π-orbital would extend over the original olefinic carbons discouraging *trans*-to-*cis* conversion, especially in view of the *trans* being the more stable form. Later, Misumi and Nakagawa (*147*) studied the same reaction and reported obtaining a mixture of the *trans-trans-* and *trans-cis*-diphenylbutadienes. This most likely indicates that the original double bond in the ylid maintained its configuration (*trans*) and that the new double bond was either *cis* or *trans,* a situation encountered in most Wittig reactions. Bohlmann (*148*) also studied this system but again could find no evidence for reaction at the γ-carbon.

Bestmann *et al.* (*149*) have provided some concrete evidence for the delocalization of negative charge portrayed by the resonance forms LXXa ↔ LXXb. Treatment of cinnamyltricyclohexylphosphonium bromide with sodium ethoxide in *O*-tritioethanol followed by reaction of the ylid in solution with benzaldehyde afforded 1,4-diphenylbutadiene labeled with tritium. Oxidation of the diene with osmium tetroxide and then sodium periodate afforded benzaldehyde and glyoxal. The detection of tritium in both of these oxidation products indicated that both the α-carbon and the γ-carbon of the phosphonium salt were susceptible to exchange under the influence of base, presumably via the respective carbanions, LXXa and LXXb [3.74].

Harrison and Lythgoe (*150*) found that the cyclohexenyl ylid (LXXI) reacted with cyclohexanone at 5° to afford a mixture of the cisoid and transoid dienes in a 20:1 ratio. Carrying out the same reaction but at 40° narrowed this ratio to 4:1. They concluded that the contribution of a resonance form in which the original double bond had single bond character led to a decrease in the resistance of that bond to rotation.

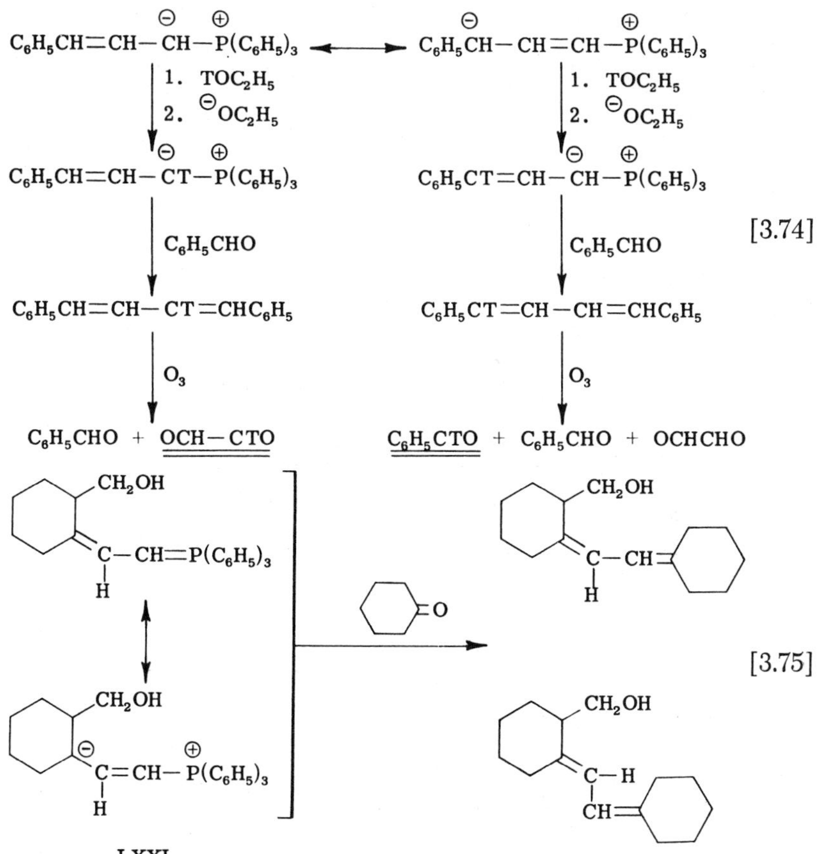

This would account for the presence of the transoid diene, the amount of which increased as the reaction temperature was raised. They showed that the cisoid diene could be isomerized to the transoid in the presence of iodine at 30–60° but did not demonstrate that the cisoid diene was stable to heat or chromatography. It would be of interest to compare the rate of isomerization of the ylid (LXXI) with the rate of isomerization of the cisoid diene since the steric repulsions should not be too different.

C. Effect of Phosphorus Substituents

The previous discussion has been concerned with the effect of the carbanion substituents on the electronic nature of phosphonium ylids. Just as important is the effect of the phosphonium group on the ylid characteristics but this aspect of ylid chemistry virtually has been neglected. The same physical and chemical tools as mentioned in the preceding section could be applied to this question.

The early work of Wittig and his students (*3, 4*) indicated that there was considerable difference in the behavior of methylene*trimethyl*phosphorane and methylene*triphenyl*phosphorane (see Section I,A of this chapter). However, in 1959 Johnson and LaCount (*6*) showed that fluorenylidene*trimethyl*phosphorane would undergo a Wittig reaction with aldehydes as had been reported earlier for fluorenylidene*triphenyl*phosphorane (*44*, Johnson). Later, the same workers (*7*) prepared and isolated fluorenylidenetri-*n*-butylphosphorane. A comparison of the two fluorenylides, the triphenyl and the tributyl, is the only one of its type on isolable ylids. The triphenyl ylid melted at 258–260° but the tributyl ylid melted at 123–124°. Both ylids were yellow, their ultraviolet spectra were nearly identical, the triphenyl ylid was perfectly stable but the tributyl ylid slowly was decolorized upon standing exposed to the atmosphere. However, it could be stored for an indefinite period in the absence of moisture. The triphenyl ylid was not hydrolyzed in water whereas the tributyl ylid was cleaved to fluorene and tri-*n*-butylphosphine oxide.

A study of the physical properties of the two fluorenylides indicated that the triphenyl ylid was the least basic and the least nucleophilic. The pK_a's of the two ylids were determined spectroscopically in 31% water-dioxane: the triphenyl ylid (pK_a 7.5) was less basic than the tributyl ylid (pK_a 8.0). In addition a study of the reactivity of the two ylids with carbonyl compounds indicated that the tributyl ylid reacted faster than

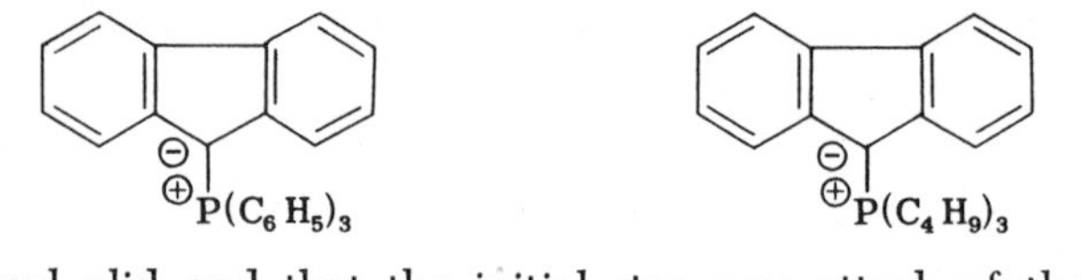

the triphenyl ylid and that the initial step was attack of the ylid carbanion on a carbonyl carbon. Therefore, it seems reasonable to conclude that the triphenyl ylid also was the least nucleophilic as well as the least basic of the two. Interestingly, the triphenyl ylid ($\mu = 7.09$ D) also had a lower dipole moment than the tributyl ylid ($\mu = 7.39$ D).

Since the hydrocarbon portions of the two fluorenylides are the same the variation in the nature of the phosphorus substituents seems the only source of explanation for the difference in the carbanionic properties. Presumably the extent of delocalization of negative charge through the fluorenyl rings should be constant for both ylids. Since the triphenyl ylid was the least basic and the least nucleophilic it is apparent that the electron density on the carbanion must be lower in that case. This implies that the triphenylphosphonium group is better able to delocalize the negative charge of the carbanion than is the tributylphosphonium group. Since the major stabilization afforded the ylid carbanions by the phosphonium group should be via acceptance of electrons into the vacant, low-energy 3*d*-orbitals of phosphorus (see Chapter 2), the 3*d*-orbitals of a triphenylphosphonium group must provide better overlap than do those of the tributylphosphonium group with the filled 2*p*-orbital of the carbanion. In other words, there must be more double bond character to the carbanion-phosphorus bond in the triphenyl ylid than in the tributyl ylid. The dipole moment data listed in the preceding paragraph are consistent with this conclusion (*7, 131*).

If the proposal of Jaffe (*151*) and of Craig *et al.* (*152*) is correct, that π bonding with an atom carrying vacant *d*-orbitals is more efficient when the atom carries a positive charge, it may be concluded that a lower electron density is induced on the phosphorus atom by its substituents in the case of the triphenyl ylid than in the tributyl ylid. It must follow that the phenyl substituents on the phosphonium group are either electron withdrawing with respect to the *n*-butyl groups or they are less electron donating. At first sight this is a startling statement since phenyl groups normally are more electron donating than alkyl groups. However, there are some instances known when the phenyl group does exert a strong electron withdrawing effect. For example, Wepster (*153*) showed that in the aniline system the total withdrawing power of the phenyl group was composed of both resonance and inductive effects, the latter contributing about one half to the total effect. When the phenyl ring was incorporated into the bicyclic quinuclidine system, the geometry prevented appreciable resonance interaction between the nitrogen atom and the phenyl ring. What remained was the inductive effect of the phenyl ring which was strongly electron withdrawing.

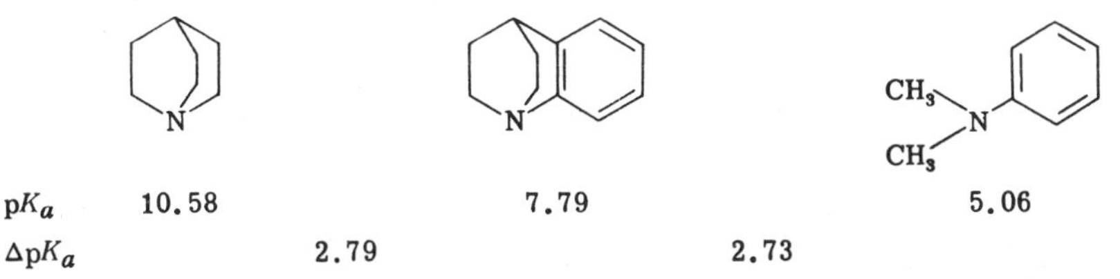

The previous arguments seem to indicate that the phenyl groups attached to the phosphorus atom in fluorenylidenetriphenylphosphorane are exerting their inductive effect (withdrawal), making the ylid less basic and with more phosphorus–carbon double bond character than is the case with fluorenylidenetributylphosphorane in which the butyl groups presumably exert their normal inductive effect (donation). Johnson *et al.* (*131*) have found that fluorenylidenetri(*p*-X-phenyl)phosphoranes where X was an electron withdrawing group were less basic and nucleophilic than where X was hydrogen. In these cases the phosphorus substituents would be more electronegative inducing more phosphorus–carbon double bond character and leaving a lower electron density on the ylid carbanion. Conversely, those ylids where X was an electron donating group were more basic and nucleophilic, the less electronegative phosphorus substituents apparently inducing less phosphorus–carbon double bond character and leaving the carbanion with a higher electron density.

Several other groups have prepared trialkylphosphonium ylids, mainly for use in the Wittig reaction, but there have been no other comparisons of physical properties between alkyl and aryl ylids. Bestmann and Kratzer (*10*), Trippett and Walker (8) and Heitman *et al.* (9) found that with stabilized ylids the trialkylphosphonium ylids were more nucleophilic than the triphenylphosphonium ylids. These observations are consistent with the preceding rationale for the fluorenylides.

A study of the P^{31} NMR chemical shifts in a series of ylids and their conjugate acids might be expected to shed light on the electron density about the phosphorus atom. Very little has been done in this regard. *p*-Nitrobenzylidenetriphenylphosphorane showed a shift of —13 ppm while triphenylphosphine showed a shift of +5.9 ppm and triphenylphosphine oxide one of —24 ppm (all in benzene solution using 85% H_3PO_4 as reference) (*119*, Grayson and Keough). As might be expected, the phosphorus atom appeared more shielded in the ylid than in the phosphine oxide.

Speziale and Ratts (*140*) found that in a series of carbonyl-substituted methylenetriphenylphosphoranes the chemical shifts all were between —19 and —22 ppm; this indicated little, if any, variation of chemical shift with structure. Somewhat surprisingly, however, they also found that the chemical shifts of the corresponding phosphonium salts (conjugate acids) were very similar to those of the ylids (see Table 3.5).

TABLE 3.5

P^{31} NMR Chemical Shifts of β-Ketophosphonium Ylids and Conjugate Acids

R	Chemical shift of $(C_6H_5)_3\overset{+}{P}CH_2COR$ (ppm)	Chemical shift of $(C_6H_5)_3P{=}CHCOR$ (ppm)
C_6H_5	-16.9	-21.6
C_2H_5O	-19.7	-19.1

More recently Driscoll *et al.* (*154*) reported that methylide-bis(triphenylphosphonium)bromide (LXVI) showed a P^{31} chemical shift of —21.2 ppm whereas its hydrobromide (i.e., the bis-phosphonium dibromide) showed a shift of —18.4 ppm. It was first concluded that the two phosphorus atoms were equivalent in the ylid and that its structure was best represented by LXVIb rather than LXVIa. A more interesting

$$\overset{\oplus}{(C_6H_5)_3P}-CH{=}P(C_6H_5)_3 \quad Br^{\ominus} \qquad (C_6H_5)_3P\overset{\oplus}{\overline{\cdots CH\cdots}}P(C_6H_5)_3 \quad Br^{\ominus}$$

(a) (b)

LXVI

and much less expected conclusion is demanded by the similarity in chemical shift between the phosphonium salt and the ylid, a situation identical to that reported by Speziale and Ratts (*140*) and mentioned above.

Denney and Smith (*13*) earlier had observed that triphenylalkyl-phosphonium salts showed P^{31} chemical shifts near —20 ppm. On the basis of the observation that treatment of triphenyl(2-carboxyethyl)phosphonium chloride ($\delta = -24$ ppm) with a base afforded a substance with a P^{31} chemical shift of —23 ppm, they concluded that a proton had been removed from the carboxyl group rather than from the carbon α to the phosphorus atom. In other words the substance was claimed to be a phosphonium betaine rather than a phosphonium ylid (VII) [3.76]. Examination of the data of Speziale and Ratts (*140*) indicates

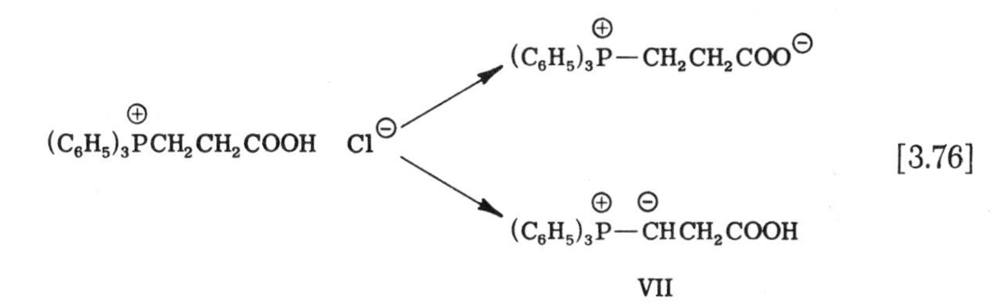

that it would be virtually impossible to differentiate between a phosphonium salt and a phosphonium ylid on the basis of P^{31} chemical shifts, and Denney's conclusion is not warranted. Corey *et al.* (*14*) recently showed that the phosphonium salt could be converted to VII with sodium hydride in tetrahydrofuran. They did not report the P^{31} NMR spectrum.

The initial conclusion that follows from these NMR observations is that there is nearly identical shielding of the phosphorus atom in phosphonium salts and the corresponding ylids. If this implies that the electron density on the phosphorus atom is the same in these two cases it also infers that there is little, if any, delocalization of the electron density from a carbanion to a phosphorus atom in an ylid. In other words, essentially no *p-d* π bonding. To draw such a conclusion would necessitate the negation of all previous evidence for octet expansion in the phosphonium atom, especially the striking difference between the stability of ammonium ylids and phosphonium ylids (see Chapter 2). This is a distasteful task in view of the available wealth of information consistent with valence shell expansion. The other alternative is to look elsewhere for an explanation of the P^{31} NMR results.

Some sanity was restored to the NMR results with the very recent report by Mark of his preparation of difluoromethylenetris(dimethylamino)phosphorane [3.77] (*37*). This ylid showed a P^{31} chemical shift

$$C_6H_5COCF_3 + [(CH_3)_2N]_3P \longrightarrow [(CH_3)_2N]_3P{=}CF_2 \qquad [3.77]$$

of +65.5 ppm while the analogous diethylamino derivative showed a shift of +58 ppm. The high phosphorus shielding was taken as indicative of a high degree of *p-d* π overlap between the carbanion and the phosphonium group. This conclusion was borne out by the observation that the ylids would not react with benzaldehydes in the Wittig reaction. The P^{31} shifts for the conjugate acids were not reported.

Examination of a series of triply connected phosphorus compounds and a comparison of their chemical shifts with those of corresponding quadruply connected but potentially pentavalent phosphorus compounds indicates that on conversion of the former to the latter an increase in chemical shift normally occurs (Δ is positive, see Table 3.6). Such is the case when the atoms connected to phosphorus are oxygen, nitrogen, sulfur or halogen. However, when the atoms connected to phosphorus are carbon (i.e., a phosphine converted to its oxide, ylid or sulfide) the chemical shift decreases (Δ is negative). Therefore, since the conversion of *any* phosphine to a pentavalent derivative results in a net decrease in P^{31} shielding, the reports of negative shifts by Speziale and Ratts (*140*) and by Driscoll *et al.* (*154*) for triphenylphosphonium ylids are not inconsistent. Likewise, the report by Mark (*37*) of the high shielding in the tris-(dimethylamino)phosphonium ylids is not inconsistent with the behavior of other nitrogen-substituted pentavalent phosphorus compounds.

It would seem that any useful discourse on the significance of the P^{31} NMR data for phosphonium ylids must await a resolution of the differences indicated by the contrasting Δ values in Table 3.6, differences which appear to be due to the nature of the three non-carbanionic ligands attached to phosphorus. Suffice it to say that the P^{31} NMR data on phosphonium ylids have not clarified the electronic description of ylids although they are not inconsistent with other physical and chemical data. The near identity of the shielding in phosphonium salts and phosphonium ylids is most mystifying and needs clarification.

TABLE 3. 6

P^{31} NMR Chemical Shifts of Tri- and Pentavalent Phosphorus[a]

Trivalent	δ^{III} (ppm)[b]	Pentavalent	δ^{V} (ppm)[b]	$\Delta(\delta^{V}-\delta^{III})$ (ppm)
$(C_2H_5)_3P$	+20.4	$(C_2H_5)_3PO$	-48.3	-68.7
		$(C_2H_5)_3PS$	-54.5	-74.5
$(C_6H_5)_3P$	+5.9	$(C_6H_5)_3PO$	-28.9	-34.8
		$(C_6H_5)_3P{=}CHCN$	-22.6[c]	-28.5
		$(C_6H_5)_3P{=}CHCOOC_2H_5$	-19.1[c]	-25.0
$(C_2H_5O)_3P$	-136.9	$(C_2H_5O)_3PO$	+0.9	+137.8
		$(C_2H_5O)_3PS$	-68.1	+68.8
$(C_6H_5O)_3P$	-126.8	$(C_6H_5O)_3PO$	+17.3	+144.1
		$(C_6H_5O)_3PS$	-53.4	+73.4
$(C_2H_5S)_3P$	-115.6	$(C_2H_5S)_3PO$	-61.3	+54.3
		$(C_2H_5S)_3PS$	-92.9	+22.7
$[(C_2H_5)_2N]_3P$	-118.2	$[(CH_3)_2N]_3PO$	-23.4	+94.8
		$[(C_2H_5)_2N]_3PS$	-77.8	+40.4
		$[(C_2H_5)_2N]_3P{=}CF_2$	+58[d]	+176.2
Cl_3P	-219.4	Cl_3PO	-1.9	+217.5
		Cl_3PS	-28.8	+190.6
		PCl_5 (vapor)	+80	+299.4
		$^{+}PCl_4$ (solid)	-91.0[e]	+128.4
		$^{-}PCl_6$ (solid)	+281[e]	+500.4
F_3P	-97	$^{-}PF_6$	+118	+215

[a] Data mainly from P. C. Lauterbur, *in* "Determination of Organic Structures by Physical Methods" (F. C. Nachod and W. D. Phillips, Eds.), Vol. 2, p. 517. Academic Press, New York, 1962.

[b] Relative to 85% phosphoric acid as reference.

[c] Speziale and Ratts (*140*).

[d] Mark (*37*).

[e] E. R. Andrews, A. Bradburg, R. G. Eades, and G. J. Jenkes, *Nature* **188**, 1096 (1960).

From the studies mentioned in the preceding discussion it is evident that phosphonium ylids essentially are carbanions and their special characteristics can be accounted for by a consideration of the various factors contributing to their stability. On the basis of information obtained from studies of their basicity, dipole moments, infrared spectra, ultraviolet spectra and their nucleophilicity it is apparent that the carbanions can be stabilized through delocalization by the phosphonium group, presumably by *p-d* π overlap, and by the carbanion substituents through *p-p* π overlap. The degree of carbanionic character seems well correlated to the degree of delocalization of the negative charge by both the phosphonium group and the other substituents on the carbanion. Any factor tending to increase the extent of delocalization also decreases the carbanionic character of the ylid. It is abundantly clear that stabilization by the phosphonium group is of the utmost importance since the analogous ammonium ylids (see Chapter 7) are, as a rule, much less stable and more nucleophilic.

D. Molecular Structure of Phosphonium Ylids

Throughout this discussion nothing has been said regarding the actual molecular structure of phosphonium ylids. The reason is that very little is known regarding their three-dimensional structure, and there is an urgent need for X-ray crystallographic work on phosphonium ylids.

The most interesting feature of ylid structure is the bond between the carbanion and the phosphorus atom, especially its length. In the course of proving the structure of the adduct of diethyl acetylenedicarboxylate and *N*-*p*-bromophenyliminotriphenylphosphorane to be LXXII, Mak and Trotter (*155*) found the phosphorus-carbanion length

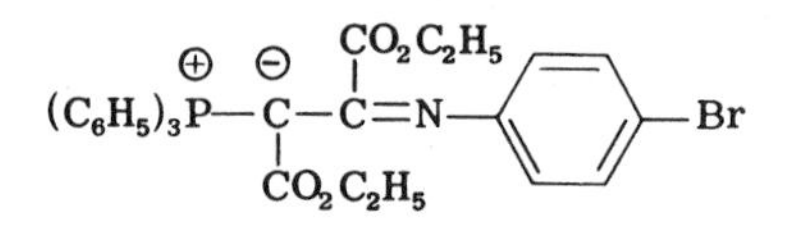

LXXII

to be 1.70 ± 0.03 Å. This is considerably shorter than the sum (1.87) of the single bond radii for phosphorus and carbon and for typical phosphorus–carbon lengths such as that in triphenylphosphine of 1.83 Å (*156*, Daly). The length is a little longer than the sum (1.67) of the double bond radii of phosphorus and carbon (*157*, Pauling) but clearly indicates considerable double bond character in agreement with most other chemical and physical evidence.

No information is available regarding the hybridization about the carbon atom in phosphonium ylids. The two possibilities are trigonal hybridization with the unshared pair of electrons in a 2*p*-orbital (Figure 3.1) or tetrahedral hybridization with the unshared pair in a sp^3-hybrid orbital (Figure 3.2). Carbanions normally tend to exist in the trigonally

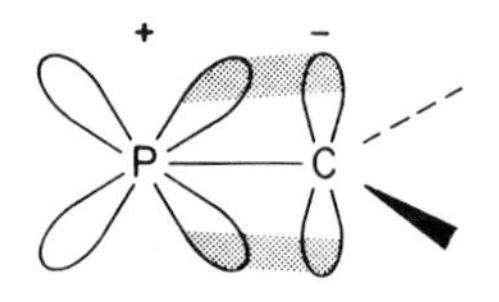

Fig. 3.1. Ylid bond using trigonal hybridization.

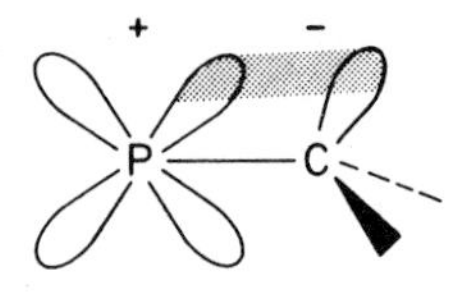

Fig. 3.2. Ylid bond using tetrahedral hybridization.

hybridized state but in recent years there have been discovered several examples in which the carbanions seemed to exist in the tetrahedral configuration. Cram *et al.* (*158*) have studied the ratio of the rates of exchange of α-hydrogen to the rates of racemization of the same substances as influenced by the adjacent heteroatom group (X in LXXIII), groups such as sulfone, sulfoxy and phosphinoxy, all of which are capable

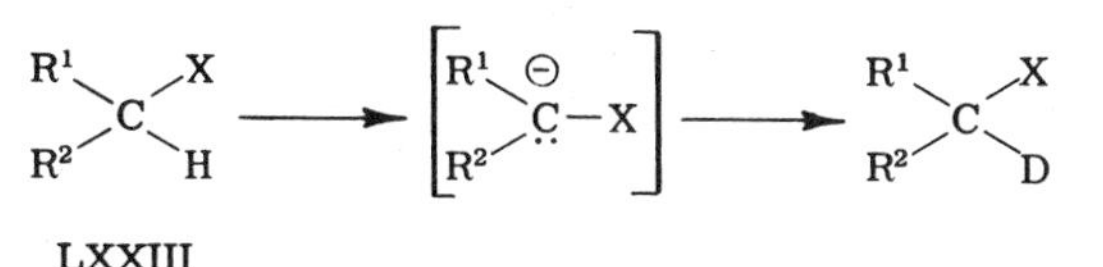

LXXIII

of expanding the outer shell to provide *p-d* π-overlap between the carbanion and the heteroatom. In some of these cases the evidence indicated that the ratio of exchange to racemization was greater than one due to the carbanion maintaining a tetrahedral configuration during the exchange process.

In accord with the theoretical predictions of Koch, Moffitt and Kimball (*159, 160*), Doering *et al.* (*141, 161*) found that there was little angular requirement for *p-d* π-overlap between carbanions and sulfonium or sulfone groups. Of special interest was the observation that a bridgehead hydrogen of the bicyclic trisulfone (LXXIV) was acidic (soluble in sodium bicarbonate) whereas the more recently reported and structurally analogous bicyclo[2.2.2]octane-2,6,7-trione (*162*, Theilacker and Wegner) showed no sign of acidity. Doering (*161*) assumed that

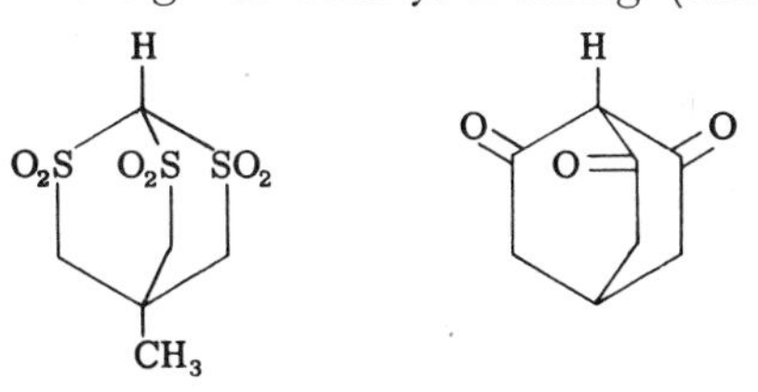

LXXIV

stabilization of the carbanion of LXXIV by the three sulfone groups implied that it was trigonally hybridized and that overlap was between a $2p$-orbital of carbon and a $3d$-orbital of sulfur. It seems equally likely that the overlap was between a sp^3-hybrid orbital of carbon and a $3d$-orbital of sulfur. Tetrahedral hybridization of the bridgehead carbanion would be more consistent with the probable molecular geometry.

On the basis of these examples it is apparent that it cannot be assumed that ylid carbanions are trigonally hybridized. Such is likely to be the case when the carbanion is able to delocalize its charge into the carbanion substituents by normal *p-p* π overlap. However, in the absence of such overlap the carbanion may well be tetrahedrally hybridized. The overlap between a sp^3-hybrid orbital and a d-orbital of phosphorus well may be more efficient than overlap of a p-orbital with the same d-orbital. Experimental evidence bearing on this point is non-existent.

Similarly, little information is available regarding the hybridization or geometry of the phosphorus atom in a phosphonium ylid. Phosphonium salts contain a tetrahedrally hybridized phosphorus atom (*163*, Van Wazer) and it is possible that this geometry is maintained in the conversion to a phosphonium ylid. This would imply that unhybridized $3d$-orbitals of phosphorus would be used in the p-d π-overlap for the ylid bond. Assuming that the σ bond between the phosphorus and the carbanion lies along an x-axis, the d_{xz}-orbital of phosphorus then would be oriented to permit efficient overlap either with a p-orbital (Figure 3.1) or with a sp^3-hybrid orbital (Figure 3.2). Rotation of the phosphorus atom and its three other substituents by 90° about the x-axis would permit overlap between the d_{xy}-orbital of phosphorus and the same orbital of the carbanion. As a result, the overlap forming the p-d π bond should not be too dependent on the conformation of the phosphorus group with respect to the carbanion.

The preceding proposal is borne out by the only evidence bearing on the question, the recent structure determination of LXXII by Mak and Trotter (*155*), which indicated that the phenyl-phosphorus-phenyl bond angles were 106, 106 and 110° and that the phenyl-phosphorus-carbanion angles were 108, 112 and 114°. The mean carbon-phosphorus-carbon angle was 109.4°, certainly consistent with the proposed tetrahedral hybridization of the phosphorus atom.

The alternative proposal of the geometry and hybridization about phosphorus would require a trigonal bipyramid structure (sp^3d hybridization) since there would be, in effect, five pairs of bonding electrons about it. Thus, it would be pentavalent in one sense but would still be tetravalent in terms of the number of substituents actually attached. In a phosphonium ylid, $R_3P{=}CHY$, two of the R groups could be in basal positions and the third in an axial position about the phosphorus atom. The $=$CHY group could be σ bonded either to a basal or axial position of the bipyramid and the π bonding could take place by overlap between the p-orbital or sp^3-hybrid orbital of the carbanion and either a basally or axially oriented sp^3d-hybrid orbital of the phosphorus atom (Figures 3.3 and 3.4). Such a description of the geometry of the substituents about the phosphorus atom probably would lead to some distortion of the pure trigonal bipyramid shape, resulting in a structure which could be classified either as a distorted trigonal bi-

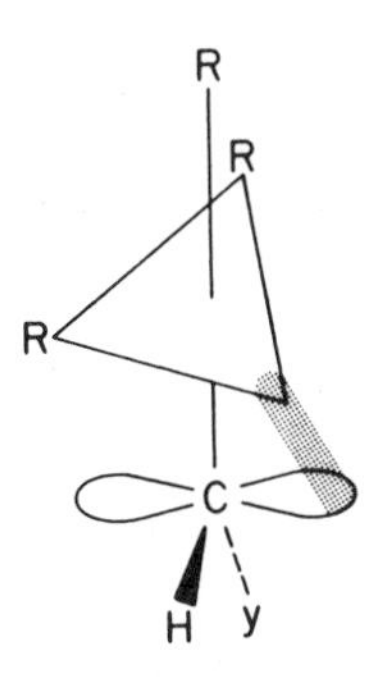

FIG. 3.3. Ylid carbanion at axial position of trigonal bipyramidal phosphorus.

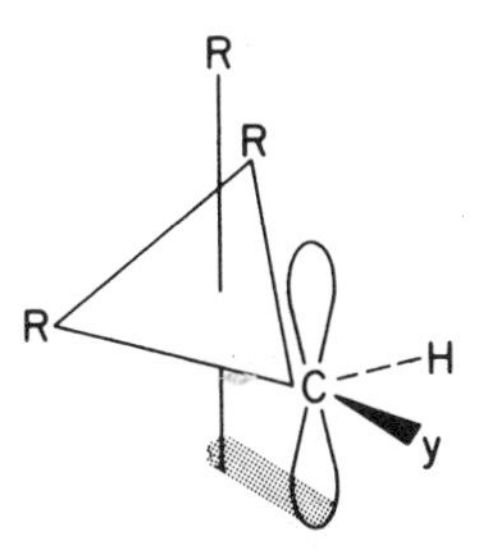

FIG. 3.4. Ylid carbanion at basal position of trigonal bipyramidal phosphorus.

pyramid or a distorted tetrahedron. A structure with two R groups in the axial positions seems unlikely since the angle between basal positions is 120°, inconsistent with appreciable π-overlap.

At present the only available chemical information pertaining to the geometry of the phosphorus atom in phosphonium ylids was provided by McEwen *et al.* (*164*) through their finding that optically active benzylidenemethylethylphenylphosphorane, prepared from the active benzylmethylethylphenylphosphonium iodide with phenyllithium, underwent the Wittig reaction with benzaldehyde to afford stilbene and optically active methylethylphenylphosphine oxide. The latter had been formed with retention of configuration, consistent with the proposal of a pentavalent transition state for the reaction with the phosphorus atom in a trigonal bipyramid configuration. This type of transition state could have been formed from an ylid in which the phosphorus atom had a tetrahedral configuration (and one face of the tetrahedron was attacked by the oxyanion) or from an ylid in which the phosphorus atom had a trigonal bipyramid configuration (Figures 3.3 or 3.4). In any event, the two possible transition states would be LXXV and LXXVI. The third possibility with both the oxygen atom and the benzylidene group in basal positions can be neglected due to angle strain. Both of the transi-

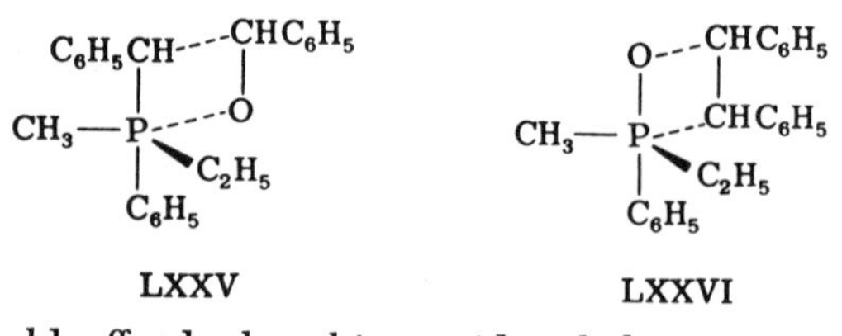

LXXV LXXVI

tion states would afford phosphine oxide of the same configuration as the phosphonium salt and ylid as observed (*165*, Hamilton *et al.*; *166*, Haake and Westheimer). Therefore, this experiment provided results which were consistent with both proposals regarding the geometry of the phosphorus atom in phosphonium ylids, whether tetrahedral or trigonal bipyramidal.

Interestingly, McEwen *et al.* (*164*) did demonstrate that the benzyl ylid would retain its optical activity in solution. This result certainly would be expected if the asymmetry was due to the phosphorus atom since there is no obvious means by which it could racemize short of a bond fracture, regardless of the phosphorus configuration.

E. Cyclic Phosphonium Ylids

With the accumulation of appreciable evidence that there is a degree of p-d π bonding in phosphonium ylids attempts have been made to incorporate such a bond into a cyclic molecule to permit a comparison of the characteristics of such a bond with a normal p-p π bond. The one approach that has been explored is the incorporation of an ylid structure into cyclic molecules which might exhibit some aromatic character. The question then became whether an ylid bond could be incorporated into an aromatic system without disruption of its aromaticity. The corollary question—How much would the nature of the ylid bond be altered by incorporation into a potential aromatic molecule?—would be much easier to answer, and the studies to date have been directed accordingly. In a rapidly appearing series of short communications Markl reported several studies on cyclic phosphonium ylids. The data reported were somewhat fragmentary but indicated that considerable moderation of the carbanionic character of an ylid could be achieved by incorporating the ylid in a potentially aromatic molecule.

The first cyclic ylid was not isolated but was prepared and studied in solution (*78*, Markl). 2,3,3-Triphenyl-3-phosphaindene (LXXVII) was prepared from the corresponding phosphonium salt with phenyllithium and was found to undergo a Wittig reaction with benzaldehyde, thereby demonstrating its nucleophilicity. The ylid appeared to behave normally indicating little if any interaction with the benzene ring.

In a similar manner, 1,1-diphenyl-3,4-dihydro-1-phosphanaphthalene

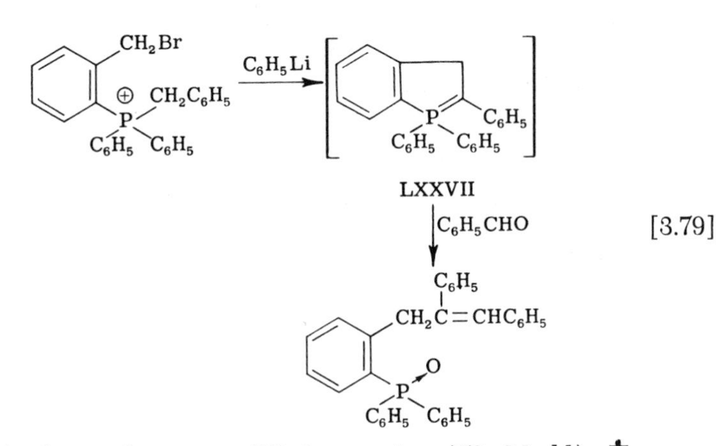

[3.79]

(LXXVIII) also underwent a Wittig reaction (*79*, Markl). However, introduction of a double bond into the 3,4-position of the naphthalene nucleus of LXXVIII afforded another ylid, 1,1-diphenyl-1-phosphanaphthalene (LXXIX), which did not undergo the Wittig reaction with benzaldehyde. This may indicate some loss of nucleophilic character

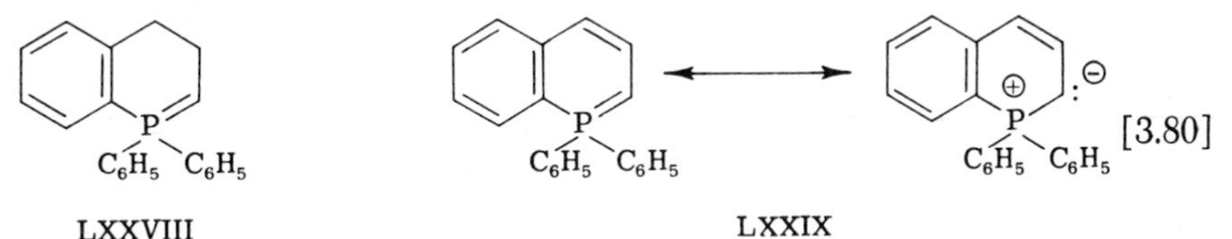

[3.80]

upon incorporation of the ylid bond into a cyclic conjugated system. However, the ylid was soluble in dilute hydrochloric acid, could be alkylated with methyl iodide and was hydrolyzed by water. It could be isolated as an amorphous, yellow mass which was sensitive to oxygen and showed ultraviolet absorption at 362 and 420 mμ. The proton NMR spectrum showed 16 protons at 2.5–3.5 τ and proton, presumably the one on the ylid carbanion, at 4.8 τ (*167*, Markl). The P^{31} NMR spectrum would be most significant.

Shortly after the above work Markl also reported the preparation of some monocyclic ylids. 1,1-Diphenyl-1-phosphabenzene (LXXX) was obtained as an amorphous, yellow solid which was subject to air oxida-

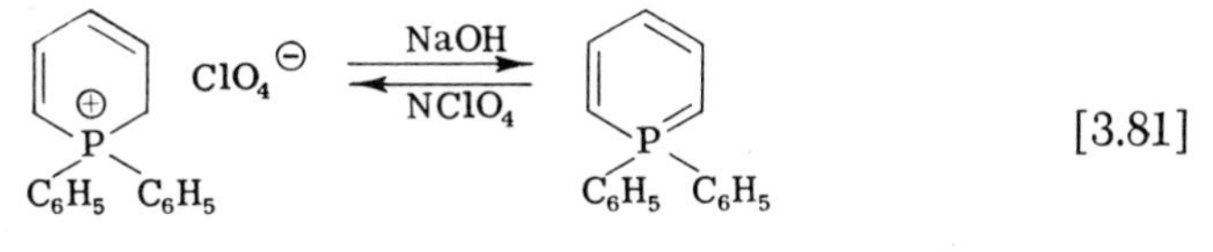

[3.81]

tion (*168*, Markl). It showed ultraviolet absorption at 409 mμ and would dissolve in dilute hydrochloric acid to reform its conjugate acid. An indication of the stability of the ylid system was evident from the observation that it could be formed from its precursor salt using aqueous sodium hydroxide. In line with previous discussion on the acidity of ylid salts, this would indicate considerable stability for the carbanion, mainly through *p-d* π-overlap. Some stability also must arise from the incorporation of the ylid in the potentially benzenoid system since the acyclic analogs could not be isolated (*106,* Bohlmann and Mannhardt). Markl (*169*) also reported the preparation of 1,1,3,3-tetraphenyl-1,3-diphosphabenzene (LXXXI) which showed ultraviolet absorption at 281

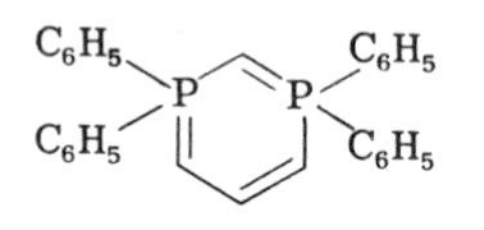

LXXI

and 384 mμ. The preparation of this ylid involved the stepwise removal of two protons from the corresponding phosphonium salt, the first with sodium carbonate but the second with sodium hydroxide, and both in aqueous media.

Cookson and Crofts (*170*) recently reported the incorporation of an ylid bond into a phenanthrene nucleus. 9,9-Diphenyl-9-phosphaphenanthrene appeared to retain its ylidic properties since it underwent a Wittig reaction with benzaldehyde. In view of the somewhat olefinic nature of a normal 9,10-bond in phenanthrene, this observation is not surprising.

The description of the bonding in such cyclic ylids presents a problem somewhat analogous to that in the phosphonitriles. In view of this similarity, Markl (*171*) proposed that the phosphorus atom was tetrahedrally hybridized to form the σ bond skeleton of the molecule. As was

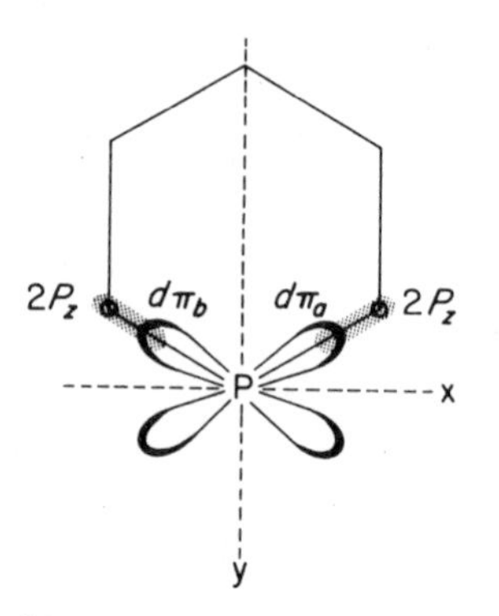

FIG. 3.5. Markl view of bonding in phosphabenzenes.

first proposed by Dewar *et al.* (*172*) for the phosphonitriles, Markl proposed that a pair of linear combination orbitals, designated $d\pi_a$ and $d\pi_b$, were formed from the d_{xz}- and d_{yz}-orbitals, the new orbitals lying in the direction of the σ bonds from phosphorus to carbon but oriented above and below the plane of the ring (Figure 3.5). Such orbitals then would be in a position to overlap with the p_z-orbital of each adjacent carbon atom. Such a proposal also implies, as did Dewar *et al.* (*172*), that the phosphorus atom is then an "isolator" atom (i.e., there can be no through conjugation) since separate orbitals of the phosphorus would be overlapping with each adjacent p_z-orbital of carbon. Thus, the system would be delocalized, perhaps "quasi-aromatic," but not aromatic. In support of this view Markl (*171*) pointed out that the phosphabenzene (LXXX) absorbed in the ultraviolet at 409 mμ whereas an acyclic analog, $(C_6H_5)_3P{=}CH{-}CH{=}CH{-}CH{=}CH{-}P^+(C_6H_5)_3$, absorbed at 432 m$\mu$.

Price (*173*) has proposed that in some sulfonium and phosphonium ylids the heteroatom enters into conjugation by overlap of a single d_{yz}-orbital with both p_z-orbitals of the adjacent carbon atoms, the direction of overlap therefore not being in the direction of the σ bond axis (Figure 3.6). The key difference between this and Markl's proposal is that the former utilizes one *d*-orbital whereas the latter utilizes two *d*-orbitals for overlap with the p_z-orbitals of the two adjacent carbon atoms. Consequently, Price's proposal predicts continuous delocalization about the ring whereas Markl's proposal predicts delocalization only over a three-atom unit of the ring. At present there is insufficient evidence available to make further discussion useful. It is clear, however, that incorporation of an ylid bond into a potentially aromatic system lowers the carbanionic character of the ylid. Whether or not the systems are "aromatic" is still open to debate and, most important, to experiment.

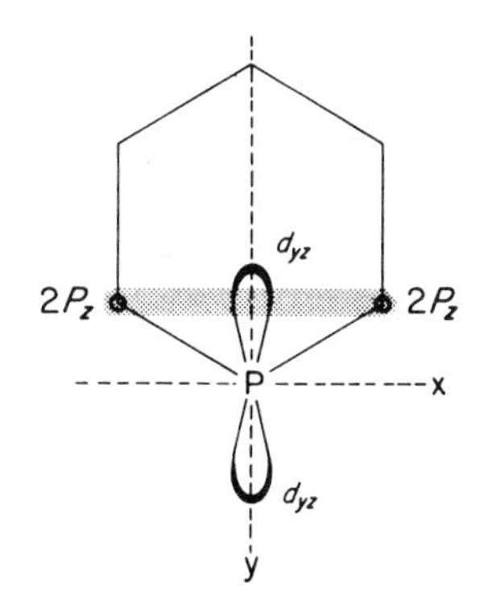

FIG. 3.6. Price view of bonding in phosphabenzenes.

III. Reactions of Phosphonium Ylids

As indicated in the previous discussion, phosphonium ylids are but a unique form of carbanion. Most of the reactions of such ylids depend on the use of this carbanionic character, usually in the initial step of a reaction. The unique reactions of these ylids, those depending on something other than simple carbanionic behavior, usually rely on the ability of the phosphorus atom to become pentavalent in an intermediate or transition state of the reaction. Such behavior usually is not exhibited in the initial step of ylid reactions but rather is shown in subsequent steps.

The best known reaction of phosphonium ylids is the Wittig reaction with carbonyl compounds to form an olefin and phosphine oxide [3.82].

$$\begin{matrix} (C_6H_5)_3P{=}CR_2 \\ + \\ R'_2C{=}O \end{matrix} \longrightarrow \left[\begin{matrix} (C_6H_5)_3\overset{\oplus}{P}-CR_2 \\ \ominus O-CR'_2 \end{matrix}\right] \longrightarrow \begin{matrix} R_2C{=}CR'_2 \\ \\ (C_6H_5)_3PO \end{matrix} \qquad [3.82]$$

The development of this reaction into an important method for the synthesis of olefins ignited a general interest in the chemistry of phosphonium ylids (*1*, Wittig and Schollkopf). The Wittig reaction is of such importance and mechanistic interest that it will be discussed separately in Chapter 4. Suffice it to mention at this point that the reaction appears to involve the nucleophilic attack by an ylid carbanion on the carbonyl carbon to form a betaine intermediate. The reaction is completed by transfer of the oxyanion to phosphorus, affording phosphine oxide and olefin. The major driving force behind the reaction appears to be the formation of the highly stable phosphorus-oxygen bond.

In this section are discussed the other reactions that are characteristic of phosphonium ylids. Several of these reactions already have been mentioned in Section I,C of this chapter as being applicable to the synthesis of complex ylids from simple ylids. Spontaneous decomposition or rearrangement reactions have been discussed in Section I,B of this chapter.

A. Hydrolysis of Phosphonium Ylids

Most phosphonium ylids are subject to hydrolysis to a hydrocarbon and a phosphine oxide but the conditions required to effect the cleavage vary widely depending on the structure of the ylid. Some ylids hydrolyze upon exposure just to water while others require long heating in the presence of hydroxide. Alkylidenetriphenylphosphoranes, such as the methylene and benzylidene ylids, hydrolyze spontaneously in the presence of moisture thereby necessitating their preparation and handling in an inert atmosphere. On the other hand, stable ylids such as the fluorenylidene-, *p*-nitrobenzylidene-, carbethoxymethylene- and phenacylidenetriphenylphosphoranes all are stable to atmospheric moisture and can be prepared in aqueous solution. The fluorenylide must be refluxed with alcoholic sodium hydroxide for more than twelve hours to effect complete hydrolysis (*44*, Johnson) while cyclopentadienylidenetriphenylphosphorane is inert to such conditions (*130*, Ramirez and Levy). The order of stability to hydrolysis seems to parallel ylid basicity. Those which are the most basic, the alkylidenephosphoranes, are the most easily hydrolyzed. This may imply that the hydrolysis involves protonation of the ylid carbanion initially to afford a quaternary phosphonium hydroxide.

The first hydrolyses of ylids were reported by Coffmann and Marvel (*117*) in 1929. The addition of water to an orange solution of ethylidenetriphenylphosphorane led to immediate loss of the color and the subsequent isolation of ethyldiphenylphosphine oxide [3.83]. Although not

$$(C_6H_5)_3P{=}CHCH_3 \xrightarrow{H_2O} [(C_6H_5)_3\overset{\oplus}{P}{-}CH_2CH_3 \quad \overset{\ominus}{O}H] \longrightarrow \begin{matrix} C_6H_6 \\ + \\ (C_6H_5)_2P(O)C_2H_5 \end{matrix} \qquad [3.83]$$

isolated at the time, benzene must have been the other reaction product. Much later, Wittig reported the hydrolysis of methylenetriphenylphosphorane to methyldiphenylphosphine oxide and benzene (*1*).

The observation that the alkyl portion of such ylids was not lost in the course of such hydrolyses struck a familiar note. Fenton and Ingold (*128*) had found that quaternary phosphonium salts would undergo hydroxide-catalyzed cleavage to a hydrocarbon and a tertiary phosphine oxide. Furthermore, the group ejected from the phosphonium salt as a hydrocarbon appeared consistently to be that group which formed the most stable carbanion. The preferential loss of groups from such salts was in the order benzyl > phenyl > alkyl. VanderWerf *et al.* (*174*) have studied this reaction in further detail and have borne out the original conclusions of Fenton and Ingold. On the basis of the similarity of these observations to those encountered in ylid hydrolyses it appears likely that hydrolysis of ylids proceeds by initial formation of a quaternary phosphonium hydroxide as portrayed above followed by hydroxide-catalyzed cleavage.

The probable carbanionic nature of the leaving group was substantiated by hydrolysis of other phosphonium ylids. Wittig and Laib (*26*) found that fluorenylidenetrimethylphosphorane was hydrolyzed to fluorene and trimethylphosphine oxide in the presence of water. Ramirez and Dershowitz (*129*) noted that benzoyl- and acetylmethylenetriphenylphosphoranes were hydrolyzed to acetophenone and acetone, respectively [3.84]. In these instances it would be expected that the fluorenyl

$$C_6H_5COCH{=}P(C_6H_5)_3 \xrightarrow{\ominus OH/H_2O} C_6H_5COCH_3 + (C_6H_5)_3PO \qquad [3.84]$$

anion and the α-ketocarbanions would be more stable than the phenyl anion and therefore be the group ejected. Schlosser (*175*) recently has shown that substituted methyltriphenylphosphonium hydroxides will undergo hydrolytic cleavage to triphenylphosphine oxide and a substituted methane provided the methyl substituent is one that can effectively stabilize a carbanion. Thus the methoxymethyl salt (LXXXIIa) was

$$(C_6H_5)_3\overset{\oplus}{P}-CH_2X \ \overset{\ominus}{O}H \begin{cases} \xrightarrow{(a)} C_6H_6 + (C_6H_5)_2P(O)CH_2OCH_3 \\ \xrightarrow{(b),(c),(d)} CH_3X + (C_6H_5)_3PO \end{cases} \qquad [3.85]$$

LXXXII a $X = OCH_3$
LXXXII b $X = SCH_3$
LXXXII c $X = Cl$
LXXXII d $X = Br$

cleaved to benzene (path *a*) but the methylthio-, chloro-, and bromomethyl salts (LXXXIIb,c,d) afforded triphenylphosphine oxide and the substituted methane. The latter three substituents probably stabilize the carbanion through use of their *d*-orbitals.

VanderWerf *et al.* (*174*) had shown that hydroxide-catalyzed decomposition of phosphonium salts followed third-order kinetics, first order in salt and second order in hydroxide. They proposed the mechanism shown in [3.86] for the decomposition. The slow step was thought to be the

$$(C_6H_5)_3\overset{\oplus}{P}-CH_2C_6H_5 + \overset{\ominus}{O}H \underset{}{\overset{\text{fast}}{\rightleftharpoons}} (C_6H_5)_3P\begin{matrix} OH \\ CH_2C_6H_5 \end{matrix}$$

$$(C_6H_5)_3P\begin{matrix} OH \\ CH_2C_6H_5 \end{matrix} + \overset{\ominus}{O}H \overset{\text{fast}}{\rightleftharpoons} (C_6H_5)_3P\begin{matrix} O^{\ominus} \\ CH_2C_6H_5 \end{matrix} + H_2O \qquad [3.86]$$

$$(C_6H_5)_3P\begin{matrix} O^{\ominus} \\ CH_2C_6H_5 \end{matrix} \overset{\text{slow}}{\rightleftharpoons} (C_6H_5)_3PO + C_6H_5CH_2^{\ominus}$$

$$C_6H_5CH_2^{\ominus} + H_2O \overset{\text{fast}}{\rightleftharpoons} C_6H_5CH_3 + \overset{\ominus}{O}H$$

ejection of the carbanionic group. As indicated earlier, the rate of decomposition varied with the nature of the substituted benzyl group being ejected. It was conceivable that decomposition of such benzyl salts was preceded by formation of a phosphonium ylid which then underwent addition of water to form a pentavalent intermediate. The latter could be decomposed to hydrocarbon and oxide as in [3.87]. Depending

$$
\begin{aligned}
&(C_6H_5)_3\overset{\oplus}{P}-CH_2C_6H_5 + \overset{\ominus}{O}H \rightleftharpoons (C_6H_5)_3P{=}CHC_6H_5 + H_2O \\
&(C_6H_5)_3P{=}CHC_6H_5 + H_2O \rightleftharpoons (C_6H_5)_3P(OH)(CH_2C_6H_5) \\
&(C_6H_5)_3P(OH)(CH_2C_6H_5) + \overset{\ominus}{O}H \rightleftharpoons (C_6H_5)_3PO + C_6H_5CH_2^{\ominus} + H_2O \\
&C_6H_5CH_2^{\ominus} + H_2O \rightleftharpoons C_6H_5CH_3 + \overset{\ominus}{O}H
\end{aligned}
\qquad [3.87]
$$

on the relative rates of the various steps such a mechanism could be either first or second order in hydroxide ion. Aksnes and Songstad (*176*) showed that tetraphenylphosphonium bromide, a salt which could not form a phosphonium ylid, and triphenylbenzylphosphonium chloride, a salt which could form an ylid, both underwent hydroxide-catalyzed decomposition in a third-order reaction (second order in hydroxide ion). The same workers, however, noted that *p*-nitrobenzyltriphenylphosphonium chloride was cleaved in a second-order reaction (first order in hydroxide ion). This indicated that either the first mechanism was followed but the formation, rather than the decomposition, of the pentavalent intermediate was the slow step or that the second (phosphorane) mechanism was followed and ylid formation was the slow step. The observation that the energies of activation remained about the same (18 kcal/mole) for the benzyl and *p*-nitrobenzyl salt cleavages while the kinetics did change from second to first order in hydroxide ion indicated that the slow step in a common mechanism must be changing or the mechanism itself must be changing. This duality of possibilities has not been resolved.

Grayson and Keough (*119*) have found that the ethoxide-catalyzed decomposition of triphenyl-*p*-nitrobenzylphosphonium bromide afforded *p*-nitrotoluene, diethyl ether and triphenylphosphine oxide. The same products were obtained by reaction of *p*-nitrobenzylidenetriphenylphosphorane with ethanol, leading to the conclusion that a pre-equilibrium condition existed for the decomposition reaction. The mechanism was portrayed as in [3.88]. Reaction of the ylid with *O*-deuteroethanol afforded di- and trideuteronitrotoluene, indicating clearly that pre-equilibrium occurred with the alcoholic deuterium being removed by the ylidic carbanion. The demonstration that phosphonium ylids undergo alcoholysis by the same mechanism as phosphonium salts undergo alkoxide-catalyzed cleavage provides good analogy for the proposal that phosphonium ylids undergo hydrolysis by a mechanism virtually identical

$$(C_6H_5)_3\overset{\oplus}{P}-CH_2C_6H_4NO_2(p) + C_2H_5O^{\ominus} \rightleftharpoons (C_6H_5)_3P{=}CHC_6H_4NO_2 + C_2H_5OH$$

$$(C_6H_5)_3\overset{\oplus}{P}-CH_2C_6H_4NO_2 + C_2H_5O^{\ominus} \underset{}{\overset{\text{slow}}{\rightleftharpoons}} (C_6H_5)_3P\begin{smallmatrix}OC_2H_5\\CH_2C_6H_4NO_2\end{smallmatrix}$$

$$(C_6H_5)_3P\begin{smallmatrix}OC_2H_5\\CH_2C_6H_4NO_2\end{smallmatrix} \xrightarrow{\text{fast}} (C_6H_5)_3\overset{\oplus}{P}OC_2H_5 + \overset{\ominus}{C}H_2C_6H_4NO_2 \quad [3.88]$$

$$(C_6H_5)_3\overset{\oplus}{P}OC_2H_5 + C_2H_5OH^{\ominus} \xrightarrow{\text{fast}} (C_2H_5)_2O + (C_6H_5)_3PO$$

$$^{\ominus}CH_2C_6H_4NO_2 + C_2H_5OH \xrightarrow{\text{fast}} CH_3C_6H_4NO_2 + C_2H_5O^{\ominus}$$

to that by which phosphonium salts undergo hydroxide-catalyzed cleavage.

There have been no studies to ascertain what effect the phosphorus substituents have on the rates and mechanism of ylid hydrolysis. Likewise, the stereochemical changes undergone by the phosphorus atom during such hydrolyses have not been studied in spite of the fact that the stereochemistry of the phosphonium salt cleavages is known (*164*, McEwen *et al.*).

The hydrolysis of phosphonium ylids is useful in two main respects. It is one of the few reactions by which a carbon–phosphorus bond can be cleaved and so finds use in the course of determining the structure of phosphonium ylids. The major use of the hydrolysis reaction, however, is to cleave that same bond after an ylid has been used in a synthetic operation. For example, Bestmann and Arnason (*81*) reported that acylation of an alkylmethylenetriphenylphosphorane afforded a keto-ylid which could be hydrolyzed to a ketone and triphenylphosphine oxide [3.89]. This sequence represents a useful ketone synthesis but

$$RCOCl + R'CH{=}P(C_6H_5)_3 \longrightarrow \begin{smallmatrix}RCO\\R'\end{smallmatrix}C{=}P(C_6H_5)_3 \xrightarrow{^{\ominus}OH} RCOCH_2R' \quad [3.89]$$

depends on having available a method for removing the phosphonium group after it has served its purpose.

B. Reduction of Phosphonium Ylids

For phosphonium ylids to be of use in organic synthesis there must be available methods for cleavage of the carbon–phosphorus bond after the synthetic operation. Hydrolysis is one such method. The reduction of phosphonium ylids is a second method that should be applicable but it appears to be fraught with difficulty in many instances, not the least of which is the presence of other reducible functional groups.

Saunders and Burchmann (*177*) reported the reduction of methylenetriphenylphosphorane with lithium aluminum hydride to methyldiphenylphosphine and benzene, the former being isolated as its methiodide [3.90]. Such a fragmentation appears consistent with the

$$(C_6H_5)_3P{=}CH_2 \xrightarrow{LiAlH_4} C_6H_6 + (C_6H_5)_2PCH_3 \qquad [3.90]$$

hydrolytic cleavages in that a phenyl carbanion is a better leaving group than an alkyl carbanion (*128, 174*). However, reduction of acetylmethylenetriphenylphosphorane also afforded benzene as the hydrocarbon fragment whereas hydrolysis of the same ylid afforded acetone. Gough and Trippett (*178*) found that benzylidenetriphenylphosphorane and isopropylidenetriphenylphosphorane also could be reduced with lithium aluminum hydride but in each case benzene and the corresponding alkyldiphenylphosphine were formed [3.91]. Here again the reduction

$$(C_6H_5)_3P{=}CHC_6H_5 \xrightarrow{LiAlH_4} C_6H_6 + (C_6H_5)_2PCH_2C_6H_5 \qquad [3.91]$$

of the ylid seemed to result in the formation of benzene, presumably via the phenyl anion, in spite of the possibility of forming a more stable carbanion (i.e., benzyl). Clearly the reduction of phosphonium ylids with lithium aluminum hydride does not result in the formation of a pentavalent intermediate in which there is equal opportunity for any group to be ejected as the anion, depending solely on its stability as an anion.

Bailey and Buckler (*179*) had shown that lithium aluminum hydride reduction of phosphonium salts containing a benzyl group invariably led to the ejection of the benzyl group as toluene and to the formation of a tertiary phosphine. Obviously this reduction did not occur via an ylid intermediate. Gough and Trippett (*178*) found that isopropyltriphenylphosphonium iodide was cleaved to propane and triphenylphosphine but the corresponding ylid, isopropylidenetriphenylphosphorane, afforded benzene and isopropyldiphenylphosphine. Methyltriphenylphosphonium and ethyltriphenylphosphonium salts both afforded benzene and the corresponding alkyldiphenylphosphine [3.92].

$$\begin{aligned}
(C_6H_5)_3\overset{\oplus}{P}{-}CH_2C_6H_5 \;\; Br^{\ominus} &\xrightarrow{LiAlH_4} (C_6H_5)_3P + C_6H_5CH_3 \\
(C_6H_5)_3\overset{\oplus}{P}{-}CH(CH_3)_2 \;\; I^{\ominus} &\xrightarrow{LiAlH_4} (C_6H_5)_3P + C_3H_8 \qquad [3.92] \\
(C_6H_5)_3\overset{\oplus}{P}{-}C_2H_5 \;\; Br^{\ominus} &\xrightarrow{LiAlH_4} (C_6H_5)_2PC_2H_5 + C_6H_6
\end{aligned}$$

In view of the marked differences in the nature of the group ejected as a hydrocarbon upon reduction of a phosphonium salt and of a phosphonium ylid it is clear that these two groups of substances must be reduced by quite different mechanisms. Reduction of the salts with lithium aluminum hydride appears to follow a mechanism similar to that proposed for the hydroxide-catalyzed cleavage of phosphonium salts or ylids. In every case of ylid reduction the carbanion group has remained attached to the final phosphine product. Therefore, ylid cleavage may take place via direct displacement of phenyl anion from the ylid, affording a new but short-lived ylid containing a phosphorus–hydrogen bond which would undergo a prototropic shift to afford phosphine [3.93].

$$(C_6H_5)_3\overset{\oplus}{P}-\overset{\ominus}{C}\langle \cdots H-Al\langle \longrightarrow \left[C_6H_5-\overset{C_6H_5}{\underset{H}{\overset{|}{P}}}\overset{\oplus}{}-\overset{\ominus}{C}\langle \; | \; Al- \right] \xrightarrow{H_2O} (C_6H_5)_2P-CH\langle \quad [3.93]$$

Trippett and Walker (*8*) found that sodium borohydride, aluminum isopropoxide, diborane and hydrogen over a platinum catalyst all were ineffective in reducing phenacylidenetriphenylphosphorane. Zinc in acetic acid would effect the reduction to acetophenone and triphenylphosphine. Schonberg *et al.* (*180*) have found that zinc in hydrochloric acid or hydrogen over a Raney nickel catalyst would reduce phosphonium ylids, fluorenylidenetriphenylphosphorane affording fluorene and triphenylphosphine. In all of these reductions, the most electronegative group was cleaved from the phosphorus.

Clearly the use of a reduction method as the terminal step of a synthetic sequence requiring cleavage of a specific carbon–phosphorus bond may lead to ambiguity. The mechanisms of the reductions have not been clearly defined. It should be mentioned, however, that reduction of phosphonium salts often is a predictable reaction. Therefore, an ylid reduction can be avoided by converting the ylid into its conjugate acid and then proceeding with a reduction.

C. Oxidation of Phosphonium Ylids

A third method for the cleavage of the carbon–phosphorus bond in a phosphonium ylid involves the use of various oxidative procedures. In this case the carbanion–phosphorus bond is the only one affected. Bestmann *et al.* (*181–183*) found that phosphonium ylids were subject to autoxidation resulting in the formation of a phosphine oxide and an olefin [3.94]. This accomplished the effective dimerization of the carbon

$$2\ RCH{=}P(C_6H_5)_3 \xrightarrow{O_2} RCH{=}CHR + 2\ (C_6H_5)_3PO \qquad [3.94]$$

portion of the ylid and constituted a means of preparing some olefins. For example, benzylidenetriphenylphosphorane afforded stilbene in 72%

yield and *n*-propylidenetriphenylphosphorane afforded 3-hexene in 68% yield. Likewise, cinnamylidenetriphenylphosphorane afforded 1,6-diphenyl-1,3,5-hexatriene. The phosphonium ylid obtained by conversion of vitamin A alcohol into its bromide followed by quaternization with triphenylphosphine and abstraction of a proton also underwent dimerization in the presence of oxygen to afford *trans*-β-carotene in 28% yield (*184*, Bestmann and Kratzer).

The autoxidation reaction also has been applied to bis-ylids resulting in the formation of cyclic olefins. Thus cycloheptene, cyclohexene and cyclopentene were formed in 60–68% yields upon autoxidation of the bis-ylids (IX) [3.95] where $n = 5$, 4 and 3, respectively (*185*, Bestmann

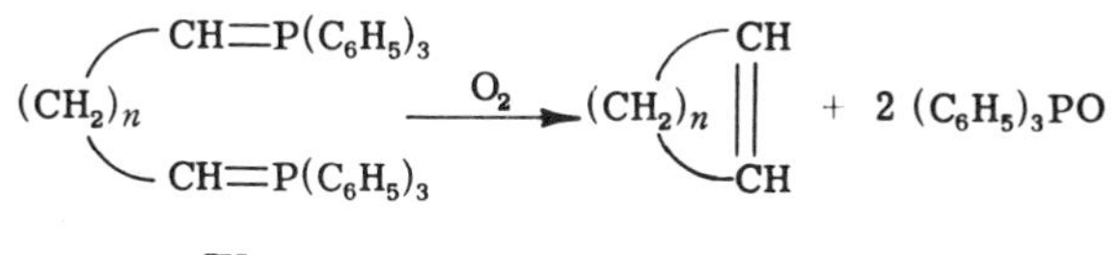

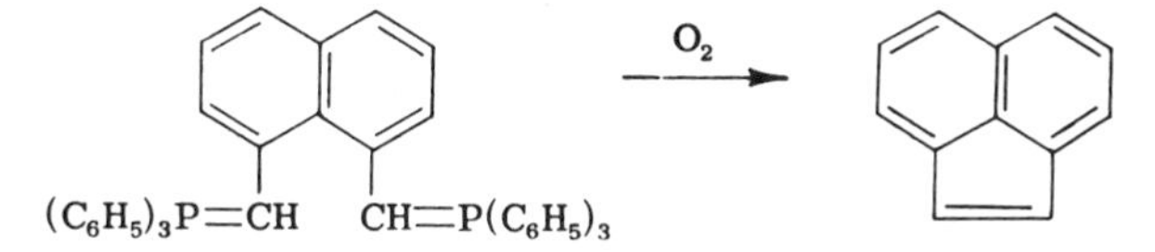

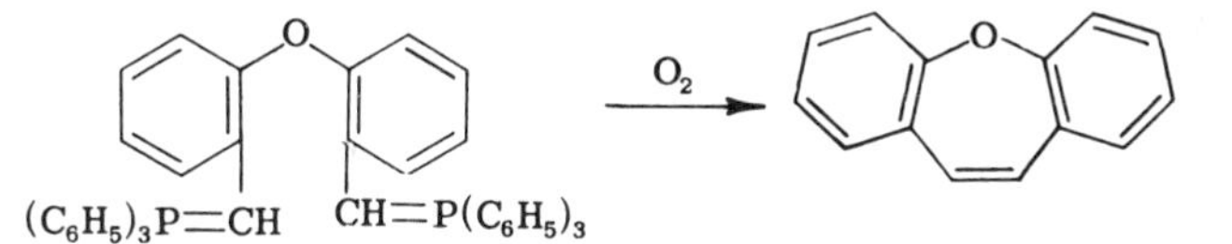

et al.). Acenaphthylene and dibenzoxepine could be formed by analogous reactions.

Bestmann and Kratzer (*183*) noted that the dimerization reaction only would proceed in the presence of a limited amount of oxygen. If excess oxygen were present a carbonyl compound and triphenylphosphine oxide were obtained in equimolar quantities. For example, benzhydrylidenetriphenylphosphorane afforded benzophenone in 70% yield, and isopropylidenetriphenylphosphorane afforded acetone in 60% yield. These workers proposed that autoxidation of phosphonium ylids proceeded by initial cleavage of the ylid to a phosphine oxide and a carbonyl component. If additional ylid were present it would react with the carbonyl component in a typical Wittig reaction to form the olefin and more phosphine oxide. For these products to prevail there must be excess ylid or insufficient oxygen present [3.96]. There is no evidence

$$RCH{=}P(C_6H_5)_3 \xrightarrow{O_2} \left[\begin{array}{c} O{-}O \\ |\quad\ | \\ RCH{-}P(C_6H_5)_3 \end{array}\right] \longrightarrow RCHO + (C_6H_5)_3PO$$

$$RCHO \xrightarrow{RCH{=}P(C_6H_5)_3} RCH{=}CHR + (C_6H_5)_3PO \qquad [3.96]$$

available to indicate in what manner the oxygen attacks the phosphonium ylid. It is conceivable that it attacks the phosphorus first but then forms a four-membered intermediate or transition state similar to that involved in the Wittig reaction. The driving force probably is the formation of phosphine oxide.

Other oxidative conditions also have been investigated. Ramirez *et al.* (*186*) found that phosphonium ylids which were not reactive (basic) enough to be cleaved by oxygen would react with ozone. Thus phenacylidenetriphenylphosphorane reacted with ozone at —70° in methylene chloride solution to afford phenylglyoxal and triphenylphosphine oxide in high yield. A cyclic intermediate also was suggested for this reaction. The ylid was inert to oxygen [3.97].

$$C_6H_5COCH{=}P(C_6H_5)_3 \xrightarrow{O_3} \left[\begin{array}{c} C_6H_5COCH{-}P(C_6H_5)_3 \\ |\qquad\quad | \\ O{-}\overset{\oplus}{O}{-}O^{\ominus} \end{array}\right]$$

$$\longrightarrow [(C_6H_5)_3PO_2] + C_6H_5COCHO; \quad [(C_6H_5)_3PO_2] \longrightarrow (C_6H_5)_3PO + \tfrac{1}{2}O_2 \qquad [3.97]$$

Denney *et al.* (*187, 188*) have investigated the oxidation of phosphonium ylids with peracetic acid in acetic acid media. Again the overall result of this oxidation, when the peracid was added to ylid, was the formation of triphenylphosphine oxide and an olefin in a ratio of 2:1. The olefin was a dimer of the carbanion portion of the ylid. Thus, phenacylidenetriphenylphosphorane afforded *trans*-dibenzoyl ethylene in 73% yield [3.98]. This dimerizing oxidation was effective only if the

$$2\ C_6H_5COCH{=}P(C_6H_5)_3 \xrightarrow{CH_3CO_3H} C_6H_5COCH{=}CHCOC_6H_5 + 2\ (C_6H_5)_3PO \qquad [3.98]$$

phosphonium ylid was not too basic. The reaction could not be applied to benzylidenetriphenylphosphorane and was minimally effective with carbethoxymethylenetriphenylphosphorane. The major reaction in these instances was simple protonation of the basic ylid. This is not a serious drawback, however, because the autoxidation method of Bestmann (*183*) was effective with the more basic ylids.

The mechanism of the peracid oxidation was proposed to be as shown in [3.99]. Evidence for this proposal rests mainly on the observa-

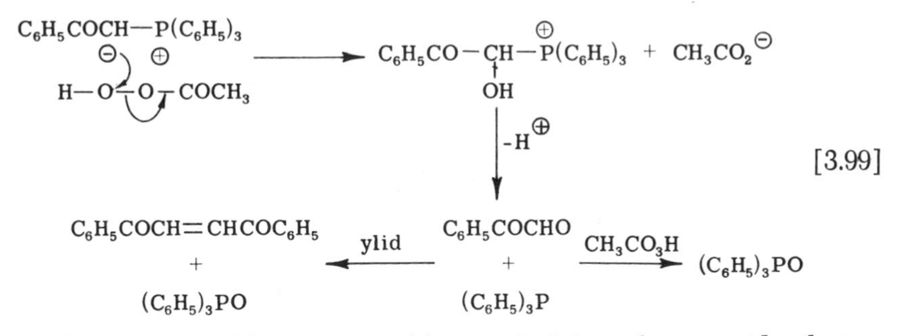

tion that inverse addition, i.e., addition of ylid to the peracid solution, resulted in the formation of phosphine oxide but not of olefin. In other words, there appeared to be no ylid left to react with the phenylglyoxal formed in the first step of the reaction. There was no evidence, however, for the initial formation of triphenylphosphine. It seems reasonable that the ylid carbanion would attack the peroxidic oxygen.

Denney and Valega (*189*) have reported the reaction of phenacylidenetriphenylphosphorane with benzoyl peroxide but they were unable to isolate the primary products of the reaction. Neither the scope nor the mechanism of this reaction has been studied but a complex series of esters and anhydrides appear to be the major products.

Oxidative cleavage of phosphonium ylids does appear to be a useful means of breaking the carbanion–phosphorus bond. The combination of Bestmann's autoxidation procedure (*183*) and Denney's peracid oxidation procedure (*188*) permits the cleavage of ylids of varying degrees of basicity. In the course of these cleavages the ylid carbon evolves in an oxidized state whereas in the other two cleavage reactions, the hydrolyses and the reductions, the ylid carbon evolved in a reduced state [3.100].

$$RCH{=}P(C_6H_5)_3 \begin{cases} \xrightarrow{\text{reduction}} RCH_3 + (C_6H_5)_3P \\ \xrightarrow{\text{oxidation}} RCHO + (C_6H_5)_3PO \\ \xrightarrow{\text{hydrolysis}} RCH_3 + (C_6H_5)_3PO \end{cases} \qquad [3.100]$$

Accordingly, there is considerable flexibility available for cleavage of carbon–phosphorus bonds in ylids.

D. Halogenation of Phosphonium Ylids

The halogenation of phosphonium ylids was discussed in detail in Section I,C of this chapter as a method for the preparation of α-halomethylenetriphenylphosphoranes. Suffice it to mention here that Markl (92) found that reaction of chlorine with one equivalent of phosphonium ylid afforded one-half equivalent of chloroylid and one-half equivalent of unchlorinated phosphonium salt [3.101]. It appeared that the ylid was

$$(C_6H_5)_3P{=}CHCOOCH_3 + Cl_2 \longrightarrow (C_6H_5)_3\overset{\oplus}{P}{-}CH(Cl)(COOCH_3)\ \ Cl^{\ominus}$$

$$\xrightarrow{(C_6H_5)_3P{=}CHCOOCH_3} (C_6H_5)_3P{=}C(Cl)(COOCH_3) + (C_6H_5)_3\overset{\oplus}{P}{-}CH_2COOCH_3\ \ Cl^{\ominus} \qquad [3.101]$$

carrying out a nucleophilic attack on the chlorine forming a chlorophosphonium salt but that half of the original ylid was acting simply as a base for the generation of chloroylid from chlorophosphonium salt.

Denney and Ross (93) found that carrying out the halogenation in pyridine solvent would permit the conversion of one equivalent of the original ylid into one equivalent of chloroylid since the pyridine would act as the base for regeneration of the chloroylid from its conjugate acid. These workers also noted that *tert*-butylhypochlorite would serve as a halogenating agent.

E. Alkylation of Phosphonium Ylids

The alkylation of phosphonium ylids to produce higher alkylated ylids was discussed in Section I,C of this chapter. The scope of this reaction and its application in other synthetic procedures will be discussed here.

In principle phosphonium ylids can be alkylated by a variety of reagents but in practice only two types have been studied. Wittig and Reiber (3), in their very early work on phosphonium ylids, found that methylenetrimethylphosphorane could be alkylated with methyl iodide to form ethyltrimethylphosphonium iodide. Much later, Bestmann *et al.* (72, 74) studied the alkylation of carbomethoxymethylenetriphenylphosphorane (XXXVIII) with a variety of alkyl halides. For example, reaction with benzyl bromide afforded a mixture of alkylated ylid (L) and conjugate acid of the original ylid. They proposed that the original

$$\underset{\text{XXXVIII}}{(C_6H_5)_3P{=}CHCOOCH_3} + C_6H_5CH_2Br \longrightarrow \underset{\text{LI}}{\left[(C_6H_5)_3\overset{\oplus}{P}{-}CH(COOCH_3)(CH_2C_6H_5) \quad Br^{\ominus}\right]}$$

$$\text{LI} + \text{XXXVIII} \longrightarrow \underset{\text{L}}{(C_6H_5)_3P{=}C(COOCH_3)(CH_2C_6H_5)} + (C_6H_5)_3\overset{\oplus}{P}{-}CH_2COOCH_3 \quad Br^{\ominus}$$

alkylation product (LI) was converted to the ylid (L) by the original ylid (XXXVIII) acting as a base. Most of the alkyl halides used have been primary halides. The use of tertiary and perhaps of secondary halides may be impractical due to competing elimination reactions. Such alkylations have not been explored, however.

The only other alkylating agent whose use has been reported to date was triethyloxonium tetrafluoborate. Markl (75) found that benzylidenetriphenylphosphorane could be alkylated to 1-phenylpropyltriphenylphosphonium tetrafluoborate in 98% yield with this reagent.

In principle any type of ylid of sufficient nucleophilicity should be capable of undergoing an alkylation. In fact the types of ylids are somewhat limited. As indicated by the work of Wittig *et al.* (3, 26) and Markl (75), alkyl- or arylmethylenetriphenylphosphoranes appear to undergo alkylation in a straightforward manner. Bestmann and Haberlein (77) indicated such also is the case with intramolecular alkylation of arylmethylenetriphenylphosphoranes. Bestmann and Schulz (72, 74) noted that carboalkoxy ylids (e.g., XXXVIII) also undergo alkylation with alkyl, allylic and benzylic halides to afford normal alkylated ylids. However, β-ketomethylene ylids do not undergo normal alkylation. For

$$\underset{\text{LXV}}{\left[(C_6H_5)_3\overset{\oplus}{P}{-}\overset{\ominus}{C}H{-}\overset{O}{\overset{\|}{C}}C_6H_5 \longleftrightarrow (C_6H_5)_3\overset{\oplus}{P}{-}CH{=}\underset{O^{\ominus}}{\underset{|}{C}}C_6H_5\right]} + C_2H_5I \longrightarrow (C_6H_5)_3\overset{\oplus}{P}{-}CH{=}C(OC_2H_5)(C_6H_5) \quad I^{\ominus} \qquad [3.102]$$

example, treatment of phenacylidenetriphenylphosphorane (LXV) with ethyl iodide afforded the O-alkylated product (*129*, Ramirez and Dershowitz). This result presumably is due to the high degree of charge delocalization from the ylid carbanion through the carbonyl group. That O-alkylation was not observed with the ester ylids (XXXVIII) must indicate that there is less charge delocalization through the ester group than through the ketone group in LXV. This conclusion agrees with the observation that the ester ylid was more basic than was the benzoyl ylid (*140*, Speziale and Ratts).

Siemiatycki and Strzelecka (*190*) found that the benzoyl ylid (LXV) reacted with phenacyl bromide to produce an interesting mixture of products which did not correspond to the expected results of a simple alkylation. They found that 1,2-dibenzoylethene (50%), phenacyltriphenylphosphonium bromide (95%) and tribenzoylcyclopropane (7%) were formed in this reaction. They proposed that a carbene intermediate was formed by one of two possible mechanisms (see (*a*) and (*b*) in [3.103]). The benzoylcarbene was thought to dimerize to form the

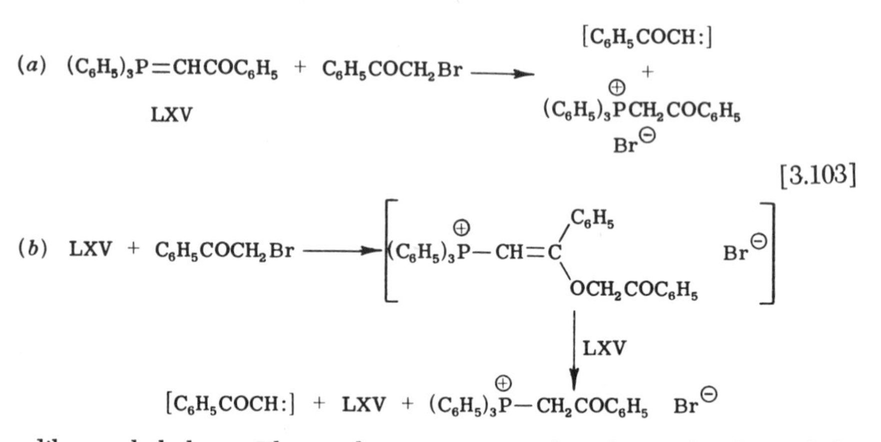

dibenzoylethylene. The cyclopropane was thought to be formed by carbene addition to the olefin. No evidence was provided for the existence of carbenes in this reaction and until some is forthcoming it is imperative to consider a completely ionic mechanism such as that outlined in [3.104]. Cyclopropane formation from the reaction of phosphonium ylids with conjugated olefins has been reported (*67*, Bestmann and Seng; *191*, Mechoulam and Sondheimer), and the Hofmann-type elimination is well known (*192*, Bestmann *et al.*). The carbenic and ionic mechanistic proposals clearly could be differentiated by determining the fate of the phosphorus in the system. The carbene mechanism predicts there should be no carbon–phosphorus bond cleavage in the reaction.

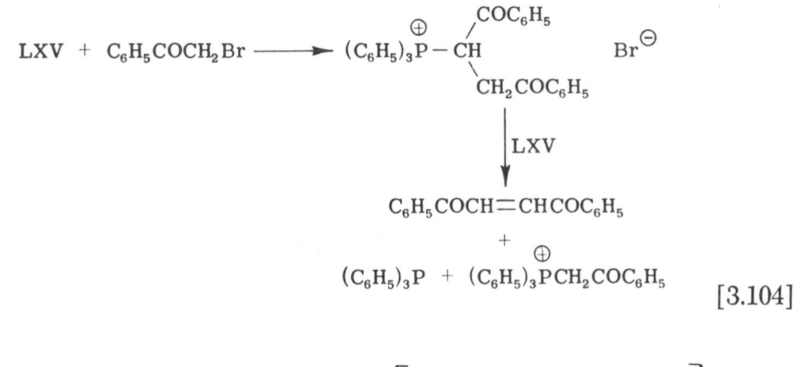

[3.104]

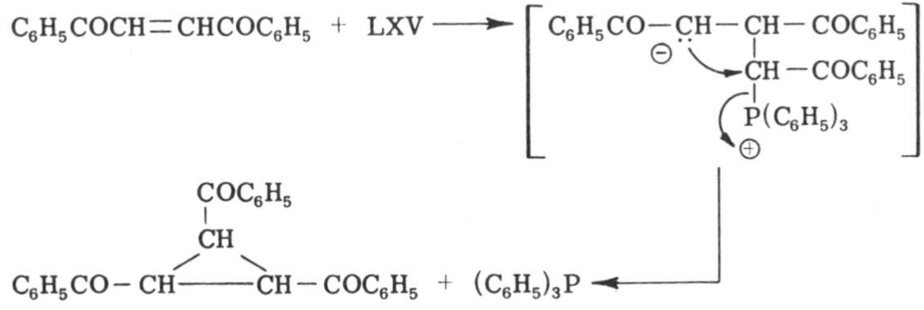

Interestingly, Bestmann *et al.* (*193, 194*) reported the reaction of the same halide, phenacyl bromide, with an ester ylid (XXXVIII). They reported the isolation of triphenylphosphine, the conjugate acid of XXXVIII, and methyl β-benzoylacrylate. They proposed that the ylid carried out a displacement of halide to form an intermediate phosphonium salt (LXXXIII) which was then subjected to a Hofmann-type elimination with another molecule of the original ylid acting as the base.

$$\underset{\text{XXXVIII}}{(C_6H_5)_3P{=}CHCOOCH_3} + C_6H_5COCH_2Br \longrightarrow \underset{\text{LXXXIII}}{(C_6H_5)_3\overset{\oplus}{P}{-}CH(COOCH_3)(CH_2COC_6H_5)\quad Br^{\ominus}}$$

$$\text{LXXXIII} + \text{XXXVIII} \longrightarrow C_6H_5COCH{=}CHCOOCH_3 + (C_6H_5)_3\overset{\oplus}{P}CH_2COOCH_3\ Br^{\ominus} + (C_6H_5)_3P$$

[3.105]

$$C_6H_5COCH_2{-}\underset{\displaystyle Br}{\underset{|}{CH}}{-}COOCH_3 + (C_6H_5)_3P \longrightarrow \text{LXXXIII}$$

The isolation of the triphenylphosphine is consistent with the proposal. In addition, they reported the preparation of the proposed intermediate phosphonium salt (LXXXIII) by another route. Treatment of this intermediate with the original ylid (XXXVIII) did effect the proposed elimination.

The alkylation of phosphonium ylids has served two main purposes. It has served as a route to more complex ylids which often are virtually unavailable by the normal salt method. Such ylids can be used in the Wittig reaction for the preparation of 1,1-disubstituted ethenes. This avoids having to alkylate triphenylphosphine with a secondary halide for the same end result [3.106].

$$R^1CH{=}P(C_6H_5)_3 + R^2X \longrightarrow (R^1)(R^2)C{=}P(C_6H_5)_3 \xrightarrow{>C=O} >C{=}C(R^1)(R^2) \qquad [3.106]$$

The second main role for the alkylation of phosphonium ylids has utilized the ylids as a convenient source of a carbanion with which to form carbon–carbon bonds followed by removal of the phosphorus group. For example, Bestmann and Schulz (*74*) prepared a series of carboxylic acids by alkylation of XXXVIII followed by hydrolysis of the resulting ylid and ester groups [3.107]. The alkylation can be used to prepare

$$\underset{\text{XXXVIII}}{(C_6H_5)_3P{=}CHCOOCH_3} + RBr \longrightarrow (C_6H_5)_3P{=}C(R)(COOCH_3) \xrightarrow{\ominus OH} RCH_2COOH \qquad [3.107]$$

almost any disubstituted methane system, R—CH_2—R′, by this general route.

F. Acylation of Phosphonium Ylids

The acylation of phosphonium ylids has been discussed in Section I,C of this chapter as a method for the conversion of simple ylids into complex carbonyl-substituted ylids. These data will be summarized briefly, and some practical uses of the reaction will be mentioned.

The acylation reaction is one that does not depend on any properties peculiar to ylids only but is characteristic of any carbanionic system. The general reaction scheme is presented in [3.108]. The initial step appears

$$RCOX + R'CH{=}P(C_6H_5)_3 \longrightarrow (C_6H_5)_3\overset{\oplus}{P}{-}CH(R')(C(=O){-}R)\ X^{\ominus} \xrightarrow{B:} (C_6H_5)_3P{=}C(R')(COR) \qquad [3.108]$$

to be the formation of an acylalkylmethyltriphenylphosphonium salt. The salt seldom is isolated, however, since the acyl group is acidifying and the salt therefore is prone to conversion into an acylalkylmethylenetriphenylphosphorane.

The efficiency of the above reaction varies with the nature of the acylating agent. In the case of acyl halides or *N*-acylimidazoles a second molecule of the original ylid serves as the base for the conversion of the acyl salt into the acyl ylid. Thus, the maximum yield is 50%. A variety of carboxylic acid chlorides (*80, 81,* Bestmann; *8, 82,* Trippett; *195,* Markl), chlorocarbonates (*83,* Bestmann; *94,* Speziale), acylimidazoles (*87,* Staab; *196,* Bestmann) and sulfonyl halides (*76,* Petragnini) have been used in the acylation reaction.

If the acylating agent is one whose anion (X^-) is a sufficiently strong base it will serve to convert the acyl phosphonium salt into the acyl ylid, thereby permitting a maximum yield of 100%. Such is the case with carboxylate esters (*1,* Wittig; *8, 86,* Trippett) and with carboxylate thioesters (*81, 85,* Bestmann). Acylation with esters has been applied to the synthesis of cyclic acyl ylids. Bergelson *et al.* (*28*) reported that generation of the five-carbon ylid from its corresponding salt (LIII, $n = 2$) led to spontaneous intramolecular acylation to form the five-membered keto-ylid (LIV, $n = 2$) [3.40]. They claimed that the cycliza-

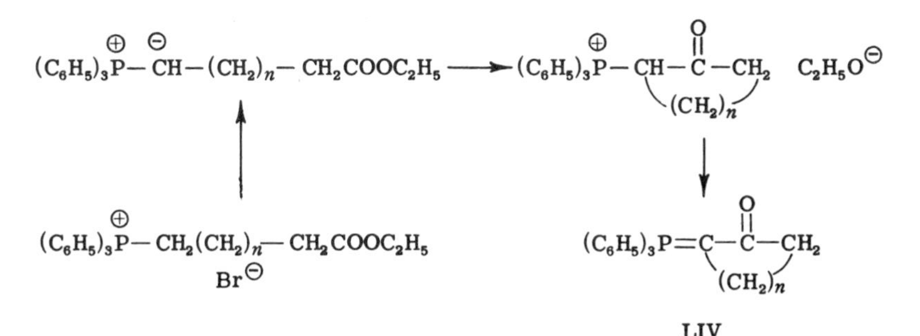

tion was effective only when $n = 2$. However, House and Babad (*88*) found that the cyclization reaction was effective for the preparation of five-, six- and seven-membered rings in from 41–84% yield.

The acylation of phosphonium ylids has found three main uses in synthetic chemistry. The most obvious application is the preparation of the acyl ylids themselves and their application in normal ylid chemistry such as use in the Wittig reaction for the preparation of α,β-unsaturated

$$RCH{=}P(C_6H_5)_3 \xrightarrow{R'COX} (C_6H_5)_3P{=}C(R)(COR) \xrightarrow{R''_2C{=}O} R''_2C{=}C(R)(COR') \qquad [3.109]$$

carbonyl compounds [3.109]. The pattern of substitution available from such a synthetic scheme virtually is unlimited.

A common application of the acylation reaction is the preparation of ketones by the hydrolysis or reduction of the acyl ylids [3.110] (*81*,

$$(C_6H_5)_3P{=}C\begin{matrix}R\\ COR'\end{matrix} \begin{matrix}\xrightarrow{\ominus OH/H_2O} RCH_2COR' + (C_6H_5)_3PO \\ \xrightarrow{Zn/H^+} RCH_2COR' + (C_6H_5)_3P\end{matrix} \qquad [3.110]$$

Bestmann). This application produces the same results as would alkylation of a methyl ketone ($R'CO—CH_3$) with an alkyl halide (RBr). The use of the ylid route permits reaction where RBr is not subject to direct displacement (i.e., C_6H_5Br) and also removes the complexity of α-*vs.* α'-substitution and the possibility of disubstitution. All of these complications are serious drawbacks to normal ketone alkylation.

Bestmann and Schulz (*83*) have applied the normal ylid acylation reaction to the synthesis of carboxylic acids by treating a phosphonium ylid with methyl chlorocarbonate and hydrolyzing the resulting ylid. This portion of the synthesis duplicates the simple carbonation reaction for acid preparation since the halide $R—CH_2—Br$ must be available for both the ylid approach and the carbonation approach [3.111]. Use of

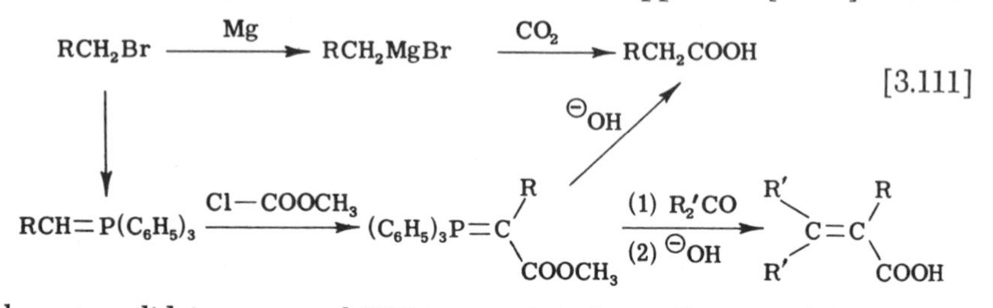

the ester ylid in a normal Wittig reaction does offer a useful approach to substituted acrylic acids.

The acylation of β-keto-ylids ($X_3P{=}CH—CO—R$) could result in C- or O-acylation. Chopard *et al.* (*90*) have found that use of acid anhydrides led to C-acylation but acid chlorides led to O-acylation. They have proposed the latter formation to be kinetic controlled and the former to be thermodynamically controlled. C-Acylation would involve initial reversible formation of the O-acylated ylid which reverts to the more stable C-acyl isomer. In agreement with this proposal they found that O-acylated ylids were isomerized to C-acylated ylids in chloroform solution in the presence of tetrabutylammonium acetate [3.112].

$$(C_6H_5)_3\overset{\oplus}{P}—CH{=}\overset{OCOCH_3}{\overset{|}{C}}—CH_3 \;\; Cl^{\ominus} \xrightarrow[CHCl_3,\ 50^\circ]{(C_4H_9)_4N^{\oplus}\ {}^{\ominus}OAc} (C_6H_5)_3P{=}C\begin{matrix}COCH_3\\ COCH_3\end{matrix} \qquad [3.112]$$

G. Pyrolysis of Keto-Ylids

In an attempt to prepare a nitromethylenetriphenylphosphorane by treatment of nitromethyltriphenylphosphonium bromide with aqueous hydroxide, Trippett and Walker (*21*) were able to isolate only triphenylphosphine oxide but they detected the presence of the fulminate ion. They proposed that an intramolecular transfer of oxygen took place through the mechanism in [3.113]. Extrapolation of this mechanism to

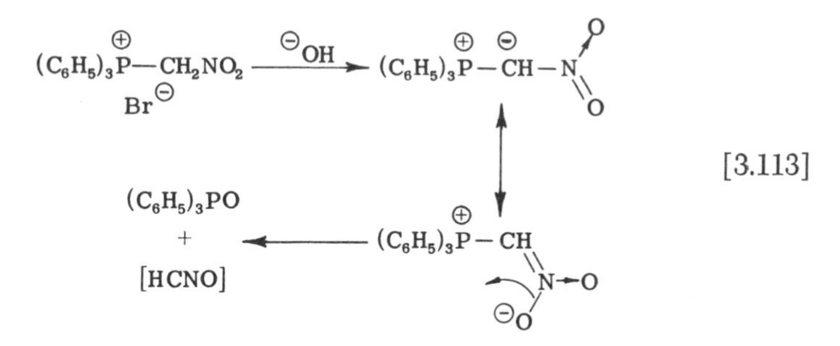

[3.113]

a purely carbon system led them to attempt the elimination of triphenylphosphine oxide from an ylid in which the carbanion was stabilized by delocalization through a carbonyl group, i.e., a keto-ylid. Pyrolysis of α-benzoylbenzylidenetriphenylphosphorane did lead to the formation of triphenylphosphine oxide along with a 59% yield of diphenylacetylene. Gough and Trippett (*82*) obtained ethyl phenylpropiolate in 91% yield

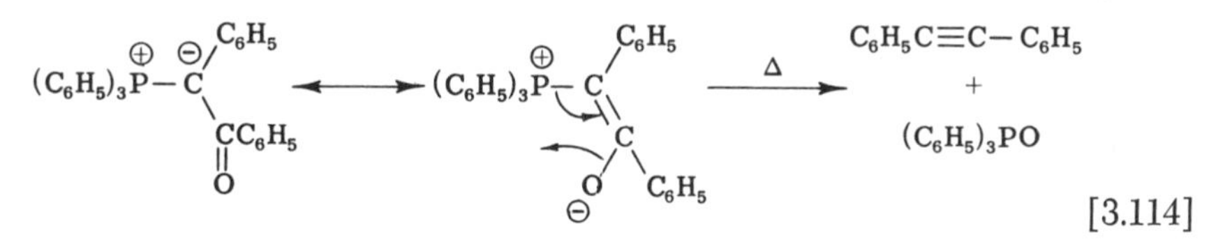

[3.114]

$$(C_6H_5)_3\overset{\oplus}{P}-\overset{\ominus}{C}(COOC_2H_5)(COC_6H_5) \xrightarrow{\Delta} C_6H_5C{\equiv}C-COOC_2H_5 + (C_6H_5)_3PO$$

by this same approach, and Markl (*195*) reported other examples. The substance obtained by Hendrickson (*69*) from the reaction of phenacylidenetriphenylphosphorane with dimethyl acetylenedicarboxylate also was found to eliminate triphenylphosphine oxide and afford an alkyne upon pyrolysis. For this reason, Hendrickson's cyclic phosphorane structure (XLII) seems less likely than a β-keto-ylid structure XLIII for this substance (see discussion on p. 42).

Bestmann and Hartung (*197*) applied the keto-ylid pyrolysis reaction to the preparation of allenic acids. They found that α-carbethoxyethylidenetriphenylphosphorane would react with a series of acid chlorides containing at least one α-hydrogen to afford an allenic ester, the conjugate acid of the original ylid and triphenylphosphine oxide. They proposed that the reaction occurred by the initial acylation of the ylid [3.115].

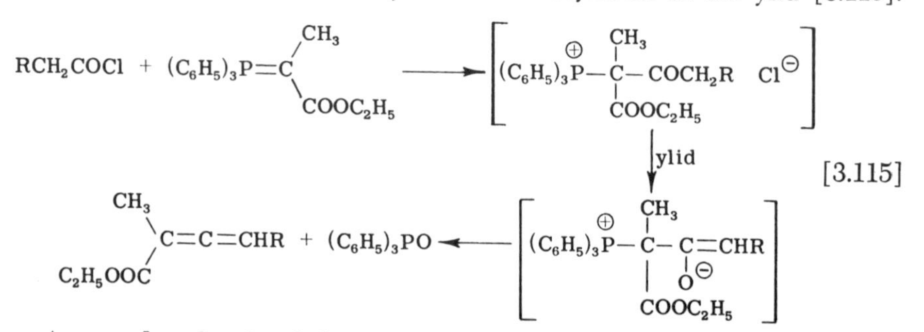

A second molecule of the original ylid was thought to form the enolate of the acylated phosphonium salt which then eliminated triphenylphosphine oxide. The conditions necessary for this reaction were not described in the brief communication but yields of 51–59% were reported for acetyl chloride, butyryl chloride and dihydrocinnamoyl chloride. From the proposed mechanism it is obvious that a disubstituted ylid must be used in order to prevent proton removal from a carbon adjacent to the phosphorus atom. More recently, Tomoskozi and Bestmann (*198*) have used the optically active α-carbo-(—)-menthoxyethylidenetriphenylphosphorane in this same synthesis and have isolated allenic esters and their acids which exhibited optical activity, presumably as a result of asymmetric induction.

All of these eliminations presumably are driven by the formation of a phosphorus–oxygen bond in the phosphine oxide.

H. Reaction of Phosphonium Ylids with Assorted Electrophilic Agents

Phosphonium ylids will react with a variety of electrophilic reagents in addition to those mentioned in preceding discussions. Reaction with the simplest electrophile, the proton, was discussed in Section II of this chapter concerned with the effect of molecular structure on ylid basicity.

Hawthorne (*199*) found that alkylidenetriphenylphosphoranes reacted with diborane to form solid complexes [3.116]. These zwitterions were stable in aqueous media but the carbon–boron and boron–hydrogen bonds were cleaved by dilute acid. Likewise, methylenetriphenylphosphorane would displace trimethylamine from its complex with phenylboron. Seyferth and Grim (*200*) reported the formation of analogous

$$(C_6H_5)_3P{=}CH_2 + B_2H_6 \longrightarrow (C_6H_5)_3\overset{\oplus}{P}{-}CH_2{-}\overset{\ominus}{B}H_3 \qquad [3.116]$$

$$(C_6H_5)_3P{=}CH_2 + C_6H_5BH_2 \cdot N(CH_3)_3 \longrightarrow (C_6H_5)_3\overset{\oplus}{P}{-}CH_2\overset{\ominus}{B}H_2C_6H_5$$

complexes with boron trifluoride, boron trichloride and triphenylboron. In a similar vein, Nesmeyanov *et al.* (*201*) found that phosphonium ylids would react with sulfur trioxide dioxanate to form a zwitterion

$$(C_6H_5)_3P{=}CHCOOCH_3 \xrightarrow{SO_3} (C_6H_5)_3\overset{\oplus}{P}{-}CH(COOCH_3)(SO_3^{\ominus}) \xrightarrow{{}^{\ominus}OH} (C_6H_5)_3P{=}C(COOCH_3)(SO_3^{\ominus}) \qquad [3.117]$$

LXXXIV

which could be converted to a new ylid (LXXXIV) in the presence of base.

Recently Oda *et al.* (*202*) found that phosphonium ylids would react with chloroform in the presence of a strong base, usually potassium *tert*-butoxide, to afford olefins and triphenylphosphine. For example, fluorenylidenetriphenylphosphorane afforded 9-dichloromethylenefluorene [3.118]. The reaction was formulated as an attack of the nucleophilic

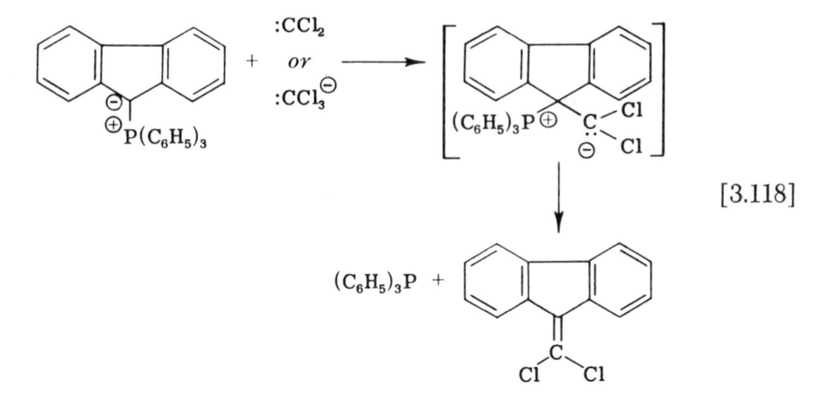

[3.118]

ylid on a carbene followed by ejection of the phosphine. It also could be envisioned as an attack of the ylid on the trichloromethyl anion. Carbenes, or incipient carbenes, have been shown to be electrophilic agents (*30, 40, 41*). The same type of reaction occurred between the fluorenylide and benzal chloride to afford 9-(α-chlorobenzal)fluorene.

Several groups have studied the nucleophilic attack by phosphonium ylids on a variety of metallic and non-metallic halides to form the corresponding substituted methyltriphenylphosphonium salts. Thus, methylenetriphenylphosphorane and trimethylbromosilane afforded a salt identical to that obtained by alkylating triphenylphosphine with bromomethyltrimethylsilane [3.119] (*203,* Seyferth and Grim). Similar reac-

$$(C_6H_5)_3P{=}CH_2 + BrSi(CH_3)_3 \longrightarrow (C_6H_5)_3\overset{\oplus}{P}{-}CH_2Si(CH_3)_3 \; Br^{\ominus} \longleftarrow (C_6H_5)_3P + BrCH_2Si(CH_3)_3 \qquad [3.119]$$

tions occurred between the same ylid and triphenylbromogermane and trimethylbromostannane (*203*). The same phosphonium ylid would replace all halogens in dimethyldichlorostannane (*203*), mercuric bromide (*203*), dibromophenylphosphine (*204*) and zinc chloride (*47*) forming diphosphonium salts in all cases. Reaction of a stable ylid, carbethoxymethylenetriphenylphosphorane, with phenylbromoselenide led to a similar displacement reaction but the resulting salt was converted, *in situ,* to a new ylid (LXXXV) by unchanged original ylid (*76,* Petragnini and Campos). The selenium atom apparently is able to stabilize the carbanion making the substituted ylid less basic than the original ylid. In similar fashion, Nesmeyanov *et al.* (*205*) obtained a mercury-

$$(C_6H_5)_3P{=}CHCOOC_2H_5 \xrightarrow{C_6H_5SeBr} \left[(C_6H_5)_3\overset{\oplus}{P}{-}CH(COOC_2H_5)(SeC_6H_5) \; Br^{\ominus}\right] \xrightarrow{\text{ylid}} (C_6H_5)_3P{=}C(COOC_2H_5)(SeC_6H_5)$$

LXXXV

[3.120]

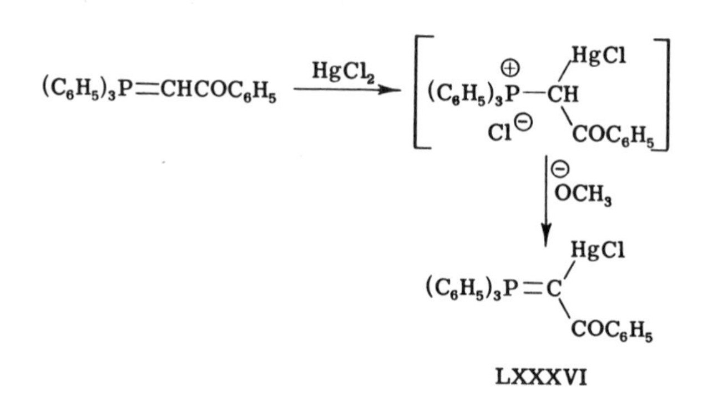

substituted ylid (LXXXVI) from phenacylidenetriphenylphosphorane and mercuric chloride.

Phosphonium ylids have been found to react with sulfur; this reaction results in the formation of a phosphine sulfide and a thioketone. Thus, Staudinger and Meyer (*42*) reported the cleavage of benzhydrylidenetriphenylphosphorane to thiobenzophenone and triphenylphosphine sulfide. Similarly, Schonberg *et al.* (*206*) cleaved fluorenylidenetriphenylphosphorane to thiofluorenone. These cleavages probably occurred via attack of the ylid carbanion on sulfur to form a zwitterionic intermediate (LXXXVII). Transfer of sulfur from sulfur to phosphorus via a four-membered transition state similar to that proposed for the Wittig reaction

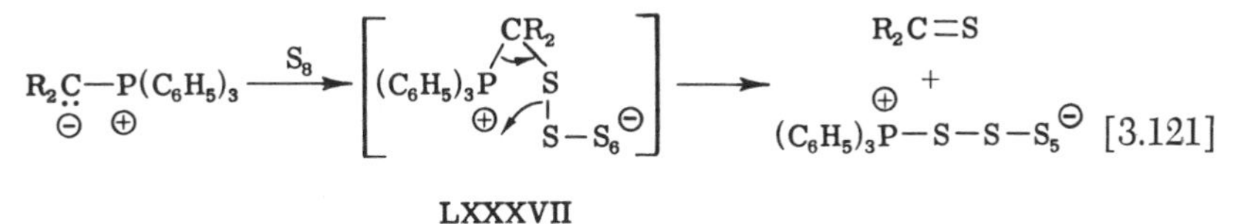

would account for the observed products. The phosphonium polysulfide would be susceptible to attack by a second molecule of ylid on the β-sulfur atom, and the whole process would be repeated. This proposal is identical to that made by Bartlett and Meguerian (*207*) for the reaction of triphenylphosphine with sulfur.

Phosphonium ylids react with a variety of electrophilic nitrogen compounds. Benzylidenetriphenylphosphorane reacted with excess phenyl azide to afford *N*-phenyliminotriphenylphosphorane and benzal aniline (*120,* Hoffmann). The nucleophilic ylid probably displaced nitrogen as do some other nucleophiles (e.g., triphenylphosphine) to afford a zwitterionic intermediate (LXXXVIII) which can collapse to benzal aniline and triphenylphosphine. The excess azide in the solution would convert the phosphine to the phosphinimine.

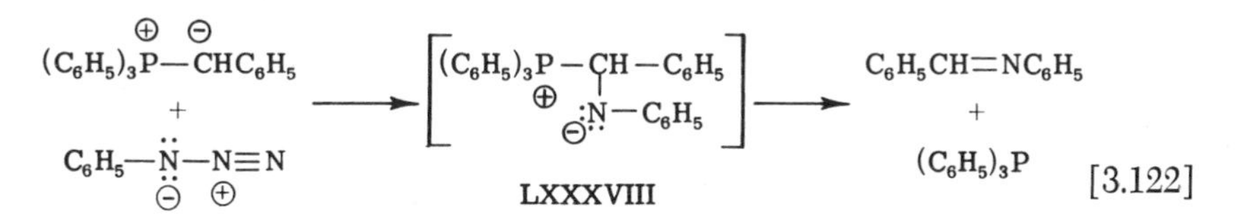

$$(C_6H_5)_3P + C_6H_5N_3 \longrightarrow (C_6H_5)_3P=NC_6H_5 + N_2\uparrow$$

Phosphonium ylids will couple directly with diazonium salts. Carbomethoxymethylenetriphenylphosphorane reacted with benzenediazonium tetrafluoborate to afford a phosphonium salt which could be converted to a new ylid (LXXXIX) upon treatment with aqueous base (*208,*

Markl). Cyclopentadienylidenetriphenylphosphorane also reacted with

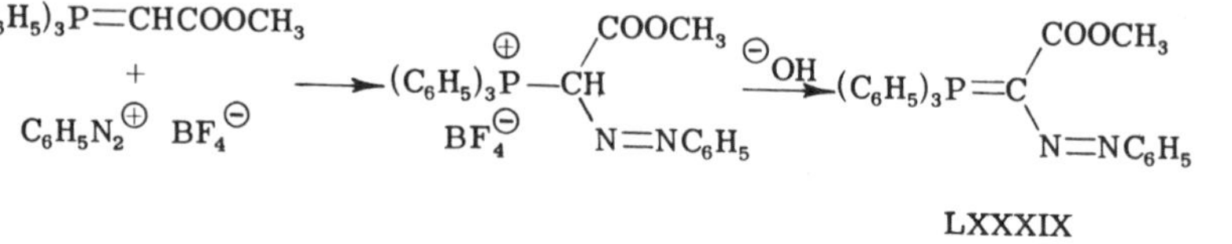

LXXXIX

[3.123]

$$\text{C}_5\text{H}_4^{\ominus}\text{–}\overset{\oplus}{\text{P}}(\text{C}_6\text{H}_5)_3 + \text{C}_6\text{H}_5\text{N}_2^{\oplus}\ \text{Cl}^{\ominus} \longrightarrow \text{C}_5\text{H}_3^{\ominus}(\text{N{=}N{-}C}_6\text{H}_5)\text{–}\overset{\oplus}{\text{P}}(\text{C}_6\text{H}_5)_3$$

benzenediazonium chloride but underwent electrophilic aromatic substitution on the cyclopentadienyl ring (*144*, Ramirez and Levy). This may be viewed as substitution at a position vinylogous to the carbanion or it may be viewed as a normal aromatic substitution. The latter view seems more realistic since the ring system must have a high degree of aromatic character.

Phosphonium ylids also have been found to react with diazo carbonyl compounds. Markl (*49*) reported that reaction of benzylidenetriphenylphosphorane with diazoacetophenone afforded a mixture of a mixed azine and a phosphinazine [3.124]. The products were accounted for by pro-

$$(\text{C}_6\text{H}_5)_3\overset{\oplus}{\text{P}}-\overset{\ominus}{\text{C}}\text{HC}_6\text{H}_5 + \text{C}_6\text{H}_5\text{COCHN}_2 \longrightarrow \left[(\text{C}_6\text{H}_5)_3\overset{\oplus}{\text{P}}-\underset{}{\overset{\text{C}_6\text{H}_5}{\text{CH}}}-\text{N{=}N}-\overset{\ominus}{\text{C}}\text{HCOC}_6\text{H}_5\right] \longrightarrow \text{C}_6\text{H}_5\text{CH{=}N{-}N{=}CHCOC}_6\text{H}_5 + (\text{C}_6\text{H}_5)_3\text{P} \quad [3.124]$$

$$(\text{C}_6\text{H}_5)_3\text{P} + \text{C}_6\text{H}_5\text{COCHN}_2 \longrightarrow (\text{C}_6\text{H}_5)_3\text{P{=}N{-}N{=}CHCOC}_6\text{H}_5$$

posing the initial formation of a direct-combination zwitterion which ejected triphenylphosphine. Wittig and Schlosser (*48*) also reported azine formation from phenyldiazomethane and the same ylid. A much more complicated reaction occurred between benzoylmethylenetriphenylphosphorane and diazoacetophenone, affording a 2,6-diphenyl-4-benzylidenepyran (*209*, Strzelecka *et al.*). The mechanism of this reaction must be quite different from Markl's case (*49*) since the nitrogen completely disappears.

I. Reaction of Phosphonium Ylids with Epoxides

Several research groups, Denney *et al.* (*210*), McEwen *et al.* (*211*, *212*) and Zbiral (*213*), have examined the reaction of stabilized phosphonium ylids with epoxides. In a recent review, Trippett (*102*) proposed a comprehensive mechanism [3.125] which seems to account for the various products of such reactions.

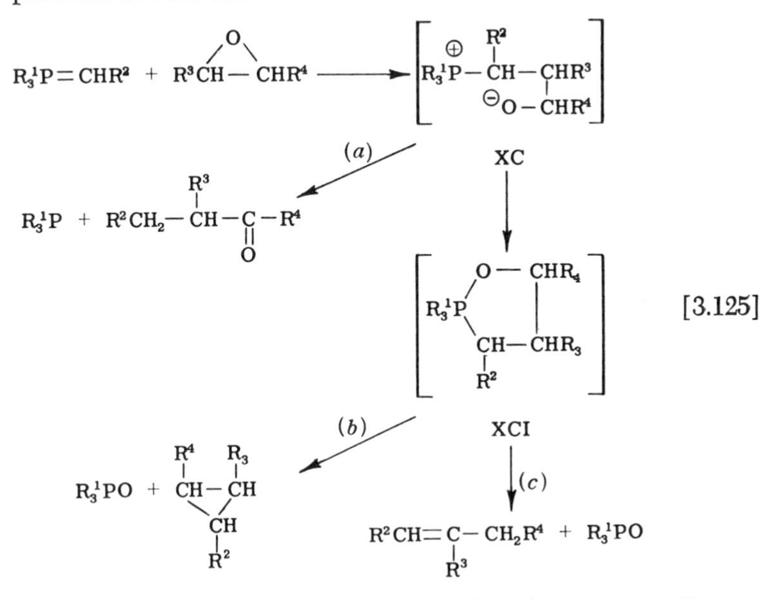

Denney *et al.* (*210*) found that the ester ylid (XXXVIII) would react with 1-octene oxide, cyclohexene oxide and styrene oxide to afford the corresponding carbethoxycyclopropane derivative and triphenylphosphine oxide (path *b* in [3.125]). They also found that use of *l*-(—)-styrene oxide afforded the optically active *trans*-cyclopropane but in low optical

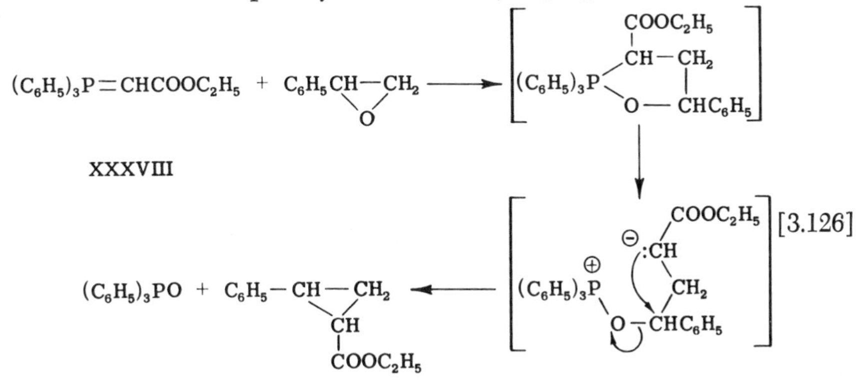

yield. Recent work by Walborsky *et al.* (*214*) has established that this reaction [3.126] took place with inversion of configuration, a fact consistent with the detailed portrayal of path *b* in [3.125].

Using styrene oxide with optically active methylethylphenylbenzylidenetriphenylphosphorane, McEwen *et al.* (*212*) found that the phosphine oxide product had undergone about 50% net inversion. The phosphorus would be expected to invert when either path *b* or path *c* were followed and the methinyl and oxygen groups both were located in basal positions about the phosphorus atom.

The same group (*211, 212*) found that reaction of the benzylidenemethylethylphenylphosphorane with styrene oxide afforded a lithium iodide adduct of the betaine intermediate (XC) which, upon pyrolysis in decalin solution, afforded a mixture of the phosphine, phosphine oxide, the cyclopropane, an olefin and a ketone [3.127]. All of these products

$$CH_3-P(C_2H_5)(C_6H_5)=CHC_6H_5 + C_6H_5-CH-CH_2(O) \longrightarrow C_6H_5-P(CH_3)(C_2H_5) + (C_6H_5)(CH_3)(C_2H_5)P{=}O + C_6H_5CH-CHC_6H_5(CH_2) + C_6H_5CH{=}CHCH_2C_6H_5 + C_6H_5CH_2CH_2COC_6H_5 \quad [3.127]$$

can be accounted for by the Trippett mechanism. The phosphine and ketone were thought to be formed via path *a*, involving a hydride shift from the R^4 carbon to the R^2 carbon in XC, displacing the phosphine (*102*). McEwen *et al.* (*211*) proposed that a proton shift occurred from the R^4 carbon to the oxyanion and then the resulting carbanion displaced the phosphine from the R^2 carbon affording a cyclopropanol which was thought to rearrange to the ketone in the basic medium. Both of these proposals are consistent with the observation that use of ylid labeled with C^{14} on the benzylic carbon afforded 1,3-diphenylpropanone labeled only on C-3, thus ruling out any symmetrical intermediates (*211*). McEwen (*212*) felt that the large amount of ketone and phosphine present indicated a reluctance of the betaine intermediate (XC) to close to a pentavalent intermediate (XCI) due to the electron-donating ability of the alkyl groups on phosphorus. As expected, use of benzylidenetriphenylphosphorane led to very little ketone formation since pentavalent intermediate formation should be much easier with three phenyl groups on phosphorus (*211, 213*). The formation of 1,3-diphenylpropene, 1,2-diphenylcyclopropane and the phosphine oxide all were proposed to proceed via the pentavalent intermediate (XCI), following paths *b* and *c* of the comprehensive mechanism. The ratio of olefin to cyclopropane should depend on whether the carbon–oxygen bond breaks first to form olefin via a carbonium ion or whether the carbon–phosphorus bond breaks first to form cyclopropane via a carbanion. Carbethoxy ylids (e.g., XXXVIII) form cyclopropanes exclusively, perhaps due to the ability of the ester group (R^2 in XCI) to stabilize a carbanion formed by cleavage of the carbon–phosphorus bond. It would be possible to design interesting tests of this hypothesis by varying the nature of the substituents R^1 through R^4, determining product ratios and perhaps obtaining ρ-σ correlations.

Trippett (*102*) proposed that the olefinic products were formed by initial cleavage of the carbon–oxygen bond in XCI, formation of a carbonium ion which underwent a hydride shift. Compelling evidence for this proposal was provided by Zbiral's observation (*213*) that cyclohexene oxide reacted with benzylidenetriphenylphosphorane to afford both the expected cyclopropane(7-phenylnorcarane) and a rearranged olefin, 1-phenyl-2-cyclopentylethene [3.128]. In this case, a hydride shift would have been impossible since it would have been forced to attack

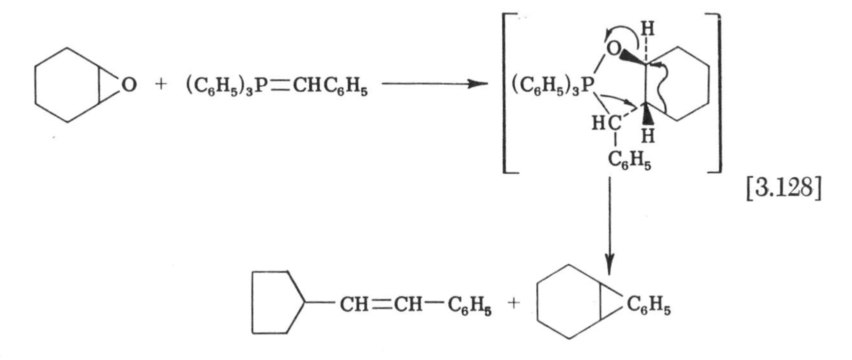

[3.128]

the carbon from the same side that the oxygen was leaving. In this case, then, a methylene group migrated as a carbanion and the ring contracted. This is the only known case of olefin formation in which a hydride shift apparently did not occur.

J. Reaction of Phosphonium Ylids with Multiple-Bonded Compounds

It has been mentioned previously in this chapter that perhaps the best known reaction of phosphonium ylids and the reason for their revival as useful synthetic tools is the reaction with aldehydes and ketones to form olefins, the Wittig reaction. This reaction will be discussed in detail in the following chapter but suffice it to mention again that the mechanism of the reaction appears to involve attack of the ylid carbanion on the carbonyl carbon of the aldehyde or ketone [3.82], a typical nucleophilic addition, as the first step. Carbonyl compounds,

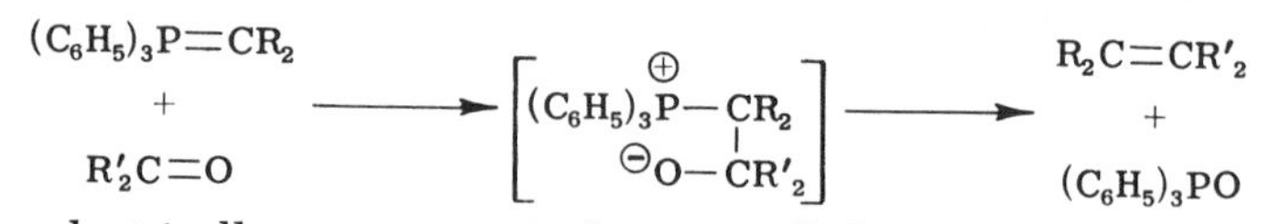

being electrically unsymmetrical, are well known to undergo nucleophilic addition and it is not surprising that they will react with ylids. It is also well known, however, that nucleophilic additions normally do not occur with electrically symmetrical multiple-bonded systems such as alkenes and alkynes. Thus, ylids will not react with an isolated alkene or alkyne function. However, a variety of nucleophiles will react with these functional groups if they are electrically unsymmetrical or have powerful electron-withdrawing groups attached. The Michael addition is a classic example. Phosphonium ylids would be expected to behave in similar fashion and undergo reaction with any multiple bond system which is electrically unsymmetrical or carries powerful electron-withdrawing groups.

This discussion will be divided into two major segments. The first will deal with those systems whose electrical dissymmetry is due to the presence of two different atoms at the termini of the multiple bond. The second group will be those systems having powerful electron-withdrawing groups at one or both ends of the multiple bond system, e.g., unsaturated esters.

1. Heteroatom-Containing Multiple Bond Systems

Schonberg and Brosowski (*215*) found that phosphonium ylids, for example, fluorenylidenetriphenylphosphorane, would react with nitrosobenzene to afford imines and triphenylphosphine oxide [3.129]. The

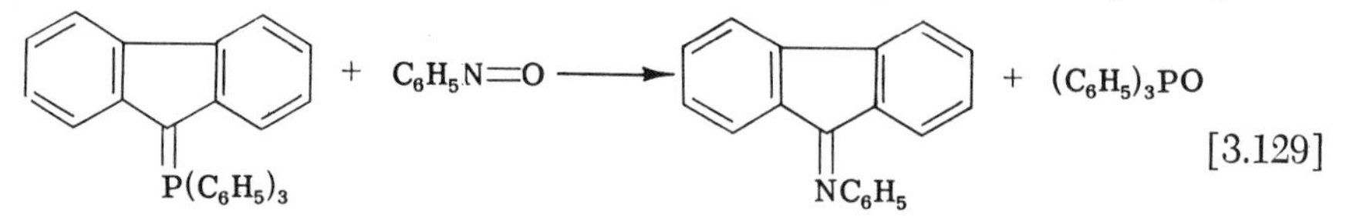

[3.129]

mechanism undoubtedly is very similar to that of the Wittig reaction. Bestmann and Seng (*216*) reported a similar reaction between benzylidenetriphenylphosphorane and benzal aniline to afford stilbene and *N*-phenyliminotriphenylphosphorane [3.130]. This reaction too probably

$$\begin{matrix}(C_6H_5)_3P{=}CHC_6H_5 \\ + \\ C_6H_5CH{=}NC_6H_5\end{matrix} \longrightarrow \left[\begin{matrix}(C_6H_5)_3\overset{\oplus}{P}-CH-C_6H_5 \\ | \\ C_6H_5-\underset{\ominus}{N}-CH-C_6H_5\end{matrix}\right] \longrightarrow \begin{matrix}C_6H_5CH{=}CHC_6H_5 \\ + \\ (C_6H_5)_3P{=}NC_6H_5\end{matrix} \qquad [3.130]$$

involved the typical betaine intermediate which collapsed by transfer of nitrogen from carbon to phosphorus.

VanderWerf *et al.* (*217*) found that optically active benzylidenemethylethylphenylphosphorane reacted with benzonitrile to afford deoxybenzoin and the phosphine oxide. The latter was not the product of a complete inversion or a complete retention of configuration so it was concluded that two routes to the products may have been followed. The mechanism in [3.131] was proposed. Path *a* would produce inverted phos-

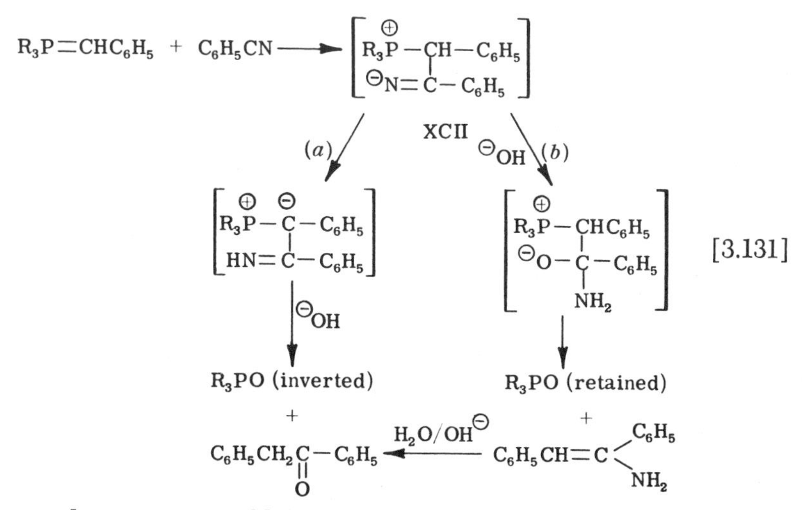

phine oxide since it would be formed by a hydrolysis whereas path *b* would produce oxide with retained configuration since it would have been formed by a Wittig reaction. Path *b* seems a somewhat unlikely possibility in that the first intermediate probably would hydrolyze directly to products rather than forming a new oxybetaine. The first intermediate (XCII) could have undergone a Wittig-type reaction to form a phosphinimine which then was hydrolyzed but this would have led to inversion of the phosphorus configuration.

Phosphonium ylids have been found to react with carbon disulfide to form thioketenes which then dimerized (*218*, Schonberg *et al.*). In addition, they react with pyrylium salts to afford benzene derivatives. Markl (*219*) found that methylenetriphenylphosphorane converted 2,4,6-triphenylpyrylium tetrafluoroborate into 2,4,6-triphenylbenzene. Use of carbomethoxymethylenetriphenylphosphorane permitted interception of the intermediate keto-ylid (XCIII) since the Wittig reaction proposed for the last step was slow.

Staudinger and Meyer (*42*) noted that benzhydrylidenetriphenylphosphorane reacted with phenylisocyanate to form a ketenimine, presumably via a Wittig reaction-type mechanism. Trippett and Walker (*21*) have found that when the ylid carried a hydrogen on the ylid

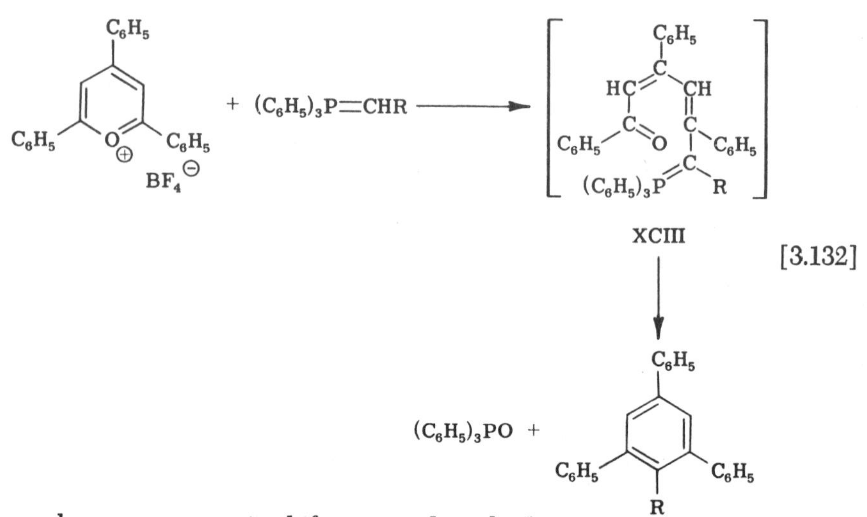

carbon, a prototropic shift occurred at the betaine stage to form a new acylated ylid [3.133].

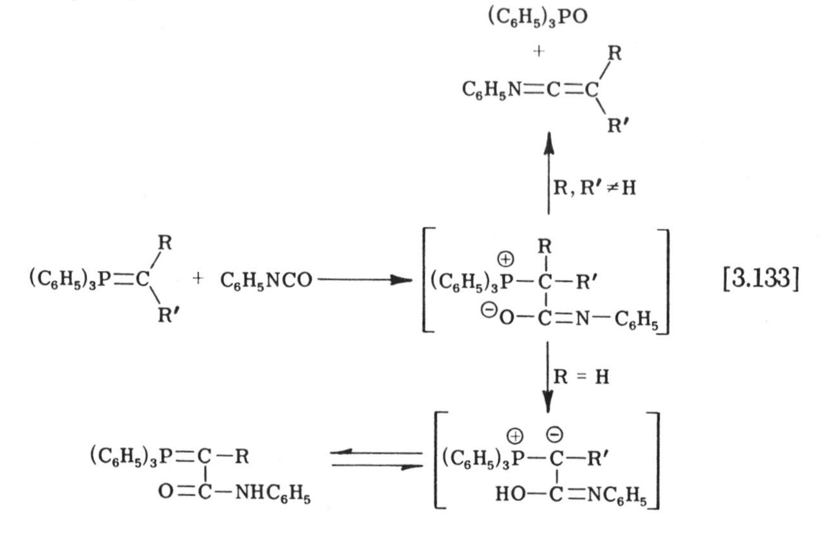

2. *Conjugated Alkenes*

There are many reports of reaction of phosphonium ylids with α,β-unsaturated ketones and aldehydes nearly all of which involved reaction at the carbonyl group to form olefins. For example, Surmatis and Ofner

(*220*) prepared β-carotene by reacting a bis-phosphonium ylid with two moles of β-ionylideneacetaldehyde with no evidence of any conjugate addition.

There have been a few reports of the addition of phosphonium ylids to the olefinic portion of conjugated ketones. Freeman (*221*) found that methylenetriphenylphosphorane added to mesitoylphenylethylene to afford a cyclopropane adduct and triphenylphosphine, presumably by way of an initial conjugate addition [3.134]. The normal carbonyl addi-

$$\begin{array}{c} \text{Mes}-\text{COCH}{=}\text{CHC}_6\text{H}_5 \\ + \\ (\text{C}_6\text{H}_5)_3\text{P}{=}\text{CH}_2 \end{array} \longrightarrow \left[\text{Mes}-\overset{\overset{\displaystyle O}{\|}}{C}-\overset{\ominus}{C}H-\underset{\underset{\displaystyle \text{CH}_2-\overset{\oplus}{P}(\text{C}_6\text{H}_5)_3}{|}}{\text{CH}}-\text{C}_6\text{H}_5 \right]$$

$$\downarrow \qquad [3.134]$$

$$(\text{C}_6\text{H}_5)_3\text{P} + \text{Mes}-\overset{\overset{\displaystyle O}{\|}}{C}-\text{CH}-\text{CH}-\text{C}_6\text{H}_5 \quad (\text{CH–CH bridged by } \text{CH}_2)$$

Mes = 2, 4, 6-trimethylphenyl

tion probably was prevented by steric hindrance from the mesityl group. Inhoffen *et al.* (*107*) claimed that addition of a phosphonium ylid to 2-ketomethylenecyclohexane afforded a conjugate triene which could result only from 1,4-addition of the ylid. The product was characterized only by its ultraviolet spectrum. The authors proposed a mechanism [3.135], which has as its sole justification the fact that it does account

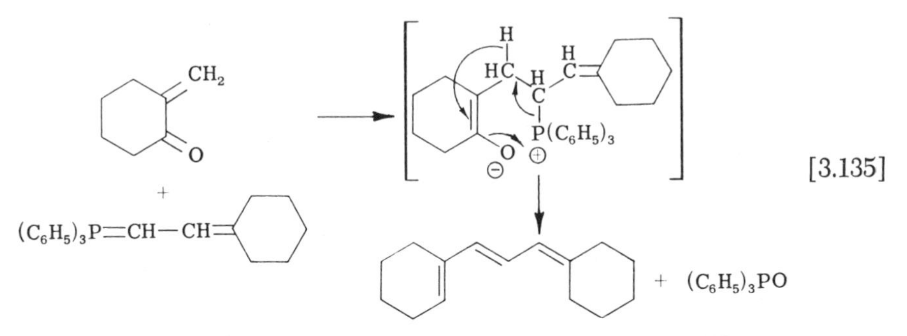

for the products. The proposed hydride shift seems unlikely to occur. Bohlmann (*148*) reported products from both 1,2- and 1,4-addition of cinnamylidenetriphenylphosphorane to cinnamylideneacetophenone. No phenylacetylene was isolated, however, and no yields were reported. The mechanism in [3.136] accounts for the products:

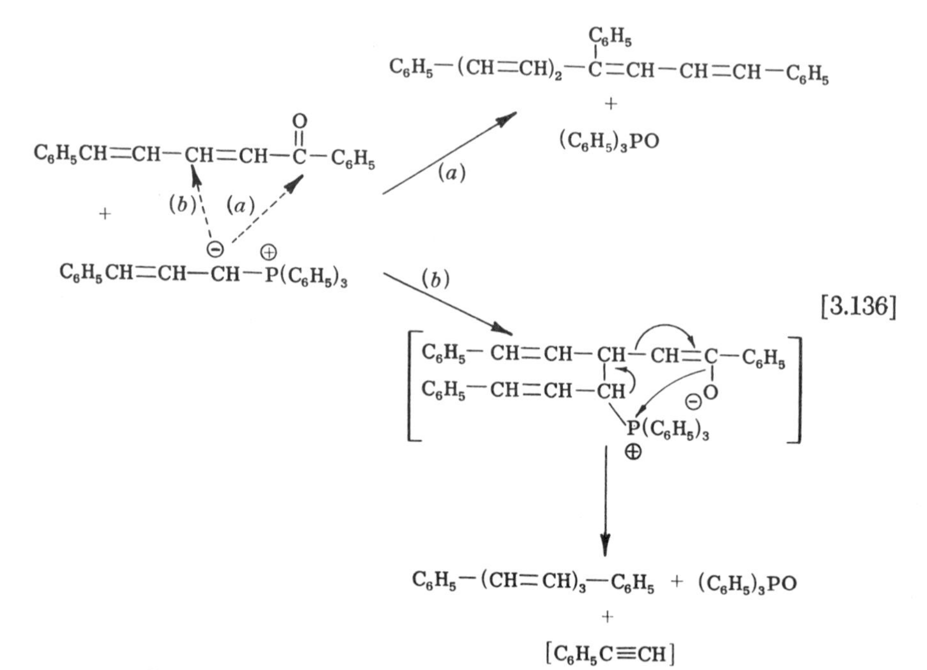

[3.136]

Mechoulam and Sondheimer (*191*) found that excess *n*-butylidenetriphenylphosphorane reacted with fluorenone to form the spirohydrocarbon (XCIV). That the hydrocarbon had been formed from 9-butyl-

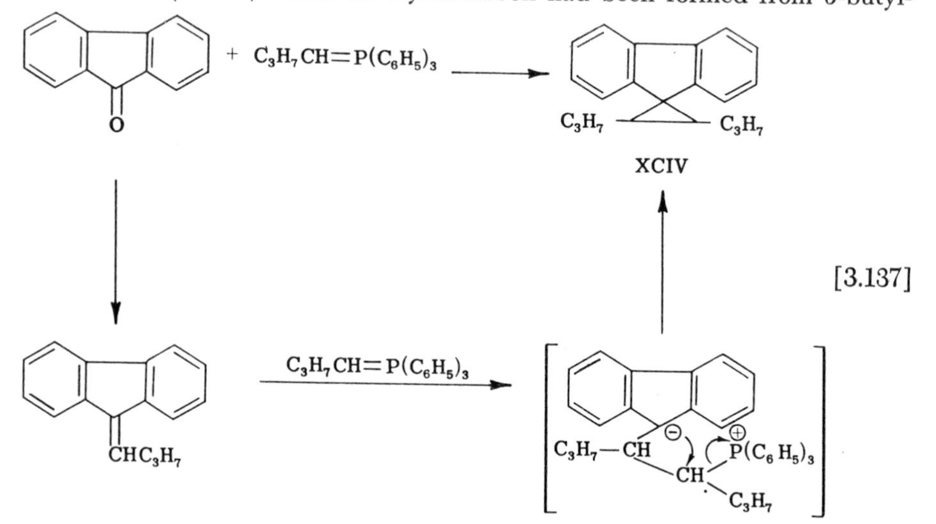

[3.137]

idenefluorene was indicated by reaction of the same ylid with the olefin to afford XCIV. This example of conjugate addition by a phosphonium ylid must depend on the stabilization of the intermediate carbanion by the fluorenyl group and the leaving ability of the phosphonium group.

Bestmann and Seng (*67*) also observed this course of reaction when ethylidenetriphenylphosphorane reacted with ethyl crotonate. However, use of a stabilized phosphonium ylid (XXXVIII) permitted a second course of reaction to prevail, the formation of a new ylid [3.138]. These

$$\underset{\text{XXXVIII}}{(C_6H_5)_3P{=}CHCOOCH_3} + C_6H_5COCH{=}CHCOOCH_3 \longrightarrow (C_6H_5)_3P{=}\overset{\overset{\displaystyle COOCH_3}{|}}{C}-\underset{\underset{\displaystyle COOCH_3}{|}}{CH}-CH_2COC_6H_5 \qquad [3.138]$$

workers proposed that a prototropic shift occurred, the carbomethoxy group permitting the formation and stabilization of the new ylid.

Trippett (*68*) uncovered yet a third course for such addition reactions. Cyanomethylenetriphenylphosphorane reacted with ethoxymethylenemalonate or tetracyanoethylene but in both cases resulted in the formation of new ylids with concommitant ejection of ethoxide in the first case and cyanide in the second case [3.139].

$$(C_6H_5)_3P{=}CHCN + C_2H_5OCH{=}C(COOC_2H_5)_2 \longrightarrow (C_6H_5)_3P{=}C(CN)CH{=}C(COOC_2H_5)_2$$

[3.139]

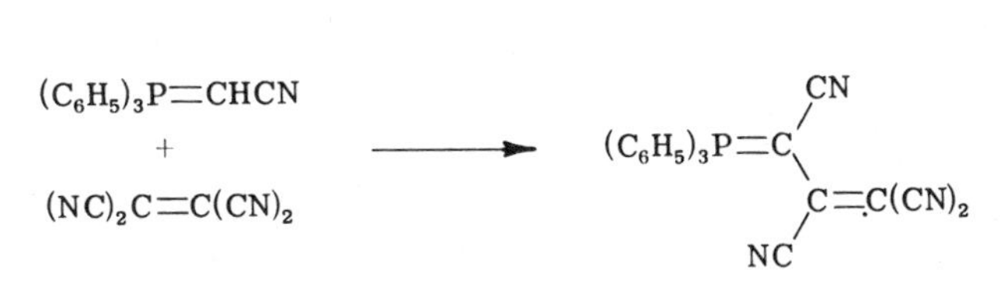

All three of these reaction paths for addition of phosphonium ylids to conjugated alkenes can be accounted for by a single comprehensive mechanism involving a Michael addition for the first step with the fate of the common zwitterionic intermediate (XCV) depending on the nature of substituent R groups [3.140]. The normal course of events would be the formation of the cyclopropane (path *a*). The examples discovered by Freeman (*221*), Mechoulam (*191*) and Bestmann (*67*) fall into this category. If the substituent R on the original ylid is capable of stabilizing the ylid a prototropic shift could occur as in path *b* (*67*). If the R group on the original ylid is capable of providing such stabiliza-

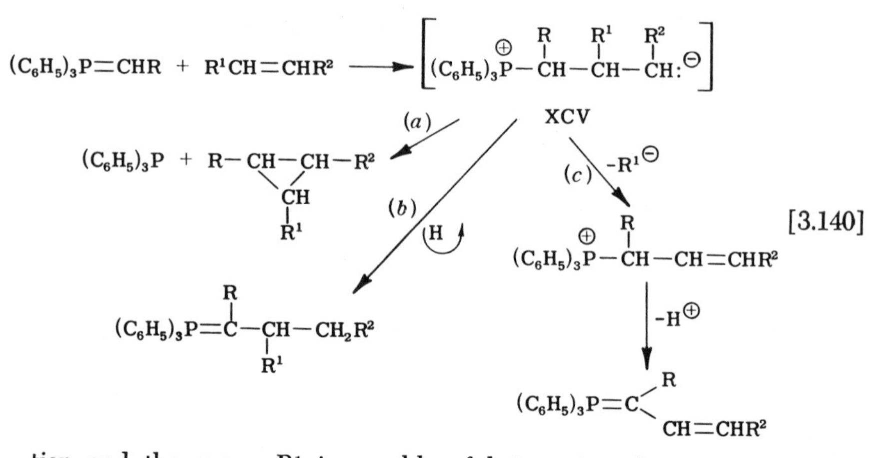

tion and the group R^1 is capable of being ejected as an anion then path *c* is followed, the ejected anion (R^1) acting as a base to reform a new ylid in the last step (*68*).

3. *Conjugated Alkynes*

Several groups recently have investigated the reaction of phosphonium ylids with conjugated alkynes, especially with dimethyl acetylenedicarboxylate. Trippett (*68*) first reported the reaction of cyanomethylenetriphenylphosphorane with this ester and claimed the formation of a

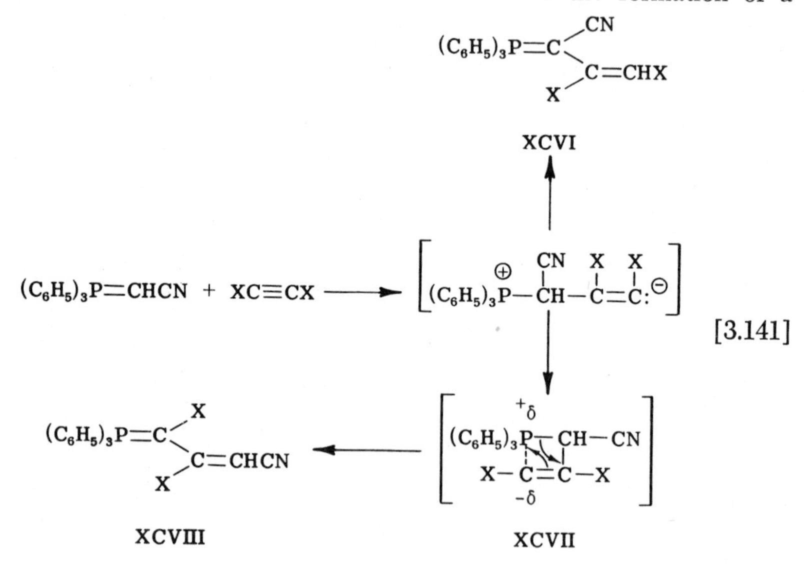

new ylid (XCVI), presumably via prototropic shift in the intermediate analogous to that in path *b* of [3.140]. However, no concrete evidence was provided for that structure.

Some doubt was shed on Trippett's assignment by the report by Brown *et al.* (*70*) that benzhydrylidenetriphenylphosphorane reacted with the same acetylenic ester to afford a rearranged ylid (XCIX). However, in this case as well no evidence was provided to permit an unambiguous structure assignment. The latter group was using analogy with the behavior of phosphinimines for the structure assignment. They

$$RR'C{=}P(C_6H_5)_3 + X{-}C{\equiv}C{-}X \longrightarrow (C_6H_5)_3P{=}C(X){-}C(X){=}CRR' \quad [3.142]$$

XCIX

$R = R' = C_6H_5$

proposed that the ylid (XCIX) was formed via a four-membered intermediate analogous to XCVII. More recently Bestmann and Rothe (*71*) examined the reaction of a series of phosphonium ylids with the same ester, and they have provided experimental evidence (degradation, IR and NMR spectra) which indicated that in all cases studied the product was the rearranged ylid (XCIX), presumably formed via an intermediate analogous to XCVII. The product structure did not change as the R and R′ groups were varied from hydrogen through alkyl to aryl. In view of these results, Trippett's structure assignment for XCVI must be treated with considerable caution since the compound may well be XCVIII, the rearranged ylid.

Hendrickson (*69*) had reported that phenacylidenetriphenylphosphorane reacted with the acetylenic diester to afford a cyclic oxaphosphorabenzene (XLII). In our earlier discussion (p. 42) it was mentioned that since this substance could be pyrolyzed to triphenylphosphine oxide and an alkyne, its structure more likely was the keto-ylid (XLIII). In view of the above results of Bestmann and Rothe (*71*), structure C, a rearranged ylid, might have been expected. Hendrickson *et al.* (*222*), in a more recent paper, has made no comment on his original proposal of XLII but has reported, in the absence of experimental data, that XLIII was formed in methanolic solution but that the rearranged ylid (C) was produced in ethereal solution. It might be expected that there would be less driving force for the formation of the rearranged ylid (C) in this case since the benzoyl group can provide effective stabilization for XLIII, the result of a simple prototropic shift in the intermediate.

It is clear that in the addition of phosphonium ylids to olefinic and

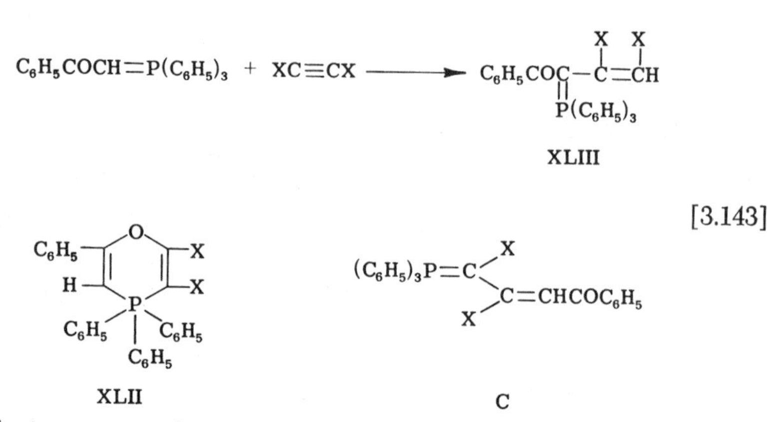

acetylenic systems there are very delicate balances between several possible products. The balance of these products has been shown to be affected by the structure of the individual reactants but there has been little examination of the effect of environment on the course of reaction.

Zbiral (*223, 224*) has found that phosphonium ylids reacted with benzyne. Benzylidenetriphenylphosphorane and bromobenzene in the presence of phenyllithium afforded a 1:1 adduct of structure CIII in 30% yield. He proposed the reaction to proceed as shown [3.144] on the

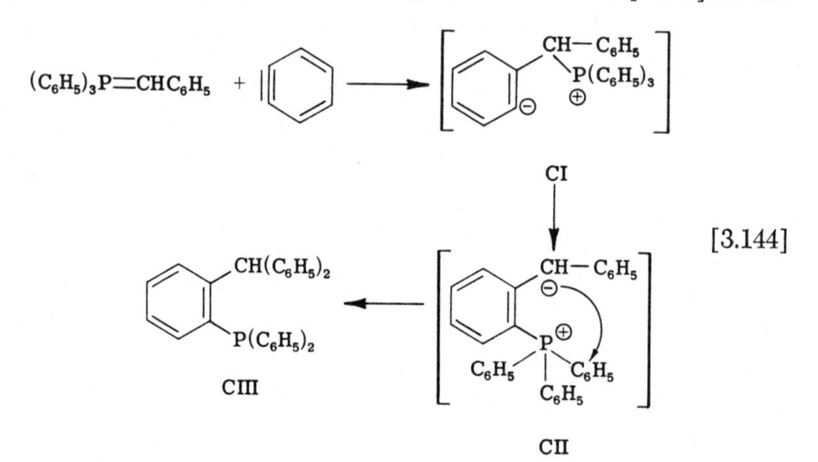

basis of several pieces of evidence. The fact that 2-ethylphenyltriphenylphosphonium bromide with phenyllithium afforded the same phosphine (CIV) as did benzyne with ethylidenetriphenylphosphorane indicated the possible intermediacy of an ylidic structure (e.g., CII). Reaction of ethylidenetris(*p*-tolyl)phosphorane with the benzyne from *m*-bromoanisole indicated that a *p*-tolyl group had migrated from phosphorus to carbon but retained the same point of attachment (i.e., in the *para*

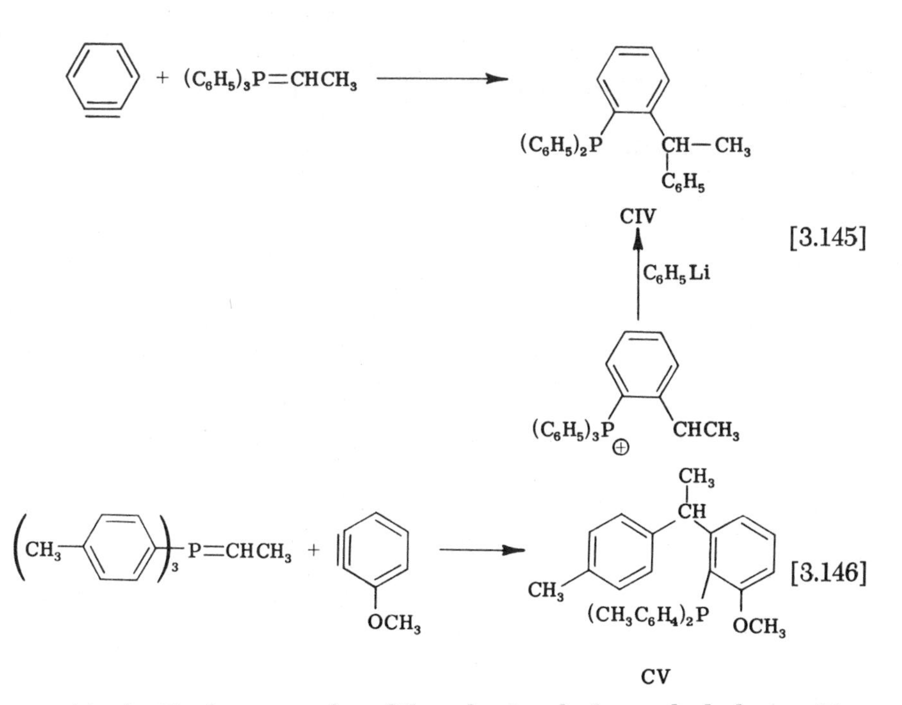

position). Furthermore, the ylid carbanion had attacked the position *meta* to the methoxy group as expected for a benzyne reaction.

The most surprising feature of this reaction is that the proposed intermediate zwitterion (CI) did not undergo a prototropic shift from the benzylic carbon to the phenyl carbanion to afford benzhydrylidene-

C_6H_5 CH $P(C_6H_5)_3$ $\ominus$ $\oplus$ —H→ $(C_6H_5)_2\overset{\ominus}{C}-\overset{\oplus}{P}(C_6H_5)_3$ [3.147]

CI

triphenylphosphorane. Franzen (*53*) and Seyferth (*54*) both had observed the latter type of reaction when benzyne was treated with sulfides and phosphines, respectively.

Appendix

Since this manuscript was written numerous publications concerning phosphonium ylids have appeared. Relatively few, however, have added significantly to our knowledge of the fundamental chemical and physical characteristics of phosphonium ylids.

Bestmann (*256*) has published a series of review articles covering some of the chemistry of phosphonium ylids. Hendrickson *et al.* (*257*) have confirmed that XLII is an incorrect structure for the adduct formed from dimethyl acetylenedicarboxylate and phenacylidenetriphenylphosphorane. The correct structure is XLIII [3.143]. In addition they obtained and provided evidence for the structure of the isomeric ylid (C) mentioned in their previous communication (*222*).

Horner and Winkler (*258*) have isolated an optically active ylid (CVI), although as a non-crystallizable oil, whose asymmetry was due to an asymmetric phosphorus atom. The ylid could be alkylated with methyl iodide to form an optically active phosphonium salt (CVII). Cathodic cleavage of the salt and then hydrolysis afforded optically active α-phenylpropionic acid indicating that there was asymmetric induction in the alkylation reaction [3.148].

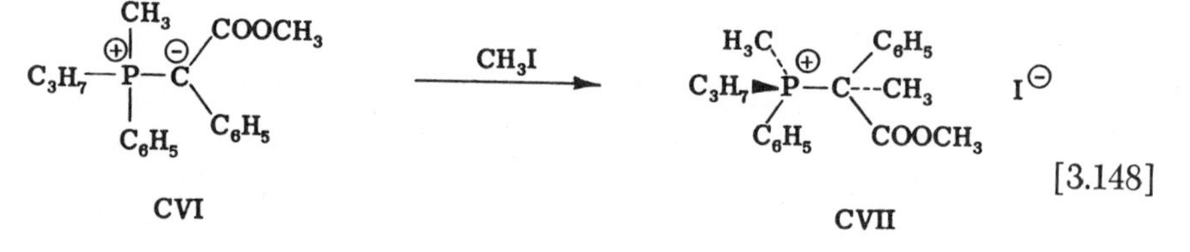

[3.148]

Significant strides have been made toward elucidating the molecular structure of phosphonium ylids. X-ray crystallographic results recently have become available for α-halophenacylidenetriphenylphosphoranes. Speziale and Ratts (*259*) (from the Monsanto Company, St. Louis, Missouri) presented a verbal report concerning the analysis of the ylids CVIII. They indicated that the ylids contained a planar skeleton with a trigonally hybridized ylidic carbon and the oxygen and phosphorus groups oriented *cis* to one another. Stephens (*260*) (also from the Monsanto Company, Zurich) has reported the experimental details of these analyses which are summarized in [3.149]. These data indicate that the

$(C_6H_5)_3\overset{\oplus}{P}—\overset{\ominus}{C}(X)COC_6H_5$

CVIII

$(C_6H_5)_3P^{\oplus}(X)C{=}C(O^{\ominus})C_6H_5$

CIX

[3.149]

Bond lengths	X = I	X = Cl
$\overset{\oplus}{P}—\overset{\ominus}{C}$	1.71 Å	1.74 Å
P—phenyl	1.77-1.82	1.80-1.82
C—CO	1.35	1.36
C—O	1.28	1.30

ylid bond is between the single bond length (1.87 Å) and the double bond length (1.67 Å), that the C—CO bond is very close to the double bond length (1.33 Å), and that the C—O bond is considerably longer than the double bond (1.23 Å). It appears that structures such as CIX contribute appreciably to the hybrid structure of the ylid. The oxyanion and phosphonium groups were oriented *cis* about the C—CO bond, probably due to both steric and electrostatic factors. On the basis of the bond angles the ylid carbon seemed to be trigonally hybridized and the phosphorus atom seemed to be tetrahedrally hybridized in accord with the representation in Fig. 3.1 (see p. 80). It is clear that there is appreciable double bond character to the ylid bond in agreement with the general conclusions of the chemical studies. It also is clear that the ylid carbanion is stabilized by conjugation with the carbonyl group.

In a brief report Luttke and Wilhelm (*261*) presented their results from an infrared spectral study of methylenetriphenylphosphorane $CH_2{=}P(C_6H_5)_3$ (III). Using C^{12}-, C^{13}-, and H^2-labeled III and the corresponding conjugate acids they found that the phosphorus-carbon force constant changed from 3.6 mdynes/Å in the salt to 4.6 mdynes/Å in the ylid. On this basis they estimated a bond order of 1.3 for the phosphorus-carbon bond of III. This result, too, is in general agreement with the results of chemical studies.

REFERENCES

1. G. Wittig and U. Schollkopf, *Chem. Ber.* **87,** 1318 (1954).
2. A. Michaelis and H. V. Gimborn, *Ber. deut. chem. Ges.* **27,** 272 (1894).
3. G. Wittig and M. Rieber, *Ann.* **562,** 177 (1949).
4. G. Wittig and G. Geissler, *Ann.* **580,** 44 (1953).
5. J. Levisalles, *Bull. Soc. Chim. Fr.* 1021 (1958).
6. A. Wm. Johnson and R. B. LaCount, *Chem. & Ind. (London)* 52 (1959).
7. A. Wm. Johnson and R. B. LaCount, *Tetrahedron* **9,** 130 (1960).
8. S. Trippett and D. M. Walker, *Chem. & Ind. (London)* 933 (1960); *J. Chem. Soc.* 1266 (1961).
9. H. Heitman, J. H. S. Wieland and H. O. Huisman, *Koninkl. Ned. Akad. Wetenschap., Proc. Ser. B.* **64,** 165 (1961); *Chem. Abstr.* **55,** 17562 (1961).
10. H. J. Bestmann and O. Kratzer, *Chem. Ber.* **95,** 1894 (1962).
11. G. Wittig, H. D. Weigmann and M. Schlosser, *Chem. Ber.* **94,** 676 (1961).
12. S. Trippett and D. M. Walker, *J. Chem. Soc.* 2130 (1961).
13. D. B. Denney and L. C. Smith, *J. Org. Chem.* **27,** 3404 (1962).
14. H. S. Corey, J. R. D. McCormick and W. E. Swensen, *J. Am. Chem. Soc.* **86,** 1884 (1964).
15. G. Wittig, H. Eggers and P. Duffner, *Ann.* **619,** 10 (1958).
16. S. Trippett, *Chem. & Ind. (London)* 80 (1956).
17. D. W. Dicker and M. C. Whiting, *J. Chem. Soc.* 1994 (1958).
18. H. Burger, Dissertation, University of Tubingen, 1957.
19. H. Heitman, U. K. Pandit and H. O. Huisman, *Tetrahedron Letters* 915 (1963).
20. J. A. Ford, Jr. and C. V. Wilson, *J. Org. Chem.* **26,** 1433 (1961).

21. S. Trippett and D. M. Walker, *J. Chem. Soc.* 3874 (1959).
22. L. Horner and H. Oediger, *Chem. Ber.* **91**, 437 (1958).
23. G. Wittig and W. Boll, *Chem. Ber.* **95**, 2526 (1962).
24. S. Trippett, *Proc. Chem. Soc.* 19 (1963).
25. L. Pinck and G. E. Hilbert, *J. Am. Chem. Soc.* **68**, 751 (1946).
26. G. Wittig and H. Laib, *Ann.* **580**, 57 (1953).
27. A. Mondon, *Ann.* **603**, 115 (1957); K. V. Scherer and R. S. Lunt, *J. Org. Chem.* **30**, 3215 (1965).
28. L. D. Bergelson, V. A. Vaver, L. I. Barsukov and M. M. Shemyakin, *Izvest. Akad. Nauk S.S.S.R.* 1134 (1963); *Bull. Acad. Sci. U.S.S.R.* 1037 (1963).
29. C. E. Griffin and G. Witschard, *J. Org. Chem.* **27**, 3334 (1962).
30. W. von E. Doering and A. K. Hoffmann, *J. Am. Chem. Soc.* **76**, 6162 (1954).
31. A. J. Speziale and K. W. Ratts, *J. Am. Chem. Soc.* **82**, 1260 (1960); **84**, 854 (1962).
32. D. Seyferth, S. O. Grim and T. O. Read, *J. Am. Chem. Soc.* **82**, 1510 (1960); **83**, 1617 (1961).
33. G. Wittig and M. Schlosser, *Angew. Chem.* **72**, 324 (1960); *Chem. Ber.* **94**, 1373 (1961).
34. V. Franzen, *Angew. Chem.* **72**, 566 (1960).
35. S. A. Fuqua, W. G. Duncan and R. M. Silverstein, *Tetrahedron Letters* 1461 (1964); *J. Org. Chem.* **30**, 2543 (1965).
36. J. Hine and D. C. Duffey, *J. Am. Chem. Soc.* **81**, 1131 (1959).
37. V. Mark, *Tetrahedron Letters* 3139 (1964).
38. D. M. Lemal and E. H. Banitt, *Tetrahedron Letters* 245 (1964).
39. V. Franzen and G. Wittig, *Angew. Chem.* **72**, 417 (1960).
40. J. Hine, "Divalent Carbon," Ronald Press Co., New York, 1964.
41. W. Kirmse, "Carbene Chemistry," Academic Press, New York, 1964.
42. H. Staudinger and J. Meyer, *Helv. Chim. Acta* **2**, 635 (1919).
43. H. Staudinger and J. Meyer, *Helv. Chim. Acta* **2**, 619 (1919).
44. A. Wm. Johnson, *J. Org. Chem.* **24**, 282 (1959).
45. F. Ramirez and S. Levy, *J. Org. Chem.* **23**, 2036 (1958).
46. G. Wittig and W. Haag, *Chem. Ber.* **88**, 1654 (1955).
47. G. Wittig and K. Schwarzenbach, *Ann.* **650**, 1 (1961).
48. G. Wittig and M. Schlosser, *Tetrahedron* **18**, 1023 (1962).
49. G. Markl, *Tetrahedron Letters* 811 (1961).
50. D. Seyferth, J. S. Fogel and J. K. Heeren, *J. Am. Chem. Soc.* **86**, 307 (1964).
51. E. E. Schweizer, *J. Am. Chem. Soc.* **86**, 2744 (1964).
52. E. E. Schweizer and K. K. Light, *J. Am. Chem. Soc.* **86**, 2963 (1964).
53. V. Franzen, H. I. Joschek and C. Mertz, *Ann.* **654**, 82 (1962).
54. D. Seyferth and J. M. Burlitch, *J. Org. Chem.* **28**, 2463 (1963).
55. A. Schonberg and R. Michaelis, *Chem. Ber.* **69**, 1080 (1936).
56. F. Ramirez and S. Dershowitz, *J. Am. Chem. Soc.* **78**, 5614 (1956).
57. A. Schonberg and A. F. A. Ismail, *J. Chem. Soc.* 1374 (1940).
58. G. Aksnes, *Acta Chem. Scand.* **15**, 692 (1961).
59. C. Osuch, J. E. Franz and F. B. Zienty, *J. Org. Chem.* **29**, 3721 (1964).
60. H. Hoffmann, *Chem. Ber.* **94**, 1331 (1961).
61. F. Ramirez and O. P. Madan, *Abstr. Papers, 148th Meeting, Am. Chem. Soc.,* p. 13S (1964).
62. F. Ramirez, O. P. Madan and C. P. Smith, *Tetrahedron Letters* 201 (1965).
63. F. Ramirez, O. P. Madan and C. P. Smith, *J. Am. Chem. Soc.* **86**, 5339 (1964).

64. L. Horner and K. Klupfel, *Ann.* **591,** 69 (1955).
65. N. Takashina and C. C. Price, *J. Am. Chem. Soc.* **84,** 489 (1962).
66. R. Oda, T. Kawabata and S. Tanimoto, *Tetrahedron Letters* 1653 (1964).
67. H. J. Bestmann and F. Seng, *Angew. Chem.* **74,** 154 (1962).
68. S. Trippett, *J. Chem. Soc.* 4733 (1962).
69. J. B. Hendrickson, *J. Am. Chem. Soc.* **83,** 2018 (1961).
70. G. W. Brown, R. C. Cookson and I. D. R. Stevens, *Tetrahedron Letters* 1263 (1964).
71. H. J. Bestmann and O. Rothe, *Angew. Chem.* **76,** 569 (1964).
72. H. J. Bestmann and H. Schulz, *Tetrahedron Letters* No. **4,** 5 (1960).
73. H. J. Bestmann, *Chem. Ber.* **95,** 58 (1962).
74. H. J. Bestmann and H. Schulz, *Chem. Ber.* **95,** 2921 (1962).
75. G. Markl, *Tetrahedron Letters* 1027 (1962).
76. N. Petragnini and M. de M. Campos, *Chem. & Ind.* (*London*) 1461 (1964).
77. H. J. Bestmann and H. Haberlein, *Z. Naturforschg.* **17b,** 787 (1962).
78. G. Markl, *Z. Naturforschg.* **18b,** 84 (1963).
79. G. Markl, *Angew. Chem.* **75,** 168 (1963); *International Edn.* **2,** 153 (1963).
80. H. J. Bestmann, *Tetrahedron Letters* No. **4,** 7 (1960).
81. H. J. Bestmann and B. Arnason, *Chem. Ber.* **95,** 1513 (1962).
82. S. T. D. Gough and S. Trippett, *Proc. Chem. Soc.* 302 (1961).
83. H. J. Bestmann and H. Schulz, *Angew. Chem.* **73,** 27 (1961); *Ann.* **674,** 11 (1964).
84. S. Trippett and D. M. Walker, *J. Chem. Soc.* 2976 (1960).
85. H. J. Bestmann and B. Arnason, *Tetrahedron Letters* 455 (1961).
86. S. Trippett and D. M. Walker, *Chem. & Ind.* (*London*) 202 (1960).
87. H. A. Staab and N. Sommer, *Angew. Chem.* **74,** 294 (1962).
88. H. O. House and H. Babad, *J. Org. Chem.* **28,** 90 (1963).
89. H. Staudinger and J. Meyer, *Ber. deut. chem. Ges.* **53,** 72 (1920).
90. P. A. Chopard, R. J. G. Searle and F. H. Devitt, *J. Org. Chem.* **30,** 1015 (1965).
91. F. Ramirez, N. B. Desai, B. Hansen and N. McKelvie, *J. Am. Chem. Soc.* **83,** 3539 (1961).
92. G. Markl, *Chem. Ber.* **94,** 2996 (1961).
93. D. B. Denney and S. T. Ross, *J. Org. Chem.* **27,** 998 (1962).
94. A. J. Speziale and K. W. Ratts, *J. Org. Chem.* **28,** 465 (1963).
95. G. Markl, *Chem. Ber.* **95,** 3003 (1962).
96. H. Hoffmann, L. Horner and G. Hassel, *Chem. Ber.* **91,** 58 (1958).
97. R. N. McDonald and T. W. Campbell, *J. Am. Chem. Soc.* **82,** 4669 (1960).
98. L. Horner, H. Hoffmann, H. G. Wippel and G. Hassel, *Chem. Ber.* **91,** 52 (1958).
99. L. Horner and H. G. Schmelzer, *Chem. Ber.* **94,** 1326 (1961).
100. N. Kornblum, P. J. Berrigan and W. J. LeNoble, *J. Am. Chem. Soc.* **85,** 1141 (1963).
101. H. Hellmann and O. Schumacher, *Ann.* **640,** 79 (1961).
102. S. Trippett, *Quart. Rev.* **17,** 406 (1964).
103. L. Horner and A. Mentrup, *Ann.* **646,** 65 (1961).
104. D. Seyferth, M. A. Eisert and J. K. Heeren, *J. Organometallic Chem.* **2,** 101 (1964).
105. H. Pommer, *Angew. Chem.* **72,** 811, 911 (1960).
106. F. Bohlmann and H. J. Mannhardt, *Chem. Ber.* **89,** 1307 (1956).

107. H. H. Inhoffen, K. Bruckner, G. F. Domagk and H. M. Erdmann, *Chem. Ber.* **88,** 1415 (1955).
108. F. W. Lichtenthaler, *Chem. Revs.* **61,** 607 (1961); P. A. Chopard, R. F. Hudson and G. Klopman, *J. Chem. Soc.* 1379 (1965).
109. I. J. Borowitz and R. Virkhaus, *J. Am. Chem. Soc.* **85,** 2183 (1963).
110. A. Michaelis and E. Kohler, *Ber. deut. chem. Ges.* **32,** 1566 (1899).
111. H. Hoffmann and H. Forster, *Tetrahedron Letters* 1547 (1963).
112. D. W. Grisley, J. C. Alm and C. N. Matthews, *Tetrahedron* **21,** 5 (1965).
113. F. Ramirez and N. McKelvie, *J. Am. Chem. Soc.* **79,** 5829 (1957).
114. R. Rabinowitz and R. Marcus, *J. Am. Chem. Soc.* **84,** 1312 (1962).
115. F. Ramirez, N. B. Desai and N. McKelvie, *J. Am. Chem. Soc.* **84,** 1745 (1962).
116. L. Pinck and G. E. Hilbert, *J. Am. Chem. Soc.* **69,** 723 (1947).
117. D. D. Coffmann and C. S. Marvel, *J. Am. Chem. Soc.* **51,** 3496 (1929).
118. R. Greenwald, M. Chaykovsky and E. J. Corey, *J. Org. Chem.* **28,** 1128 (1963).
119. M. Grayson and P. T. Keough, *J. Am. Chem. Soc.* **82,** 3919 (1960).
120. H. Hoffmann, *Chem. Ber.* **95,** 2563 (1962).
121. G. Köbrich, *Angew. Chem.* **74,** 33 (1962).
122. R. P. Welcher and N. E. Day, *J. Org. Chem.* **27,** 1824 (1962).
123. L. D. Bergelson, V. A. Vaver and M. M. Shemyakin, *Izvest. Akad. Nauk S.S.S.R.* 1900 (1960); *Bull. Acad. Sci. U.S.S.R.* 1779 (1960).
124. D. Seyferth, J. K. Heeren and S. O. Grim, *J. Org. Chem.* **26,** 4783 (1961).
125. R. G. Jones and H. Gilman, *in* "Organic Reactions" (R. Adams, ed.), Vol. 6, p. 341. John Wiley & Sons, Inc., New York, 1951.
126. G. Wittig and M. Rieber, *Ann.* **562,** 187 (1949).
127. D. Seyferth, J. K. Heeren and W. B. Hughes, Jr., *J. Am. Chem. Soc.* **84,** 1764 (1962); **87,** 2847, 3467 (1965).
128. G. W. Fenton and C. K. Ingold, *J. Chem. Soc.* 2342 (1929).
129. F. Ramirez and S. Dershowitz, *J. Org. Chem.* **22,** 41 (1957).
130. F. Ramirez and S. Levy, *J. Am. Chem. Soc.* **79,** 67 (1957).
131. A. Wm. Johnson, S. Y. Lee, R. A. Swor and L. D. Royer, *J. Am. Chem. Soc.* **88,** in press (1966).
132. C. N. Matthews, J. S. Driscoll, J. E. Harris and R. J. Wineman, *J. Am. Chem. Soc.* **84,** 4349 (1962).
133. A. Streitwieser and J. I. Brauman, *J. Am. Chem. Soc.* **85,** 2633 (1963).
134. L. D. Bergelson and M. M. Shemyakin, *Tetrahedron* **19,** 149 (1963).
135. L. D. Bergelson, V. A. Vaver, L. I. Barsukov and M. M. Shemyakin, *Izvest. Akad. Nauk S.S.S.R.* 1053 (1963); *Bull. Acad. Sci. U.S.S.R.* 957 (1963).
136. N. E. Miller, *J. Am. Chem. Soc.* **87,** 390 (1965); *Inorg. Chem.* **4,** 1458 (1965).
137. A. Streitwieser, *Tetrahedron Letters* No. **6,** 23 (1960).
138. H. H. Jaffe, *Chem. Revs.* **53,** 191 (1953).
139. S. Fliszar, R. F. Hudson and G. Salvadori, *Helv. Chim. Acta* **46,** 1580 (1963).
140. A. J. Speziale and K. W. Ratts, *J. Am. Chem. Soc.* **85,** 2790 (1963).
141. W. von E. Doering and A. K. Hoffmann, *J. Am. Chem. Soc.* **77,** 521 (1955).
142. J. Hine, M. W. Burske, M. Hine and P. B. Langford, *J. Am. Chem. Soc.* **79,** 1406 (1957).
143. P. Chopard and G. Salvadori, *Gazz. Chim. Ital.* **73,** 668 (1963).
144. F. Ramirez and S. Levy, *J. Am. Chem. Soc.* **79,** 6167 (1957).
145. H. Depoorter, J. Nys and A. Van Dormael, *Tetrahedron Letters* 199 (1961); *Bull. Soc. Chim. Belge* **73,** 921 (1964).
146. T. W. Campbell and R. N. McDonald, *J. Org. Chem.* **24,** 1969 (1959).

147. S. Misumi and M. Nakagawa, *Bull. Chem. Soc. Jap.* **36,** 399 (1963).
148. F. Bohlmann, *Chem. Ber.* **89,** 2191 (1956).
149. H. J. Bestmann, O. Kratzer and H. Simon, *Chem. Ber.* **95,** 2750 (1962).
150. I. T. Harrison and B. Lythgoe, *J. Chem. Soc.* 843 (1958).
151. H. H. Jaffe, *J. Chem. Phys.* **58,** 185 (1954).
152. D. P. Craig, A. Maccoll, R. S. Nyholm, L. E. Orgel and L. E. Sutton, *J. Chem. Soc.* 332 (1954); D. P. Craig and E. A. Magnusson, *J. Chem. Soc.* 4895 (1956).
153. B. M. Wepster, *Rec. Trav. Chim.* **71,** 1159, 1171 (1952).
154. J. S. Driscoll, D. W. Grisley, J. V. Pustinger, J. E. Harris and C. N. Matthews, *J. Org. Chem.* **29,** 2427 (1964).
155. T. C. W. Mak and J. Trotter, *Acta Cryst.* **18,** 81 (1965).
156. J. J. Daly, *J. Chem. Soc.* 3799 (1964).
157. L. Pauling, "Nature of the Chemical Bond," 3rd Edition, p. 224. Cornell University Press, Ithaca, New York, 1960.
158. D. J. Cram, R. D. Trepka and P. St. Janiak, *J. Am. Chem. Soc.* **86,** 2731 (1964) and earlier papers in the series.
159. G. E. Kimball, *J. Chem. Phys.* **8,** 188 (1940).
160. H. P. Koch and W. E. Moffitt, *Trans. Far. Soc.* **47,** 7 (1951).
161. W. von E. Doering and L. K. Levy, *J. Am. Chem. Soc.* **77,** 509 (1955).
162. W. Theilacker and E. Wegner, *Ann.* **664,** 125 (1963).
163. J. R. Van Wazer, "Phosphorus and Its Compounds," Vol. 1, p. 14, Interscience Publishers, Inc., New York, 1958.
164. W. E. McEwen, K. F. Kumli, A. Blade-Font, M. Zanger and C. A. VanderWerf, *J. Am. Chem. Soc.* **86,** 2378 (1964).
165. W. C. Hamilton, S. J. LaPlaca and F. Ramirez, *J. Am. Chem. Soc.* **87,** 127 (1965).
166. P. C. Haake and F. H. Westheimer, *J. Am. Chem. Soc.* **83,** 1102 (1961).
167. G. Markl, private communication.
168. G. Markl, *Angew. Chem.* **75,** 669 (1963).
169. G. Markl, *Z. Naturforschg.* **18b,** 1136 (1963).
170. E. A. Cookson and P. C. Crofts, *Angew. Chem.* **76,** 755 (1964).
171. G. Markl, *Angew. Chem.* **75,** 1121 (1963).
172. M. J. S. Dewar, E. A. C. Lucken and M. A. Whitehead, *J. Chem. Soc.* 2423 (1960).
173. C. C. Price, *Abstr. Papers, 147th Meeting Am. Chem. Soc.,* p. 6N (1964).
174. C. A. VanderWerf, W. E. McEwen and M. Zanger, *J. Am. Chem. Soc.* **81,** 3806 (1959).
175. M. Schlosser, *Angew. Chem.* **74,** 291 (1962).
176. G. Aksnes and J. Songstad, *Acta Chem. Scand.* **16,** 1426 (1962).
177. M. Saunders and G. Burchmann, *Tetrahedron Letters* No. **1,** 8 (1959).
178. S. T. D. Gough and S. Trippett, *J. Chem. Soc.* 4263 (1961).
179. W. J. Bailey and S. A. Buckler, *J. Am. Chem. Soc.* **79,** 3567 (1957).
180. A. Schonberg, K. H. Brosowski and E. Singer, *Chem. Ber.* **95,** 2984 (1962).
181. H. J. Bestmann, *Angew. Chem.* **72,** 34 (1960).
182. H. J. Bestmann and O. Kratzer, *Angew. Chem.* **74,** 494 (1962).
183. H. J. Bestmann and O. Kratzer, *Chem. Ber.* **96,** 1899 (1963).
184. H. J. Bestmann and O. Kratzer, *Angew. Chem.* **73,** 757 (1961).
185. H. J. Bestmann, H. Haberlein and O. Kratzer, *Angew. Chem.* **76,** 226 (1964).
186. F. Ramirez, R. B. Mitra and N. B. Desai, *J. Am. Chem. Soc.* **82,** 5763 (1960).
187. D. B. Denney and L. S. Smith, *J. Am. Chem. Soc.* **82,** 2396 (1960).

188. D. B. Denney, L. S. Smith, J. Song, C. J. Rossi and C. D. Hall, *J. Org. Chem.* **28**, 778 (1963).
189. D. B. Denney and T. M. Valega, *J. Org. Chem.* **29**, 440 (1964).
190. M. S. Siemiatycki and H. Strzelecka, *Compt. rend.* **250**, 3469 (1960).
191. R. Mechoulam and F. Sondheimer, *J. Am. Chem. Soc.* **80**, 4386 (1958).
192. H. J. Bestmann, H. Haberlein and I. Pils, *Tetrahedron* **20**, 2079 (1964).
193. H. J. Bestmann and H. Schulz, *Angew. Chem.* **73**, 620 (1961).
194. H. J. Bestmann, F. Seng and H. Schulz, *Chem. Ber.* **96**, 465 (1963).
195. G. Markl, *Chem. Ber.* **94**, 3005 (1961).
196. H. J. Bestmann, N. Sommer and H. A. Staab, *Angew. Chem.* **74**, 293 (1962).
197. H. J. Bestmann and H. Hartung, *Angew. Chem.* **75**, 297 (1963).
198. I. Tomoskozi and H. J. Bestmann, *Tetrahedron Letters* 1293 (1964).
199. M. F. Hawthorne, *J. Am. Chem. Soc.* **80**, 3480 (1958); **83**, 367 (1961).
200. D. Seyferth and S. O. Grim, *J. Am. Chem. Soc.* **83**, 1613 (1961).
201. N. A. Nesmeyanov, S. T. Zhuzhlikova and O. A. Reutov, *Doklady Akad. Nauk S.S.S.R.* **151**, 856 (1963).
202. R. Oda, Y. Ito and M. Okano, *Tetrahedron Letters* 7 (1964).
203. D. Seyferth and S. O. Grim, *J. Am. Chem. Soc.* **83**, 1610 (1961).
204. D. Seyferth and K. A. Brandle, *J. Am. Chem. Soc.* **83**, 2055 (1961).
205. N. A. Nesmeyanov, V. M. Novikov and O. A. Reutov, *Izvest. Akad. Nauk S.S.S.R.* 772 (1964); *Bull. Acad. Sci. U.S.S.R.* 724 (1964).
206. A. Schonberg, K. H. Brosowski and E. Singer, *Chem. Ber.* **95**, 2144 (1962).
207. P. D. Bartlett and G. Meguerian, *J. Am. Chem. Soc.* **78**, 3710 (1956).
208. G. Markl, *Tetrahedron Letters* 807 (1961).
209. H. Strzelecka, M. S. Siemiatycki and C. Prevost, *Compt. rend.* **254**, 696 (1962); **255**, 731 (1962).
210. D. B. Denney, J. J. Vill and M. J. Boskin, *J. Am. Chem. Soc.* **84**, 3944 (1962).
211. W. E. McEwen and A. P. Wolf, *J. Am. Chem. Soc.* **84**, 676 (1962).
212. W. E. McEwen, A. Blade-Font and C. A. VanderWerf, *J. Am. Chem. Soc.* **84**, 677 (1962).
213. E. Zbiral, *Monatshefte Chemie* **94**, 78 (1963).
214. Y. Inouye, T. Sugita and H. M. Walborsky, *Tetrahedron* **20**, 1695 (1964).
215. A. Schonberg and K. H. Brosowski, *Chem. Ber.* **92**, 2602 (1959).
216. H. J. Bestmann and F. Seng, *Angew. Chem.* **75**, 475 (1963); *Tetrahedron* **21**, 1373 (1965).
217. A. Blade-Font, W. E. McEwen and C. A. VanderWerf, *J. Am. Chem. Soc.* **82**, 2646 (1960).
218. A. Schonberg, E. Frese and K. H. Brosowski, *Chem. Ber.* **95**, 3077 (1962).
219. G. Markl, *Angew. Chem.* **74**, 696 (1962).
220. J. D. Surmatis and A. Ofner, *J. Org. Chem.* **26**, 1171 (1961).
221. J. P. Freeman, *Chem. & Ind. (London)* 1254 (1959).
222. J. B. Hendrickson, R. Rees and J. F. Templeton, *J. Am. Chem. Soc.* **86**, 107 (1964).
223. E. Zbiral, *Monatshefte Chemie* **95**, 1759 (1964).
224. E. Zbiral, *Tetrahedron Letters* 3963 (1964).
225. S. G. Levine, *J. Am. Chem. Soc.* **80**, 6150 (1958).
226. G. Drefahl, G. Plotner and R. Scholz, *Z. Chemie* **1**, 93 (1961).
227. A. T. Blomquist and V. J. Hruby, *J. Am. Chem. Soc.* **86**, 5041 (1964).
228. G. Markl, *Z. Naturforschg.* **17b**, 782 (1962).
229. D. E. Worrall, *J. Am. Chem. Soc.* **52**, 2933 (1930).

230. T. I. Bieber and E. H. Eisman, *J. Org. Chem.* **27,** 678 (1962).
231. F. Bohlmann and E. Inhoffen, *Chem. Ber.* **89,** 1276 (1956).
232. T. W. Campbell and R. N. McDonald, *J. Org. Chem.* **24,** 1246 (1959).
233. K. Dimroth and G. Pohl, *Angew. Chem.* **73,** 436 (1961).
234. H. Gilman and R. A. Tomasi, *J. Org. Chem.* **27,** 3647 (1962).
235. C. E. Griffin, K. R. Martin and B. E. Douglas, *J. Org. Chem.* **27,** 1627 (1962).
236. C. E. Griffin and G. Witschard, *J. Org. Chem.* **29,** 1001 (1964).
237. R. F. Hudson and P. A. Chopard, *J. Org. Chem.* **28,** 2446 (1963).
238. O. Isler, G. Gutmann, H. Lindlar, M. Montavon, R. Buegy, C. Ryser and P. Zeller, *Helv. Chim. Acta* **39,** 463 (1956).
239. R. Ketcham, D. Jambatkar and L. Martinelli, *J. Org. Chem.* **27,** 4666 (1962).
240. E. E. Schweizer and R. Schepers, *Tetrahedron Letters* 979 (1963).
241. M. S. Siemiatycki, J. Carretto and F. Malbec, *Bull. Soc. Chim. Fr.* 125 (1962).
242. G. Wittig, W. Boll and K. H. Kruck, *Chem. Ber.* **95,** 2514 (1962).
243. F. Bohlmann and H. G. Viehe, *Chem. Ber.* **88,** 1245, 1347 (1955).
244. F. Dallacker, K. Ulrichs and M. Lipp, *Ann.* **667,** 50 (1963).
245. H. H. Inhoffen and K. Irmacher, *Chem. Ber.* **89,** 1833 (1956).
246. H. H. Inhoffen, H. Burkhardt and G. Quinkert, *Chem. Ber.* **92,** 1564 (1959).
247. F. Bohlmann, *Chem. Ber.* **90,** 1519 (1957).
248. G. P. Schiemenz and H. Engelhard, *Chem. Ber.* **94,** 578 (1961).
249. R. F. Hudson, P. A. Chopard and G. Salvadori, *Helv. Chim. Acta* **47,** 632 (1964).
250. K. Eiter and H. Oediger, *Ann.* **682,** 62 (1965).
251. H. Oediger and K. Eiter, *Ann.* **682,** 58 (1965).
252. E. Buchta and F. Andree, *Chem. Ber.* **92,** 3111 (1959).
253. M. Schlosser and K. F. Christmann, *Angew. Chem.* **76,** 683 (1964).
254. D. E. Bissing, *J. Org. Chem.* **30,** 1296 (1965).
255. T. L. Fletcher, M. J. Namkung, J. R. Price and S. K. Schaefer, *J. Med. Chem.* **8,** 347 (1965).
256. H. J. Bestmann, *Angew. Chem.* **77,** 609, 651, 850 (1965); *International Edn.* **4,** 583, 645, 830 (1965).
257. J. B. Hendrickson, C. Hall, R. Rees and J. F. Templeton, *J. Org. Chem.* **30,** 3312 (1965).
258. L. Horner and H. Winkler, *Ann.* **685,** 1 (1965).
259. A. J. Speziale and K. W. Ratts, *Abstr. Papers, 150th Meeting, Am. Chem. Soc.* p. 57S (1965); *J. Am. Chem. Soc.* **87,** 5603 (1965).
260. F. S. Stephens, *J. Chem. Soc.* 5640, 5658 (1965).
261. W. Luttke and K. Wilhelm, *Angew. Chem., International Edn.* **4,** 875 (1965).

4

THE WITTIG REACTION

The Wittig reaction [4.1], involving a condensation-elimination between a phosphonium ylid and an aldehyde or ketone to form an olefin and a phosphine oxide, is named after Professor George Wittig of the University of Heidelberg. This reaction has been reviewed several times

$$R_3P{=}CR_2' + R_2''CO \longrightarrow R_3PO + R_2'C{=}CR_2'' \qquad [4.1]$$

in recent years by Wittig himself (*1*), Levisalles in 1958 (*2*), Schollkopf in 1959 and in 1964 (*3*), Trippett in 1960 (*4*), Yanovskaya in 1961 (*5*) and Trippett again in 1963 (*6*). These reviews all have been concerned only with the Wittig reaction whereas this book is an attempt to deal with all the chemistry of ylids. The reactions of phosphonium ylids have been discussed in Section III of the preceding chapter, but the Wittig reaction is being discussed by itself in this chapter due to its importance in synthetic organic chemistry and since the discovery and subsequent development of this reaction actually incited the explosive interest in the chemistry of ylids in their own right.

The history of the development of the Wittig reaction was mentioned briefly by Wittig in his second review article in 1956 (*1*) but has been discussed in more detail by him in a recent paper presented before the IUPAC Symposium on Organophosphorus Compounds in May, 1964 (*7*). In that paper, entitled "Variationen zu einem Thema von Staudinger; Ein Beitrag zur Geschichte der phosphororganischen Carbonyl-Olefinierung," Wittig drew a delightful and elegant analogy between the evolution of a piece of music and the development of the Wittig reaction.

The first condensation between a carbonyl compound and a phosphonium ylid was reported in 1919 by Staudinger and Meyer (*8*) who found that benzhydrylidenetriphenylphosphorane would react with phenylisocyanate to eliminate triphenylphosphine oxide and form triphenylketenimine. Three years later, Luscher (*9*) found that heating the same ylid with diphenylketene afforded tetraphenylallene and triphenyl-

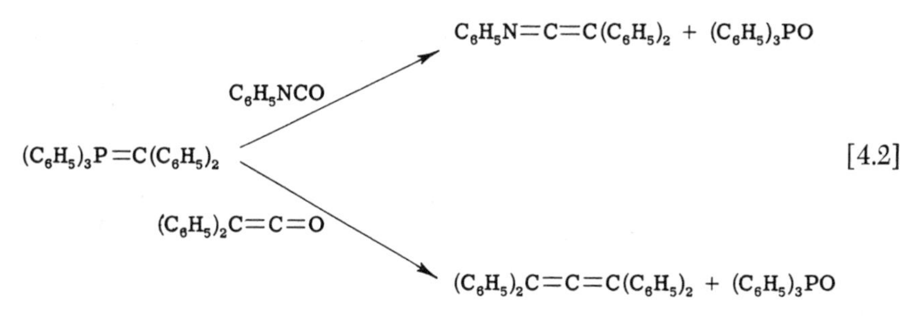

phosphine oxide. This reaction then lay dormant in the literature for twenty-seven years even though there were sporadic isolated investigations into the chemistry of phosphonium ylids in the interim.

In 1949 Wittig and Rieber (*10*) treated tetramethylphosphonium iodide with methyllithium in hopes of obtaining pentamethylphosphorane, $(CH_3)_5P$. Instead, methylenetrimethylphosphorane, $(CH_3)_3P{=}CH_2$, was formed, and reaction with benzophenone afforded a betaine which was trapped as (2-hydroxy-2,2,diphenyl)ethyltrimethylphosphonium iodide after quenching the reaction with acid and potassium iodide. Four years later, Wittig and Geissler (*11*) found that methyltriphenylphosphonium iodide likewise could be converted to methylenetriphenylphosphorane, $(C_6H_5)_3P{=}CH_2$, with phenyllithium, and that it would react with benzophenone to afford triphenylphosphine oxide and diphenylethylene in 84% yield [4.3]. This observation signaled

$$(CH_3)_3P{=}CH_2 + (C_6H_5)_2CO \xrightarrow[(2)\ H^{\oplus},\ KI]{(1)\ \text{ether}} (CH_3)_3\overset{\oplus}{P}{-}CH_2{-}\underset{OH}{C}(C_6H_5)_2 \quad I^{\ominus}$$

$$(C_6H_5)_3P{=}CH_2 + (C_6H_5)_2CO \xrightarrow{\text{ether}} (C_6H_5)_2C{=}CH_2 + (C_6H_5)_3PO \qquad [4.3]$$

the birth of the Wittig reaction. The scope and usefulness of the reaction were exploited in Wittig's subsequent papers, especially those written with Schollkopf (*12*) and Haag (*13*). Therefore, whereas the original example of this reaction was not discovered by Wittig but by Staudinger, the former certainly developed and elaborated the important synthetic applications of the reaction.

The role of the Wittig reaction has been established firmly in the arsenal of synthetic organic chemistry as an important method for the preparation of olefins. One of the advantages of the reaction is that it permits the introduction of an olefinic group into a molecule simultaneously with a homologation operation. In other words, the carbon skeleton of a molecule can be elaborated and a functional group introduced in one step. The most important feature of the Wittig reaction is the lack of ambiguity in the position of the new double bond. Olefin synthesis by dehydration or dehydrohalogenation often permits the formation of positional isomers or rearrangement of the original olefinic product under the influence of the acid or base catalyst. Under ideal conditions, the availability of a stable, isolable ylid, the Wittig reaction can be carried out in neutral media in the absence of extraneous ions or molecules. Often, however, the ylid must be formed *in situ* and halide ions and excess base may be present. Nevertheless, structural ambiguity in the olefinic product virtually is non-existent.

I. Experimental Conditions of the Wittig Reaction

The conditions used in many Wittig reactions have been patterned after those originally employed by Wittig's group. This generally has involved the formation of the ylid from the phosphonium salt in ethereal solution using an appropriate base, often phenyllithium, followed by addition of the carbonyl component and heating for periods of from a few hours to several days in a Schlenck tube (*12*). However, when using "reactive" (i.e., non-stabilized) ylids even these conditions are more severe than necessary. For example, Hauser *et al.* (*14*) found that benzylidenetriphenylphosphorane and benzaldehyde afforded a 78% yield of stilbene after five minutes at 10° whereas Wittig and Schollkopf (*12*) carried out the reaction over a period of two days and obtained an 82% yield. More severe conditions often are used with "stabilized" (i.e., unreactive, isolable) ylids to counterbalance their reduced nucleophilicity. Carbethoxymethylenetriphenylphosphorane and cyclohexanone afforded cyclohexylideneacetic acid ethyl ester in 44% yield when reacted at room temperature for long periods of time, but heating to 100° for short periods afforded the ester in 60% yield (*15*, Sugasawa and Matsuo; *16*, Fodor and Tomoskozi). Similarly, fluorenylidenetributylphosphorane and 4,4′-dinitrobenzophenone in refluxing chloroform afforded a 15% yield of olefin after three hours but afforded a 93% yield after twenty-four hours (*17*, Johnson and LaCount).

The solvents employed in the Wittig reaction are dependent on the ylid being used. The reader is referred to Chapter 3, Section I for a discussion of useful solvents. The Wittig reaction itself can be run in solvents ranging in polarity from hexane and benzene through ethanol and tetrahydrofuran to dimethylsulfoxide.

There is only one report of any catalysis of a Wittig reaction. Ruchardt *et al.* (*18*) found that addition of benzoic acid in catalytic amounts provided striking increases in olefin yields. For example, carbethoxymethylenetriphenylphosphorane and acetone reacted in 6% yield under normal conditions but in 80% yield in the presence of traces of benzoic acid. A similar increase was noted with cyclohexanone. The catalyst probably protonated the carbonyl group making it more susceptible to attack of the nucleophilic ylid. Such catalysis could not be used, however, in the presence of a reactive ylid, but neither should it be needed.

When using stable ylids in a Wittig reaction they can be prepared in a separate operation, purified and then reacted with a carbonyl compound in a solvent in the absence of extraneous ions (e.g., *17*). Several groups have noted that extraneous anions or cations in the reaction media, usually from the generation of a reactive ylid from its salt *in situ*, have an effect on the course of the Wittig reaction. The metal cation from the base used to generate the ylid clearly affects the yield. Wittig and Schollkopf (*12*) obtained a 48% yield of methylenecyclohexane from cyclohexanone and methylenetriphenylphosphorane, the latter having been generated from its hydrobromide with phenyllithium. Use of potassium *tert*-butoxide as the base in place of phenyllithium raised the yield of olefin to 91% (*19*, Schlosser and Christmann). The latter group proposed that the lithium cation complexed with the betaine intermediate (I) of the reaction, thereby hindering its decomposition to products, whereas the potassium ion would not form as strong a complex. This

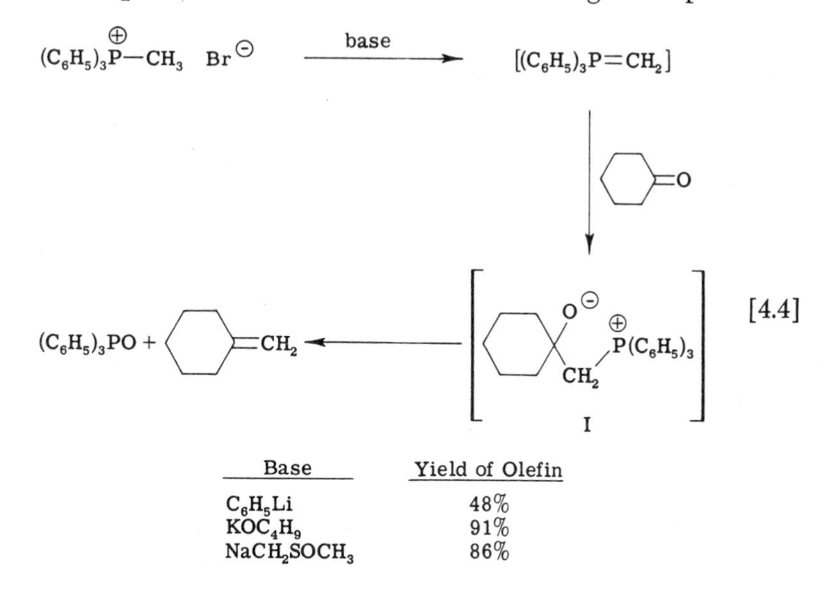

Base	Yield of Olefin
C_6H_5Li	48%
KOC_4H_9	91%
$NaCH_2SOCH_3$	86%

proposal agreed with Horner's observations for the reaction of phosphinoxy carbanions with carbonyl compounds where it was possible actually to isolate the intermediate betaines in the presence of lithium ions but not in the presence of potassium ions (*20*). The same phenomenon may account for Greenwald's observation that use of methylsulfinylcarbanion as the base for generation of methylenetriphenylphosphorane led to an 86% yield of the olefin, since soduim hydride was used to generate the carbanion from dimethylsulfoxide (*21*).

Several groups have detected apparent effects exerted on the course of the Wittig reaction by the halide anion that often is present. The halide always is present in the case of reactive ylids since the phosphonium salt is usually converted into ylid *in situ.* Collins and Hammond (*22*) reported that cyclopentanone would not react with methylenetriphenylphosphorane when the latter was generated from its hydriodide salt but that reaction would occur when the hydrobromide salt was used. This appears to be a specific anomaly since the iodides of many other phosphonium salts have been used successfully (*19,* Schlosser and Christmann). Hauser *et al.* (*14*) found that benzylidenetriphenylphosphorane reacted much faster with a variety of carbonyl compounds when it was generated from its chloride than from its bromide. The bromide may have been complexing with the phosphonium group of the betaine, thereby slowing down the collapse of the betaine to products. Bergelson *et al.* (*23*) and House *et al.* (*24*) have studied the effect of added anions, cations and neutral molecules on the stereochemistry of the Wittig reaction but this topic will be discussed later in this chapter.

Schiemenz and Engelhard (*25*) reported an interesting series of reactions to test the effectiveness of three experimental approaches to a Wittig synthesis. Following the normal procedure, methyl bromoacetate and triphenylphosphine afforded the expected phosphonium salt in 90% yield. This was converted into the ylid with aqueous sodium hydroxide in 97% yield, and the latter reacted with benzaldehyde to afford methyl cinnamate. From this three-step sequence the cinnamate was obtained in 63% overall yield based on methyl bromoacetate. Cutting the sequence to two steps by preparing the salt in one step and then heating a mixture of the salt, benzaldehyde and sodium hydroxide in the second step afforded the same product in 50% overall yield. Cutting the sequence further to just one step by heating a mixture of methyl bromoacetate, triphenylphosphine, benzaldehyde and sodium hydroxide in methanol to 150° for three hours led to a 54% overall yield of methyl cinnamate! There obviously is little advantage to the three-step procedure in this instance but the scope of this three-in-one reaction probably is not too wide.

Wittig and Haag (*13*) attempted to apply Staudinger's method to the preparation of simple phosphonium ylids but without success. Triphenylphosphine and diazomethane afforded an azine but the latter could not be converted into an ylid. More recently, however, Wittig and Schlosser (*26*) found that cuprous chloride would catalyze the decomposition of the azine to methylenetriphenylphosphorane (see Chapter 3, Section I). The most interesting feature of this work was the observation that a normal two-step procedure, preparation of the azine and then its decomposition in the presence of a carbonyl, afforded a 35% yield of 1,1-diphenylethene using benzophenone. A one-step procedure, the addition of diazomethane to a mixture of triphenylphosphine, cuprous chloride and benzophenone, afforded the same olefin in 23% yield [4.5].

$$CH_2N_2 + (C_6H_5)_3P + (C_6H_5)_2CO \xrightarrow{CuCl} (C_6H_5)_2C{=}CH_2 + N_2 + (C_6H_5)_3PO \quad [4.5]$$

To permit the easy separation of the phosphine oxide from the olefin in the product mixture from a Wittig reaction, Trippett and Walker (*27*) have suggested replacing triphenylphosphonium salts with diphenyl (*p*-dimethylaminophenyl)phosphonium salts since the phosphine oxide would then carry a basic amine group which would permit its removal by acid extraction and avoid the sometimes tedious chromatographic procedures often used.

There are many complications that can interfere with a successful Wittig reaction. Most of these have to do with the formation or stability of the phosphonium ylid required for the reaction, such as those arising from reaction of the ylid with solvent, spontaneous fragmentation of the ylid, intramolecular reaction of the ylid, the presence of two or more possible sites for attack of the base on the phosphonium salt, etc. These all have been discussed in Chapter 3, Section I,B. There are relatively few complications in the Wittig reaction itself when it is divorced from the ylid formation process. There always is the problem of reaction site when there are two or more possible sites of nucleophilic attack. This problem will be examined in Section II of this chapter dealing with the scope of the Wittig reaction. There have been a few cases reported where ylids themselves catalyze condensation or polymerization reactions of the carbonyl component of a Wittig reaction, usually aldol-type reactions (*14,* Hauser *et al.*; *28,* Witschard and Griffin). For example, 1-tetralone underwent such a condensation in the presence of benzylidenetriphenylphosphorane (*28*) [4.6].

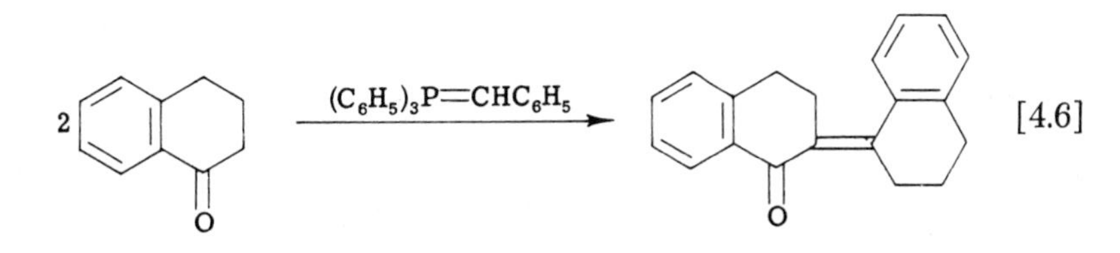

II. Scope and Limitations of the Wittig Reaction

The scope of the Wittig reaction is determined by the chemical characteristics of both the carbonyl and ylid components of the reaction. The reactivity of both components is a major factor but the individual stability and singularity in mode of reaction of each also is important.

A. Nature of the Ylid

Of the approximately two hundred ylids that are listed in Table 3.1 through 3.4 a vast majority undergoes the Wittig reaction with a variety of carbonyl compounds with no complications. The inability of the remaining ylids to undergo this reaction usually is attributable to their independent instability or their lack of reactivity (nucleophilicity). Such lowered nucleophilicity affects the intial step of the Wittig reaction, the attack of ylid carbanion on carbonyl carbon to form a betaine intermediate [4.7]. The following are examples of ylids with their

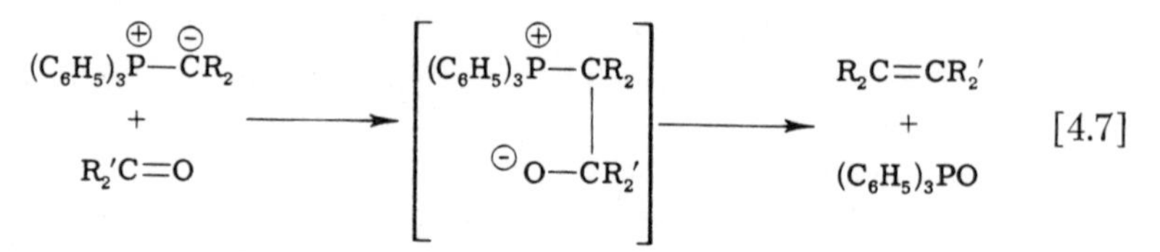

nucleophilicity lowered to the extent that their participation in Wittig reactions is hindered:

(a) Cyclopentadienylidenetriphenylphosphorane will not react with aldehydes or ketones (*29*, Ramirez and Levy),

(b) Fluorenylidenetriphenylphosphorane will react with aldehydes but not with ketones (*30*, Johnson),

(c) Benzoylmethylenetriphenylphosphorane reacted with benzaldehyde but not with cyclohexanone (*31*, Ramirez and Dershowitz),

(d) Carbethoxymethylenetriphenylphosphorane reacted much faster with aldehydes than with ketones (*15*, Sugasawa and Matsuo),

(e) 2,3,3-Diphenyl-3-phosphaindene reacted with benzaldehyde to form an olefin but 1,1-diphenyl-1-phosphanaphthalene would not react (*32*, Markl).

The nucleophilicity of ylids has been discussed in Chapter 3, Section II but suffice it to mention here that such reductions in nucleophilicity usually are due to extensive delocalization of the electron density on the carbanion by its substituents. The examples just quoted all deal with variation in ylid structure at the ylid carbon.

As pointed out in Chapter 3 the nature of the phosphorus substituents also affects the nucleophilicity of an ylid. Johnson and LaCount (*17*) have shown that replacement of a triphenylphosphonium group in an ylid by a trialkylphosphonium group increased the nucleophilicity of the ylid. On the other hand, Johnson (*33*) found that attachment of electron withdrawing substituents to the three phenyl groups does decrease the rate of the first step of the Wittig reaction. Mark (*34*) recently reported that difluoromethylenetris(dimethylamino)phosphorane would not undergo the Wittig reaction with benzaldehyde, presumably due to the decreased nucleophilicity of the carbanion as detected by the P^{31} NMR shift.

The nature of the phosphorus substituents has a marked effect on the second step of the Wittig reaction, the decomposition of betaine to olefin and phosphine oxide. Wittig *et al.* (*35*) have found that when using methylenephosphoranes, electron-donating groups on the phosphorus atom may slow the second step to the point where the betaine can be isolated and is the dominant product. For example, methylenetris(*p*-methoxyphenyl)phosphorane and benzaldehyde afforded a 69% yield of the betaine, and no olefin could be found. Under the same conditions methylenetris (*p*-tolyl)phosphorane afforded a 30% yield of betaine and a 55% yield of styrene. Such effects appear only when the second step of the Wittig reaction is slow since fluorenylidenetris(*p*-methoxyphenyl)phosphorane reacted normally with benzaldehyde to afford only benzalfluorene (*33*, Johnson). The interrelationships of these various effects will be discussed in full in Section IV of this chapter.

The inherent instability of some ylids precludes their use in the Wittig reaction. Characteristics such as fragmentations, spontaneous intramolecular alkylations, acylations and Wittig reactions, all of which have been discussed in Chapter 3, Section I,B, imposes a limitation on the use of certain ylids.

A very specific but rather interesting limitation on the use of the Wittig reaction was discovered by Gilman and Tomasi (*36*). Reaction of trimethylsilylmethyltriphenylphosphonium bromide with phenyllithium and then benzophenone was expected to afford 1-trimethylsilyl-2,2-diphenylethene and triphenylphosphine oxide. However, none of this olefin was present although triphenylphosphine oxide was isolated. Instead, tetraphenylallene was obtained in 36% yield. The mechanism presented in [4.8] was advanced. Some verification of this proposal was provided by the observation that 2,2-diphenylvinyltriphenylphosphonium bromide could be converted into an ylid (II) which reacted with benzophenone to afford tetraphenylallene in 54% yield. The complication in the original reaction apparently was due to the presence of two groups in the betaine (III) to which the oxygen could be transferred, the trimethylsilyl group and the triphenylphosphorus group. This specific example illustrates

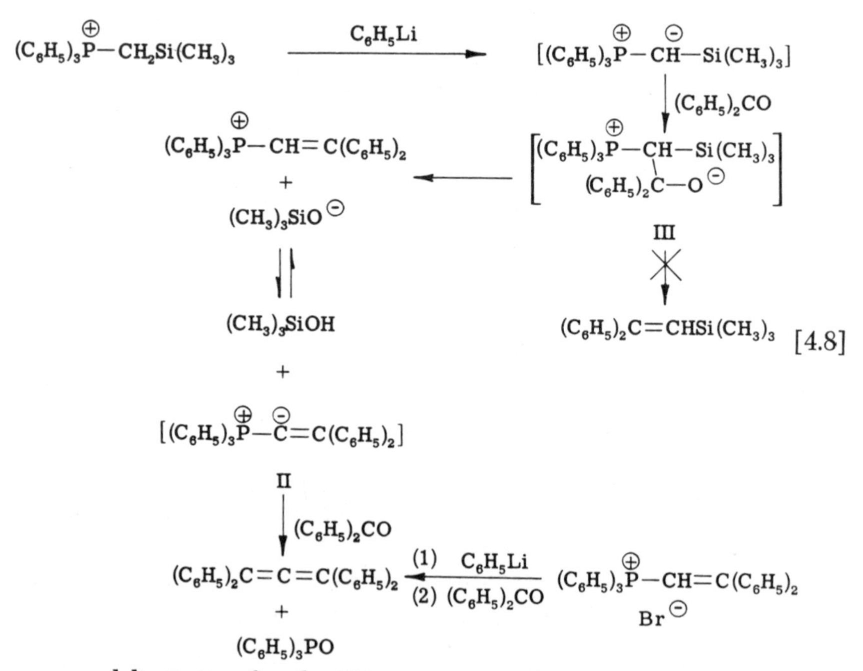

a general limitation for the Wittig reaction that product ambiguity will result whenever there are two or more possible sites for oxygen transfer. To date, however, this limitation has been more theoretical than practical.

B. Nature of the Carbonyl Compound

The Wittig reaction will take place with a wide variety of carbonyl compounds without complications. Included are saturated and unsaturated aliphatic aldehydes (*14*), aromatic aldehydes (*12, 30*), aliphatic and aromatic ketones (*12*), thioketones (*37*), ketenes (*9, 38*) and isocyanates (*8*). It is apparent from the general scheme in [4.9] that two

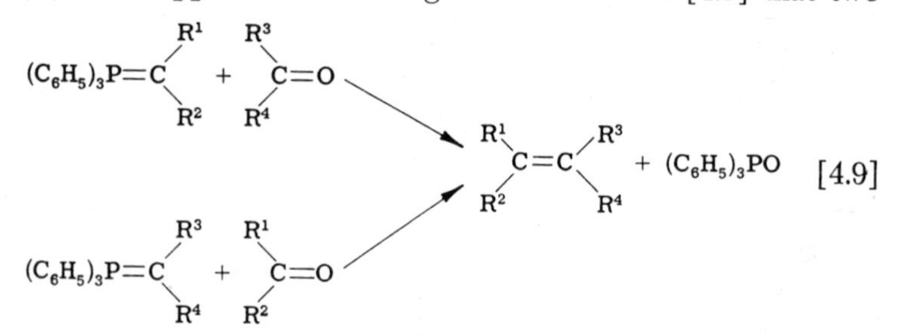

routes are available for the synthesis of any given olefin by the Wittig reaction. Given the flexibility in choice of the carbonyl component as indicated above and the few limitations on the nature of the ylid component as indicated in the previous section, almost any olefin is capable of being synthesized using the Wittig reaction. Some of the more unique applications of this reaction will be discussed in the following section.

Only in rare instances does the presence of another functional group in the carbonyl component of the Wittig reaction interfere with the normal progress of the reaction. Sondheimer and Mechoulam (*39*) formed a series of methylene-substituted steroids by reacting methylenetriphenylphosphorane with the appropriate keto-steroid. They found that the presence of hydroxyl, acetoxyl, conjugated and unconjugated double bonds did not interfere. Nuclear halogen, alkoxyl, nitro and dimethylamino groups do not interfere. Kochetkov and Dmitriev (*40*) have found that phosphonium ylids will react with free glucose or its pentaacetate to form the expected olefin.

In cases where there are other groups present in the carbonyl component which might be expected to react with a phosphonium ylid the use of inverse addition usually avoids complications. Thus, Bohlmann and Inhoffen (*41*) were able to carry out a Wittig reaction with an unsaturated ester-aldehyde exclusively at the aldehyde function. Similarly, House and Rasmusson (*42*) effected reaction of ethylidenetriphenylphosphorane at the ketonic group of methyl pyruvate with no evidence of attack at the ester function [4.10]. Drefahl and Plotner

$$C_3H_7{-}(C{\equiv}C)_2{-}(CH_2)_2{-}CH{=}P(C_6H_5)_3 + CH_3OOC{-}CH{=}CHCHO$$

$$\longrightarrow C_3H_7{-}(C{\equiv}C)_2{-}(CH_2)_2{-}(CH{=}CH)_2{-}COOCH_3 \qquad [4.10]$$

$$(C_6H_5)_3P{=}CHCH_3 + CH_3COCOOCH_3 \longrightarrow CH_3CH{=}C\begin{matrix} CH_3 \\ COOCH_3 \end{matrix}$$

(*43*) found that benzylidenetriphenylphosphorane reacted with *p*-chloromethylbenzaldehyde to afford the expected olefin with no interference from a competing alkylation.

In most cases the presence of unsaturation in the carbonyl component does not affect the course of the Wittig reaction. Campbell and McDonald (*44*) found that benzylidenetriphenylphosphorane reacted normally with cinnamaldehyde to afford 1,4-diphenyl-1,3-butadiene. However, Freeman (*45*) found that methylenetriphenylphosphorane added in a Michael addition fashion to mesitoylphenylethylene to afford 1-phenyl-2-mesitoylcyclopropane in 52% yield. Attack of the ylid carbanion at the carbonyl carbon presumably was hindered by the bulk of the mesityl group [4.11]. Bohlmann (*46*) reported an example in which the products

$$C_6H_5CH{=}CHCHO + C_6H_5CH{=}P(C_6H_5)_3 \longrightarrow C_6H_5(CH{=}CH)_2C_6H_5 + (C_6H_5)_3PO$$

$$C_6H_5CH{=}CH{-}\underset{\underset{O}{\|}}{C}{-}Mes + CH_2{=}P(C_6H_5)_3 \longrightarrow C_6H_5{-}CH{-}CH{-}\underset{\underset{O}{\|}}{C}{-}Mes \quad (\text{cyclopropane ring closed by } CH_2) \qquad [4.11]$$

(Mes = 2, 4, 6-trimethylphenyl)

of both a normal 1,2-addition and an abnormal 1,4-addition were produced. Cinnamylideneacetophenone and cinnamylidenetriphenylphosphorane afforded the expected tetraene, 1,4,8-triphenyloctatetraene, but in addition they obtained some 1,6-diphenylhexatriene. The latter was the result of a 1,4-addition which also must have produced phenylacetylene [4.12]. These two are the only reports of conjugate additions to

$$C_6H_5(CH{=}CH)_2\overset{\overset{C_6H_5}{|}}{C}{=}CH{-}CH{=}CHC_6H_5 + (C_6H_5)_3PO$$

$$\uparrow (a)$$

$$C_6H_5CH{=}CH{-}CH{=}CH{-}\overset{\overset{O}{\|}}{C}C_6H_5$$

$$+ \qquad (b) \qquad (a)$$

$$C_6H_5CH{=}CH{-}\overset{\ominus}{C}H{-}\overset{\oplus}{P}(C_6H_5)_3$$

$$\downarrow (b) \qquad [4.12]$$

$$\left[\begin{array}{l} C_6H_5CH{=}CH{-}CH{-}CH \quad C_6H_5 \\ \qquad\qquad\qquad\qquad\quad C \\ C_6H_5CH{=}CH{-}CH \qquad O^{\ominus} \\ \qquad\qquad\quad \oplus P(C_6H_5)_3 \end{array}\right]$$

$$\downarrow$$

$$C_6H_5(CH{=}CH)_3C_6H_5 + (C_6H_5)_3PO + [C_6H_5C{\equiv}CH]$$

unsaturated aldehydes or ketones. Many other conjugated carbonyls have been successfully employed in a Wittig synthesis, especially in the terpene and carotenoid fields. An excellent review of these reactions has been compiled by Yanovskaya (5).

Two instances have been reported recently in which the Wittig reaction has taken an abnormal course due to the structure of the carbonyl component. In both cases, however, the first step of the reaction, the betaine formation, appears to have taken place in a normal manner. Wittig and Haag (*38*) found that under mild conditions reaction of isopropylidenetriphenylphosphorane with diphenylketene stopped at the betaine stage (IV). The betaine could be isolated as its hydrobromide but by further heating it could be forced to eliminate triphenylphosphine oxide and produce the expected allene. The stability of the betaine

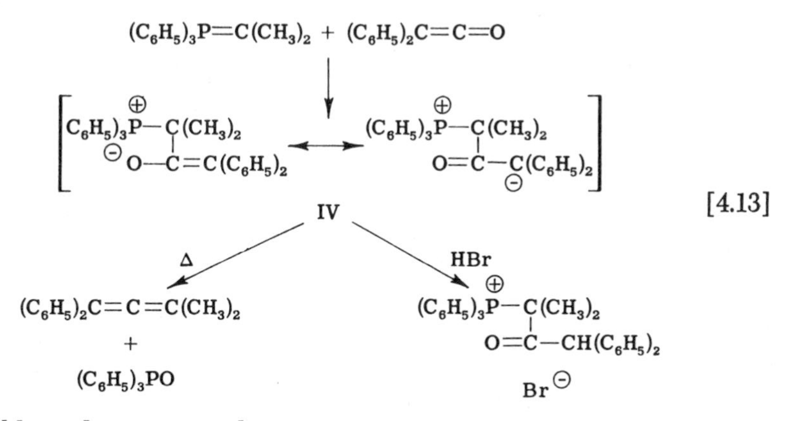

[4.13]

probably is due to its enolate structure.

The abnormal reaction discovered by LaLancette (*47*) also must depend on the stability of an enolate ion. Methylenetriphenylphosphorane and tetramethylcyclobutane-1,3-dione afforded a new ylid (V) rather than a mono- or dimethylenecyclobutane. The first step in this reaction was thought to be the normal attack of the ylid carbanion on a carbonyl

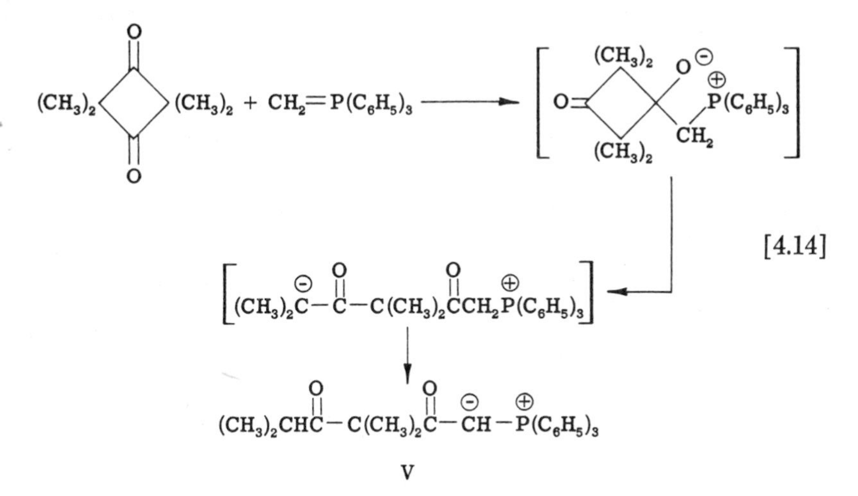

[4.14]

carbon to form a betaine. The latter probably underwent a ring opening at this point to afford an enolate ion which, in turn, underwent a prototropic shift to form a more stable enolate ion, a keto-ylid (V). The ring opening perhaps was to be expected since such was the case in the presence of other nucleophiles.

Occasionally self-condensations among the carbonyl components seem to be catalyzed by phosphonium ylids, precluding that particular combination for olefin synthesis. Hauser *et al.* (*14*) claimed that several instances of low yields of olefins when using saturated or unsaturated aliphatic aldehydes were due to the ylid catalyzing aldol condensations or polymerizations. Witschard and Griffin (*28*) reported that 1-tetralone underwent self-condensation in the presence of benzylidenetriphenylphosphorane. This complication could be avoided, however, by condensation of a tetralenylide with benzaldehyde, i.e., interchange of the ylid and carbonyl component.

A final consideration regarding the nature of the carbonyl component of the Wittig reaction must be the reactivity of the carbonyl compound. In reactions of carbonyl compounds with stabilized ylids where the first step, the attack of the carbanion on the carbonyl carbon, is the slow step, the carbonyl compound with the most electrophilic carbon should be the most reactive. In agreement with this supposition, Johnson and LaCount (*17*) found that aldehydes were more reactive than ketones with fluorenylidenetriphenylphosphorane. In addition, the reactivity of *para*-substituted benzaldehydes decreased in the following order: $NO_2 > Cl > H > OCH_3 > N(CH_3)_2$. Goetz *et al.* (*48*) later confirmed this sequence with phenacylidenetriphenylphosphorane, and Speziale and Bissing (*49*) found it to prevail with carbomethoxymethylenetriphenylphosphorane.

With relatively unreactive ylids such as fluorenylidenetriphenylphosphorane the electrophilicity of some carbonyl compounds is insufficient to permit reaction. This particular ylid would not react with any ketones, whether diaryl, dialkyl or arylalkyl (*17*). That this lack of reaction might be steric in origin was indicated by failure of the ylid to react with mesitaldehyde although it would react with *p*-tolaldehyde (*33*). On the other hand, the more nucleophilic fluorenylidenetri-*n*-butylphosphorane would react with ketones such as *p*-nitroacetophenone. It is still not clear what role steric hindrance plays in the observed reactivity sequences for carbonyl compounds.

In reactions of "non-stabilized" ylids with carbonyl compounds the second step of the reaction, the decomposition of betaine to olefin and phosphine oxide, appears to be the slow step. In these instances ketones usually are more reactive than aldehydes (*35*, Wittig *et al.*). For example, methylenetriphenylphosphorane reacted faster with benzophenone

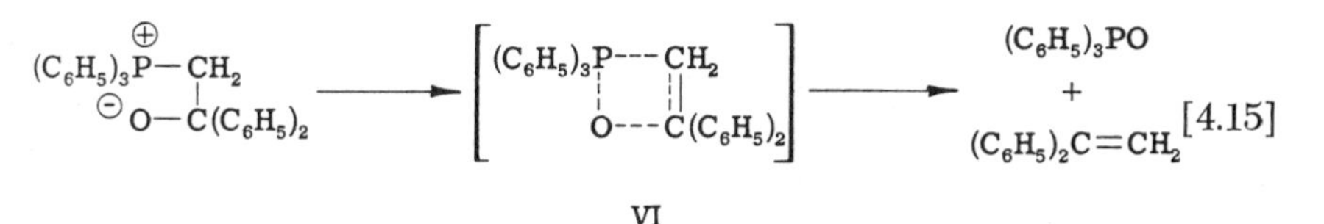

than with benzaldehyde. This difference has been attributed to the stabilization afforded by the two phenyl rings to the incipient double bond in the transition state (VI) between betaine and olefin (*35*, Wittig *et al.*; *50*, Fliszar *et al.*).

III. Applications of the Wittig Reaction

Subject to the limitations discussed in the preceding section the Wittig reaction generally is applicable to the synthesis of any olefin. In this section a few of the more novel examples of the Wittig reaction will be discussed. No attempt will be made to be encyclopedic. The reader is referred to the reviews listed at the beginning of this chapter (references *1* through *6*) for exhaustive compilations.

Because of the mild conditions usually employed in the Wittig reaction it often has been used for the synthesis of olefins which are especially prone to isomerization or rearrangement. Thus, one of the first striking examples of the power of the Wittig reaction was the synthesis of methylenecyclohexane from methylenetriphenylphosphorane and cyclohexanone (*12*, Wittig and Schollkopf). The exocyclic isomer was free from contamination by the endocyclic isomer. In a similar manner, Battiste (*51*) synthesized one of the first examples of a derivative of the elusive methylenecyclopropane by reaction of carbethoxymethylenetriphenylphosphorane with diphenylcyclopropenone [4.16]. Cava and Pohl (*52*)

$$\begin{matrix} C_6H_5-C \\ \| \\ C_6H_5-C \end{matrix}\!\!>C=O + (C_6H_5)_3P=CHCOOC_2H_5 \longrightarrow \begin{matrix} C_6H_5-C \\ \| \\ C_6H_5-C \end{matrix}\!\!>C=CHCOOC_2H_5 \quad [4.16]$$

and Blomquist and Hruby (*53*) have prepared derivatives of 1,2-bismethylene-3,4-benzocyclobutene by the Wittig reaction. The former group reacted a diketone with two equivalents of ylid whereas the latter group reacted a bis-ylid with two equivalents of carbonyl compound [4.17]. This duality of approach to a single system further illustrates the flexibility of the Wittig reaction. Interestingly, Parrick (*54*) has reported that some α-diketones, such as benzil and 9,10-phenanthroquinone, will not react with two equivalents of ylid. In both of these cases, α,β-unsaturated ketones were obtained from reaction of the ylid only at one carbonyl group.

Many groups have synthesized a large number of naturally occurring

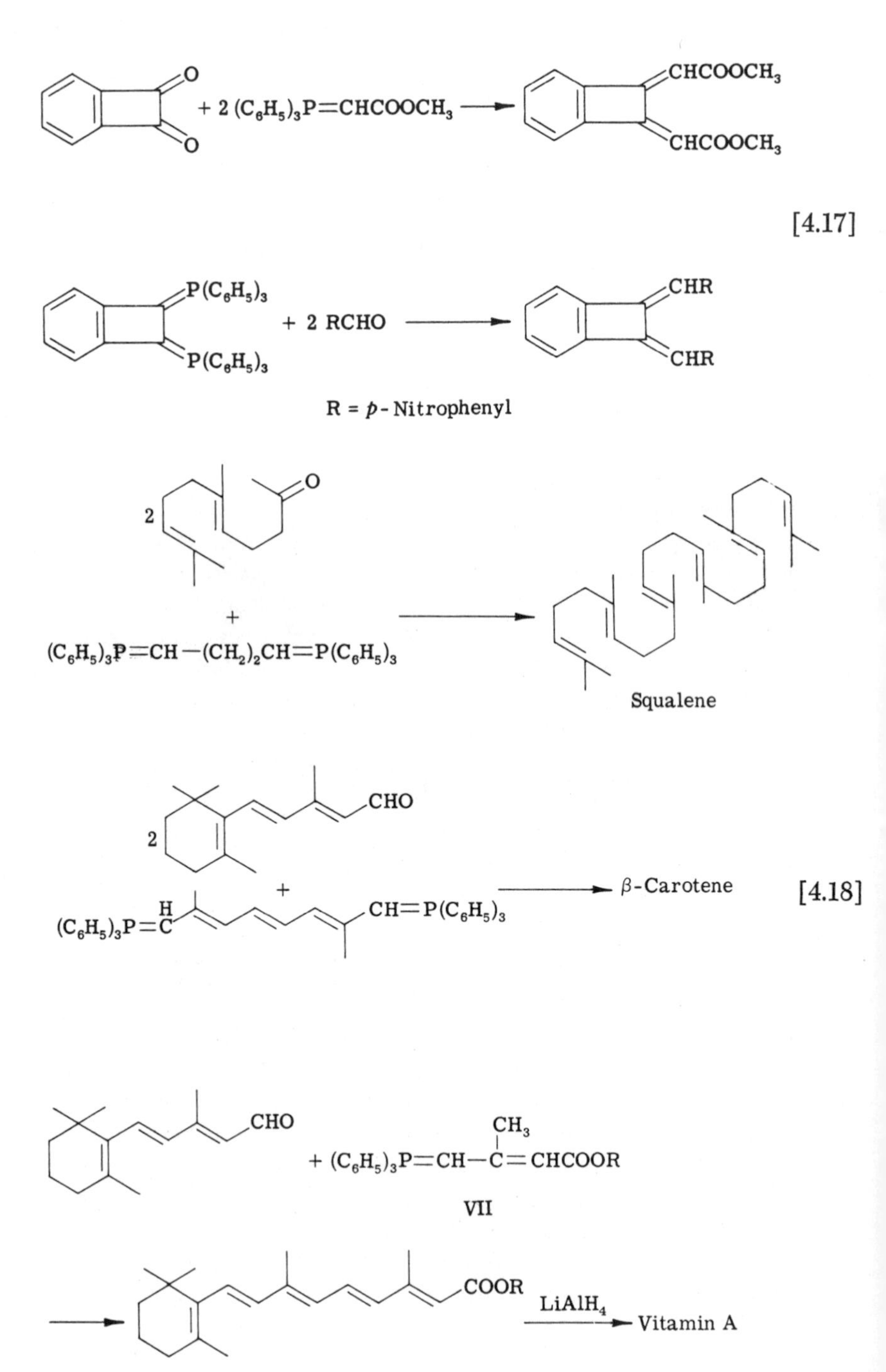

+ 2 $(C_6H_5)_3P{=}CHCOOCH_3$
$CHCOOCH_3$
$CHCOOCH_3$
[4.17]
$P(C_6H_5)_3$
$P(C_6H_5)_3$
+ 2 RCHO
CHR
CHR
R = p-Nitrophenyl
2
+
$(C_6H_5)_3P{=}CH-(CH_2)_2CH{=}P(C_6H_5)_3$
Squalene
CHO
2
+
$(C_6H_5)_3P{=}C$
H
$CH{=}P(C_6H_5)_3$
β-Carotene
[4.18]
CHO
CH_3
+ $(C_6H_5)_3P{=}CH-C{=}CHCOOR$
VII
COOR
$LiAlH_4$
Vitamin A

polyenes using the Wittig reaction. Previous to its discovery dehydration or elimination reactions usually were used but these often led to mixtures of isomers, difficulty in isolating and identifying the desired isomer and isomerization of the polyene once prepared. Trippett (*55*) appears to have prepared the first pure sample of synthetic all-*trans*-squalene by the condensation of two moles of geranyl acetone with one equivalent of the bis-ylid derived from 1,4-bis-triphenylphosphoniobutane dibromide. Similarly, Surmatis and Ofner (*56*) prepared pure all-*trans*-β-carotene using a bis-ylid and β-ionylideneacetaldehyde [4.18]. The same aldehyde but with the ester-ylid (VII) afforded a polyene ester which, upon reduction with lithium aluminum hydride, afforded vitamin A (*57*, Wittig and Pommer).

As indicated in the previous two paragraphs there are many examples of "double" Wittig reactions, those employing a bis-ylid in reaction with two equivalents of a carbonyl component (*53, 55, 56*) and those employing a bis-carbonyl compound in reaction with two equivalents of ylid (*52, 58*). Campbell and McDonald (*59*) have shown that a series of substituted distyrrylbenzenes could be prepared using both approaches [4.19].

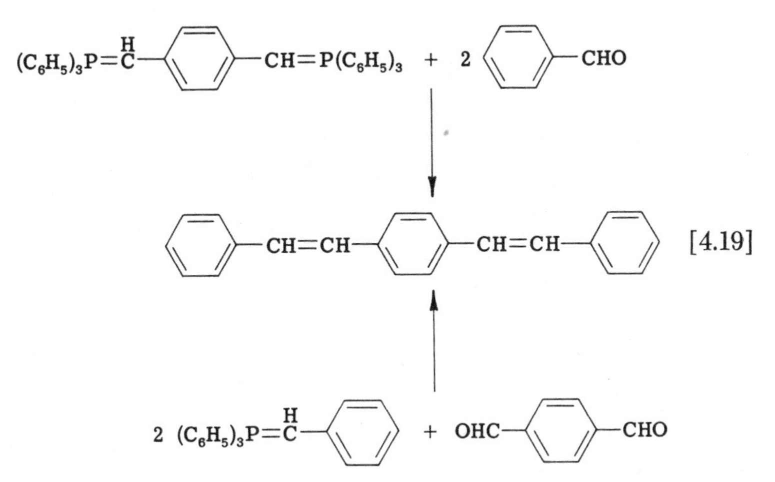

[4.19]

In some instances, use of the Wittig reaction may supplant older established procedures for standard synthetic conversions. For example, Bohlmann (*60*) has compared the condensation of methyl γ-bromocrotonate with substituted benzaldehydes via the Reformatsky route (formation of the organozinc reagent, condensation with carbonyl and dehydration of the alcohol) with the same overall result by the Wittig reaction route (formation of the phosphonium salt and ylid then reaction with the carbonyl). The Wittig reaction is the simpler experimentally but gave about the same overall yield.

Levine (*61*) developed a method which may supplant the Darzen's condensation for the one-carbon homologation of a carbonyl compound to an aldehyde. Reaction of the ylid derived from methoxymethyltriphenylphosphonium bromide with cyclohexanone afforded an enol ether which was hydrolyzed to cyclohexanecarboxaldehyde [4.20]. Later, Wit-

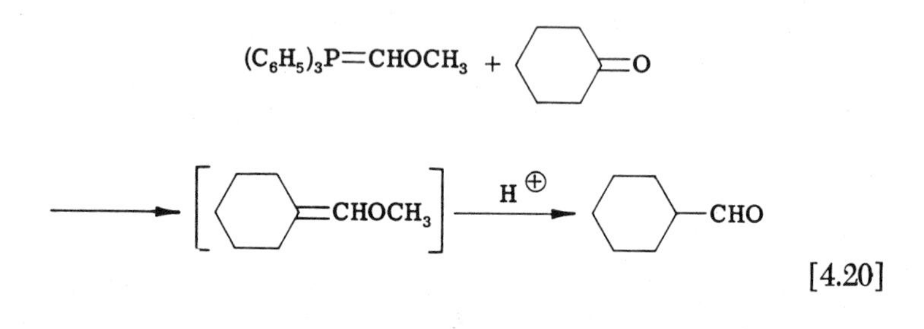

$$(C_6H_5)_3P{=}CHOCH_3 + C_6H_5CHO$$

$$\longrightarrow [C_6H_5CH{=}CHOCH_3] \xrightarrow{H^{\oplus}} C_6H_5CH_2CHO$$

tig and Knauss (*62*) reported similar homologations with benzophenone and benzaldehyde. This procedure is much simpler and less prone to side reactions than the Darzen's glycidic ester route.

Bestmann *et al.* (*63*) and, more recently, Schlosser (*64*) have used the Wittig reaction as a means of synthesis of specifically labeled olefins. For example, preparation of benzylidenetriphenylphosphorane from its

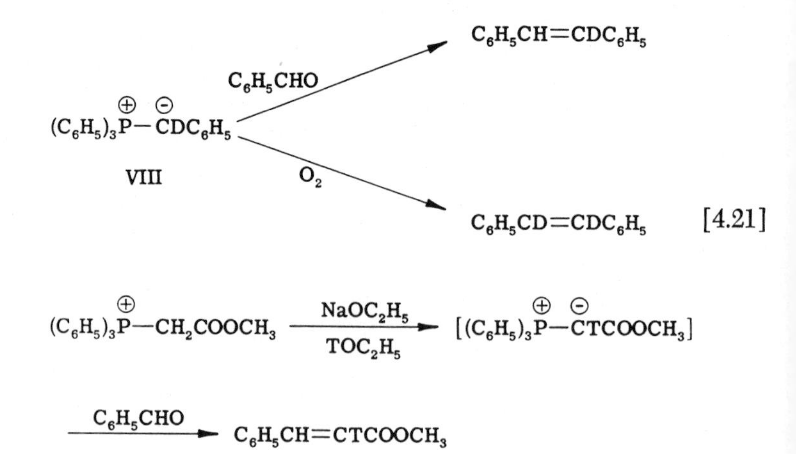

conjugate acid in the presence of a deuterium source resulted in deuterated ylid (VIII). Reaction of this ylid with benzaldehyde afforded monodeuterostilbene whereas oxidation of the ylid with molecular oxygen resulted in effective dimerization of the ylid to afford dideuterostilbene (*64*). Similarly, Bestmann *et al.* (*63*) prepared α-tritiocinnamate by treating carbomethoxymethyltriphenylphosphonium bromide with sodium ethoxide in O-tritiated ethanol and then adding benzaldehyde.

McDonald and Campbell (*65*) have attempted to apply the Wittig reaction to the preparation of high polymers. Both approaches, one of using two different bifunctional molecules (a bis-ylid and a bis-carbonyl) and the other of using one bifunctional molecule (an aldehydo-ylid) resulted in the formation of polymers [4.22].

$$(C_6H_5)_3P{=}CHC_6H_4CH{=}P(C_6H_5)_3 + OHC{-}C_6H_4{-}CHO \longrightarrow OHC{-}C_6H_4{-}(CH{=}CH{-}C_6H_4)_n{-}CHO \quad n \simeq 9$$

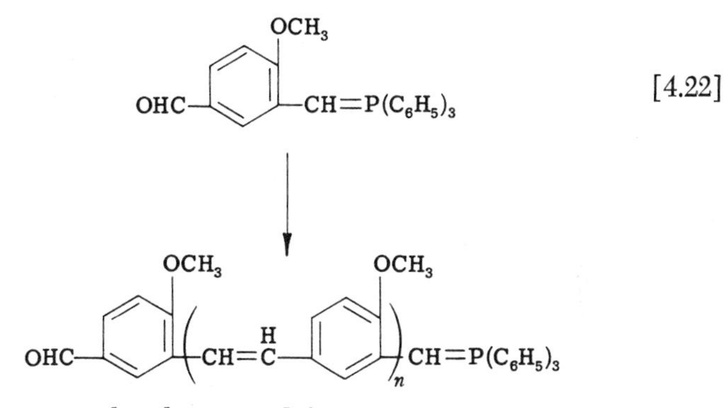

[4.22]

The Wittig reaction has been used for the synthesis of ring systems. Wittig *et al.* (*66*) first demonstrated this use with the reaction between *o*-phthalaldehyde and the bis-ylid derived from 1,3-bis(triphenylphosphonio)-propane dibromide, resulting in the formation of 3,4-benzocycloheptatriene in 28% yield [4.23]. Dimroth and Pohl (*67*) used the same approach for the preparation of benzoxepine, and Griffin and Peters (*68*) used an analogous route to 1,2,5,6-dibenzocyclooctatetraene. In the synthesis of such compounds two approaches are available as for the polymerization reactions. Reaction of bis-ylids with bis-carbonyl compounds is the most common route. The only example of the other possible route came about in an unsuccessful attempt to carry out an intramolecular Wittig reaction. Griffin and Witschard (*69*) found that

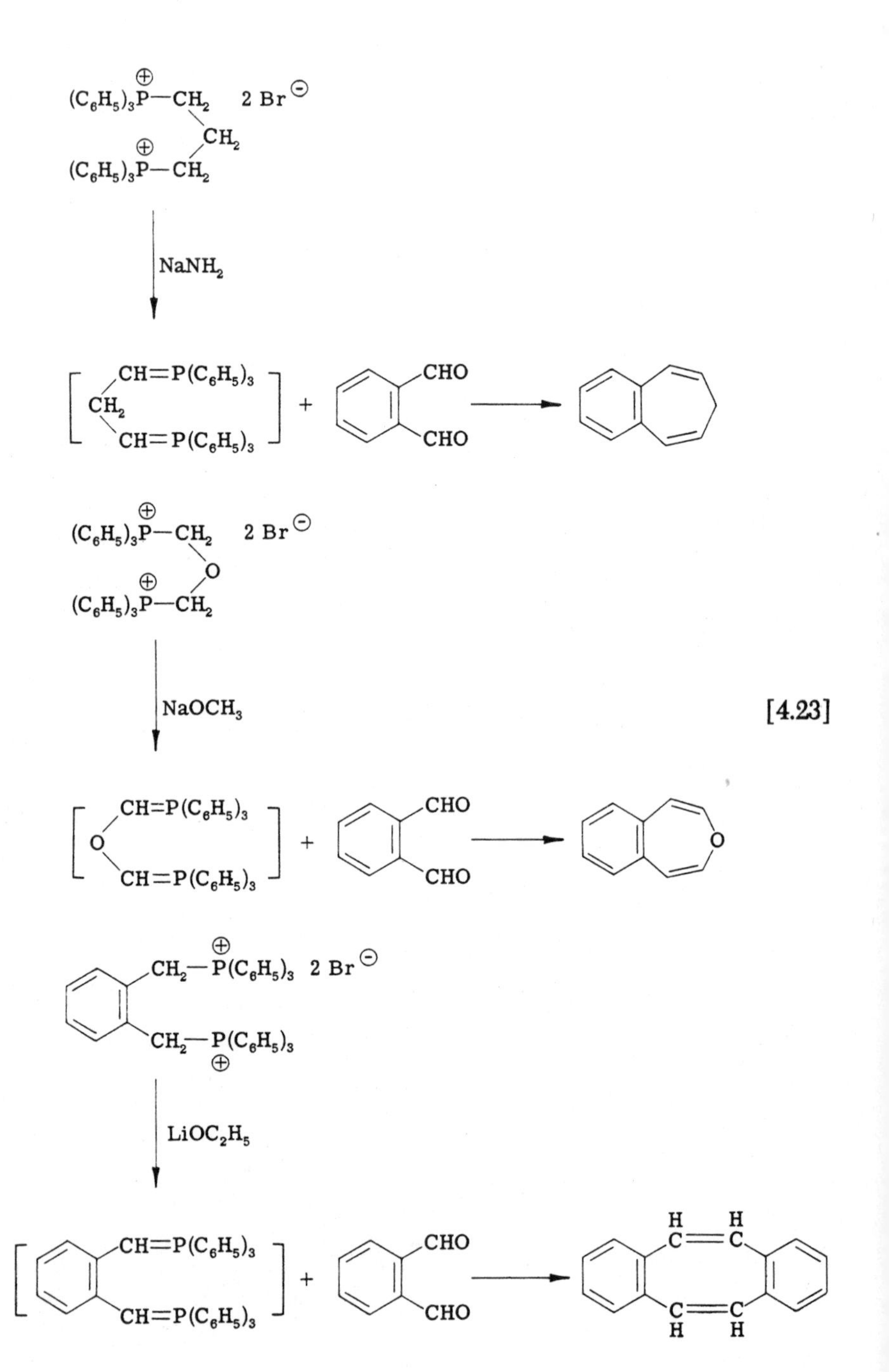
(C6H5)3P—CH2 2 Br⊖
CH2
(C6H5)3P—CH2
NaNH2
CH=P(C6H5)3
CH2
CH=P(C6H5)3
CHO
CHO
(C6H5)3P—CH2 2 Br⊖
O
(C6H5)3P—CH2
NaOCH3
[4.23]
CH=P(C6H5)3
O
CH=P(C6H5)3
CHO
CHO
O
CH2—P(C6H5)3 2 Br⊖
CH2—P(C6H5)3
LiOC2H5
CH=P(C6H5)3
CH=P(C6H5)3
CHO
CHO
H H
C=C
C=C
H H

phenacylmethylenetriphenylphosphorane underwent an intermolecular condensation to afford 1,4-diphenyl-1,4-cyclohexadiene but only in 12% yield [4.24]. An intermolecular condensation between *o*-formylbenzyl-

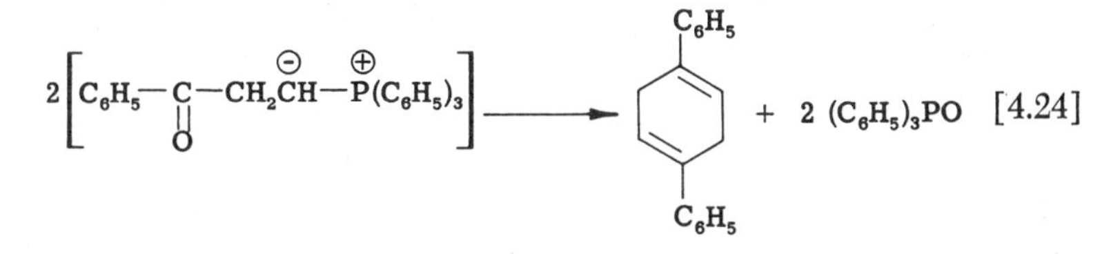

idenetriphenylphosphorane also ought to afford the dibenzocyclooctatetraene obtained by Griffin and Peters (*68*).

Several groups have examined possible intramolecular Wittig reactions. Had the phenacylmethylide (IX, $n = 1$) undergone an intramolecular reaction it would have afforded phenylcyclopropene (*69*).

$$C_6H_5{-}\underset{\underset{O}{\|}}{C}{-}(CH_2)_n{-}\overset{\ominus}{C}H{-}\overset{\oplus}{P}(C_6H_5)_3 \longrightarrow C_6H_5{-}\underbrace{C{=}CH}_{(CH_2)_n} + (C_6H_5)_3PO \qquad [4.25]$$

IX

Later work by Griffin and Witschard (*70*) with the next higher homolog (IX, $n = 2$) revealed that it also was impossible to close a four-membered ring by this reaction. However, Bieber and Eisman (*71*) were able to form 1-phenylcyclopentene in 24% yield, and Griffin and Witschard (*70*) obtained 1-phenylcyclohexene in 9% yield, both by the intramolecular reaction. The failure to obtain any four- or three-membered olefins by this intramolecular reaction may be attributed to initial failure to close the ring in the betaine-forming step as a result of the ring strain. It is likely, on the basis of an examination of Dreiding models, that had the betaine formed the oxygen and the triphenylphosphorus groups would have been *cis* to one another and thus able to complete the elimination of triphenylphosphine oxide. The same stereochemistry should prevail in the higher homologs and they did, in fact, afford olefins.

Schweizer and Schepers (*72*) have applied the intramolecular Wittig reaction to the synthesis of dihydrooxepine but the reaction was complicated by the formation, by an as yet unknown route, of 2-methyl-3,4-chromene. Schweizer later found that addition of the sodium salt of salicylaldehyde to vinyltriphenylphosphonium bromide led to an ylid intermediate which cyclized to 3,4-chromene in 62% yield [4.26] (*73*). The same principle has been used for the synthesis of 3*H*-pyrrolizine

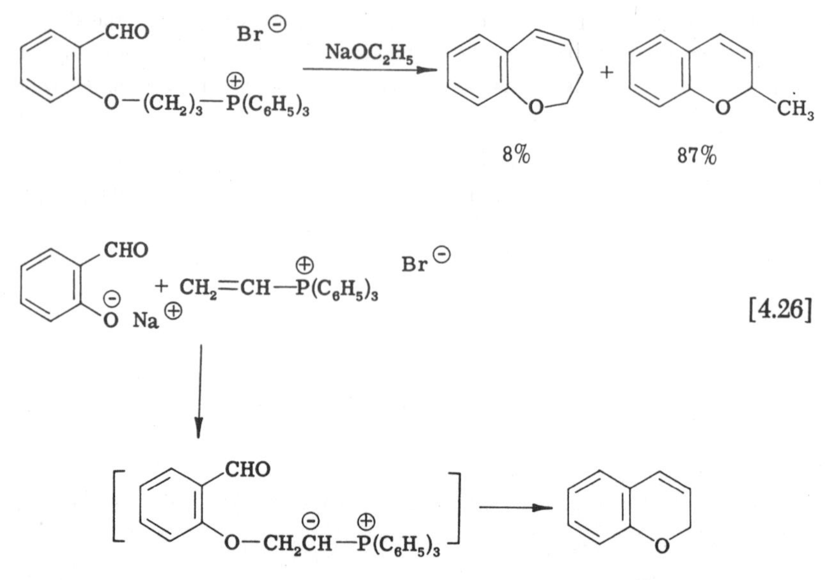

from 2-pyrrolaldehyde and vinyltriphenylphosphonium bromide [4.27]

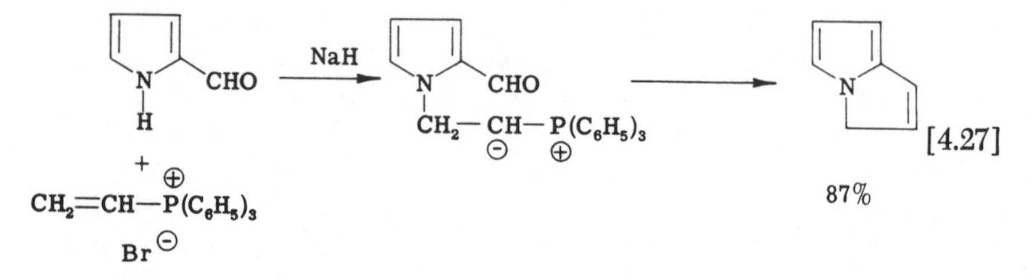

(*74*, Schweizer and Light). This reaction proceeded in 87% yield in spite of the fact that it involved the closure of a five-membered ring.

These are but a few of the more interesting examples of the application of the Wittig reaction to organic synthesis. It should be evident that the availability of this reaction has permitted or at least made much easier syntheses that otherwise would not have been attempted. Clearly this reaction has established itself as one of the standard methods of organic synthesis.

IV. Mechanism of the Wittig Reaction

In their first report of a successful reaction between a phosphonium ylid and a carbonyl compound, that between methylenetriphenylphosphorane and benzophenone, Wittig and Geissler (*11*) proposed the stepwise formation of a betaine and a four-membered intermediate to account for the products of the reaction, 1,1-diphenylethene and triphenylphosphine oxide [4.28]. This proposal was consistent with the

$$(C_6H_5)_2C{=}O + \overset{\ominus}{C}H_2{-}\overset{\oplus}{P}(C_6H_5)_3 \longrightarrow \left[(C_6H_5)_2\underset{|}{C}{-}O^{\ominus} \;\; CH_2{-}\overset{\oplus}{P}(C_6H_5)_3 \right] \longrightarrow \left[(C_6H_5)_2C{-}O \;\; CH_2{-}P(C_6H_5)_3 \right] \tag{4.28}$$

$$\longrightarrow (C_6H_5)_2C{=}CH_2 + (C_6H_5)_3PO$$

previous observation (Wittig and Rieber, *10*) that the unsuccessful reaction between methylenetrimethylphosphorane and benzophenone did afford a hydroxyphosphonium salt after treatment of the reaction mixture with mineral acid. This salt presumably was formed by protonation of a betaine intermediate. The mechanism proposed by Wittig has been accepted as a working hypothesis by most people in the field. To date its general implications have not been disproven although many refinements have been added.

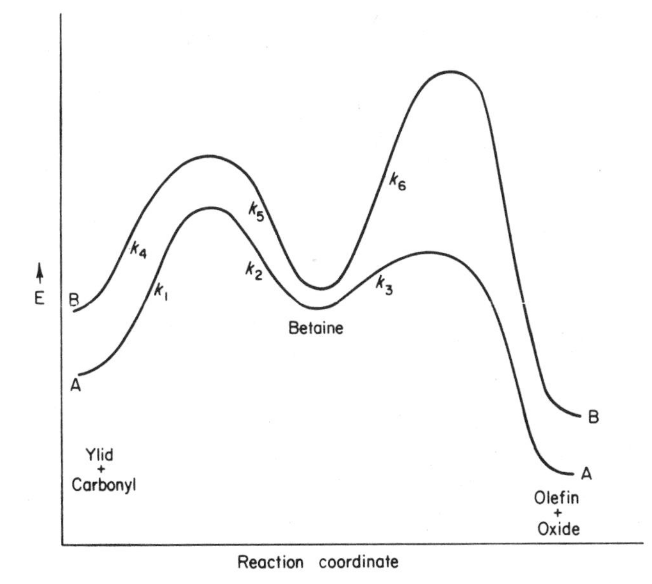

FIG. 4.1. Wittig reaction profile. A = Stabilized ylid route; B = non-stabilized ylid route.

The current status of the mechanism of the Wittig reaction perhaps is best summarized by the energy profiles shown in Figure 4.1 (*17, 35, 49, 50, 75*). As indicated by that profile and as inferred in previous discussion, phosphonium ylids have been broadly classified into "stabilized" and "non-stabilized" ylids. The line of demarcation between these two categories has not been sharply defined but a stabilized ylid usually is considered to be one which can be isolated and used in a subsequent experiment, therefore one of reduced basicity. Coincidentally it also has developed that there is a difference in the mechanistic path for the reaction of these two categories of ylids with carbonyl compounds. Until the realization of these inherent mechanistic differences crystallized two conflicting mechanistic proposals had been advanced (*35*, Wittig *et al.*; *17*, Johnson and LaCount). It can be forseen that further difficulties are bound to arise in the future with ylids of borderline "stabilization."

The approach in this section will be to present the mechanism for the Wittig reaction which currently is accepted as the most useful hypothesis and then to present the experimental data pertaining to the individual steps. Finally, alternate mechanistic proposals will be presented. It is hoped that this approach will permit a more coherent presentation than would a developmental or chronological approach.

On the basis of their studies using carbethoxymethylenetriphenyl- and tri-*n*-butylphosphorane in reactions with substituted benzaldehydes, Speziale and Bissing (*49*) proposed that stabilized ylids reacted with carbonyl compounds in a slow, reversible first step (betaine formation) which was followed by a fast decomposition of the betaine to olefin and phosphine oxide (i.e., profile A, Figure 4.1). The rate expression for this

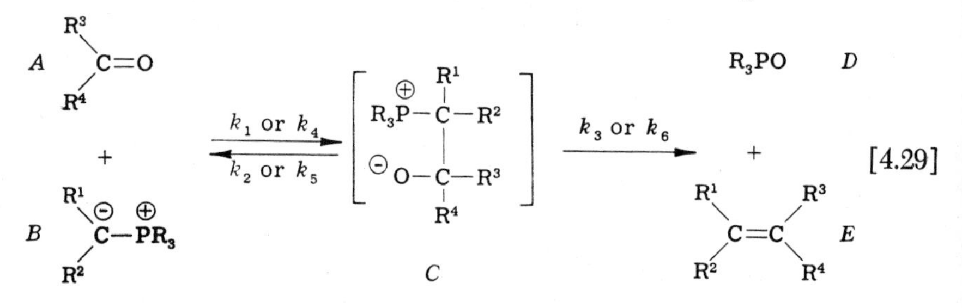

reaction using the steady-state approximation for the betaine intermediate, would be as in equation (1):

$$\frac{dE}{dt} = \frac{k_1k_3(A)(B)}{k_2 + k_3} \tag{1}$$

If their proposal were correct it would imply that $k_3 > k_2$ or that $k_2 = nk_3$ where n is a small constant. Therefore, $n = k_2/k_3$, and the authors assumed that changes in the structure of the ylid and carbonyl component affected k_2 and k_3 less than k_1. This supposition was borne out by their later data. Substituting in equation (1) would give equation (2) which

$$\frac{dE}{dt} = \frac{k_1(A)(B)}{n+1} \tag{2}$$

implies that the reaction of a stable ylid with carbonyls should essentially be a second order reaction with $k_3 > k_2 > k_1$. As k_3 increases relative to k_2, n becomes smaller and the expression approaches $dE/dt = k_1(A)(B)$.

Speziale and Bissing (*49*) found that the reaction between carbethoxymethylenetriphenylphosphorane (compound *B*, R = phenyl, R^1 = H, $R^2 = COOC_2H_5$) and substituted benzaldehydes (compound *A*, R^3 = H, R^4 = *p*-substituted phenyl) was second order, first order in each component. They established that the rate of ylid disappearance was the same as the rate of olefin formation, an observation consistent with equation (2) above. They also noted that solvent affected the rate of the reaction. The reaction was about six times faster in chloroform than in benzene and about one thousand times faster in methanol than in benzene. These observations, too, are consistent with the proposed formation of a polar betaine in the slow step of the reaction. The activation energies were relatively small, in the order of 9–11 kcal/mole but there were large negative entropies of activation, in the order of —40 eu at 25°. This is consistent with a highly oriented transition state for betaine formation, probably the result of the preferred pairing up of the ylid dipole and the carbonyl dipole.

At about the same time as the above work was reported, Fliszar *et al.* (*50*) provided some kinetic data on the reaction of various acylmethylenetriphenylphosphoranes (X) with benzaldehyde. They found that the order of reaction rates was R = *p*-methoxyphenyl > phenyl > *p*-nitrophenyl. This was the same as the order of basicity of these ylids and this system followed a Bronsted relationship. Thus it was indicated that with these stabilized ylids also the first step was the slow one.

$$\underset{\text{X}}{(C_6H_5)_3P{=}CH{-}\overset{O}{\overset{\|}{C}}{-}R} + C_6H_5CHO \longrightarrow C_6H_5CH{=}CH{-}\overset{O}{\overset{\|}{C}}{-}R + (C_6H_5)_3PO \tag{4.30}$$

The same group (*50*) studied the reaction of a non-stabilized phosphonium ylid, methylenetriphenylphosphorane, with benzaldehyde (*A*, R^3 = H, $R^4 = C_6H_5$; *B*, R = C_6H_5, $R^1 = R^2$ = H). Since a betaine had been isolated from this system (*12, 50*) it was apparent that $k_4 > k_6$.

With the betaine available, it was possible to obtain a measure of k_5 (by following the benzaldehyde formation) and also to obtain a value for K (i.e., $K = k_4/k_5 = 9 \times 10^2$ liter/mole). Thus, they were able to calculate $k_4 = 4.6 \times 10^{-3}$ liter/mole/sec at 40°. On the basis of the observation that heating the betaine to a low temperature resulted in the reversal of its formation but that heating to a much higher temperature was required before olefin and phosphine oxide were obtained, the authors concluded that $k_5 > k_6$. As a result, it appeared that for non-stabilized ylids the decomposition of the betaine to products was the slow step and $k_4 > k_5 > k_6$, just the reverse of the case for stabilized ylids proposed by Speziale and Bissing (*49*). It also might be mentioned that the quantitative studies by Speziale (*49*) and by Fliszar (*50*) are in agreement with the earlier qualitative proposals by Johnson (*17*) and Wittig (*35*) covering stabilized and non-stabilized ylids, respectively.

To summarize the current views of the mechanism of the Wittig reaction with reference to the terminology of the energy profiles in Fig. 4.1, for stabilized ylids (path *A*):

$$k_3 > k_2 > k_1$$

for non-stabilized ylids (path *B*):

$$k_4 > k_5 > k_6$$

To establish the relationship between the two different cases it is also likely that:

$$k_4 > k_1; \qquad k_2 > k_5; \qquad k_3 > k_6$$

The specific evidence pertaining to these conclusions will be discussed in the following subsections, step by step.

A. Effect of Carbonyl Structure

From all of the early work with methylenetriphenylphosphorane it was clear that benzophenone afforded a higher yield of olefin than did benzaldehyde in spite of the latter being more susceptible to nucleophilic attack (*35*, Wittig *et al.*). Wittig proposed that the structural difference between the two carbonyl compounds was being reflected in the last and slow step of the reaction wherein the two phenyl groups would stabilize the transition state (VI) between betaine and olefin

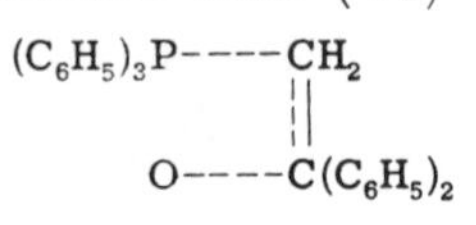

VI

more effectively than would one by conjugation with the incipient double bond. This proposal is consistent with the more recent kinetic studies of Fliszar *et al.* (*50*) which indicated that this last step was in fact the slow step of the reaction.

Before any kinetic data were available, Johnson (*17, 30*) had shown that fluorenylidenetriphenylphosphorane would not react with ketones but would react with benzaldehydes to form benzalfluorene. The reactivity of a series of *para*-substituted benzaldehydes decreased in the order $NO_2 > Cl > H > OCH_3 > N(CH_3)_2$. This is the order expected when the aldehyde is undergoing nucleophilic attack on the carbonyl carbon, and this led Johnson to propose that the first step of the Wittig reaction, the betaine formation, was the slow step. Later, Goetz *et al.* (*48*) reported a similar reactivity order for benzaldehydes reacting with phenacylidenetriphenylphosphorane. In addition, Speziale and Ratts (*76*) noted that dichloromethylenetriphenylphosphorane, in the presence of equimolar quantities of *p*-nitrobenzaldehyde and *p*-dimethylaminobenzaldehyde, afforded only 1-(*p*-nitrophenyl)-2,2-dichloroethene.

The recent kinetic data of Speziale and Bissing (*49*), obtained from the reaction of carbethoxymethylenetriphenylphosphorane with substituted benzaldehydes and plotted against the substituent σ values, gave a ρ value of $+2.7$. This value is comparable to those from other reactions known to involve nucleophilic addition to a carbonyl group (*77*, Jaffe). These results also are consistent with equation (2) which contains only a single rate constant, that for betaine formation (k_1). The ρ value is inconsistent with an alternate proposal that with stabilized ylids the first step is rapid and reversible and the second step, betaine decomposition, is slow. In such a case the ρ value would be expected to be small and variable. The acid catalysis of the reaction between the same ester ylid and acetone leading to an increase in olefin yield from 6 to 80% also is consistent with the proposal that betaine formation is the slow step with stabilized ylids. The acid would be expected to protonate the carbonyl oxygen, thereby making the carbonyl carbon more susceptible to nucleophilic attack by the ylid carbanion (*18*, Ruchardt *et al.*)

Thus, with stabilized ylids the structure of the carbonyl compound is reflected in the rate of the first step of the Wittig reaction (betaine formation) whereas with non-stabilized ylids it is reflected in the last step (betaine decomposition).

B. Effect of Ylid Structure

As pointed out in Section II of this chapter, some ylids will undergo reaction with all types of aldehydes and ketones, others will react only with aldehydes and still others will not react with either. This usually is a reflection of the nucleophilicity of the particular ylid. As pointed out in Chapter 3, the nucleophilicity of triphenylphosphoranyl ylids is directly related to the ability of the carbanion substituents to delocalize the negative charge. It would be expected, therefore, that the reactivity of a given ylid in the Wittig reaction should be related to the basicity of that ylid. Qualitatively this obviously is the case when it is noted that methylenetriphenylphosphorane will react with most carbonyls (*12*), fluorenylidenetriphenylphosphorane will react only with aldehydes (*30*) and cyclopentadienylidenetriphenylphosphorane will not react with either aldehydes or ketones (*29*).

Fliszar *et al.* (*50*) obtained a linear plot of ylid basicity versus rate of a Wittig reaction with benzaldehyde for a series of acylmethylenetriphenylphosphoranes. These ylids would be classed as "stabilized" and would be expected to follow equation (2), the rate of the overall reaction depending on the rate of betaine formation. This result bore out the earlier qualitative observations and subsequent predictions by Johnson (*17*) that the Bronsted relationship should hold for the Wittig reaction of stabilized ylids, the most basic ylid also being the most reactive.

Considerably less information is available regarding the effect of the structure of the phosphorus group on the course and mechanism of the Wittig reaction. Wittig and Rieber (*10*) had observed that methylene*trimethyl*phosphorane would not afford olefin when reacted with benzophenone. They were, however, able to isolate the conjugate acid (XI) of a betaine after treatment of the solution with mineral acid and potas-

$$(CH_3)_3\overset{+}{P}-\overset{-}{C}H_2 + (C_6H_5)_2CO \longrightarrow \left[(CH_3)_3\overset{\oplus}{P}-CH_2-C(C_6H_5)_2-O^{\ominus}\right] \xrightarrow{H^+,\ KI} (CH_3)_3\overset{\oplus}{P}-CH_2-C(C_6H_5)_2(OH)\ I^{\ominus}\ \ \text{XI} \qquad [4.31]$$

$$\left[(CH_3)_3\overset{\oplus}{P}-CH_2-C(C_6H_5)_2-O^{\ominus}\right] \not\longrightarrow (C_6H_5)_2C=CH_2$$

$$(C_6H_5)_3\overset{\oplus}{P}-\overset{\ominus}{C}H_2 + (C_6H_5)_2CO \longrightarrow \left[(C_6H_5)_3\overset{\oplus}{P}-CH_2-C(C_6H_5)_2-O^{\ominus}\right] \longrightarrow (C_6H_5)_2C=CH_2$$

sium iodide. They attributed the failure of this reaction to the phosphorus atom carrying too high an electron density to accept the transfer of an oxyanion. In other words, the betaine would not decompose to olefin and phosphine oxide. The three methyl substituents on phosphorus were thought to increase the electron density on phosphorus via an inductive effect. On the other hand, methylene*triphenyl*phosphorane would react with benzophenone to afford the expected olefin. Trippett and Walker (78) later found that methylenetrimethylphosphorane and benzophenone could be forced to produce 1,1-diphenylethene but the reaction still was much slower than for the triphenylphosphoranyl case.

More recently, Wittig *et al.* (35) have shown that methylenephosphoranes carrying other electron-donating groups on phosphorus, groups such as tri(*p*-methoxyphenyl), tripiperidyl and trimorpholino, all would undergo the initial step of the Wittig reaction to form a betaine but there was considerable difficulty in effecting its decomposition to olefin and phosphine oxide. In all this work Wittig used methylenephosphoranes which fall into the category of "non-stabilized" ylids. These observations are consistent with the energy profile B (Figure 4.1) in which the betaine decomposition is represented as the step with the largest activation energy. Changes in the structure of the ylid which should affect k_6 should be reflected directly in the overall rate of the reaction. Therefore, substitution on phosphorus by electron-donating groups would be expected to increase the electron density on phosphorus making subsequent nucleophilic attack by the oxyanion of the betaine more difficult and permitting build-up and subsequent isolation of betaine from the reaction.

The effect of phosphorus substitution on the reaction of stabilized ylids with carbonyls forms an interesting and very significant contrast to the above discussion. Examination of energy profile A (Figure 4.1) and equation (2) indicates that k_1 is the only rate constant that enters into the simple overall rate expression for the reaction of stabilized ylids with carbonyl compounds. From the above discussion it is clear that replacement of the phenyl groups on phosphorus with alkyl groups led to a decrease in k_6 (betaine decomposition), and it is reasonable to assume that with stabilized ylids k_3 (betaine decomposition) also ought to be lowered upon similar substitution. However, k_3 does not appear in the simple rate expression [equation (2)], and changes in k_3 ought not to be reflected in the overall rate, at least until it becomes very small. The only reflection of a change in the phosphorus substituent should be in the value of k_1 (betaine formation).

From previous discussion in Chapter 3 it was apparent that changes in the phosphorus substituents affected the nucleophilicity of an ylid.

Because of their electron-donating characteristics alkyl groups increased the nucleophilicity of the phosphonium ylids as reflected by an increase in their basicity. Johnson and LaCount (*17*) showed that fluorenylidene-tri-*n*-butylphosphorane was more basic than fluorenylidenetriphenyl-phosphorane and, as expected from the above arguments, the former also was found to be more reactive in the Wittig reaction. More recently, Speziale and Bissing (*49*) showed that for reaction with *p*-anisaldehyde in benzene at 25° carbethoxymethylene*triphenyl*phosphorane had a rate constant of 7.1×10^{-5} whereas carbethoxymethylene*tri-n-butyl*phosphor-ane had a rate constant of 5.6×10^{-3}, a difference of nearly two powers of ten. Therefore, it is apparent that while k_3 certainly must be smaller for a stabilized trialkylphosphonium ylid than for a triphenylphosphonium ylid, it still is not small enough to become rate controlling, and the structural change mainly is reflected in the betaine formation step (k_1).

In summary then, changes in the structure of stabilized ylids have been reflected in k_1, the betaine formation step which depends on the nucleophilicity of the ylid and the electrophilicity of the carbonyl group. In contrast, changes in the structure of non-stabilized ylids have been reflected in k_6, the betaine decomposition step which depends on the ease of oxyanion transfer to phosphorus and on the stability of the ole-finic product. As yet there have been no cases reported for non-stabilized ylids where structural changes have been sufficient to either decrease k_4 or increase k_6 to the point where the betaine-forming step becomes rate controlling. Likewise, there have been no cases reported for stabi-lized ylids where k_1 has been increased or k_3 decreased to the point where betaine formation no longer is the rate-controlling step of the Wittig reaction. To effect the latter situation one would have to construct an ylid which very rapidly would form a betaine, i.e., be strongly nucleo-philic, but in which the phosphorus atom only reluctantly would undergo attack by the oxyanion.

C. Betaine Formation

There is considerable evidence available to support the proposal, originally advanced by Wittig (*10*), that betaines are intermediate in the Wittig reaction. Reaction of methylenetriphenylphosphorane with benz-aldehyde led to the formation of a betaine (XII) which was isolated and identified as its hydrobromide (*12*, Wittig and Schollkopf). Treat-ment of the betaine with phenyllithium at a temperature higher than that used in its formation led to the completion of the Wittig reaction and the formation of styrene and triphenylphosphine oxide. In subse-quent experiments with other methylenephosphoranes carrying electron-donating groups on phosphorus, Wittig *et al.* (*35*) characterized the corresponding betaines as their conjugate acids.

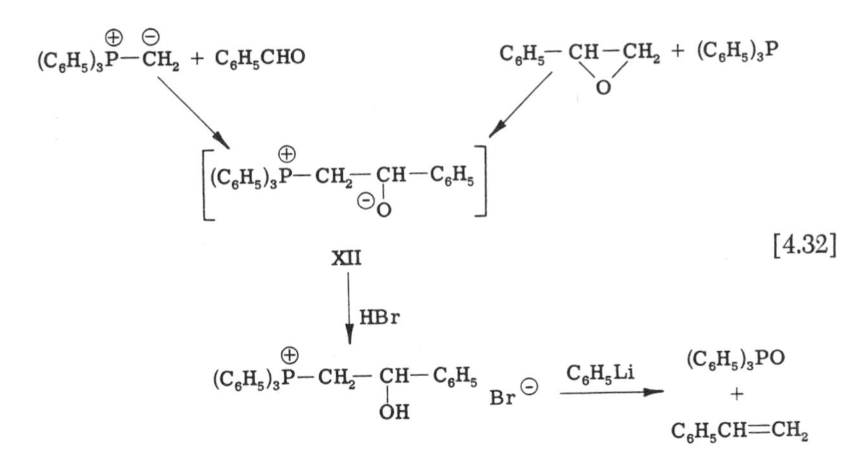

Additional support for structure XII for the betaine was provided by the observation that heating a mixture of triphenylphosphine and styrene oxide to 165° afforded a 50% yield of styrene and an 86% yield of triphenylphosphine oxide (*13,* Wittig and Haag). They proposed that the phosphine attacked styrene oxide on the primary carbon to afford XII. Subsequent examples of this reaction have been reported by Denney and Boskin (*79*) and Speziale and Bissing (*49*). More recently Trippett (*80*)

$$\underset{\textbf{XIII}}{(C_6H_5)_2\overset{\oplus}{P}(CH_3)-CH(C_6H_5)-CH(OH)(C_6H_5)}\ \ I^{\ominus} \xrightarrow{NaOC_2H_5} C_6H_5CH{=}CHC_6H_5 + \left[(C_6H_5)_2P(CH_3){=}O\right] \qquad [4.33]$$

has shown that the hydroxyphosphonium salt (XIII), prepared via an alternate route, would afford a mixture of the stereoisomeric stilbenes upon treatment with ethanolic sodium ethoxide, presumably via a betaine intermediate.

Wittig and Haag (*38*) have reported the isolation and characterization of a betaine (IV) obtained from reaction of isopropylidenetriphenylphosphorane and diphenylketene. It melted at 139–140°, afforded satisfactory microanalytical data and appeared to be monomeric in benzene solution. This betaine, which is an enolate ion, was alkylated on carbon with methyl iodide and also would form a conjugate acid with hydrogen bromide, both products showing typical ketonic absorption in the infrared at 1700 cm^{-1} [4.34]. The infrared absorption of the betaine was not reported. Heating the betaine to 140° resulted in the completion of the Wittig reaction (decomposition of the betaine) and the formation of triphenylphosphine oxide and 1,1-dimethyl-3,3-diphenyl-1,2-propadiene.

The isolation of the betaine (IV) has provided the first opportunity

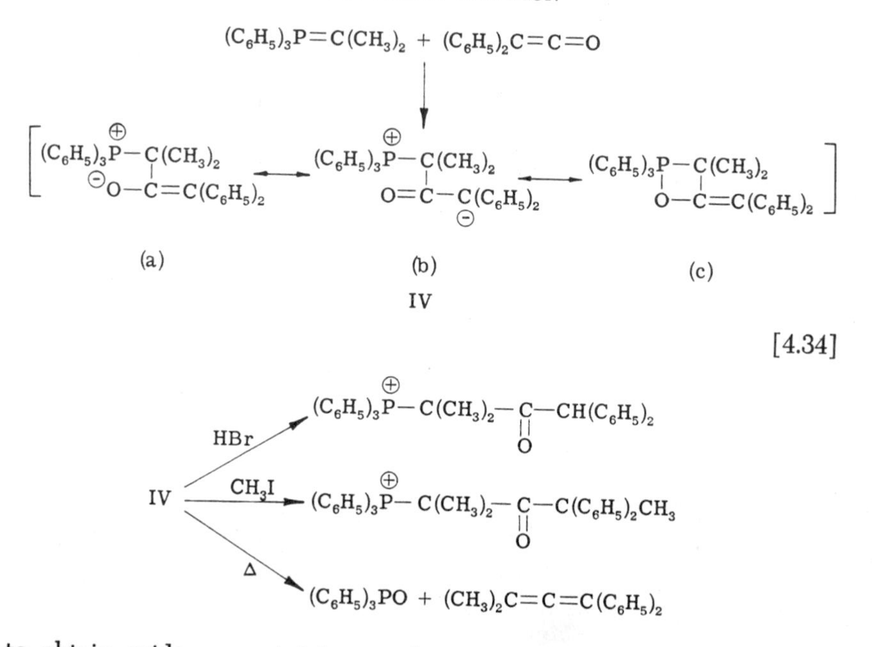

to obtain evidence pertaining to the structure of betaines—mainly to the question of the open-chain structure (IVa and IVb) versus the cyclic structure (IVc). This betaine had a dipole moment of 4.34 D, not very different from that of triphenylphosphine oxide. This value is somewhat low for that expected for a betaine in which there is no interaction between the oxygen and the phosphorus atoms. The P^{31} NMR chemical shift was +36 ppm (85% H_3PO_4 reference), too high for a phosphonium salt (usually about —20 ppm) and somewhat lower than expected for a pentavalent phosphorus compound. Denney and Relles (*81*) have just reported pentaethoxyphosphorane to show a chemical shift of +71 ppm, but Ramirez *et al.* (*82*) claim that a tetraalkoxyalkylphosphorane showed a shift of only +27.9 ppm. Thus, it might tentatively be concluded that the betaine structure involves considerable association between the oxyanion and the phosphonium group, structure IVc contributing appreciably. The fact that the betaine was stable enough to be isolated, however, indicates that the phosphorus atom could not be fully pentavalent.

In every reaction from which a betaine has been isolated the ylid used has been a non-stabilized ylid, usually a methylenetriphenylphosphorane or a C-alkyl derivative thereof. To date no betaines have been isolated from the reaction of a stabilized ylid with a carbonyl compound. Johnson and LaCount (*17*) reported the only serious attempt to do so when they reacted fluorenylidenetri-*n*-butylphosphorane with three carbonyl compounds of differing electrophilicity, acetone, *m*-nitroacetophenone and benzaldehyde. They intercepted the reactions after short periods of time by bubbling gaseous hydrogen bromide through the system hoping to trap any betaine as a conjugate acid. With acetone, only the hydrobromide of the unreacted ylid could be recovered while with benzaldehyde the product of a Wittig reaction, benzalfluorene, was isolated. With *m*-nitroacetophenone 14% of the olefinic product and 76% of the ylid conjugate acid were recovered. In no instance was there any evidence for the existence of a hydroxyphosphonium bromide. These results are to be expected if the mechanism of the Wittig reaction is as implied in the energy profiles in Figure 4.1. With stabilized ylids the decomposition of the betaine should be faster than its formation so there should be no build-up of betaine in solution. On the other hand, with non-stabilized ylids the rate of decomposition of the betaine should be slower than its rate of formation so there should be a finite concentration of betaine in the reaction medium, thereby permitting its entrapment.

The rate of betaine formation (k_1 and k_4 in Figure 4.1) should be affected by the nucleophilicity of the ylid and the electrophilicity of the carbonyl compound. In a reaction using a stabilized ylid, k_1 is the only rate constant of major import [equation (2)] so alterations either in the ylid structure or the carbonyl structure ought to be reflected in the overall, immediately observable rate of the reaction. In accord, it was observed that fluorenylidenetriphenylphosphorane would react with *p*-nitrobenzaldehyde but not with *p*-dimethylaminobenzaldehyde. For analogous reasons, the effect on k_1, *m*-nitroacetophenone would react with fluorenylidenetri-*n*-butylphosphorane but not with the triphenyl analog (*17*).

On the other hand in a reaction using non-stabilized ylids, k_4, the rate of formation of the betaine, should not directly be observable in the overall rate of the reaction for the following reasons. If one uses the energy profile B (Figure 4.1) and the rate constants indicated thereon as a working hypothesis, the general expression for the Wittig reaction with non-stabilized ylids is as shown in equation (3) [analogous to equation (1) for the stabilized ylids and assuming that the steady-state treatment still is valid (*80*)]:

$$A + B \underset{k_5}{\overset{k_4}{\rightleftharpoons}} C \overset{k_6}{\rightarrow} D + E$$

$$\frac{dE}{dt} = \frac{k_4 k_6 (A)(B)}{k_5 + k_6} \tag{3}$$

From the profile B and the previous discussion it appears that $k_4 > k_5 > k_6$ and equation (3) can be simplified to equation (4) ($K = k_4/k_5$):

$$\frac{dE}{dt} = \frac{k_4 k_6}{k_5}(A)(B)$$

$$= Kk_6(A)(B) \quad (4)$$

Thus, k_4 is involved in the overall rate expression only as a ratio playing a part in determining the value of the equilibrium constant, K, between ylid and betaine. As a result, the effect of changes in the nucleophilicity of a non-stabilized ylid and the electrophilicity of a carbonyl compound should not directly be reflected in the overall rate of the resulting Wittig reaction. The effect of structural changes on k_4 should be masked by the more important changes in k_6. These predictions are in accord with the facts.

The question of the reversibility of betaine formation is important in its own right for an adequate description of the mechanism of this reaction but it also bears directly on the stereochemistry of the reaction. Wittig and Schollkopf (*12*) reported that the conjugate acid of the betaine (XII) afforded an odor of benzaldehyde upon treatment with aqueous sodium hydroxide, perhaps indicating some reversal of the betaine to aldehyde and ylid. However, in a later paper (*35*) Wittig *et al.* reported that heating the free betaine (XII) with benzophenone led to the recovery of nearly all of the benzophenone and the formation only of "traces" of diphenylethylene. If the betaine had reverted to carbonyl and ylid the latter should have been trapped by reaction with benzophenone to form a new betaine which would have decomposed to olefin and phosphine oxide faster than would the original betaine [4.35]. Wittig concluded that betaine forma-

$$(C_6H_5)_3\overset{\oplus}{P}-CH_2-\underset{O^{\ominus}}{CH}C_6H_5 \underset{k_5}{\overset{k_4}{\rightleftharpoons}} (C_6H_5)_3\overset{\oplus}{P}-\overset{\ominus}{C}H_2 + C_6H_5CHO$$

$$\xrightarrow{(C_6H_5)_2CO} \left[(C_6H_5)_3\overset{\oplus}{P}-CH_2-\underset{^{\ominus}O}{C}(C_6H_5)_2\right] \longrightarrow (C_6H_5)_2C{=}CH_2 + (C_6H_5)_3PO \quad [4.35]$$

tion was irreversible. More recently, however, Fliszar *et al.* (*50*) have followed the reversal of the betaine (XII) by monitoring the benzaldehyde concentration by infrared spectroscopy. In fact, at 40° in chloroform solution the equilibrium constant K ($= k_4/k_5$) was found to be 9×10^2 liter/mole. Trippett (*80*) recently has reported generating a betaine (XIV) by treatment of (2-phenyl-2-hydroxy)ethyldiphenylmethylphosphonium iodide with either sodium ethoxide in ethanol or aqueous sodium hydroxide. In both instances only dimethylphenylphosphine oxide

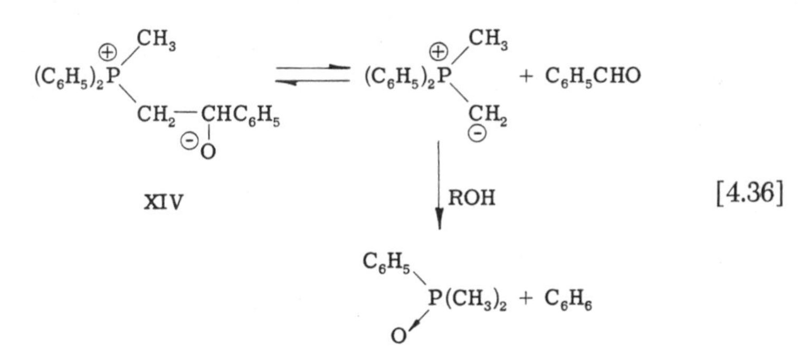

and benzaldehyde could be isolated, perhaps indicating that the betaine reverted to ylid and aldehyde, the former then undergoing hydrolysis.

The situation on betaine reversibility is much less ambiguous with betaines derived from stabilized ylids. Wittig and Haag (*13*) had shown that ethyl phenylglycidate was converted into ethyl cinnamate by heating with triphenylphosphine. They proposed that the reaction proceeded by attack of the phosphine at the α-carbon to form a betaine which then underwent normal oxyanion transfer to phosphorus as in a Wittig reaction. Speziale and Bissing (*49*) demonstrated that when the same reaction was carried out in the presence of *m*-chlorobenzaldehyde a mixture of olefins containing ethyl cinnamate and ethyl *m*-chlorocinnamate was obtained

$$(C_6H_5)_3PO + C_6H_5CH{=}CHCOOC_2H_5$$

$$\uparrow k_3$$

$$(C_6H_5)_3P + C_6H_5CH\underset{O}{-}CH{-}COOC_2H_5 \longrightarrow \left[\begin{array}{c} (C_6H_5)_3\overset{\oplus}{P}{-}CH{-}COOC_2H_5 \\ \ominus O{-}CH{-}C_6H_5 \end{array}\right] \quad [4.37]$$

$$k_2 \downarrow\uparrow k_1$$

$$(C_6H_5)_3P{=}CHCOOC_2H_5 + C_6H_5CHO$$

$$\downarrow m\text{-}Cl{-}C_6H_4CHO$$

$$m\text{-}Cl{-}C_6H_4CH{=}CHCOOC_2H_5 + (C_6H_5)_3PO$$

[4.37]. The incorporation of *m*-chlorobenzaldehyde into the olefinic product conclusively demonstrated that ylid must have been formed in the system and that betaine therefore must have been in equilibrium with ylid and aldehyde. Incidentally, they found k_3 to be larger than k_2.

In conclusion, it is clear that betaine formation from stabilized ylids and carbonyl compounds is reversible. The same conclusion appears to be valid for the non-stabilized case. Several groups, House (*42*), Bestmann (*75*) and Ketcham (*83*), much earlier had proposed that betaine formation from stabilized ylids was reversible in order to account for the stereochemistry of the olefins formed in the Wittig reaction. These proposals will be discussed in a later section.

D. Betaine Decomposition

The decomposition of the betaine to olefin and phosphine oxide must occur by attack of the oxyanion on the phosphonium atom, and the transition state between the betaine and products must have some olefinic character to it. The driving force for the decomposition most likely is the formation of the phosphine oxide, a highly stable substance. The "oxygen affinity" of tertiary phosphines is well recognized and indicated by the variety of deoxygenations effected by them. The phosphorus-oxygen bond is a particularly strong bond, having a dissociation energy of 130–140 kcal/mole (*84*, Hartley *et al.*).

The most easily obtained information on the betaine decomposition step should be from those betaines derived from non-stabilized ylids. Comparison of equations (2) and (4), the overall expressions for the rate of Wittig reactions with stabilized and non-stabilized ylids, respectively, reveals that only in the latter case does the rate of the decomposition reaction play an important role in the overall rate. With stabilized ylids the decomposition rate is submerged in a ratio. Apart from any stereochemical considerations, which will be discussed in the next section, there ought to be at least two electronic factors which affect the conversion of betaine into olefin and phosphine oxide. These are the stabilization of the incipient double bond in the transition state (XV) and the rate at which the oxyanion attacks the phosphonium atom.

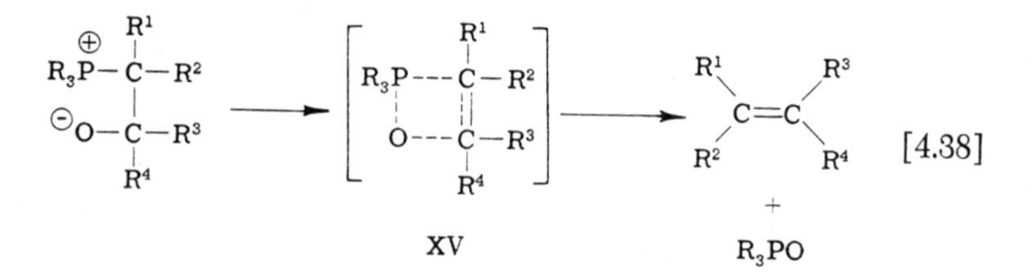

[4.38]

In the transition state (XV) the groups R^1, R^2, R^3 and R^4 all could provide stabilization for the forming double bond. However, it virtually is impossible to study the effect of the groups R^1 and R^2 since if they were chosen so as to stabilize the double bond they also would have provided delocalization for the original ylid and rendered any results invalid by making it a "stabilized" ylid. This may not be the case if these two were electron-donating groups. However, the effect of the structure of the groups R^3 and R^4 on the betaine decomposition has been studied. Wittig *et al.* (*35*) noted that in a comparison of the reaction between methylenetriphenylphosphorane and benzaldehyde ($R^1 = R^2 = R^3 = H$, $R^4 =$ phenyl) or benzophenone ($R^1 = R^2 = H$, $R^3 = R^4 =$ phenyl), the former invariably afforded a low yield of olefin along with some betaine whereas the latter afforded a high yield of olefin and no betaine. The implication from these observations is that the two phenyl groups in the latter case provided more stabilization for the olefinic double bond than did the one phenyl group of the former. In other words, k_6 was larger in the benzophenone case than it was in the benzaldehyde case. Since it is possible to isolate the betaines, the synthesis of a series of betaines and a study of their kinetics of decomposition seem most desirable.

There is some evidence available regarding the second factor mentioned above. In the reaction of two ylids, differing only in the phosphorus substituents, with a common carbonyl compound the rate of the betaine decomposition ought to depend on the ease with which the oxyanion carries out a nucleophilic attack on the phosphonium atom. Again restricting the discussion to the case of non-stabilized ylids in which the rate of decomposition (k_6) would be reflected in the overall rate of the reaction, it is apparent that phosphorus substituents which increase the electron density on the phosphorus atom hinder the decomposition of the betaine. For example, Wittig and Rieber (*10*) showed that methylene*trimethyl*phosphorane and benzophenone afforded only the betaine whereas methylene*triphenyl*phosphorane and benzopheone afforded only the olefin with no evidence for a prolonged existence of the corresponding betaine (*11,* Wittig and Geissler). The alkyl groups are known to be electron-donating and expected to hinder the decomposition and the phenyl groups appear to be electron-withdrawing (*17*), aiding the decomposition by making the phosphorus atom more susceptible to nucleophilic attack.

In an unrelated but pertinent experiment, McEwen and Wolf (*85*) found that reaction of styrene oxide with benzylidene*methylethylphenyl*phosphorane afforded 1-benzoyl-2-phenylethane (path *a*) but reaction of the same oxide with benzylidene*triphenyl*phosphorane afforded 1,2-diphenylcyclopropane as the major product (path *b*) [4.39]. The difference in the course of these two reactions can be accounted for by the reluctance of the dialkylarylphosphorus group to undergo oxyanion attack in the first case and by the willingness of the triarylphosphorus group to undergo oxyanion attack in the second case.

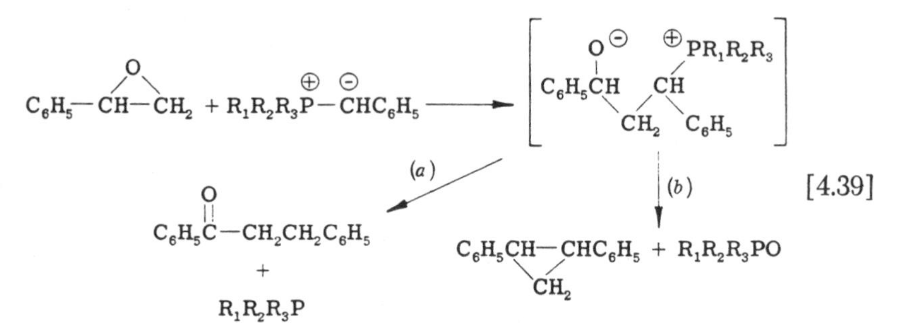

The high temperature required to effect the elimination of triphenylphosphine oxide from the betaine (IV) formed by isopropylidenetriphenylphosphorane and diphenylketene (*38*, Wittig and Haag) may be attributed to the lowered nucleophilicity of the oxyanion as a result of its enolate structure. This lowered nucleophilicity on oxygen was clearly indicated by the reactions of the betaine occurring on carbon (see p. 162).

As mentioned in the preceding section, all changes in the structure of a stabilized ylid or carbonyl component in a Wittig reaction seem to be reflected in the initial step of the reaction. Therefore, the effect of these structural variations on the rate of decomposition of betaine (k_3) have not been directly observable since this is the fastest step of the reaction.

E. Alternate Mechanisms for the Wittig Reaction

Only one other mechanism has been proposed for the Wittig reaction. Bergelson and Shemyakin (*86*) have proposed that in certain cases, especially those in which the ylid carbanion carries stabilizing groups, the electron density on the ylid carbanion is too low to permit nucleophilic attack on the carbonyl carbon of an aldehyde or ketone and that the initial step in the Wittig reaction is the attack of the nucleophilic carbonyl oxygen on the phosphonium atom of the ylid. This would result in

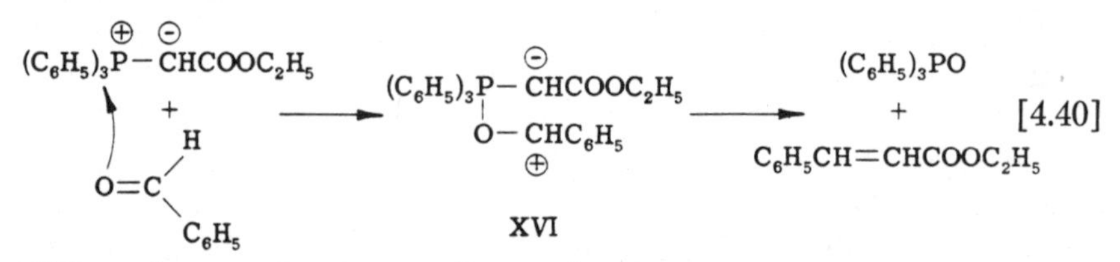

XVI as the reaction intermediate. The Russian authors agreed that for non-stabilized ylids the mechanism as proposed by Wittig probably operates.

Several arguments might be raised against the above proposal but the most pertinent argument was provided by the kinetic data of Speziale and Bissing (*49*) on the very reaction shown above. If the Bergelson proposal were correct, the rate of the reaction ought to be increased when an aldehyde carrying a more nucleophilic oxygen was used, regardless of whether the first or the second step were rate controlling. According to Johnson's qualitative data (*17*) and Speziale's quantitative data (*49*), the carbonyl compounds with the more nucleophilic oxygen are the least reactive. In other words, *p*-nitrobenzaldehyde is much more reactive than *p*-dimethylaminobenzaldehyde or *p*-anisaldehyde. Yates and Stewart (*87*) have shown that *p*-nitrobenzaldehyde is less basic than *p*-anisaldehyde. In addition, the linear Hammett plot with a ρ value of $+2.7$ (*49*) is convincing evidence that the carbonyl group of the benzaldehydes must be undergoing nucleophilic attack on carbon.

Bergelson and Shemyakin (*88*) recently have claimed there is no experimental proof for the oxyanion-phosphonium structure of betaines. However, examination of the data regarding the betaine (IV) obtained from isopropylidenetriphenylphosphorane and diphenylketene (p. 162) can leave little doubt that IV is a far better representation of the characteristics of the substance than would be a structure corresponding to XVI.

There are several variations of the original Wittig proposal that need to be considered for the mechanism of the Wittig reaction. These variations [4.41], however, involve the possibility of different ratios of the

$$\text{Ylid} + \text{Aldehyde} \underset{k_2 \text{ or } k_5}{\overset{k_1 \text{ or } k_4}{\rightleftharpoons}} \text{Betaine} \xrightarrow{k_3 \text{ or } k_6} \text{Olefin} + \text{Oxide} \qquad [4.41]$$

various rate constants than those ratios implied by the energy profiles in Figure 4.1. Four variations are possible:

(a) Rapid and irreversible betaine formation, slow decomposition
(b) Slow and irreversible betaine formation, rapid decomposition
(c) Rapid and reversible betaine formation, slow decomposition
(d) Slow and reversible betaine formation, rapid decomposition

For stabilized ylids mechanisms *a* and *b* clearly are impossible since Speziale and Bissing (*49*) demonstrated the reversibility of the betaine formation. Mechanism *c* also is ruled out by the observation that the more nucleophilic trialkylphosphonium ylids increased the overall rate of the reaction (*17, 49*), rather than decreasing it by slowing the betaine decomposition as mechanism *c* would have predicted. The rate expression for the reaction of stabilized ylids with carbonyl compounds consistent with mechanism d is in equation (2). The rate of the first step of the reaction, the formation of the betaine (k_1), is the dominant feature of that

$$\frac{d(\text{olefin})}{dt} = \frac{k_1(\text{ylid})(\text{aldehyde})}{n+1} \tag{2}$$

$$n = k_2/k_3$$

expression, the rates of the second step and the reversal of the first step appearing only as a ratio in the denominator. In the cases studied to date the variation in n has been small. In any given stereochemical series the change from triphenylphosphonium ylid to tri-n-butylphosphonium ylid has led to nearly a hundredfold increase in the overall rate of the reaction but led to a one-half- or twofold increase in the value of n [from 0.50 to 0.87 in one case and from 0.15 to 0.082 in another case (*49*, Speziale and Bissing)]. Therefore, profile A of Figure 4.1 appears to be followed, structural changes are reflected mainly in k_1 and the rate constants decrease in the order $k_3 > k_2 > k_1$.

The case is a little different for the reaction of non-stabilized ylids with carbonyl compounds. From the previous discussion in Section IV,D it is clear that betaine decomposition is the slow step in the reaction; thus mechanisms b and d are eliminated. On the basis of the work of Fliszar *et al.* (*50*) it must be concluded that the reaction is reversible so mechanism a also must be eliminated. Thus, the mechanism of the reaction of non-stabilized ylids with carbonyl compounds is best represented by mechanism c, rapid and reversible betaine formation and slow betaine decomposition to olefin and phosphine oxide. The rate expression for this situation is that of equation (4) in which the dominant feature is the rate of betaine decomposition (k_6), the other rates entering in only as a ratio, the equilibrium constant. The only indication of the size of that

$$\frac{d(\text{olefin})}{dt} = Kk_6(\text{ylid})(\text{aldehyde}) \tag{4}$$

$$K = k_4/k_5$$

ratio is from the reaction of methylenetriphenylphosphorane with benzaldehyde (*50*, Fliszar *et al.*) which gave a value of 9×10^2 liter/mole. The variation in K with changes in structure of the ylid and the carbonyl component has not been explored. It appears, however, that profile B of Figure 4.1 is followed, structural changes are reflected mainly in k_6 and the rate constants decrease in the order $k_4 > k_5 > k_6$.

In summary, the factors which favor betaine formation (nucleophilic ylid and electrophilic carbonyl compound) often retard betaine decomposition and to date these factors have been relatively clear cut. A most interesting case will evolve when the appropriate substances are used to carry out a Wittig reaction which mechanistically bridges these two rather well-defined mechanistic pathways.

V. Stereochemistry of the Wittig Reaction

The Wittig reaction of phosphonium ylids with aldehydes or ketones can give rise to *cis*- or *trans*-olefins. The first observation of this possibility was the formation of an 82% yield of stilbenes, consisting of 20% *cis* isomer and 62% *trans* isomer (i.e. a 30:70 ratio), from the reaction of benzylidenetriphenylphosphorane and benzaldehyde [4.42] (*12,* Wittig and Schollkopf). In this instance the ylid had been generated from its

$$(C_6H_5)_3P{=}CHC_6H_5 + C_6H_5CHO \longrightarrow \overset{C_6H_5}{\underset{H}{}}C{=}C\overset{H}{\underset{C_6H_5}{}} + \overset{C_6H_5}{\underset{H}{}}C{=}C\overset{C_6H_5}{\underset{H}{}} \qquad [4.42]$$

hydrochloride with phenyllithium. In their next publication Wittig and Haag (*13*) reported the formation of the same stilbenes, but in a *cis-trans* ratio of 47:53, when the ylid was generated in alcoholic solution with sodium ethoxide. Reaction of carbethoxymethylenetriphenylphosphorane with benzaldehyde was reported to afford a 77% yield of ethyl *trans*-cinnamate with no mention made of the *cis* isomer.

Most of the Wittig reactions reported to date have led to the formation of the *trans*-olefin as the dominant product (see review articles, references *1–6*). However, the selectivity of the reaction has varied widely. Nonetheless, in the early days of the Wittig reaction the dominant formation of the *trans* isomer was put to synthetic use. For example, Trippett (*55*) was able to prepare what appeared to be pure all-*trans*-squalene from the reaction of the bis-ylid prepared from 1,4-bis(triphenylphosphonio)butane dibromide with two equivalents of geranyl acetone. In a similar fashion Surmatis and Ofner (*56*) were able to prepare pure all-*trans*-β-carotene from a bis-ylid and two equivalents of β-ionylideneacetaldehyde.

At this point a note of caution should be sounded in the use of raw product analysis data as a measure of the stereochemistry of the Wittig reaction. In many cases *cis*-olefins are isomerized to the *trans* isomers by purely thermal means. In addition, however, the reactants or products of a Wittig reaction may catalyze the isomerization. For example, Blomquist and Hruby (*53*) found that *cis*-*p*-nitrobenzylidenebenzocyclobutene was 67% isomerized to the *trans* isomer after refluxing for eleven hours in dimethylformamide with triphenylphosphine oxide, and Speziale and Bissing (*49*) found that ethyl *cis*-cinnamate was isomerized to the *trans* isomer in the presence of tertiary phosphines.

The stereochemistry of the Wittig reaction can be accounted for on the basis of the mechanisms discussed in Section IV of this chapter. The formation of the isomers and the pertinent rate constants are indicated in Chart 4.1. These are a new set of rate constants. The forma-

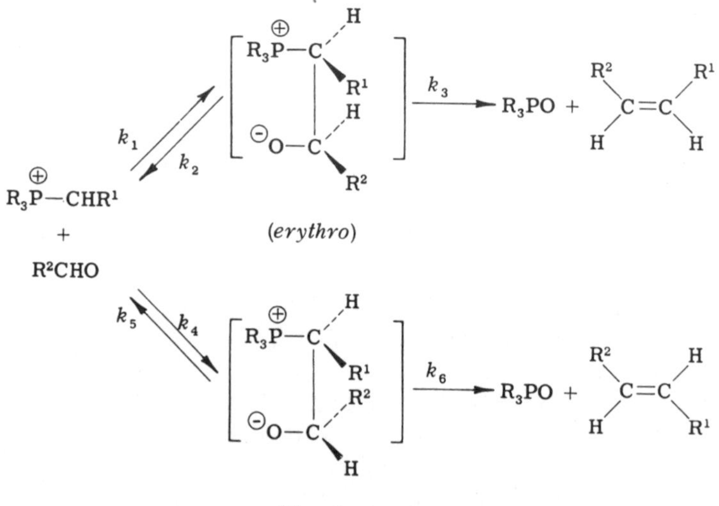

Chart 4.1

tion of the *cis*- and *trans*-olefins is thought to arise via the intermediacy of *erythro*- and *threo*-betaines, respectively. It is not surprising that *trans*-olefins are often the dominant product since, in most instances, the *trans* isomers are the most stable. More important however, is the likelihood of the transition state for the formation of the *trans*-isomer from the *threo*-betaine being more stable than that from the *erythro*-betaine. The former should have less steric hindrance since the large groups (R^1 and R^2) would not be eclipsed and the incipient double bond should be stabilized by conjugation with these groups. Such conjugation should be more difficult in the transition state from the *erythro*-betaine since the R groups could not become coplanar with the forming double bond.

Trippett (*80*) has shown that there is no direct conversion from one betaine to another. Treatment of the phosphonium iodide (XIII) with sodium ethoxide in ethanol afforded a mixture of *cis*- and *trans*-stilbenes along with diphenylmethylphosphine oxide. Repetition of this reaction but in the presence of *m*-chlorobenzaldehyde afforded a mixture of *cis*-stilbene and mixed chlorostilbenes but no *trans*-stilbene. From this data it is apparent that the originally formed *erythro*-betaine could ultimately afford a mixture of *erythro*- and *threo*-betaines, presumably via the inter-

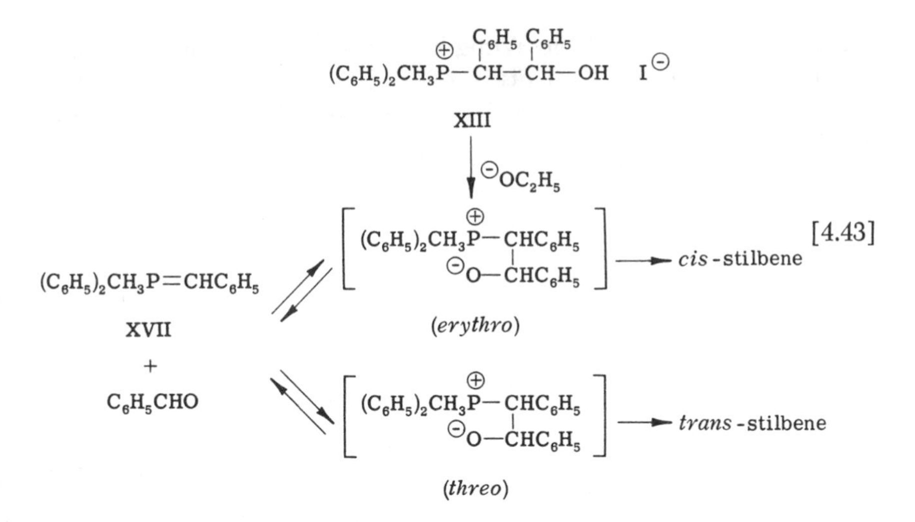

mediacy of the ylid (XVII), since these two betaines are the precursors to the *cis*- and *trans*-olefins, respectively. Addition of the more electrophilic *m*-chlorobenzaldehyde served to trap any ylid (XVII) that was formed in the system, preventing the formation of any *trans*-olefin via the *threo*-betaine. Since no *trans*-stilbene was formed in this instance, all of that formed in the original reaction must have been formed via the ylid rather than by any direct interconversion from *erythro*-betaine to *threo*-betaine (*6*, Trippett). This experiment, and somewhat similar but inconclusive results obtained by Speziale and Bissing (*49*), have confirmed what often has been assumed, that the interconversion of two stereoisomeric betaines must occur via reversion to the phosphonium ylid and the carbonyl compound.

The Wittig reaction has been assumed to involve the *cis* elimination of triphenylphosphine oxide from a betaine intermediate as portrayed by the mechanism in Chart 4.1. McEwen *et al.* (*89*), based on the stereochemical discussions by Haake and Westheimer (*90*), have argued convincingly that the Wittig reaction was a *cis* elimination and would lead to formation of a phosphine oxide whose phosphorus atom had the same configuration as the phosphonium salt. However, it was not until 1964 that Horner and Winkler (*91*) proved the point. They found that cathodic cleavage (retention of configuration) of levorotatory methylpropylphenylbenzylphosphonium bromide afforded optically active methylpropylphenylphosphine. The latter was oxidized with hydrogen peroxide (retention of configuration) to levorotatory methylpropylphenylphosphine oxide [4.44]. Treatment of the original levorotatory phosphonium salt with phenyllithium afforded a solution of optically active benzylidene-

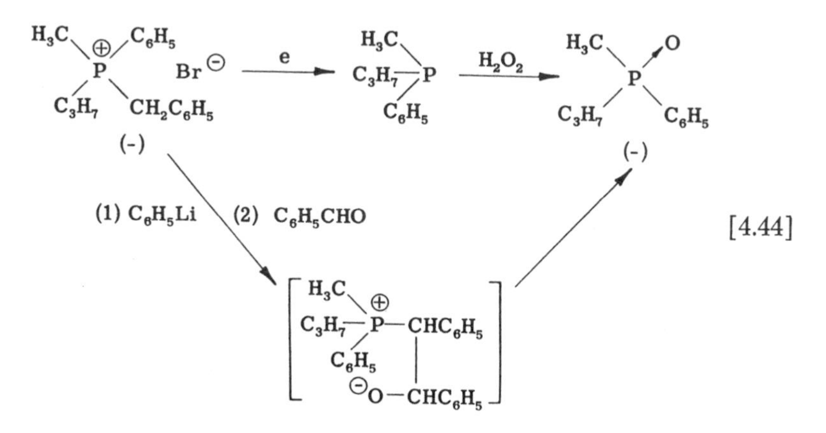

[4.44]

methylpropylphenylphosphorane [previously shown to be optically stable (*89*)] which reacted with benzaldehyde to afford stilbenes and levorotatory phosphine oxide. Thus, the phosphine oxide was produced with retention of configuration and the betaine decomposition must have occurred via a *cis* elimination.

The nature of the substituents on the carbanion portion of phosphonium ylids has been shown to affect the stereochemistry of the Wittig reaction. House and Rasmusson (*42*) found that methylcarbomethoxymethylenetriphenylphosphorane (XVIII) reacted with acetaldehyde in 90% yield to afford a 96:4 ratio of the angelic and tiglic esters. In organic nomenclature the former is called the *cis* isomer and the latter the *trans* isomer, but for purposes of this discussion the nomenclature will be reversed because the angelate is the result of a *trans* reaction, the bulky group of the carbonyl compound being *trans* to the bulky group from the ylid portion. The same products could be obtained by using a different combination of reactants but the stereochemical results were different too. Ethylidenetriphenylphosphorane (XIX) and methyl pyruvate afforded only a 21% yield of olefins with a ratio of 68:32 in favor of angelate (*trans* reaction) [4.45].

House and Rasmusson (*42*) proposed that in the first reaction the formation of an intermediate betaine was reversible since the ylid (XVIII) was stabilized. Since the rate-controlling step in the reaction was thought to be the decomposition of the betaines to olefin, the stereochemistry was thought to be determined in that step by the relative energies of the two transition states leading to *cis*- and *trans*-olefins. The *trans* transition state was expected to have the lower energy for the reasons discussed above and was expected to form the fastest.

The second reaction using ethylidenetriphenylphosphorane (XIX) was thought to be less stereospecific for two reasons. First, it was assumed

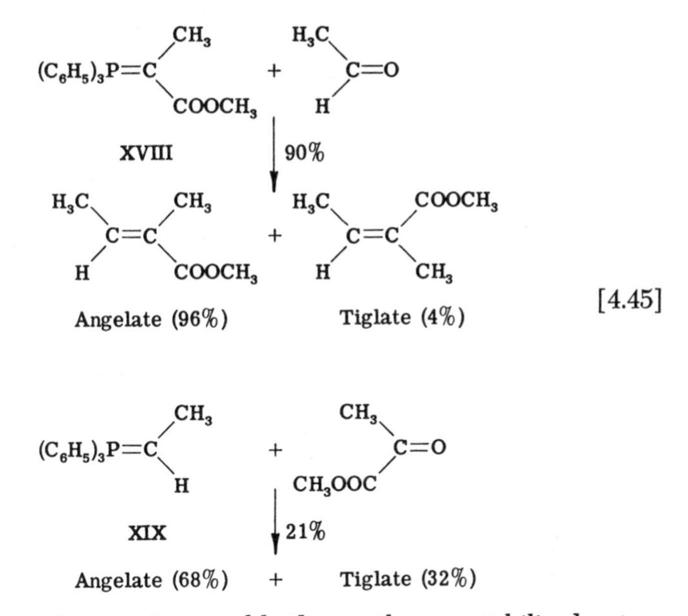

that betaine formation was irreversible due to the non-stabilized nature of the ylid. The formation of the betaine was thought to be the step in which the stereochemistry was determined. Thus, the differences in the transition states for the *formation* of the betaines was thought to be reflected in the isomeric ratio. The *threo*-betaine was expected to be the least sterically hindered of the two, and the olefin resulting from its ultimate elimination, the *trans* isomer (angelate), was expected to dominate the reaction products but by a smaller margin than in the first reaction. Since it has since been shown that betaine formation is reversible even for non-stabilized ylids this argument must be modified.

Ketcham *et al.* (83) used the same arguments to explain their observations that *p*-nitrobenzylidenetriphenylphosphorane (a stabilized ylid) reacted with *p*-anisaldehyde to afford an 89% yield of pure *trans*-4-methoxy-4′-nitrostilbene whereas the reverse reaction, that of *p*-methoxybenzylidenetriphenylphosphorane (a non-stabilized ylid) with *p*-nitrobenzaldehyde, afforded an 89% yield of the same olefins but with a *cis/trans* ratio of 48:52. Here again, the reaction of the stabilized ylid was assumed to be reversible with the stereochemistry determined by betaine decomposition whereas reaction of the non-stabilized ylid with aldehyde was assumed to be irreversible with the stereochemistry determined in the betaine-forming step [4.46].

Bestmann and Kratzer (75) have reported the effect of replacing a triphenylphosphonium group with a tricyclohexylphosphonium group on the stereochemistry of the olefins from a Wittig reaction. A series of sub-

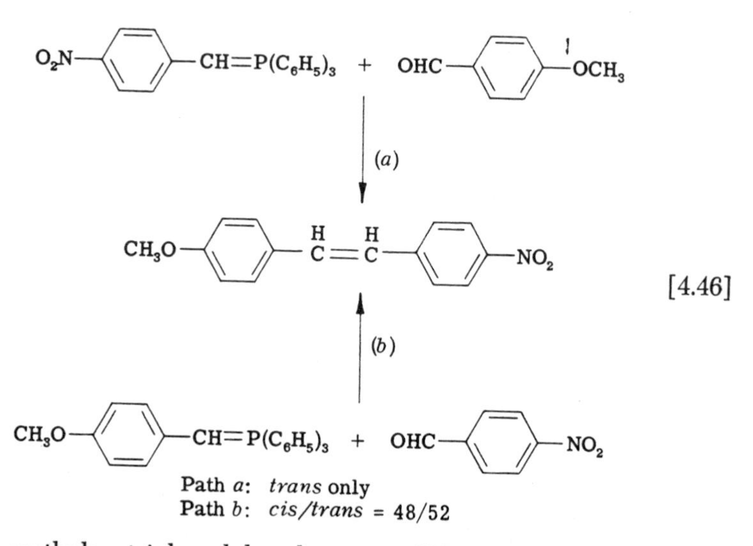

stituted methylenetriphenylphosphoranes (XX, R = phenyl, carbomethoxy, styryl; R′ = phenyl) reacted with benzaldehyde or cinnamaldehyde to afford a mixture of the *cis*- and *trans*-olefins. In contrast the corresponding substituted methylenetricyclohexylphosphoranes (XX, R = phenyl, carbomethoxy, styryl; R′ = cyclohexyl) reacted with the same

$$R-\overset{\ominus}{C}H-\overset{\oplus}{P}R_3'$$

XX

aldehydes but afforded only the *trans*-olefins. These workers proposed that with both the tricyclohexyl- and triphenylphosphonium stabilized ylids the betaine formation was slow and would afford a mixture of the two isomeric betaines. The major difference was thought to arise in the betaine decomposition step and they proposed this was so rapid in the case of the triphenylphosphonium ylids that there was little opportunity for reversal of an *erythro*-betaine to the more stable *threo*-betaine. Hence, both would be converted to their respective olefinic products, the *cis*- and *trans*-olefins. The betaine decomposition step was thought to be much slower in the case of the tricyclohexylphosphonium ylids due to the increased electron density on phosphorus. This would permit ready equilibrium between the *erythro*- and *threo*-betaines, the latter being most readily converted into olefin (*trans*) because of conjugative and steric effects. Speziale and Bissing (49) have reported similar results. Carbethoxymethylene*tributyl*phosphorane gave a higher *trans/cis* ratio than did carbethoxymethylene*triphenyl*phosphorane in reaction with benzaldehyde.

A. For Stabilized Ylids

From the qualitative observations recorded in the preceding paragraphs it is apparent that many factors are affecting the stereochemistry of the Wittig reaction. There would appear to be a delicate balance between the rates of formation of the betaines, their rates of dissociation and their rates of decomposition to olefins. If the mechanistic studies and the conclusions discussed in Section IV of this chapter are valid they also ought to account for the stereochemistry of the reaction. The interrelationships of the rates of the various steps will be discussed, dealing first with stabilized ylids in the absence of polar solvents and additives.

Speziale and Bissing (*49*) have studied the Wittig reaction between

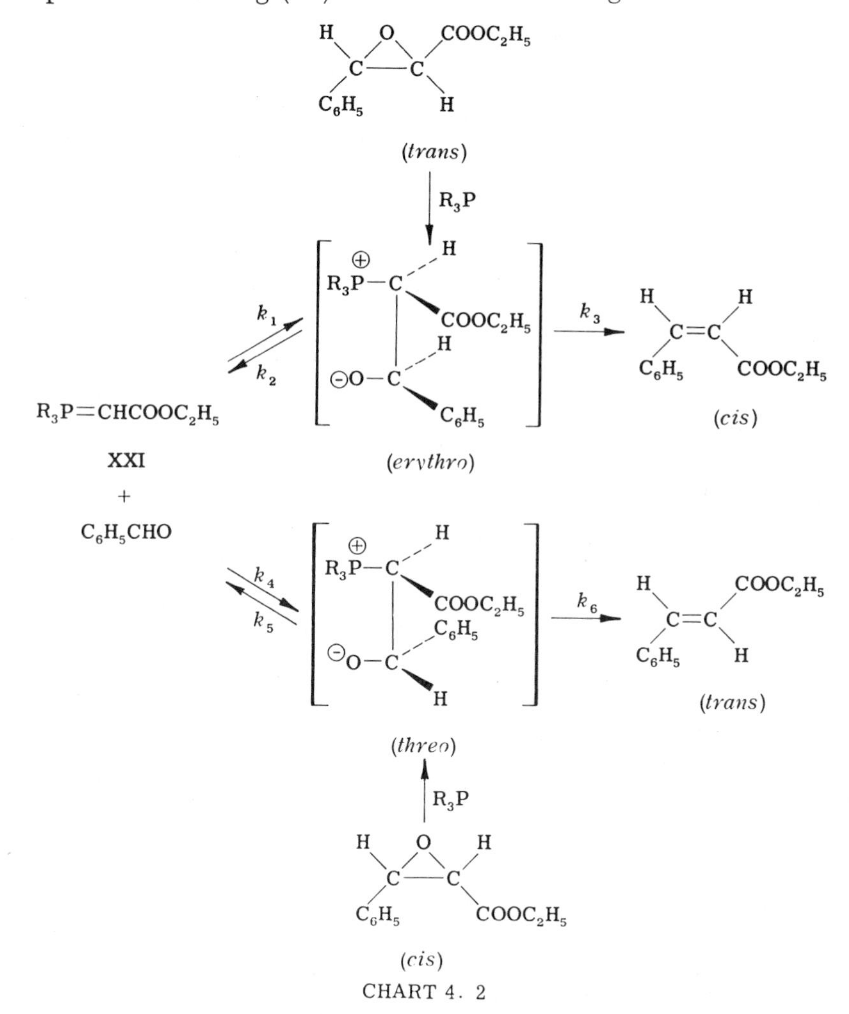

CHART 4. 2

carbethoxymethylenetriphenylphosphorane (XXI, R = phenyl) and benzaldehyde in sufficient detail that some relationships between the various rate constants could be obtained. This reaction afforded the ethyl cinnamates in a *trans/cis* ratio of 84:16 (= 5.25), presumably via the path outlined in Chart 4.2. This indicated that the overall pathway, involving the rate constants k_4, k_5, and k_6, was faster than the alternate pathway via the *erythro*-betaine and involving the rate constants k_1, k_2 and k_3. Generation of the *threo*-betaine by heating ethyl *cis*-phenylglycidate with triphenylphosphine afforded the olefins in a *trans/cis* ratio of 97:3. However, heating the *trans* epoxide in the same manner with triphenylphosphine led to the olefins in a *trans/cis* ratio of 59:41 [4.47].

$$C_6H_5CH\overset{O}{-}CHCOOC_2H_5 + (C_6H_5)_3P \longrightarrow C_6H_5CH{=}CHCOOC_2H_5 + (C_6H_5)_3PO \quad [4.47]$$

cis ----------→ *trans/cis* = 97/3

trans ----------→ *trans/cis* = 59/41

While these two ratios must be used with caution because of olefin isomerization, it is nonetheless clear that more *erythro*-betaine is converted to *threo*-betaine than vice-versa.

Heating ethyl *trans*-phenylglycidate with triphenylphosphine but in the presence of a three molar excess of *m*-chlorobenzaldehyde afforded a mixture of ethyl chlorocinnamates and ethyl cinnamates in the ratio of 33.5:66.5 (= 0.50) [4.48]. The isolation of the chlorocinnamates indi-

$$C_6H_5CH\overset{O}{-}CHCOOC_2H_5 + (C_6H_5)_3P \xrightarrow{ClC_6H_4CHO} C_6H_5CH{=}CHCOOC_2H_5 + ClC_6H_4CH{=}CHCOOC_2H_5 + (C_6H_5)_3PO \quad [4.48]$$

trans-oxirane ------→ Cl/H = 33.5/66.5

cis-oxirane ------→ Cl/H = 13/87

cated that the *erythro*-betaine was in equilibrium with the ylid (XXI, R = phenyl) and benzaldehyde, and it was assumed that any ylid formed by this route was immediately trapped by reaction with the more electrophilic *m*-chlorobenzaldehyde. This assumption was verified to a reasonable degree by the observation that reaction of the ylid (XXI, R = phenyl) with a mixture of one equivalent of benzaldehyde and three equivalents of *m*-chlorobenzaldehyde afforded the chlorocinnamates and cinnamates in a ratio of 94:6. Accordingly, the ratio of chlorocinnamates to cinnamates in the reaction between *trans*-phenylglycidate, triphenylphosphine and excess *m*-chlorobenzaldehye could be used as a measure of the ratio of k_2/k_3. Therefore, $k_2/k_3 = 0.50$ and for the *erythro*-betaine the rate of its decomposition to olefin is only twice as fast as its rate of dissociation to ylid and aldehyde. The analogous experiment with ethyl *cis*-phenylglycidate gave a k_5/k_6 ratio of 0.15. Therefore, for the *threo*-betaine the rate of decomposition to olefin is 6.7 times the rate of dissociation.

It is difficult to dissect the stereochemical implications of the mechanism portrayed in Chart 4.2 much more, but some general trends can be detected by making a few assumptions. Using the overall rate expression [equation (1), p. 154] derived in the previous section for stabilized ylids, the following two expressions can be written for the formation of the *cis*(*C*)- and *trans*(*T*)-olefin:

$$\frac{dC}{dt} = \frac{k_1k_3(\text{ylid})(\text{aldehyde})}{k_2 + k_3} \tag{5}$$

$$\frac{dT}{dt} = \frac{k_4k_6(\text{ylid})(\text{aldehyde})}{k_5 + k_6} \tag{6}$$

The ratio of isomeric olefins obtained from a Wittig reaction could then be expressed by dividing equation (5) by equation (6) and assuming that $C/T = dC/dT$:

$$\frac{C}{T} = \frac{k_1k_3(k_5 + k_6)}{k_4k_6(k_2 + k_3)} \tag{7}$$

By substituting the following experimentally determined values in equation (7), $C/T = 1/5.25$, $k_5 = 0.15k_6$, $k_2 = 0.50k_3$, it can be shown that $k_4 = 4k_1$. In other words, betaine formation is faster for the *threo* isomer than for the *erythro* isomer. If it is assumed that $k_2 = k_5$, then it would follow that $k_6 = 3.3k_3$. In other words, betaine decomposition was faster for the *threo* betaine than for the *erythro* betaine. Conversely, if it is assumed that $k_3 = k_6$, then $k_2 = 3.3k_5$, i.e. betaine dissociation is faster for the *erythro*-betaine than for the *threo*-betaine. The absolute relationships between k_2 and k_5 or between k_3 and k_6 cannot be obtained from the available data. However, the relationship between the two ratios, k_2/k_5 and k_6/k_3, can be obtained, and examples are shown in Table 4.1. Any combination of these two ratios would account for

TABLE 4.1

Relationship of k_2/k_5 and k_6/k_3 for Stabilized Ylids

If k_2/k_5 =	1	2	5	10
then k_6/k_3 =	3.3	1.6	0.7	0.3

the observed *trans*/*cis*-olefin ratio for the reaction of XXI (R = phenyl) with benzaldehyde.

It is apparent that the observed ratio of the *trans*-cinnamate to the *cis*-cinnamate, 5.25, was considerably higher than any one of the ratios of k_4/k_1, k_2/k_5 and k_6/k_3. Therefore, no single step of the reaction, the betaine formation, the betaine dissociation or the betaine decomposition, appears to account for the dominance of the *trans* over the *cis* isomer. It is likewise clear, however, that the ratios of the rate constants of all three steps are in the direction that favors the dominance of the *trans*-olefin in the Wittig reaction. In other words, the *threo*-betaine forms faster than the *erythro*, perhaps because of less steric hindrance in the transition state. Probably for the same reason the *erythro*-betaine dissociated to aldehyde and ylid faster than did the *threo*-betaine. In addition, the *threo*-betaine decomposed to olefin faster than did the *erythro*-betaine, probably because of less steric hindrance and better conjugative stabilization of the incipient double bond in the transition state. As a result of the differences in all three steps, the Wittig reaction is best classed as a stereoselective but not a stereospecific reaction.

Analogous arguments can be used in the case of the reaction of carbethoxymethylenetri-*n*-butylphosphorane (XXI, R = *n*-butyl) with benzaldehyde, also studied by Speziale and Bissing (*49*). In this case the relationships among the rate constants are as follows:

$$k_3 = 1.15k_2; \qquad k_2 = 10.6k_5; \qquad k_6 = 10.6k_3; \qquad k_4 = 11k_1$$

No assumptions need be made to obtain the first and the last of these relationships. For the other two the same assumptions are made as for the

TABLE 4. 2

Relative Rate Constants for Wittig Reaction of Carbethoxymethylenetributyl- and Triphenylphosphoranes with Benzaldehyde

$(C_4H_9)_3P{=}CHCOOC_2H_5$	$(C_6H_5)_3P{=}CHCOOC_2H_5$
$k_3 = 1.15\ k_2$	$k_3 = 2\ k_2$
$k_4 = 11\ k_1$	$k_4 = 4\ k_1$
$k_2 = 10.6\ k_5$	$k_2 = 3.3\ k_5$
$k_6 = 10.6\ k_3$	$k_6 = 3.3\ k_3$

triphenylphosphonium ylid. From these relationships, as illustrated in Table 4.2, the following conclusions can be drawn:

(a) The directions of these ratios are the same for the triphenylphosphonium ylid as for the tri-*n*-butylphosphonium ylid, both favoring the dominance of the *trans*-olefin.

(b) The relative magnitude of all four ratios is in a direction which indicates that the alkylphosphonium ylid should permit a higher dominance of *trans*- over *cis*-olefin (observed: 5.25 for the triphenyl case vs. 19.0 for the tri-*n*-butyl case). In other words, *threo*-betaine forms faster with alkyl ylid than with phenyl ylid, relative to the formation of the *erythro*-betaine. The *erythro*-betaine dissociates faster in the alkyl case than in the phenyl case relative to the dissociation of the *threo*-betaine. The *threo*-betaine decomposes to olefin faster in the alkyl case than in the phenyl case, again relative to the rates of *erythro*-betaine decomposition.

In short, every factor operating in this reaction as portrayed by the mechanism in Chart 4.2 changes to favor more *trans*-olefin formation from a trialkylphosphonium ylid than from an analogous triphenylphosphonium ylid. This explanation accounts for the data presented by Bestmann and Kratzer (*75*) and discussed earlier in this section. It would appear that an increased proportion of the *trans* isomer obtained with a trialkylphosphonium ylid is due in approximately equal measure to the increase in the ratios of the rates of betaine formation and the rates of betaine decomposition but slightly less to the decrease in ratios of betaine decomposition to betaine dissociation. In structural terms, the increase in the proportion of *trans* isomer with trialkylphosphonium ylids is due both to the increased nucleophilicity of the ylid (faster betaine formation) and the increased importance of conjugative stabilization of the incipient double bond in the transition state for betaine decomposition.

B. For Non-stabilized Ylids

The situation regarding the stereochemistry of the Wittig reaction with non-stabilized ylids is not nearly as clear as for stabilized ylids. As indicated in the previous section of this chapter, the betaine formation step appears to be the fast step of the reaction and it is probable that the steric influences are not pronounced in the transition state for betaine formation. In that case, it would be expected that k_1 and k_4, the rates of formation of the *erythro*- and *threo*-betaines, respectively, would be larger than the other rates in the reaction and that they would be similar in magnitude. If this is the case it would infer that $k_2 > k_5$, the *erythro*-betaine dissociating faster than the *threo*-betaine, since the *threo*-betaine surely must be more stable than the *erythro*-betaine by analogy with the betaines of the stabilized ylids.

Trippett (*80*) recently reported that benzylidenediphenylmethylphosphorane (XVII) and benzaldehyde afforded stilbenes in a *cis/trans* ratio of 33:66. Generation of the *erythro*-betaine from its conjugate acid (XIII) but in the presence of an excess of *m*-chlorobenzaldehyde led to the isolation of *cis*-stilbene and mixed *m*-chlorostilbenes in a ratio of 2:3. Thus, the ratio of $k_2/k_3 = 1.5$, and the *erythro*-betaine dissociated 1.5 times as fast as it decomposed. They were unable to obtain the pre-

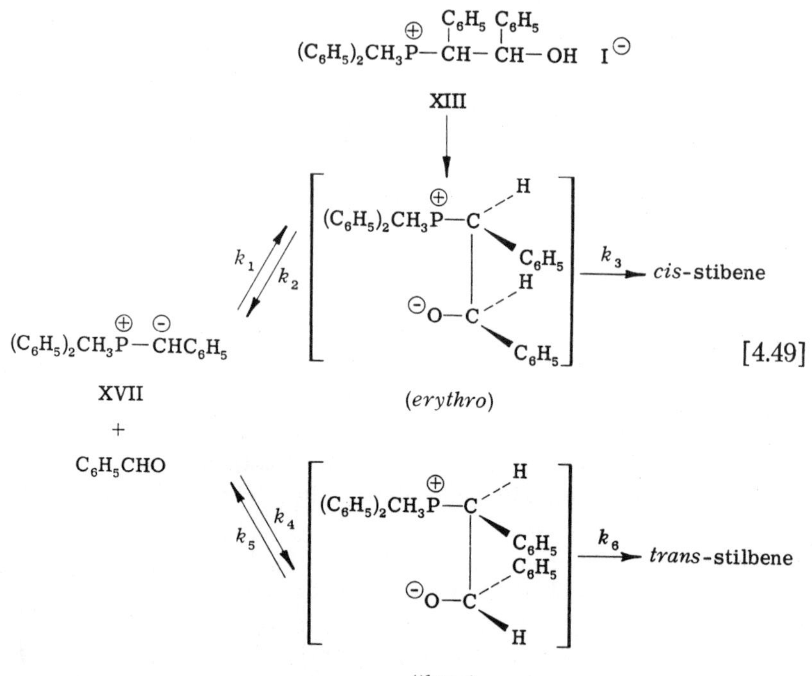

cursor to the *threo*-betaine and therefore could not obtain a measure of k_5/k_6. Nonetheless, using the general equation (7), the relationships $C/T = 0.50$ and $k_2 = 1.5k_3$, and assuming that $k_1 = k_4$, it can be shown that $k_6 = 4k_5$. Thus, the characteristics of the reaction of non-stabilized ylids with carbonyl compounds, at least as exemplified by a benzylidene-phosphorane, are that the two betaines may form at the same rate, the *erythro*-betaine dissociates faster than it decomposes to *cis*-olefin and the *threo*-betaine decomposes to *trans*-olefin faster than it dissociates to ylid and carbonyl.

Once again, the absolute relationship between k_3 and k_6 or k_2 and k_5 cannot be obtained but a relationship between their ratios is represented by equation (8):

$$k_6/k_3 = 6k_5/k_2 \tag{8}$$

Thus, as the ratio of k_2/k_5, the ratio of betaine dissociation, increases, the ratio of k_6/k_3, the ratio of the rates of betaine decomposition, can decrease and yet maintain the same stereochemical result. Since k_2 would be expected to be larger than k_5 due to a difference in the steric hindrance in the betaines, the ratio of k_6/k_3 probably is not very different from unity and in the case of non-stabilized ylids, the major factor governing the stereochemical course of the Wittig reaction would be the dissociation of the betaines.

The results obtained by Denney and Boskin (79) for the reaction of tributylphosphine with *cis*- and *trans*-2-butene oxides are accounted for by the above mechanistic proposal. *trans*-2-Butene oxide, when heated with tributylphosphine, afforded the *cis*- and *trans*-2-butenes in a ratio of 72:28 [4.50]. Since this reaction must initially proceed via the *erythro*-

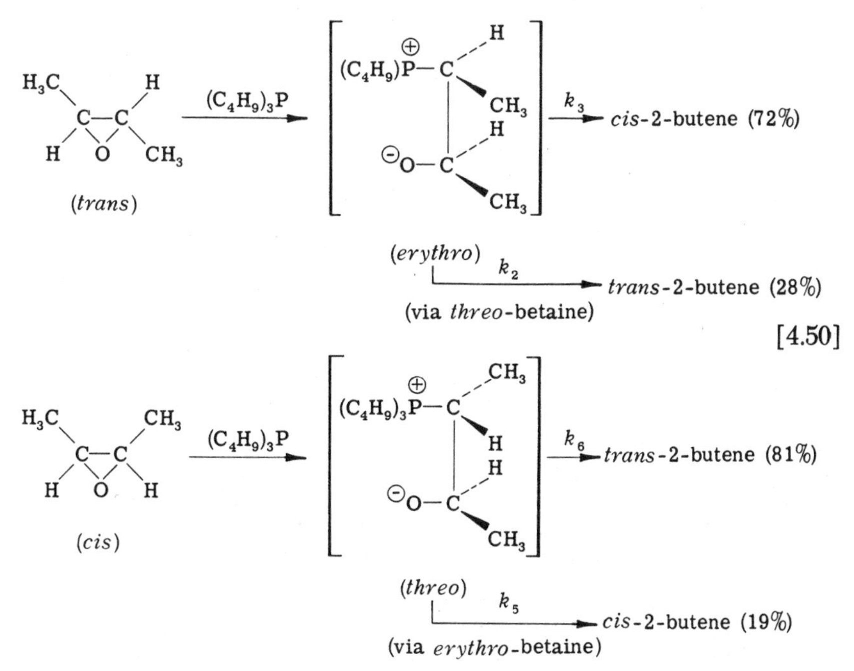

betaine, the value of $k_3/k_2 = 2.6$. In other words, the *erythro*-betaine decomposes to *cis*-olefin 2.6 times as fast as it dissociates to ylid and aldehyde. The same reaction, but with the *cis*-2-butene oxide afforded a mixture of *cis*- and *trans*-2-butene in a ratio of 19:81. Since this reaction must initially proceed via the *threo*-betaine, the value of $k_6/k_5 = 4.3$. In other words, the *threo*-betaine also decomposes to olefin (but *trans*) faster than it dissociates into ylid and aldehyde. The fact that reaction of the *cis*-oxide with tributylphosphine is more stereoselective is due to the higher decomposition/dissociation ratio for the *threo*-betaine as expected. It is impossible to ascertain at this time whether the overall difference is due more to differences in the decomposition or the dissociation rates.

On the basis of the preceding discussion, it is apparent that the stereochemistry of the Wittig reaction, whether involving stabilized or non-stabilized ylids, depends on very delicate balances between the six rate constants. It is safe to conclude that in most instances, in the absence of additives or special solvent interactions, the *trans* isomers tend to dominate the mixture of olefins obtained from the reaction. There is experimental evidence to support this conclusion and it has been possible to propose a mechanism, complete with structural and kinetic implications, which accounts for the observed data. It appears that in the case of stabilized ylids, the stereoselectivity of the reaction is due to factors entering into the betaine formation and betaine decomposition steps while in the case of non-stabilized ylids the stereoselectivity is due mainly to the balance between the betaine dissociation and betaine decomposition steps. It must be emphasized that the mechanistic discussions are in the form of hypotheses which have been developed on the basis of very limited data. These discussions are meant to incite further inquiry into the subtleties and even basic concepts of the Wittig reaction in order either to put these proposals on a firmer base or to devise new ones to take their place.

C. Effect of Reaction Conditions

The conditions under which Wittig reactions are carried out may have considerable bearing on the stereoselectivity and even the course of the reaction. All of the previous discussion has concerned those reactions carried out in non-polar solvents in the absence of additives. Recent work by a number of investigators has indicated very clearly the alterations of stereoselectivity effected by a variety of environmental changes.

As mentioned above the *trans*-olefins usually dominate the product composition in most Wittig reactions. This is due mainly to the steric effects involved in betaine formation, induced by matching of the carbonyl and ylid dipoles, and the steric and electronic effects operating in betaine decomposition. It would be expected that environmental changes which tend to reduce the steric and electronic differences between the isomeric betaines would decrease the stereoselectivity of the reaction. In other words, the *cis/trans* ratio would be expected to increase.

Hauser *et al.* (*14*) have found that raising the temperature of the reaction between *n*-hexylidenetriphenylphosphorane and acetaldehyde resulted in an increase in the *cis/trans* ratio. A higher temperature would be expected to decrease the effect of different stabilities of the *threo*- and *erythro*-betaines. However, Drefahl *et al.* (*92*) claimed that temperature changes exerted only a minor effect on the benzylidenetriphenylphosphorane reaction with α-naphthaldehyde.

Bergelson and Shemyakin (*88, 93*) have found that the presence of either excess aldehyde or excess ylid tends to increase the *cis/trans* ratio. They found that *n*-propylidenetriphenylphosphorane reacted with one equivalent of benzaldehyde in benzene solution in the presence of lithium iodide to afford a 34:66 ratio of *cis*- and *trans*-ethylstyrene. When a 1:2 ratio of aldehyde to ylid was used under the same conditions the *cis/trans* ratio was raised to 65:35 and when a 2:1 ratio of aldehyde to ylid was used the olefin ratio was raised to 61:39. These workers attributed the effect of excess reagents to a suppression of the dissociation of the betaines. In view of the previous stereochemical arguments, such suppression would be expected to raise the *cis/trans* ratio. Drefahl *et al.* (*92*) have challenged these results, however.

Two different types of solvent effects have been discerned. The use of non-protonic solvents of widely varying polarity with stabilized ylids virtually has no effect on the *cis/trans* ratio of the olefins (*86,* Bergelson and Shemyakin; *24,* House *et al.*). This would imply that general solvation of the betaine is not affecting its formation or that any effect is absorbed and counterbalanced by electronic effects exerted on the decomposition step. However, the use of protonic polar solvents such as methanol have been found to increase the *cis/trans* ratio (*24, 92*). House has proposed that either the reversibility of solvated betaines to ylid and aldehyde is slowed or the relative stability of the stereoisomeric betaines may be altered by the removal of electrostatic attraction. In fact, it has been argued (*80,* Trippett) that in the presence of powerful solvation or complexation the *erythro* isomer is the more stable and exists in the conformation represented by XXII. If such were the case, then the relative rates of betaine formation and dissociation character-

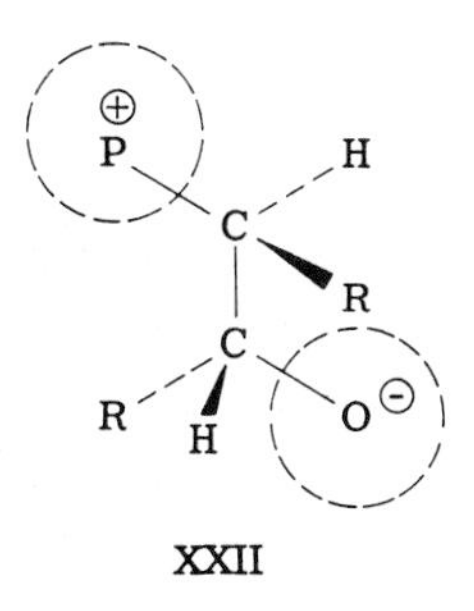

XXII

istic of the normal Wittig reaction would be reversed. The relative rates of decomposition of the isomeric betaines probably would not be altered but due to the importance of the formation and dissociation steps in the determination of the stereoselectivity of the reaction, the *cis/trans* ratio probably would be raised.

Bergelson *et al.* (*86, 88*), House *et al.* (*24*) and Drefahl *et al.* (*92*) all have found that both polar non-protonic solvents and protonic solvents or proton-donating catalysts increase the *cis/trans* ratio of the olefins obtained from the reaction of a carbonyl compound with a non-stabilized ylid. This increased ratio probably is due to the same factors mentioned above, solvation or complexation with the betaine, especially the oxyanion portion. In the case of non-stabilized ylids the normal rates of formation of the two isomeric betaines are about the same. Under these conditions any change in the isomeric ratio of olefins would be due to the effect of complexation on the relative rates of betaine dissociation and betaine decomposition. The normally high k_2/k_5 ratio would be decreased as the *erythro*-betaine approached or surpassed the *threo*-betaine in stability and the *cis/trans* ratio would then increase as well. However, a major change in the rates of formation of the betaines effected by solvation also could favor the formation of the *erythro*-betaine, raising the *cis/trans* ratio of the olefinic products.

Bergelson *et al.* (*86, 88*) originally investigated the effect of various additives on the stereochemical course of the Wittig reaction, especially with non-stabilized ylids. They concluded that suspended or dissolved halide ions tended to increase the *cis/trans* ratio of olefins. However, more recent work by House *et al.* (*24*) has indicated that the major effect is that of the cation, especially lithium, and only when it is actually dissolved in the reaction medium. In these instances they confirmed that an unusually high *cis/trans* ratio of olefins was obtained in the reaction of benzylidenetriphenylphosphorane with propionaldehyde. The effect of the lithium cation appeared to be less than that of either protonic solvents or protonic additives. The effect of these various Lewis acidic species most likely is the same as discussed above, affecting the dissociation or decomposition of the isomeric betaines more so than their formation. The Russian workers have agreed with the more recent experimental results of House *et al.* (see footnote 8 in reference *24*).

Bergelson and Shemyakin (*86*) had proposed that halide complexation with the phosphonium group of a non-stabilized ylid would reduce the reactivity of that ylid. If the phosphonium group were complexed, electronic considerations would dictate that the ylid ought to be more nucleophilic. This might be counterbalanced by steric considerations which would hinder attack by the carbanion on the carbonyl group. In that event reduction in the reactivity of the ylid would make the betaine formation steps more important in determining the stereoselectivity of the reaction. If the steric differences between the betaines also were reduced, the formation step then would not discriminate between the *erythro*- and *threo*-betaines, thus reducing the *cis/trans* ratio of the olefinic products.

From this brief discussion it must be apparent that conclusions regarding the stereochemical course of the Wittig reaction must be considered in light of the environment in which the reaction was conducted. In particular, much of the early work with non-stabilized ylids which was conducted in a variety of solvents, often with lithium halides present, must be reassessed in light of the recent work of Bergelson (*88*) and House (*24*). It must also be apparent, however, that important advantage may be gained in synthesis by a judicious choice of Wittig reaction conditions. As pointed out earlier, the normal course of the Wittig reaction results in the dominant formation of the *trans*-olefin. However, use of polar protonic solvents, protonic catalysts, Lewis acids or higher temperatures can lead to the formation of major amounts of the *cis* isomers from the same reaction. Bergelson *et al.* (*88, 94*) have used these conditions to effect striking syntheses of unsaturated fatty acids having *cis*-olefinic groups in the molecules. For example, reaction of ω-carbethoxy-*n*-octylidenetriphenylphosphorane with pelargonic aldehyde in dimethylformamide solution afforded the expected olefin which was 95% ethyl oleate [4.51] (*95*).

$$(C_6H_5)_3P{=}CH{-}(CH_2)_7{-}COOC_2H_5 + C_8H_{17}CHO \longrightarrow \underset{(95\%\ cis)}{C_8H_{17}CH{=}CH{-}(CH_2)_7COOC_2H_5} \qquad [4.51]$$

In conclusion, the Wittig reaction is a most useful synthetic tool for the preparation of olefins in general and the stereoselective synthesis of stereoisomeric olefins in particular. The reaction is noticeably free from obvious complications and it can be applied with a wide variety of ylids and carbonyl compounds. The mechanistic details of the reaction, in their intricate interrelationships, more than balance the synthetic simplicities of the reaction and make for most interesting study. A coherent mechanistic picture of the reaction has been presented and an attempt has been made to account for the stereochemical observations on the basis of this mechanism. Considerably more work is necessary, especially in the case of non-stabilized ylids, before the factors governing the stereochemical course of the reaction become clear. It is hoped that the proposals incorporated herein will lead to further experimental testing of these and other hypotheses covering the mechanism of the Wittig reaction.

Appendix

Numerous additional applications of the Wittig reaction in synthetic chemistry have appeared during the last half of 1965. None of these have been sufficiently novel to warrant special mention, however. The applications of the Wittig reaction have been thoroughly reviewed by

Maercker (*96*). The BASF group have published details of their application of phosphonium chemistry in the vitamin A field, one feature of which was the evolution of new synthetic approaches to vitamin A utilizing the Wittig reaction (*97*). Trippett and Walker (*98*) have isolated a rearrangement product (XXIV) formed in 64% yield during the reaction of benzaldehyde with methylenetriphenylphosphorane. They proposed that XXIV was formed by rearrangement of the normal betaine (XXIII) [4.52]. The former also was obtained in 40% yield when the betaine was generated by opening styrene oxide with triphenylphosphine.

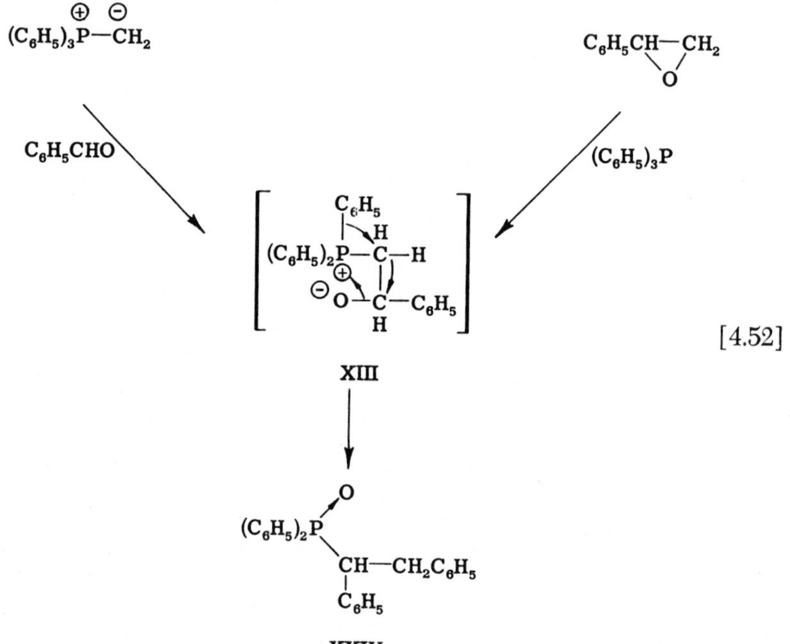

Johnson and Kyllingstad (*99*) have shown that facilitating decomposition of a betaine to olefin and phosphine oxide permits the formation of larger *cis/trans* ratios of isomeric olefins in the Wittig reaction. Thus, using a tris(*p*-chlorophenyl)phosphonium ylid (XXVa) gave a *cis/trans* ratio of stilbenes of 4.10 while using a triphenylphosphonium ylid (XXVb) gave a *cis/trans* ratio of 0.79. The effect of facilitating betaine decomposition is explicable in terms of equation (7) (see p. 179), k_1 and

$$\overset{\oplus}{R_3P}-\overset{\ominus}{CH}-C_6H_4OCH_3 + O_2NC_6H_4CHO \longrightarrow CH_3OC_6H_4CH{=}CHC_6H_4NO_2 \qquad [4.53]$$

XXVa $R = p\text{-}ClC_6H_4-$
XXVb $R = C_6H_5-$

k_3 being increased and k_2 being decreased, all relative to k_4, k_5, and k_6, respectively, and leading to a larger *cis/trans* ratio of olefins. In the synthesis of 4-nitro-4′-methoxystilbene they were able to alter the *cis/trans* ratio from 17/83 to 80/20 by proper choice of carbanion and phosphorus substituents.

Schlosser and Christmann (*100*) have provided convincing evidence that betaines formed from non-stabilized ylids are capable of dissociating to ylid and carbonyl compound. Heating the betaine formed from ethylidenetriphenylphosphorane and benzaldehyde with *m*-chlorobenzaldehyde afforded a mixture of β-methylstyrene and *p*-chloro-β-methylstyrene. These workers also demonstrated that the *threo*-betaine was more stable than the *erythro*-betaine as expected on steric grounds.

Speziale and Bissing (*101*) have elaborated on their earlier work (*49*) on the mechanism of the Wittig reaction. Using the betaines derived from the reaction of triphenylphosphine and tributylphosphine with 4-octene oxide they have demonstrated their dissociation to the non-stabilized *n*-butylidenephosphoranes and butyraldehyde. They have shown that the dissociation/decomposition ratio is larger for the tributylphosphonium betaine than for the triphenylphosphonium betaine, probably due to the easier oxyanion attack on phosphorus in the latter case. These workers were unable to provide any evidence for direct interconversion between *erythro*- and *threo*-betaines in agreement with the earlier results of Trippett (*80*). Speziale and Bissing (*101*) confirmed their earlier conclusions that the dissociation/decomposition ratio is larger for *erythro*-betaines than for *threo*-betaines. In other words, the *threo*-betaines decomposed faster than did the *erythro*-betaines as expected on both steric and electronic grounds. The differences in these ratios diminished at higher temperatures. Both ratios were found to increase with higher temperature, indicating that dissociation to ylid and aldehyde was favored and leading to the conclusion that partitioning of a betaine between the dissociation and decomposition routes was controlled mainly by the activation entropy rather than by the activation enthalpy. It was evident that a betaine formed from a stabilized ylid tended to dissociate faster than did a betaine formed from a non-stabilized ylid, verifying an assumption made much earlier by House and Rasmusson (*42*).

REFERENCES

1. G. Wittig, *Experientia* **12**, 41 (1956); *Angew. Chem.* **68**, 505 (1956).
2. J. Levisalles, *Bull. Soc. Chim. Fr.* 1021 (1958).
3. U. Schollkopf, *Angew. Chem.* **71**, 260 (1959); *in* "Newer Methods of Preparative Organic Chemistry" (W. Foerst, ed.), Vol. III, pp. 111–150, Academic Press, New York, 1964.

4. S. Trippett, *Advan. Org. Chem.* **1,** 83 (1960).
5. L. A. Yanovskaya, *Russ. Chem. Revs.* **30,** 347 (1961).
6. S. Trippett, *Quart. Rev.* **17,** 406 (1963).
7. G. Wittig, *Pure and Applied Chem.* **9,** 245 (1964).
8. H. Staudinger and J. Meyer, *Helv. Chim. Acta* **2,** 635 (1919).
9. G. Luscher, Dissertat. Eidg. Techn. Hochschule, Zurich, 1922.
10. G. Wittig and M. Rieber, *Ann.* **562,** 177 (1949).
11. G. Wittig and G. Geissler, *Ann.* **580,** 44 (1953).
12. G. Wittig and U. Schollkopf, *Chem. Ber.* **87,** 1318 (1954).
13. G. Wittig and W. Haag, *Chem. Ber.* **88,** 1654 (1955).
14. C. F. Hauser, T. W. Brooks, M. L. Miles, R. A. Raymond and G. B. Butler, *J. Org. Chem.* **28,** 372 (1963).
15. S. Sugasawa and H. Matsuo, *Chem. Pharm. Bull.* (*Jap.*) **8,** 819 (1960).
16. G. Fodor and I. Tomoskozi, *Tetrahedron Letters* 579 (1961).
17. A. Wm. Johnson and R. B. LaCount, *Tetrahedron* **9,** 130 (1960).
18. Ch. Ruchardt, S. Eichler and P. Panse, *Angew. Chem.* **75,** 858 (1963).
19. M. Schlosser and K. F. Christmann, *Angew. Chem.* **76,** 683 (1964).
20. L. Horner, H. Hoffmann, H. G. Wippel and G. Klahre, *Chem. Ber.* **92,** 2499 (1959).
21. R. Greenwald, M. Chaykovsky and E. J. Corey, *J. Org. Chem.* **28,** 1128 (1963).
22. C. H. Collins and G. S. Hammond, *J. Org. Chem.* **25,** 1434 (1960).
23. L. D. Bergelson, V. A. Vaver, L. I. Barsukov and M. M. Shemyakin, *Izvest. Akad. Nauk. S.S.S.R.* 1053 (1963); *Chem. Abstr.* **59,** 8783 (1963).
24. H. O. House, V. K. Jones and G. A. Frank, *J. Org. Chem.* **29,** 3327 (1964).
25. G. P. Schiemenz and H. Engelhard, *Chem. Ber.* **94,** 578 (1961).
26. G. Wittig and M. Schlosser, *Tetrahedron* **18,** 1023 (1962).
27. S. Trippett and D. M. Walker, *J. Chem. Soc.* 2130 (1961).
28. G. Witschard and C. E. Griffin, *J. Org. Chem.* **29,** 2335 (1964).
29. F. Ramirez and S. Levy, *J. Am. Chem. Soc.* **79,** 67 (1957).
30. A. Wm. Johnson, *J. Org. Chem.* **24,** 282 (1959).
31. F. Ramirez and S. Dershowitz, *J. Org. Chem.* **22,** 41 (1957).
32. G. Markl, *Z. Naturforschg.* **18b,** 84 (1963); *Angew. Chem.* **75,** 168 (1963).
33. A. Wm. Johnson, unpublished observations.
34. V. Mark, *Tetrahedron Letters* 3139 (1964).
35. G. Wittig, H. D. Weigmann and M. Schlosser, *Chem. Ber.* **94,** 676 (1961).
36. H. Gilman and R. A. Tomasi, *J. Org. Chem.* **27,** 3647 (1962).
37. P. Duffner, Dissertation, Tubingen, 1957.
38. G. Wittig and A. Haag, *Chem. Ber.* **96,** 1535 (1963).
39. F. Sondheimer and R. Mechoulam, *J. Am. Chem. Soc.* **79,** 5029 (1957).
40. N. K. Kochetkov and B. A. Dmitriev, *Doklady Akad. Nauk. S.S.S.R.* **151,** 106 (1963); *Chem. & Ind.* (*London*) 864 (1963).
41. F. Bohlmann and E. Inhoffen, *Chem. Ber.* **89,** 1276 (1956).
42. H. O. House and G. H. Rasmusson, *J. Org. Chem.* **26,** 4278 (1961).
43. G. Drefahl and G. Plotner, *Chem. Ber.* **94,** 907 (1961).
44. T. W. Campbell and R. N. McDonald, *J. Org. Chem.* **24,** 1969 (1959).
45. J. P. Freeman, *Chem. & Ind.* (*London*) 1254 (1959).
46. F. Bohlmann, *Chem. Ber.* **89,** 2191 (1956).
47. E. A. LaLancette, *J. Org. Chem.* **29,** 2957 (1964).
48. H. Goetz, F. Nerdel and H. Michaelis, *Naturwiss.* **50,** 496 (1963).
49. A. J. Speziale and D. E. Bissing, *J. Am. Chem. Soc.* **85,** 3878 (1963).
50. S. Fliszar, R. F. Hudson and G. Salvadori, *Helv. Chim. Acta* **46,** 1580 (1963).

51. M. A. Battiste, *J. Am. Chem. Soc.* **86**, 942 (1964).
52. M. P. Cava and R. J. Pohl, *J. Am. Chem. Soc.* **82**, 5242 (1960).
53. A. T. Blomquist and V. J. Hruby, *J. Am. Chem. Soc.* **86**, 5041 (1964).
54. J. Parrick, *Can. J. Chem.* **42**, 190 (1964).
55. S. Trippett, *Chem. & Ind.* (*London*) 80 (1956).
56. J. D. Surmatis and A. Ofner, *J. Org. Chem.* **26**, 1171 (1961).
57. G. Wittig and H. Pommer, Ger. Patent, 32741 IVB/120, Sept. 25, 1954.
58. E. Buchta and F. Andree, *Chem. Ber.* **92**, 3111 (1959).
59. T. W. Campbell and R. N. McDonald, *J. Org. Chem.* **24**, 1246 (1959).
60. F. Bohlmann, *Chem. Ber.* **90**, 1519 (1957).
61. S. G. Levine, *J. Am. Chem. Soc.* **80**, 6150 (1958).
62. G. Wittig and E. Knauss, *Angew. Chem.* **71**, 127 (1959).
63. H. J. Bestmann, O. Kratzer and H. Simon, *Chem. Ber.* **95**, 2750 (1962).
64. M. Schlosser, *Chem. Ber.* **97**, 3219 (1964).
65. R. N. McDonald and T. W. Campbell, *J. Am. Chem. Soc.* **82**, 4669 (1960).
66. G. Wittig, H. Eggers and P. Duffner, *Ann*, **619**, 10 (1958).
67. K. Dimroth and G. Pohl, *Angew. Chem.* **73**, 436 (1961).
68. C. E. Griffin and J. A. Peters, *J. Org. Chem.* **28**, 1715 (1963).
69. C. E. Griffin and G. Witschard, *J. Org. Chem.* **27**, 3334 (1962).
70. C. E. Griffin and G. Witschard, *J. Org. Chem.* **29**, 1001 (1964).
71. T. I. Bieber and E. H. Eisman, *J. Org. Chem.* **27**, 678 (1962).
72. E. E. Schweizer and R. Schepers, *Tetrahedron Letters* 979 (1963).
73. E. E. Schweizer, *J. Am. Chem. Soc.* **86**, 2744 (1964).
74. E. E. Schweizer and K. K. Light, *J. Am. Chem. Soc.* **86**, 2963 (1964).
75. H. J. Bestmann and O. Kratzer, *Chem. Ber.* **95**, 1894 (1962).
76. A. J. Speziale and K. W. Ratts, *J. Am. Chem. Soc.* **84**, 854 (1962).
77. H. H. Jaffe, *Chem. Revs.* **53**, 191 (1953).
78. S. Trippett and D. M. Walker, *Chem. & Ind.* (*London*) 933 (1960); *J. Chem. Soc.* 1266 (1961).
79. D. B. Denney and M. J. Boskin, *Chem. & Ind.* (*London*) 330 (1959).
80. S. Trippett, *Pure and Applied Chem.* **9**, 255 (1964).
81. D. B. Denney and H. M. Relles, *J. Am. Chem. Soc.* **86**, 3897 (1964).
82. F. Ramirez, O. P. Madan and C. P. Smith, *Tetrahedron Letters* 201 (1965).
83. R. Ketcham, D. Jambatkar and L. Martinelli, *J. Org. Chem.* **27**, 4666 (1962).
84. S. B. Hartley, W. S. Holmes, J. K. Jacques, M. F. Mole and J. C. McCoubrey, *Quart. Rev.* **17**, 204 (1963).
85. W. E. McEwen and A. P. Wolf, *J. Am. Chem. Soc.* **84**, 676 (1962).
86. L. D. Bergelson and M. M. Shemyakin, *Tetrahedron* **19**, 149 (1963).
87. K. Yates and R. Stewart, *Can. J. Chem.* **37**, 664 (1959).
88. L. D. Bergelson and M. M. Shemyakin, *Pure and Applied Chem.* **9**, 271 (1964).
89. W. E. McEwen, K. F. Kumli, A. Blade-Font, M. Zanger and C. A. VanderWerf, *J. Am. Chem. Soc.* **86**, 2378 (1964).
90. P. C. Haake and F. H. Westheimer, *J. Am. Chem. Soc.* **83**, 1102 (1961).
91. L. Horner and H. Winkler, *Tetrahedron Letters* 3265 (1964).
92. G. Drefahl, D. Lorenz and G. Schnitt, *J. Prakt. Chem.* **23**, 143 (1964).
93. L. D. Bergelson, V. A. Vaver, L. I. Barsukov and M. M. Shemyakin, *Tetrahedron Letters* 2669 (1964).
94. L. D. Bergelson and M. M. Shemyakin, *Angew. Chem.* **76**, 113 (1964); *International Edn.* **3**, 250 (1964).
95. L. D. Bergelson, V. A. Vaver and M. M. Shemyakin, *Izvest. Akad. Nauk. S.S.S.R.* 729 (1961).

96. A. Maercker, *in* "Organic Reactions" (R. Adams, ed.), Vol. 14, p. 270. John Wiley and Sons, Inc., New York, 1965.
97. H. Freyschlag, H. Grassner, A. Nurrenbach, H. Pommer, W. Reif and W. Sarneck, *Angew. Chem.* **77**, 277 (1965); *International Edn.* **4**, 287 (1965).
98. S. Trippett and D. J. Walker, *Chem. Comms.* 106 (1965).
99. A. Wm. Johnson and V. L. Kyllingstad, *J. Org. Chem.* **31**, 334 (1966).
100. M. Schlosser and K. F. Christmann, *Angew. Chem.* **77**, 682 (1965); *International Edn.* **4**, 689 (1965).
101. D. E. Bissing and A. J. Speziale, *J. Am. Chem. Soc.* **87**, 2683 (1965).

5

OTHER PHOSPHORUS YLIDS

As has been discussed in Chapter 2, current theory predicts that a carbanion, when located adjacent to a phosphorus atom carrying a high degree of positive charge, ought to be stabilized by overlap of the orbital carrying the carbanion electrons with a vacant 3*d*-orbital of the phosphorus atom. The nature of the substitution on the phosphorus atom would be expected to alter only the degree of this stabilization, not the kind. As indicated by the discussion in the preceeding two chapters, phosphonium groups are especially proficient at providing such stabilization for carbanions since the phosphorus atom carries a large positive charge. Accordingly it has been possible even to isolate some phosphonium ylids due to their decreased basicity.

There are many other types of organophosphorus compounds which ought to lend themselves to carbanion formation. It is the purpose of this chapter to discuss the chemistry of these molecular systems. In most instances the interest in these systems has arisen either from a search for carbanion-stabilizing groups or from a search for reagents which would convert carbonyl compounds into olefins, i.e. complement the Wittig synthesis of olefins. Accordingly, one of the first reactions to be studied when such a system is at hand often is the reaction of the carbanion with a carbonyl compound, an olefin being the desired product.

I. Phosphinoxy Carbanions

Horner *et al.* (*1*) were the first to explore the generation of a phosphinoxy carbanion and its subsequent reaction with a carbonyl compound. Treatment of diphenylmethylphosphine oxide with sodamide and then benzophenone afforded a 70% yield of 1, 1-diphenylethene and a 95% yield of diphenylphosphinic acid. They proposed that a phosphinoxy carbanion (I, methylenediphenyloxyphosphorane) was formed which then attacked the carbonyl carbon of the ketone to afford a betaine

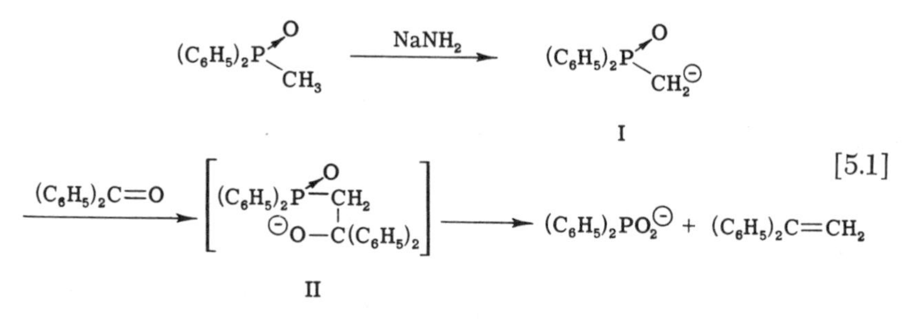

intermediate (II), analogous to that proposed for the Wittig reaction. This betaine was thought to undergo oxyanion transfer to the phosphorus atom resulting in olefin formation. The driving force for the reaction was thought to be the formation of the new phosphorus-oxygen bond, the diphenylphosphinate anion.

Seyferth *et al.* (*2*) recently have shown that it is possible to prepare certain phosphinoxy carbanions by reaction of an alkyllithium with triphenylphosphine oxide; in effect an exchange reaction is carried out. Thus, triphenylphosphine oxide and methyllithium appeared initially to afford diphenylmethylphosphine oxide and phenyllithium since quenching of the solution with deuterium oxide afforded largely monodeuterated benzene [5.2]. Continued heating of the solution, however,

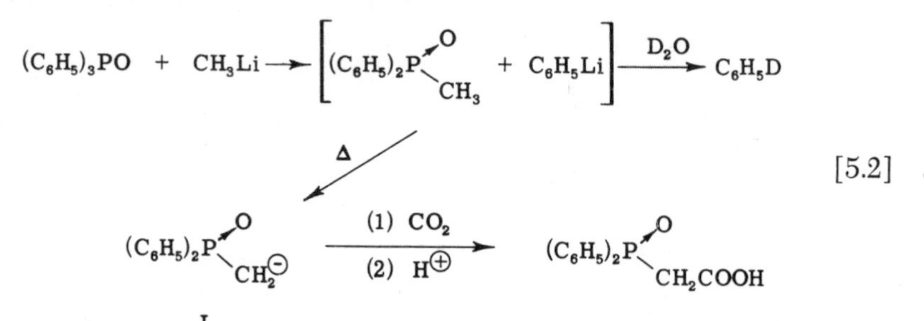

resulted in the metalation of the new phosphine oxide forming a phosphinoxy carbanion (I). The existence of I was indicated by its reaction with carbon dioxide. Analogous results have been reported by Seyferth and Welch (*3*) with triphenylphosphine sulfide.

The scope of the reaction of phosphinoxy carbanions with carbonyl compounds to afford olefins has not been studied in great detail but thus far seems to parallel the normal Wittig reaction of phosphonium carbanions (ylids). Both Horner (*1*) and Hoffmann (*4*) have studied the reaction and Horner *et al.* (*5*) have shown that bis-carbanions also will partake in the reaction [5.3]. The analogous reaction in the phos-

$$(C_6H_5)_2P(O)\text{—}CH_2CH_2\text{—}P(O)(C_6H_5)_2 + (C_6H_5)_2C{=}O \xrightarrow{KOC_4H_9^t} (C_6H_5)_2C{=}CH\text{—}CH{=}C(C_6H_5)_2 \quad [5.3]$$

phonium carbanion series was not possible since the intermediate ylid underwent fragmentation into triphenylphosphine and vinyltriphenylphosphonium bromide (*6*, Wittig *et al.*).

The conversion of phosphine oxides into their corresponding carbanions and subsequent reaction with carbonyl compounds to afford olefins is quite sensitive to the nature of the base used for carbanion formation. No olefin was obtained when diphenylethylphosphine oxide was treated with benzophenone in the presence of sodium methoxide, sodium ethoxide, methylmagnesium bromide, metallic sodium or metallic potassium. On the other hand, good yields of olefin resulted when the same mixture was treated with potassium *tert*-butoxide, sodamide, potassium amide or potassium phenyl. Perhaps the most interesting and pertinent observation was made by Horner *et al.* (*7*) in 1959 and later by Richards and Banks in 1963 (*8*). They found that reaction of diphenylbenzylphosphine oxide with benzaldehyde in the presence of phenyllithium afforded no olefin. Instead, the hydroxyphosphine oxide (IV) was obtained in good yield. However, reaction of the same two

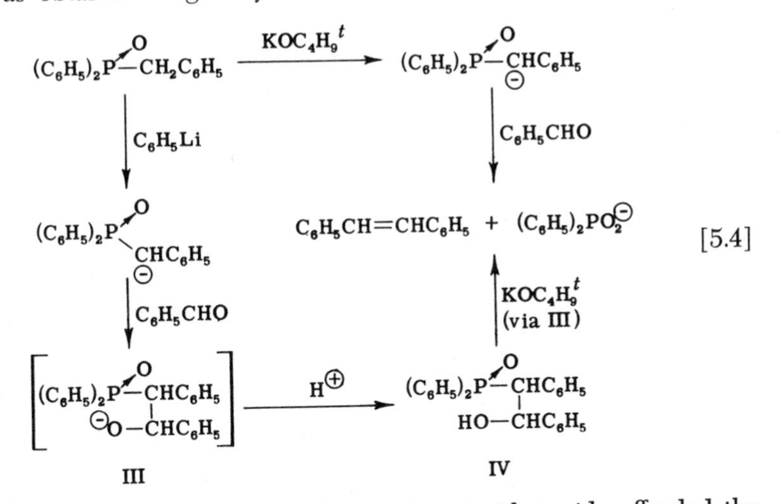

components but in the presence of potassium *tert*-butoxide afforded the expected olefin, stilbene, in good yield. The isolation of the hydroxyphosphine oxide (IV) provided convincing evidence that the betaine

(III) was an intermediate in the reaction. Furthermore, regeneration of the betaine from IV using potassium *tert*-butoxide led to the formation of stilbene in near quantitative yield. Horner concluded that lithium cations complexed rather strongly with the oxyanion of the betaine (III), thereby preventing its decomposition whereas the potassium ion did not. These observations parallel those of Schlosser and Christmann (*9*) with betaines from phosphonium ylids.

Horner and Winkler (*10*) showed, by reaction of the carbanion from (+)-methylphenylbenzylphosphine oxide with benzal aniline, that the resulting (−)-methylphenylphosphinic acid anilide had the same configuration as the starting phosphine oxide [5.5]. Therefore, the olefin-

$$(+)\text{-}CH_3(C_6H_5)P(\rightarrow O)CH_2C_6H_5 + C_6H_5CH{=}NC_6H_5 \xrightarrow{KOC_4H_9{}^t} (-)\text{-}CH_3(C_6H_5)P(\rightarrow O)NHC_6H_5 + C_6H_5CH{=}CHC_6H_5 \qquad [5.5]$$

forming reaction must have proceeded with retention of configuration about the phosphorus atom and appeared to be a *cis* elimination, identical to the normal Wittig reaction of phosphonium ylids.

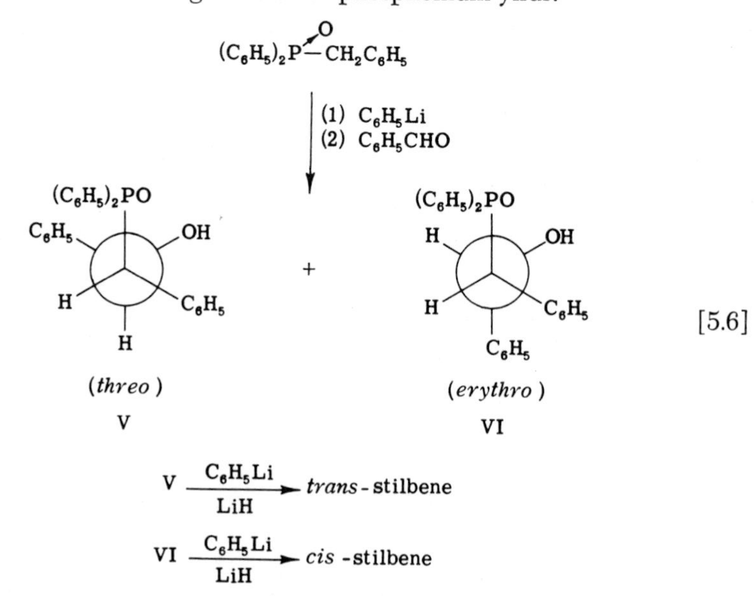

Horner and Klink (*11*) have investigated the stereochemistry of the olefin-forming aspect of the reaction of phosphinoxy carbanions with carbonyl compounds. Reaction of diphenylbenzylphosphine oxide with benzaldehyde in the presence of phenyllithium afforded a mixture of the *threo* (V)- and *erythro* (VI)-hydroxyphosphine oxides in 90% and 5% yields, respectively. They were separated by thin-layer chromatography. The stereochemistry of these intermediates was assigned on the basis of the observation that the supposed *threo* form (V), when treated with phenyllithium and lithium hydride, was converted into *trans*-stilbene. Likewise, the *erythro* form (VI) was converted into *cis*-stilbene. These are the products expected from a *cis* elimination but the overall logic fails to prove the stereochemistry [5.6].

Both the *threo*- and *erythro*-hydroxyphosphine oxides also were prepared by catalytic hydrogenation of diphenyldesylphosphine oxide [5.7]. However, this reaction provided no confirmation of the assignment of stereochemistry to V and VI.

$$(C_6H_5)_2P(\rightarrow O){-}CH(C_6H_5){-}\overset{O}{\overset{\|}{C}}C_6H_5 \xrightarrow[H_2]{Pd/C} 70\%\ \text{V} + 10\%\ \text{VI} \qquad [5.7]$$

A third preparation of the two hydroxyphosphine oxides (V and VI) has led to considerable doubt regarding the original stereochemical assignment. Reaction of lithiodiphenylphosphine oxide with *cis*- and *trans*-stilbene oxides led to the formation, in very low yields, of the *erythro* (VI)- and *threo* (V)-hydroxyphosphine oxides, respectively, along with small quantities of diphenylbenzylphosphine oxide and benzaldehyde. By analogy with the reactions of other nucleophiles with epoxides (*12, 13*) a *trans* ring opening would be expected and the *cis*-epoxide should have afforded the *threo* form (V) and the *trans*-epoxide should have afforded the *erythro form* (VI) [5.8]. Two conclusions are

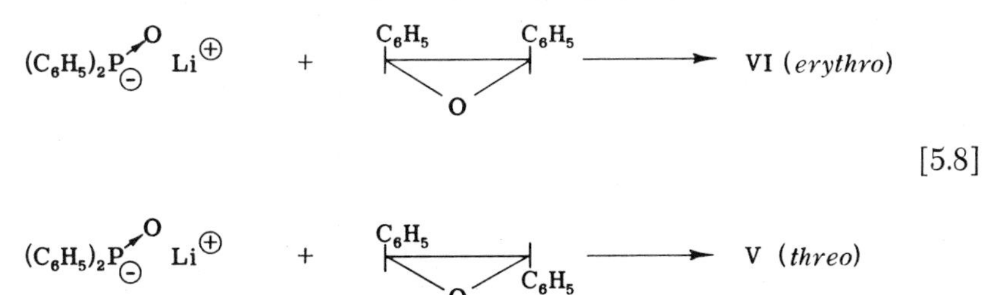

possible. If the original assignment of stereochemistry for V and VI was correct, then the above reaction with the stilbene oxides must have proceeded by a *cis* ring opening. There is little precedent for such a proposal. On the other hand, if the original assignment for V and VI was incorrect, then the mechanism of the conversion of the betaine (III) into olefin and phosphinate must not have involved a *cis* elimination. However, Horner and Winkler (*10*) had previously shown the reaction to involve a *cis* elimination. Further experimental clarification obviously is in order.

The isolation of small amounts of diphenylbenzylphosphine oxide and benzaldehyde from the reaction of lithiodiphenylphosphine oxide with the stilbene oxides indicated that the two stereoisomeric betaines corresponding to V and VI were capable of following two reaction paths: either decomposition to olefin and phosphinate or dissociation to starting materials, the aldehyde and carbanion. Confirmation was provided by the observation that the supposed *erythro*-hydroxyphosphine oxide (VI), when treated with potassium *tert*-butoxide, afforded *trans*-stilbene [5.9] (*11*). Previously it had been shown that with phenyl-

$$\text{VI } (erythro) \xrightarrow[\text{LiH}]{C_6H_5Li} cis\text{-stilbene}$$

$$\text{VI} \xrightarrow{KOC_4H_9{}^t} trans\text{-stilbene}$$

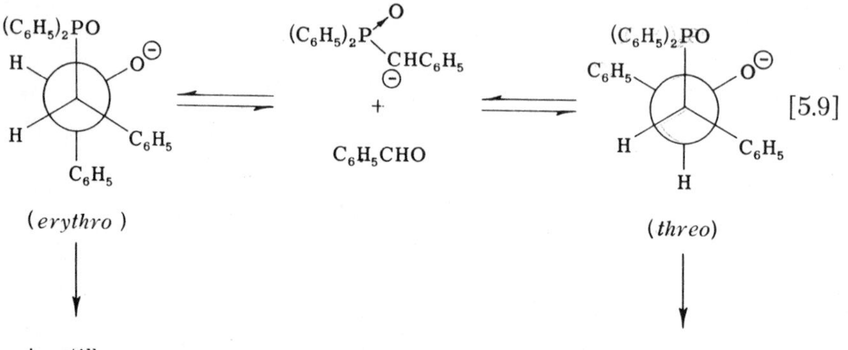

lithium-lithium hydride VI was converted into *cis*-stilbene. It appeared possible that in the presence of the potassium cation the betaine was capable of dissociating to carbanion and benzaldehyde. The presence of the latter was proven by its entrapment with 2,4-dinitrophenylhydrazine. The carbanion and benzaldehyde probably recombined to form mainly the *threo*-betaine (V) which decomposed into *trans*-stilbene and phosphinate. This argument is consistent with the less sterically hindered nature of the transition state for the formation of *trans*-olefin from *threo*-betaine. In the other transition state, that for the formation of *cis*-olefin from *erythro*-betaine, the two phenyl groups would be eclipsed and unable to stabilize the forming double bond.

In summary then, on the basis of the limited data available to date it appears that the stereochemistry and mechanism of the reaction of phosphinoxy carbanions with carbonyl compounds to form olefins and phosphinates are very similar, in a gross sense, to the mechanism and stereochemistry of the Wittig reaction of phosphonium ylids with carbonyl compounds to form olefins and phosphine oxides. Betaine formation seems to be involved and it appears to be a reversible reaction. The elimination seems to be of a *cis* type in which the *threo*-betaines afford *trans*-olefins and the *erythro*-betaines afford *cis*-olefins, the former being the route of lowest energy.

In view of this interesting parallel between the reaction of phosphonium and phosphinoxy carbanions with carbonyl compounds, Horner *et al.* (*14*) were led to compare the two in competition reactions. Treatment of a mixture of one equivalent of diphenylbenzylphosphine oxide and one equivalent of triphenylbenzylphosphonium bromide with a mixture of one equivalent of potassium *tert*-butoxide and one equivalent of benzaldehyde led to the formation of stilbene (68%), triphenylphosphine oxide (53%) and diphenylphosphinic acid (10%) along with the recovery of 77% unreacted diphenylbenzylphosphine oxide and only 39% unre-

$$(C_6H_5)_2P(\rightarrow O)CH_2C_6H_5 \quad + \quad (C_6H_5)_3\overset{\oplus}{P}CH_2C_6H_5 \quad Br^{\ominus}$$

$$\xrightarrow[\;]{KOC_4H_9{}^t \quad | \quad C_6H_5CHO} \qquad [5.10]$$

$$C_6H_5CH{=}CHC_6H_5 \quad + \quad (C_6H_5)_3PO \quad + \quad (C_6H_5)_2PO_2H$$

acted triphenylbenzylphosphonium bromide [5.10]. Thus, about 84% of the olefin was formed by reaction of the phosphonium ylid with aldehyde, resulting in the formation of triphenylphosphine oxide. On this basis, the phosphonium carbanion (ylid) appeared either to be formed faster than the phosphinoxy carbanion or to be more reactive. Repetition of the experiment but using nearly four equivalents of potassium *tert*-butoxide led to the formation of stilbene (89%), triphenylphosphine oxide (7%) and diphenylphosphinic acid (65%) along with the recovery of unchanged diphenylbenzylphosphine oxide (7%) and triphenylphosphonium bromide (83%). Thus, when excess base was present so as to permit the formation of both phosphonium and phosphinoxy carbanions, about 90% of the stilbene was formed by reaction of the phosphinoxy carbanion with aldehyde.

These two experiments lead to the conclusion that a phosphonium group provides more stabilization for a carbanion than does a phosphinoxy group, a phosphonium salt therefore being more acidic than a phosphine oxide [5.11]. This is the relationship to be expected if

$$(C_6H_5)_2P(\rightarrow O)\overset{\ominus}{C}HC_6H_5 + (C_6H_5)_3\overset{\oplus}{P}CH_2C_6H_5 \rightleftharpoons (C_6H_5)_2P(\rightarrow O)CH_2C_6H_5 + (C_6H_5)_3\overset{\oplus}{P}-\overset{\ominus}{C}HC_6H_5 \qquad [5.11]$$

$$K > 1$$

the stabilization of the carbanion depends upon the valence shell expansion of the phosphorus atom. The more positive charge localized on the phosphorus atom the better should be the overlap of the filled orbital of the carbanion with the phosphorus empty 3*d*-orbital. The phosphorus atom should have a larger positive charge in the phosphonium salt than in the phosphine oxide due to back donation from the oxygen in the latter. Since the phosphonium carbanion appears to be the most stabilized of the two, it also would be expected to be the least nucleophilic of the two. This prediction was verified by the observation that in the presence of both carbanions, both benzaldehyde and benzophenone reacted preferentially with the phosphinoxy carbanion. The difference in the reactivity between the two carbanions apparently was reflected in their relative rates of nucleophilic attack on the carbonyl carbon, i.e. the betaine formation step.

Confirmation of the above reactivity relationships was provided by the reaction of the combined phosphine oxide-phosphonium salt (VII) with benzophenone in the presence of excess potassium *tert*-butoxide. The phosphonium-olefin (VIII) and diphenylphosphinic acid were the only isolable products from the reaction, there being no sign of triphenylphosphine oxide. Here again the phosphinoxy carbanion proved to be more reactive than the phosphonium carbanion (ylid) [5.12].

Although the most interesting reaction of phosphinoxy carbanions may be their reaction with carbonyl compounds to form olefins, so-called "PO-activated olefin formation" (*14*), these carbanions participate in a number of other interesting reactions which are also characteristic of

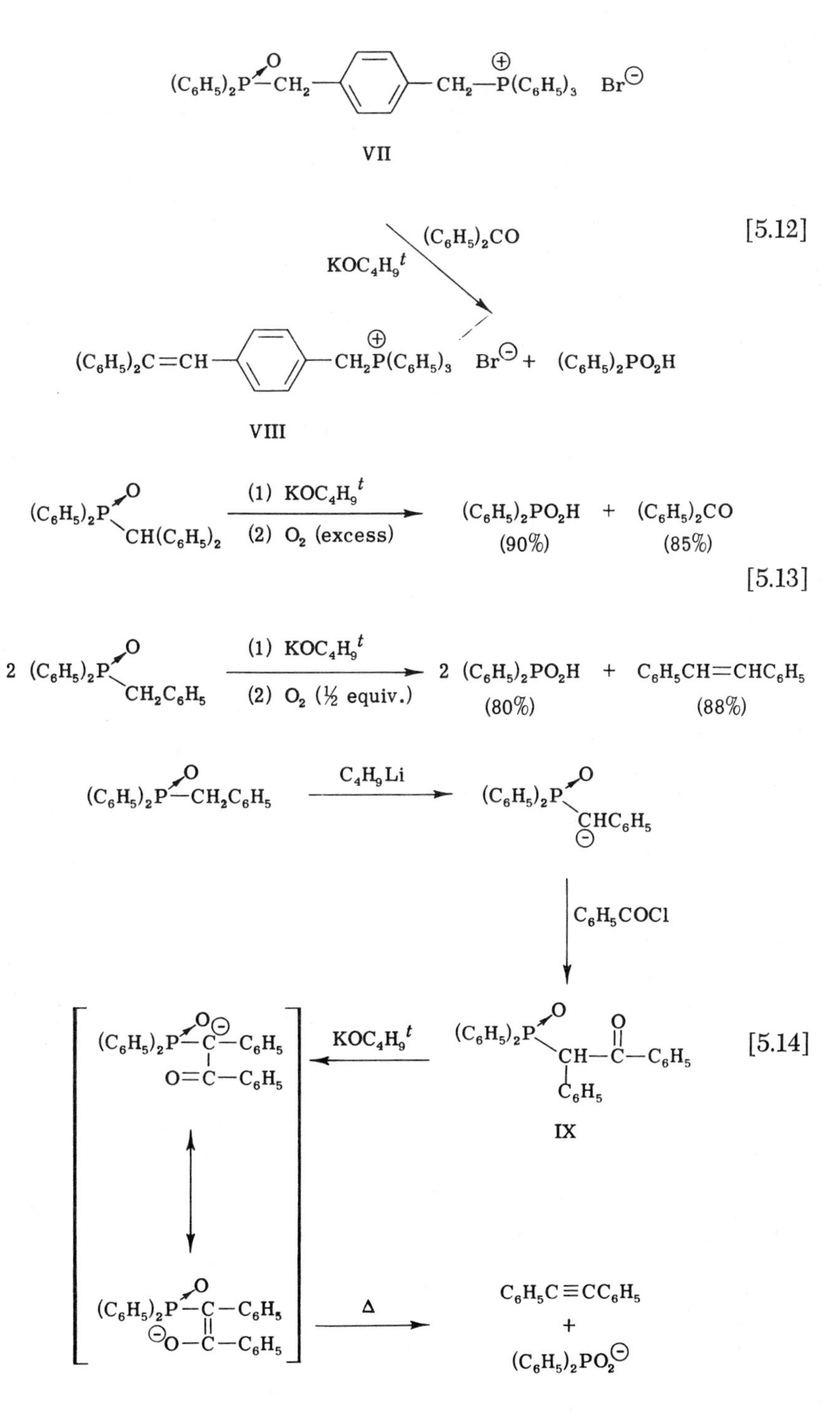
(C6H5)2P(O)—CH2—C6H4—CH2—P(C6H5)3 Br⊖
VII
KOC4H9t
(C6H5)2CO
[5.12]
(C6H5)2C=CH—C6H4—CH2P(C6H5)3 Br⊖ + (C6H5)2PO2H
VIII
(C6H5)2P(O)CH(C6H5)2
(1) KOC4H9t
(2) O2 (excess)
(C6H5)2PO2H + (C6H5)2CO
(90%)
(85%)
[5.13]
2 (C6H5)2P(O)CH2C6H5
(1) KOC4H9t
(2) O2 (½ equiv.)
2 (C6H5)2PO2H + C6H5CH=CHC6H5
(80%)
(88%)
(C6H5)2P(O)—CH2C6H5
C4H9Li
(C6H5)2P(O)CHC6H5
C6H5COCl
(C6H5)2P(O)CH(C6H5)—C(O)—C6H5
IX
KOC4H9t
[5.14]
(C6H5)2P(O)—C⊖(C6H5)—C(=O)—C6H5
(C6H5)2P(O)—C(C6H5)=C(O⊖)—C6H5
Δ
C6H5C≡CC6H5
+
(C6H5)2PO2⊖

the closely related phosphonium carbanions (ylids). Horner *et al.* (*15*) found that phosphinoxy carbanions will react with oxygen undergoing cleavage to phosphinates and carbonyl compounds [5.13]. In the presence of one-half equivalent of oxygen, the resulting carbonyl compound reacts with the remaining carbanion to form an olefin, the overall result being the superficial dimerization of the carbon portion of the original carbanion. These observations are analogous to those made by Bestmann and Kratzer (*16*) on phosphonium ylids.

Hoffmann (*4*) reported the acylation of phosphine oxides by the reaction of phosphinoxy carbanions with esters. Later, Gough and Trippett (*17*) acylated diphenylbenzylphosphine oxide with benzoyl chloride in the presence of butyllithium. The resulting desyldiphenylphosphine oxide (IX) could be converted into its carbanion whose enolate form was, in essence, a phosphinoxy betaine and would eliminate phosphinate to form diphenylacetylene. Both the acylation and the alkyne formation are analogous to reactions undergone by phosphonium ylids [5.14].

Horner *et al.* (*18*) found that phosphinoxy carbanions also would react with oxiranes to afford cyclopropanes in a manner analogous to that exhibited by phosphonium ylids (*19*, Denney *et al.*). Diphenylbenzylphosphine oxide reacted with styrene oxide in the presence of potassium *tert*-butoxide to afford 1,2-diphenylcyclopropane in 37% yield. In the presence of phenyllithium the same phosphine oxide reacted with stilbene oxide but the reaction stopped at the betaine stage, a hydroxyphosphine oxide analogous to XI actually being isolated. Here again the lithium cation appears to have prevented oxyanion transfer to phosphorus in the betaine stage (i.e. XII) as was the case for the betaine (III) from the same phosphinoxy carbanion and benzaldehyde (*7*, Horner *et al.*). The isolation of a betaine conjugate acid provides evidence that the reaction between the carbanion and the oxide normally proceeds via nucleophilic attack on carbon.

Horner *et al.* (*18*) indicated the hydroxyphosphine oxide from the reaction between diphenylbenzylphosphine oxide and styrene oxide in the presence of phenyllithium to be the primary alcohol (X). No specific evidence was provided for this structure and it seems more likely that the actual structure is that of a secondary alcohol (XI) since the carbanion would be expected to attack the primary carbon of the styrene oxide. Reaction of benzylidenetriphenylphosphorane with styrene oxide also afforded 1,2-diphenylcyclopropane and appeared to proceed by attack of the phosphonium carbanion on the primary carbon of styrene oxide to form an intermediate analogous to XII (*20*, McEwen and Wolf). Tomoskozi (*21*) recently has shown that as for the phosphonium

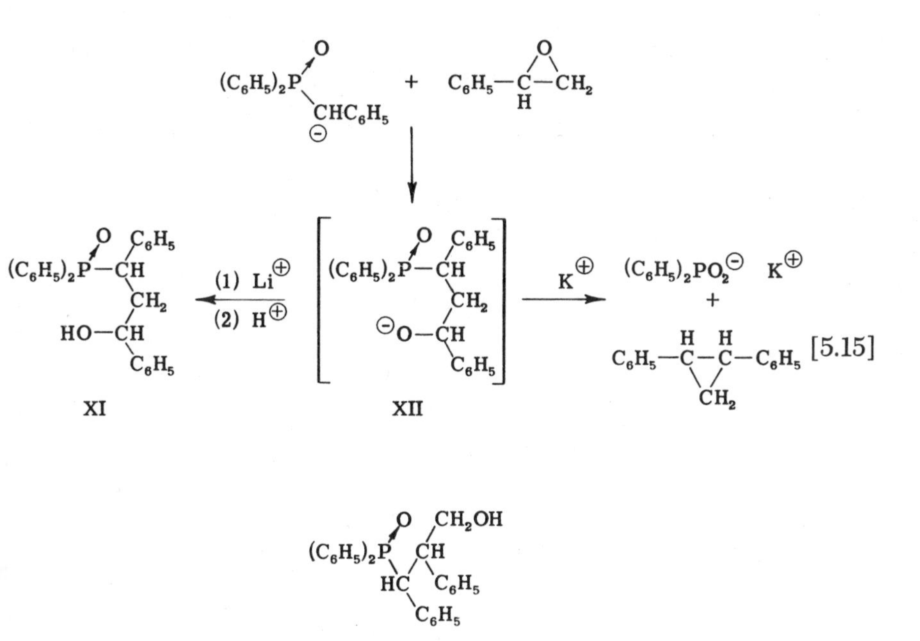

ylid case, reaction of phosphinoxy carbanions with optically active styrene oxide proceeded with inversion of configuration of the asymmetric carbon.

On the basis of the discussion in this section it is apparent that phosphinoxy carbanions are very similar to phosphonium ylids in their chemical properties. None of the former have been isolated so there is no physical data available. The phosphinoxy groups appears to be somewhat less effective than a phosphonium group in delocalizing carbanion electrons, presumably by valence shell expansion and use of the 3*d*-orbitals of phosphorus.

II. Phosphonate Carbanions

The history of phosphonate carbanions dates back to 1927. The major interest at that time was whether or not a phosphonate group could replace one of the carbonyl groups in a *β*-dicarbonyl compound with retention of the active methylenic properties. Arbuzov *et al.* (*22, 23*) found that carbethoxymethyldiethylphosphonate (XIII) could be converted into its carbanion by treatment with metallic sodium or potassium. This carbanion could be alkylated on carbon with a variety of alkyl halides.

Subsequent to this work other groups explored the scope of the

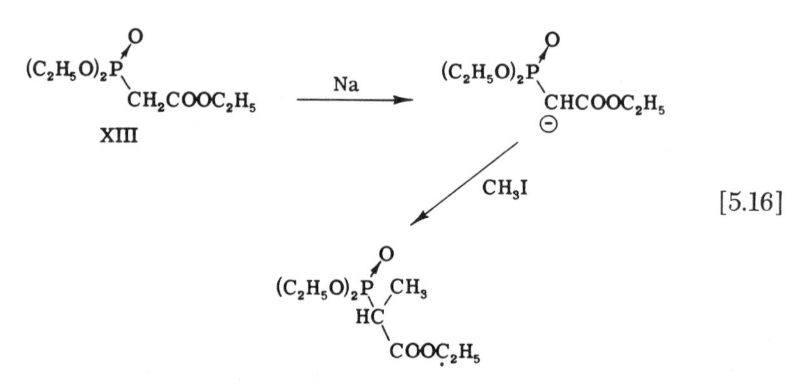

[5.16]

alkylation reaction. Pudovik and Lebedeva (*24*) were able to alkylate acetonyldiethylphosphonate (XIV) on carbon and Kosolapoff (*25*) was able to prepare and alkylate a bis-phosphonyl carbanion [5.17].

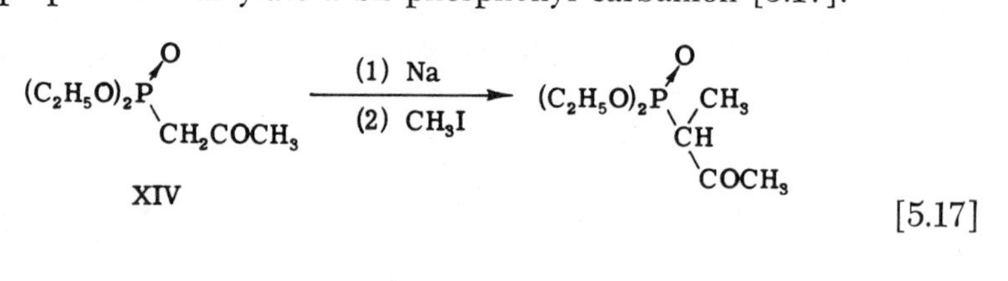

[5.17]

Phosphonate carbanions also have been found to undergo Michael-type additions to conjugated carbonyl systems. Thus, the carbanion prepared from XIV with sodium ethoxide in ethanol added to methyl acrylate to afford the expected adduct (XV) (*26*, Pudovik and Lebedeva). In the presence of excess base, the original adduct was converted to its carbanion and underwent a second addition to methyl acrylate (*24*). Michalski and Musierowicz (*27*) recently have found that the sulfur analog of XIII (XVI) also would form a carbanion and undergo a Michael addition to acrylonitrile.

On the basis of the above examples it is apparent that alkyldiethylphosphonates will form carbanions under standard conditions and that these carbanions will undergo normal carbanionic reactions. In view of the fact that these carbanions probably are stabilized to some extent by the vacant 3*d*-orbitals of the adjacent phosphorus atom, reaction of the carbanions with carbonyl compounds to form betaines would be expected and these betaines also would be expected to undergo oxyanion transfer to afford olefins and diethylphosphates. In other words, phosphonate carbanions should undergo reaction with carbonyl compounds

$$\text{XIV} + CH_2{=}CHCOOCH_3 \xrightarrow{NaOC_2H_5} \underset{\text{XV}}{(C_2H_5O)_2P(\rightarrow O){-}CH(COCH_3){-}CH_2CH_2COOCH_3} + (C_2H_5O)_2P(\rightarrow O){-}C(COCH_3){-}(CH_2CH_2COOCH_3)_2 \qquad [5.18]$$

$$\underset{\text{XVI}}{(C_2H_5O)_2P(\rightarrow S){-}CH_2CO_2C_2H_5} \xrightarrow[(2)\ CH_2{=}CHCN]{(1)\ Na} (C_2H_5O)_2P(\rightarrow S){-}CH(COOC_2H_5)CH_2CH_2CN$$

to afford olefins in a manner analogous to that of phosphinoxy and phosphonium carbanions.

Horner *et al.* (*1*) in 1958 were the first to observe this reaction. Diethylbenzylphosphonate reacted with benzophenone in the presence of sodamide to afford triphenylethylene in 88% yield [5.19]. Subsequent

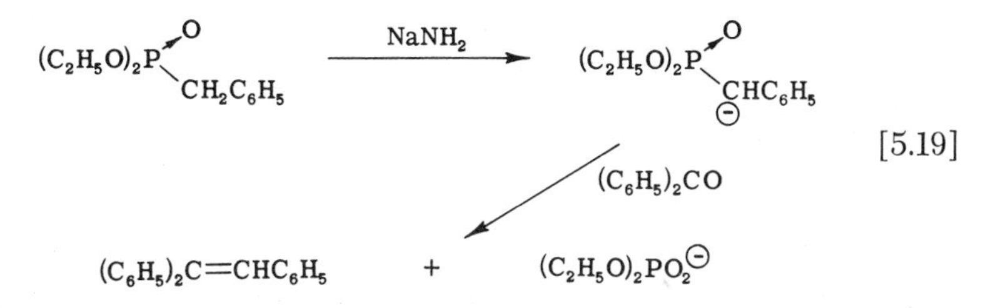

to this observation Wadsworth and Emmons (*28*) carried out a broad study of the use of phosphonate carbanions in olefins synthesis. Using a slurry of sodium hydride in dimethoxyethane to generate the carbanion from a variety of alkyl diethylphosphonates, they were able to effect reaction with both aldehydes and ketones under mild conditions to obtain high yields of olefins. Many of the reactions occurred at or only slightly above room temperature and the sodium diethylphosphate precipitated almost immediately.

Since the above reports many have used phosphonate carbanions in olefin synthesis. The nature of the bases and the solvents used have varied from sodium hydride to sodamide and sodium ethoxide and from tetrahydrofuran and dimethylformamide to benzene. Kovalev *et al.* (*29*), Takahashi *et al.* (*30*) and Bose and Dahill (*31*) have explored the use of XIII as a replacement for the Reformatsky synthesis of acrylates. Normant and Sturtz (*32*) have used XIV for the synthesis of α,β-unsaturated ketones, and Drefahl *et al.* (*33*) have used cyanomethyl diethylphosphonate to prepare acrylonitrile derivatives [5.20]. Horner

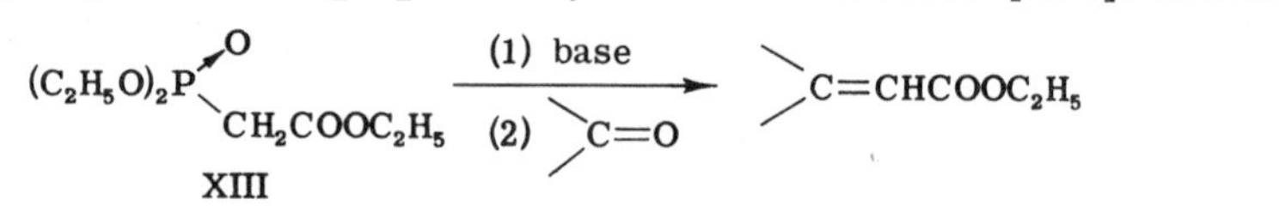

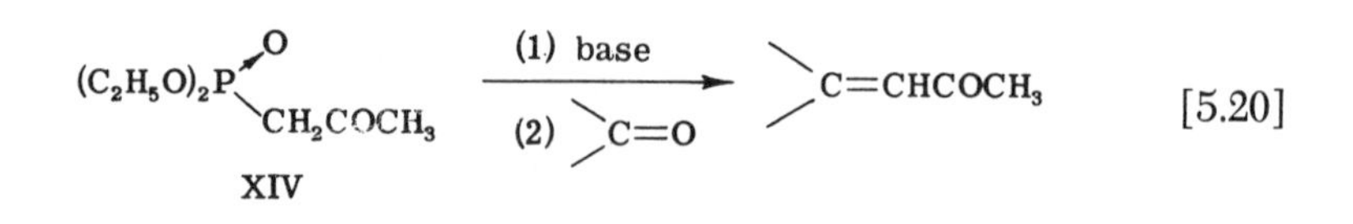

$$(C_2H_5O)_2P(O)CH_2CN \xrightarrow[(2)\ >C=O]{(1)\ \text{base}} >C=CHCN$$

et al. (*5*) have used bis-phosphonates for the ultimate preparation of diolefins, and Zimmer and Bercz (*34*) have used aminomethyl diethylphosphonates for the preparation of enamines. Whitlock (*35*) has used *o*-xylylene-bis-diethylphosphonate in a reaction with an α-diketone for the synthesis of a cyclic diene. Michalski and Musierowicz (*27*) have found that XVI also will form a carbanion in the presence of potassium *tert*-butoxide which will afford olefins upon reaction with carbonyl compounds.

Tomoskozi and Janzso (*36*) reported an example of asymmetric induction in the synthesis of olefins by the phosphonate carbanion method. Formation of the carbanion from (−)-carbomenthoxymethyl diethylphosphonate using sodium hydride and then its reaction with 4-methylcyclohexanone afforded an optically active ester which was hydrolyzed to an optically active acid (XVII) of about 50% optical purity.

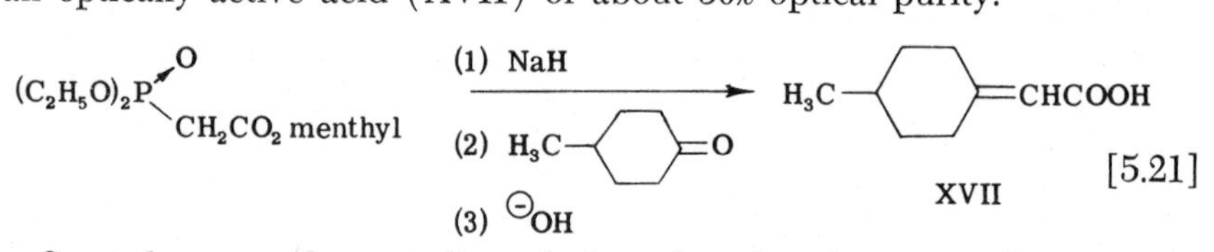

Several groups have indicated that the phosphonate carbanions are more reactive than are the corresponding phosphonium carbanions (ylids) in the olefin-forming reaction. For example, Trippett and Walker (*37*) indicated that carbethoxymethyloxydiethoxyphosphorane (the carbanion of XIII) would react with most ketones whereas carbethoxymethylenetriphenylphosphorane would not. Similarly, Wadsworth and Emmons (*28*) reported that the carbanion from phenacyl diethylphosphonate was much more reactive than phenacylidenetriphenylphosphorane. The clearest demonstration of the relative reactivity of the phosphonium and phosphonate carbanions is the report by Horner *et al.* (*14*) that treatment of a mixture of one equivalent of benzyl diethylphosphonate and one equivalent of benzyltriphenylphosphonium bromide with one equivalent of benzophenone and 3.6 equivalents of potassium *tert*-butoxide led to the formation of a 67% yield of triphenylethylene with a recovery of 70% of the phosphonium salt unchanged from the reaction mixture. There was no indication of the amounts, if any, of triphenylphosphine oxide and potassium diethylphosphate recovered. However, the amount of phosphonium salt recovered indicates that a minor amount of the olefin probably was formed by reaction of the phosphonium carbanion with benzophenone. A phosphonate carbanion would be expected to be more nucleophilic than a phosphonium carbanion, and hence more reactive with carbonyl compounds, since the phosphonate group should have a lower net positive charge and thereby afford less stabilization for the carbanion by valence shell expansion. This relative nucleophilicity was observed by Horner *et al.* (*14*).

The mechanism of the reaction of phosphonate carbanions with carbonyl compounds probably is analogous to that for the Wittig reaction. The betaine probably is formed by nucleophilic attack of the carbanion on the carbonyl carbon of an aldehyde or ketone. This betaine may be able to dissociate to carbanion and carbonyl compound—there is no evidence on this point. The betaine probably undergoes decomposition to olefin and phosphate by oxygen transfer to phosphorus via a *cis* elimination. The driving force for the elimination must be the formation of a new phosphorus-oxygen bond in the phosphate.

The stereochemistry of the olefin formation has not been clarified adequately but even more so than for the Wittig reaction, the *trans*-olefin usually dominates the reaction mixture (*38*, Wadsworth *et al.*). Attempts at structural and environmental modifications in the olefin formation have failed to produce appreciable amounts of *cis*-olefin (*38*). Bergelson and Shemyakin (*39*) also have reported, but without experimental data, that the dominant formation of *trans*-olefin from reaction of carbethoxymethyl diethylphosphonate carbanion with aldehydes is not subject to environmental changes such as in solvent, temperature and added Lewis acids or bases. All of these fragmentary stereochemical observations must mean that if betaine formation is irreversible and olefin isomerization is absent (a point often *not* checked), the *threo*-betaine (XVIII) must be formed the fastest. If betaine formation is reversible the results would imply that the combined rates of *threo*-betaine formation and decomposition are greater than the combined rates of *threo*-betaine dissociation, *erythro*-betaine formation and *erythro*-betaine decomposition to *cis*-olefin. Clearly there is a need for considerable experimental investigation into the stereochemistry of this reaction.

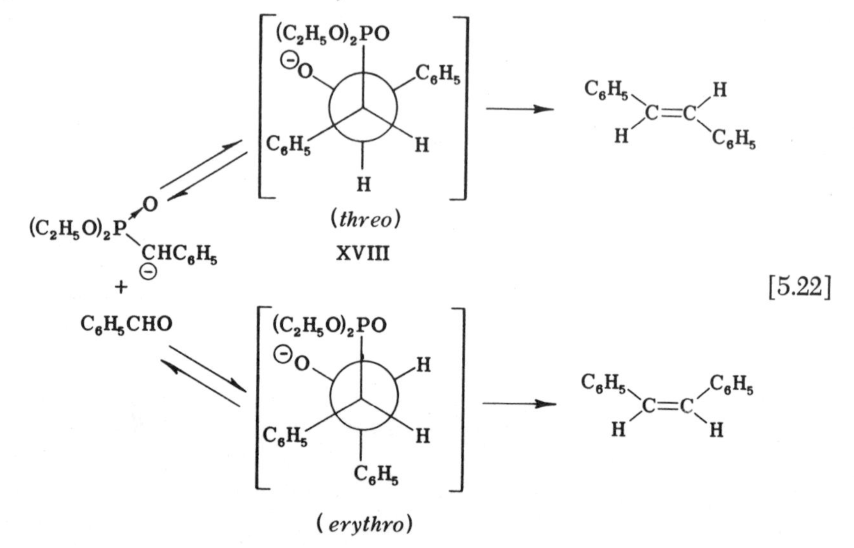

Patai and Schwartz (*40*) have reported that reaction of XIII with benzaldehyde in the presence of piperidine and in ethanolic solution afforded the benzylidene ester (XIX) in 70% yield. The same substance had been reported earlier by Pudovik and Lebedeva (*26*) from heating XIII and benzaldehyde in the presence of acetic anhydride; these are essentially the conditions of the Perkin reaction. The former group found that the diester (XIX) was hydrolyzed to cinnamic acid in the presence of potassium hydroxide, indicating the loss of the phosphonate group. The same workers found that the original reaction, but with *p*-nitro-

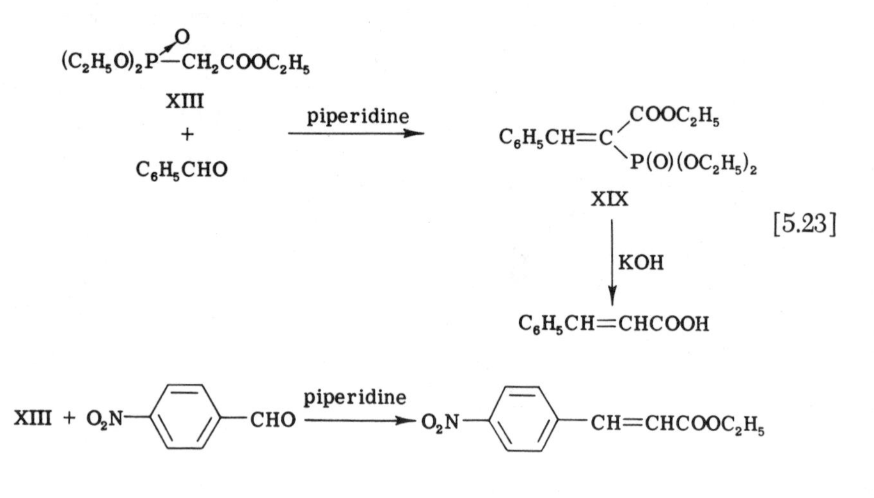

benzaldehyde as the carbonyl component, afforded an 83% yield of ethyl *p*-nitrocinnamate directly without any hydrolysis. They proposed that the reaction proceeded by the same route, the initial formation of a phosphonate-carboxylate diester which was dephosphonylated by the piperidine. They could obtain no reaction with benzophenone.

The direct formation of the *p*-nitrocinnamate in the above reaction was attributed to the stabilization of an intermediate carbanion in the de-phosphonylation step by the nitro group (*40*). It is very difficult to see how the nitro group could provide effective stabilization in the proposed carbanion (XX) if the latter is actually involved in the reaction. It seems more likely that the *p*-nitrocinnamate was formed by a Wittig-type reaction between the phosphonate carbanion and *p*-nitrobenzaldehyde.

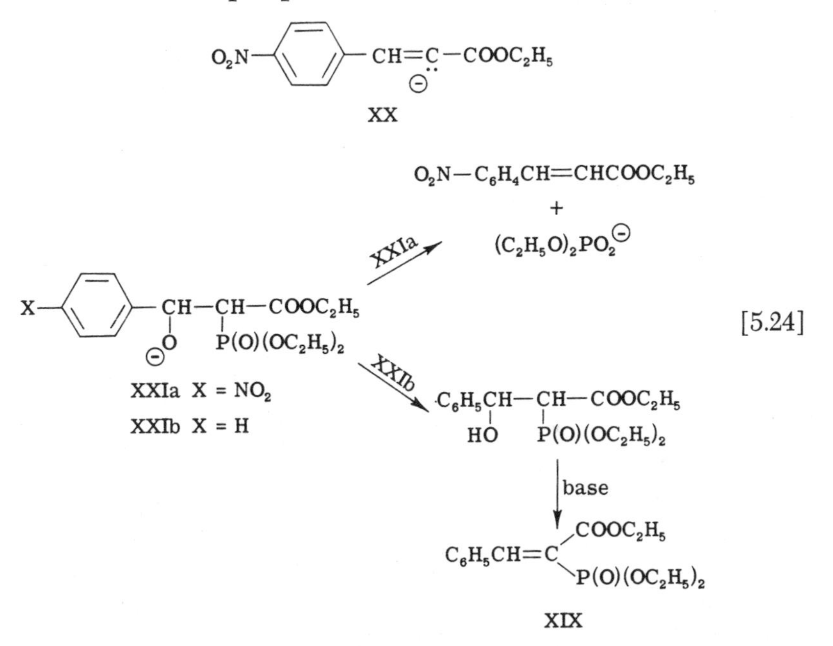

The failure of the reaction with benzaldehyde to proceed by the same route may be attributed to the fact that the reactions were carried out in protonic solvents with neutral bases. The transition state from the nitro-substituted betaine (XXIa) to olefin should be of lower energy than that from the unsubstituted betaine (XXIb) due to stabilization of the incipient double bond by the nitro group. Thus, the former would proceed directly and rapidly to olefin whereas the latter probably would be protonated to form alcohol which then would be dehydrated as for a normal Perkin reaction. This proposal remains to be tested, however.

Phosphonate carbanions have been observed to react with oxiranes in a manner analogous to that found for phosphonium and phosphinoxy carbanions. Wadsworth and Emmons (*28*) initially reported the reaction of carbethoxymethyl diethylphosphonate with styrene oxide in the presence of sodium hydride in dimethoxyethane at 85° to result in a 42% yield of 1-phenyl-2-carbethoxycyclopropane. Other examples were reported, and in general the conditions were considerably milder than

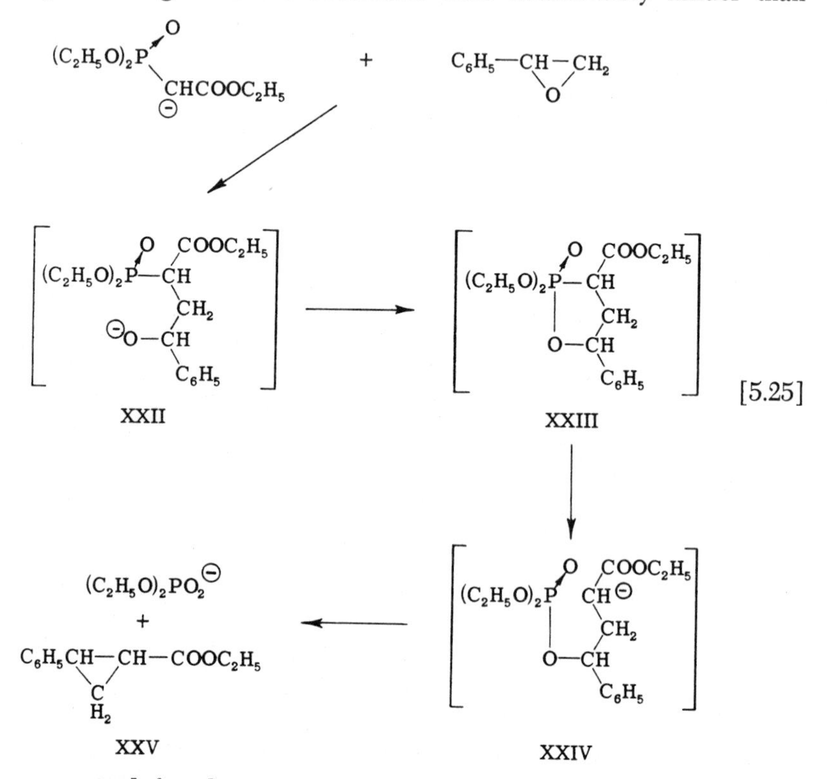

those required for the reaction of analogous phosphonium carbanions with the same oxiranes (*19*, Denney *et al.*). The mechanism of this reaction probably is analogous to that proposed by McEwen and Wolf. (*20*), involving the intermediacy of a betaine. The milder conditions can be attributed to the increased nucleophilicity of the phosphonate carbanions.

Denney *et al.* (*19*) had shown that the formation of XXV from reaction of carbethoxymethylenetriphenylphosphorane with styrene oxide proceeded with retention of optical activity. Furthermore, recent work by Walborsky *et al.* (*41*) showed that the reaction occurred with inversion of the configuration of the asymmetric carbon in styrene oxide. Tomoskozi (*42*) now has found that reaction of the carbanion of carbethoxymethyl diethylphosphonate with styrene oxide also proceeds with retention of optical activity and inversion of configuration. Accordingly, the reaction probably involves the intermediate formation of a betaine (XXII) which cyclizes to a pentavalent phosphorus derivative (XXIII). The phosphorus-carbon bond of the latter probably cleaves to afford a new betaine (XXIV), and the ejection of diethylphosphate probably gives rise to the inversion.

Cotton and Schunn (*43*) were able to isolate the sodium and zinc salts of the carbanion of acetonyl diethylphosphonate by treatment of XIV with metallic sodium in benzene and then with zinc chloride for the second salt. These salts undoubtedly existed in chelated forms (XXVI). The P^{31} NMR chemical shifts indicated that the phosphorus atom was less shielded in the chelates than in the protonated form (XIV)

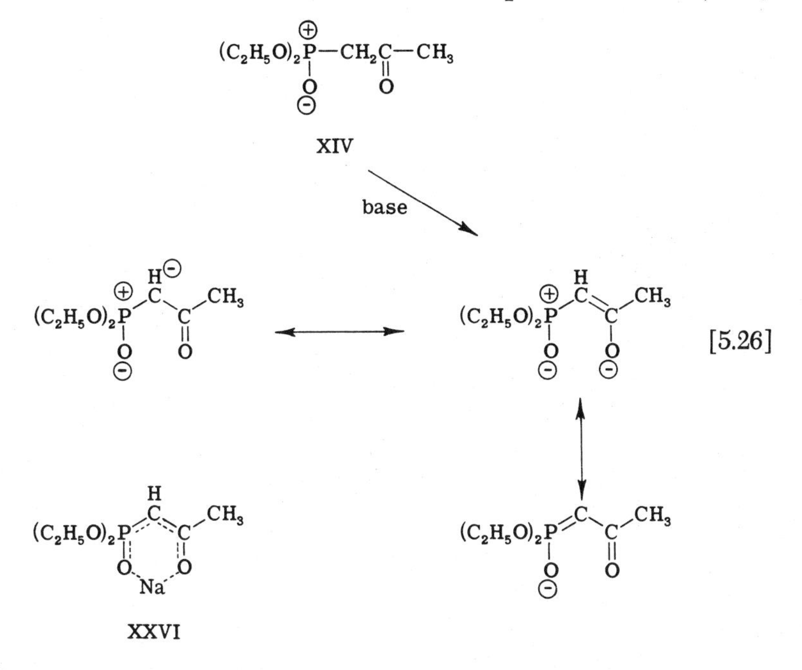

(— 38 ppm *vs.* —23 ppm). Ultraviolet and infrared spectra of the chelates indicated that there was some delocalization of the carbanion electrons into the phosphorus atom, presumably by overlap with the 3*d*-orbitals. They concluded that the phosphonate group was approximately as effective as a carbonyl group in providing stabilization for the carbanion.

In summary, phosphonate carbanions have a chemistry similar to that of phosphonium carbanions (ylids) and phosphinoxy carbanions.

They undergo normal carbanionic reactions and also undergo reactions characteristic of phosphonium ylids, especially reactions with carbonyl compounds to form olefins and reactions with oxiranes to form cyclopropanes. Phosphonate carbanions appear to be more nucleophilic than phosphonium carbanions, probably due to decreased stabilization of the carbanion by valence shell expansion of the phosphorus atom.

III. Miscellaneous Phosphorus Carbanions

Several other phosphorus-containing carbanions have been studied briefly in recent years. In all of these cases the phosphorus atom is adjacent to the carbanion and, presumably through valence shell expansion, provides stabilization for the carbanion. The extent of this stabilization should be governed by the degree of overlap between the filled orbital of the carbanion and the vacant 3*d*-orbital of the phosphorus atom. This overlap should, in turn, be governed by the degree of positive charge localized on the phosphorus atom. In the cases that are mentioned in this section very little experimental data currently are available but there is sufficient data to indicate the gross nature of the carbanions.

Horner *et al.* (*5*) have found that ethyl phenylbenzylphosphinate (XXVII) would form a carbanion in the presence of potassium *tert*-butoxide. The carbanion reacted with a series of aldehydes and ketones to afford olefins and ethyl phenylphosphonic acid. Thus XXVII and benzaldehyde afforded *trans*-stilbene in 85% yield while XXVII and

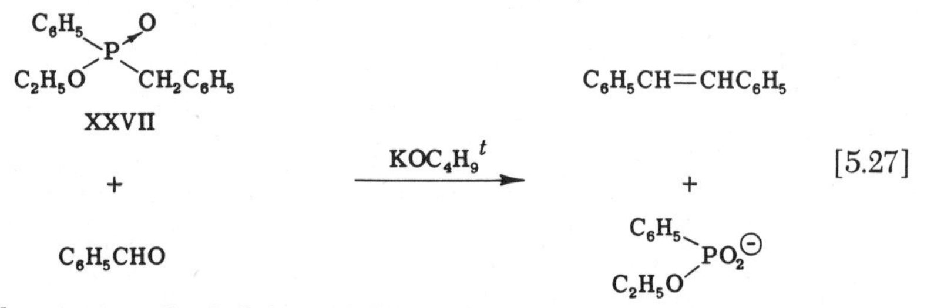

cyclohexanone afforded benzylidenecyclohexane in 55% yield. This reaction probably proceeded by a mechanism analogous to that of the Wittig reaction. The scope and stereochemistry of the reactions of phosphinate carbanions have not been investigated.

Ramirez *et al.* (*44*) reported that addition of ethyl diphenylphosphinite to *trans*-dibenzoylethylene afforded the phosphinite carbanion (XXVIII) in 42% yield. This ylid showed a P^{31} chemical shift of −54.2 ppm whereas the alternate structure (XXIX) proposed by Harvey and Jensen (*45*) would have been expected to show a positive chemical shift. The proton NMR spectrum also was in accord with structure

XXVIII. The ylid was sensitive to moisture and it did react with benzyl

$$(C_6H_5)_2POC_2H_5 + C_6H_5COCH{=}CHCOC_6H_5 \longrightarrow C_6H_5\underset{\underset{O}{\|}}{C}-\overset{\ominus}{C}(\overset{\oplus}{P}(C_6H_5)_2-OC_2H_5)-CH_2COC_6H_5 \quad \text{XXVIII}$$

$$\text{XXVIII} \xrightarrow{C_6H_5CH_2Br} C_6H_5-\underset{(C_6H_5)_2PO}{C}{=}\overset{OCH_2C_6H_5}{C}-CH_2COC_6H_5 \qquad [5.28]$$

$$\text{XXIX: cyclic } C_6H_5C{=}CH-CHCOC_6H_5 \text{ ring closed through O and } P(C_6H_5)_2(OC_2H_5)$$

XXIX

bromide. The latter reaction afforded the O-alkylated product which, in the course of the reaction, also underwent a displacement of the ethyl group. The O-alkylation is analogous to that undergone by acylmethylenetriphenylphosphoranes (*46,* Ramirez and Dershowitz).

There have been three reports of the preparation of trialkoxymethylenephosphoranes. In a U.S. patent, Birum (*47*) reported the formation of a series of complex substituted phosphitemethylenes (XXX) from the reaction between thiophosgene and trialkylphosphites. No properties were mentioned. At about the same time, Middleton, also in a U.S. patent (*48*), reported that reaction between certain thioketones and trialkylphosphites also afforded phosphitemethylenes. For example, trimethylphosphite and thionohexafluoroacetone afforded hexafluoroiso-

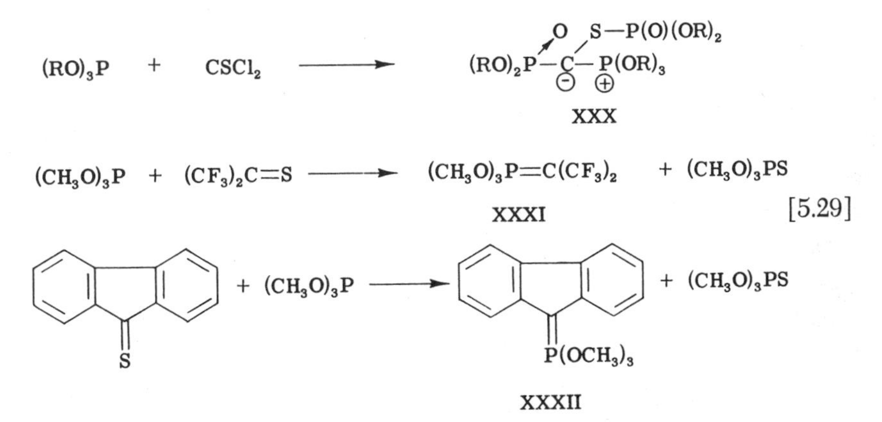

propylidenetrimethoxyphosphorane (XXXI) and thiofluorenone with the same phosphite afforded the crystalline ylid (XXXII). The former compound exhibited characteristic carbanionic properties in that it was protonated and demethylated by dilute hydrochloric acid and it could be halogenated and demethylated by bromine. Pyrolysis resulted in methyl transfer from oxygen to carbon, presumably via an Arbuzov-like reaction [5.30]. There were no reports of any simple alkylation reactions

$$\underset{\text{XXXI}}{(CH_3O)_3\overset{\oplus}{P}-\overset{\ominus}{C}(CF_3)_2} \xrightarrow[H_2O]{HCl} (CF_3)_2CH-P(\rightarrow O)(OCH_3)_2$$

$$\text{XXXI} \xrightarrow{Br_2} (CF_3)_2CBr-P(\rightarrow O)(OCH_3)_2 \qquad [5.30]$$

$$\text{XXXI} \xrightarrow{\Delta} (CF_3)_2C(CH_3)-P(\rightarrow O)(OCH_3)_2$$

or of any attempt to effect reaction with carbonyl compounds.

Ramirez and Madan (*49*) reported that addition of trimethylphosphite to *trans*-dibenzoylethylene afforded a phosphitemethylene (XXXIII), presumably via initial formation of a zwitterion which underwent a proton shift analogous to that proposed for similar additions of phosphines to conjugated olefins (see Chapter 3, Section I,C for ex-

$$(CH_3O)_3P + C_6H_5COCH{=}CHCOC_6H_5 \longrightarrow \underset{\text{XXXIII}}{C_6H_5-\underset{\underset{O}{\|}}{C}-\underset{\underset{\overset{\oplus}{P}(OCH_3)_3}{|}}{\overset{\ominus}{C}}-CH_2COC_6H_5}$$

$$\longrightarrow \; (C_6H_5)(CH_3O)C{=}C(CH_2COC_6H_5)-P(\rightarrow O)(OCH_3)_2 \qquad [5.31]$$

amples and discussion). The phosphitemethylene showed a P^{31} chemical shift of —56.2 ppm and it underwent a spontaneous methyl transfer analogous to that reported for XXXI to afford an enol-phosphonate methyl ether. Reaction of XXXIII with benzyl bromide also resulted in O-alkylation. Hydrolysis of the ylid with water resulted in the cleavage of the ylid bond to afford methyl phosphate and dibenzoylethane. There was no attempt to investigate the reaction of XXXIII with carbonyl compounds or to compare its basicity with other ylids or carbanions.

Addition of trimethylphosphite to conjugated systems results in the formation of cyclic tetraalkoxyalkylphosphoranes when a simple prototropic shift cannot afford a highly stabilized ylid. Such was the case in the

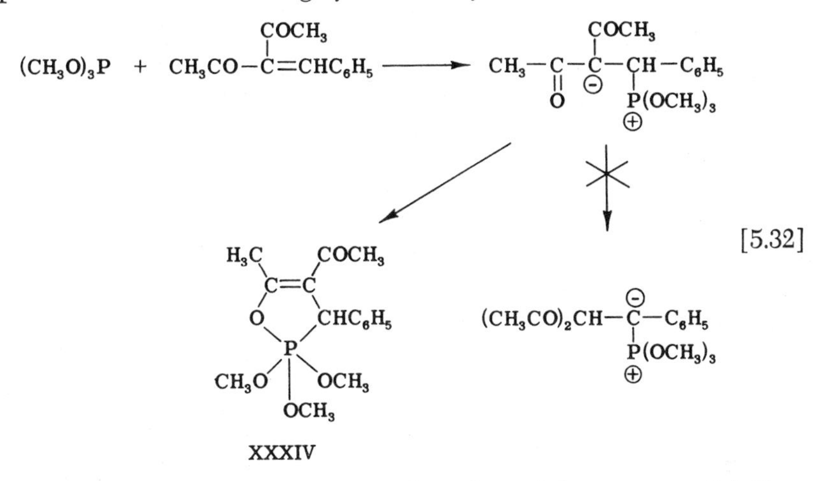

reaction of trimethylphosphite with 3-benzylidenepentane-2,4-dione which afforded a cyclic phosphorane (XXXIV). That it was not the phosphitemethylene was indicated by its P^{31} chemical shift of +27.9 ppm (*50*, Ramirez *et al.*).

In conclusion, it is apparent that there are many different types of phosphorus-stabilized carbanions in addition to the best known example, the phosphonium ylids. The overall chemical properties of the various carbanions are similar, differing mainly in degree rather than kind. From the sparce physical data available it appears there are differences in the degree of stabilization afforded the carbanions by the various phosphorus groups. This stabilization roughly is in proportion to the degree of positive charge carried by the phosphorus atom. Much more experimental investigation of these and other phosphorus carbanions is needed, especially with a view to obtaining some quantitative chemical and physical data which would pertain to the degree of valence shell expansion in the different phosphorus groups.

REFERENCES

1. L. Horner, H. Hoffmann and H. G. Wippel, *Chem. Ber.* **91**, 61 (1958).
2. D. Seyferth, D. E. Welch and J. K. Heeren, *J. Am. Chem. Soc.* **86**, 1100 (1964).
3. D. Seyferth and D. E. Welch, *J. Organometallic Chem.* **2**, 1 (1964).
4. H. Hoffmann, *Angew. Chem.* **71**, 379 (1959).
5. L. Horner, H. Hoffmann, W. Klink, H. Ertel and V. G. Toscano, *Chem. Ber.* **95**, 581 (1962).
6. G. Wittig, H. Eggers and P. Duffner, *Ann.* **619**, 10 (1958).
7. L. Horner, H. Hoffmann, H. G. Wippel and G. Klahre, *Chem. Ber.* **92**, 2499 (1959).

8. J. J. Richards and C. V. Banks, *J. Org. Chem.* **28**, 123 (1963).
9. M. Schlosser and K. F. Christmann, *Angew. Chem.* **76**, 683 (1964).
10. L. Horner and H. Winkler, *Tetrahedron Letters* 3265 (1964).
11. L. Horner and W. Klink, *Tetrahedron Letters* 2467 (1964).
12. A. J. Speziale and D. E. Bissing, *J. Am. Chem. Soc.* **85**, 3878 (1963).
13. S. Trippett, *Pure and Applied Chem.* **9**, 255 (1964).
14. L. Horner, W. Klink and H. Hoffmann, *Chem. Ber.* **96**, 3133 (1963).
15. L. Horner, H. Hoffmann, G. Klahre, V. G. Toscano and H. Ertel, *Chem. Ber.* **94**, 1987 (1961).
16. H. J. Bestmann and O. Kratzer, *Chem. Ber.* **96**, 1899 (1963).
17. S. T. D. Gough and S. Trippett, *J. Chem. Soc.* 2333 (1962).
18. L. Horner, H. Hoffmann and V. G. Toscano, *Chem. Ber.* **95**, 536 (1962).
19. D. B. Denney, J. J. Vill and M. J. Boskin, *J. Am. Chem. Soc.* **84**, 3944 (1962).
20. W. E. McEwen and A. P. Wolf, *J. Am. Chem. Soc.* **84**, 676 (1962).
21. I. Tomoskozi, *Chem. & Ind.* (*London*) 689 (1965).
22. A. E. Arbuzov and A. A. Dunin, *Ber. deut. chem. Ges.* **60**, 291 (1927).
23. A. E. Arbuzov and A. I. Razumov, *J. Russ. Phys. Chem. Soc.* **61**, 623 (1929); *Chem. Abstr.* **23**, 4444 (1929).
24. A. N. Pudovik and N. M. Lebedeva, *Zhur. Obshchei Khim.* **25**, 1920 (1955); *Chem. Abstr.* **50**, 8442 (1956).
25. G. M. Kosolapoff, *J. Am. Chem. Soc.* **75**, 1500 (1953).
26. A. N. Pudovik and N. M. Lebedeva, *Doklady Akad. Nauk. S.S.S.R.* **90**, 799 (1953); *Chem. Abstr.* **50**, 2429 (1956).
27. J. Michalski and S. Musierowicz, *Tetrahedron Letters* 1187 (1964).
28. W. S. Wadsworth and W. D. Emmons, *J. Am. Chem. Soc.* **83**, 1733 (1961).
29. B. G. Kovalev, L. A. Yanovskaya and V. F. Kucherov, *Izvest. Akad. Nauk. S.S.S.R.* 1876 (1962); *Bull. Acad. Sci. U.S.S.R.* 1788 (1962).
30. H. Takahashi, K. Fujiwara and M. Ohta, *Bull. Chem. Soc. Jap.* **35**, 1498 (1962).
31. A. K. Bose and R. T. Dahill, *J. Org. Chem.* **30**, 505 (1965).
32. H. Normant and G. Sturtz, *Compt. rend.* **256**, 1800 (1963).
33. G. Drefahl, K. Ponsold and H. Schick, *Chem. Ber.* **97**, 2011 (1964).
34. H. Zimmer and J. P. Bercz, *Abstr. Papers, 146th Meeting, Am. Chem. Soc.* p. 28C (1964).
35. H. W. Whitlock, Jr., *J. Org. Chem.* **29**, 3129 (1964).
36. I. Tomoskozi and G. Janzso, *Chem. & Ind.* (*London*) 2085 (1962).
37. S. Trippett and D. M. Walker, *Chem. & Ind.* (*London*) 990 (1961).
38. D. H. Wadsworth, O. E. Schupp, E. J. Seus and J. A. Ford, Jr., *J. Org. Chem.* **30**, 680 (1965).
39. L. D. Bergelson and M. M. Shemyakin, *Pure and Applied Chem.* **9**, 271 (1964).
40. S. Patai and A. Schwartz, *J. Org. Chem.* **25**, 1232 (1960).
41. Y. Inouye, T. Sugita and H. M. Walborsky, *Tetrahedron* **20**, 1695 (1964).
42. I. Tomoskozi, *Angew. Chem.* **75**, 294 (1963); *Tetrahedron* **19**, 1969 (1963).
43. F. A. Cotton and R. A. Schunn, *J. Am. Chem. Soc.* **85**, 2394 (1963).
44. F. Ramirez, O. P. Madan and C. P. Smith, *J. Am. Chem. Soc.* **86**, 5339 (1964).
45. R. G. Harvey and E. V. Jensen, *Tetrahedron Letters* 1801 (1963).
46. F. Ramirez and S. Dershowitz, *J. Org. Chem.* **22**, 41 (1957).
47. G. H. Birum, U.S. Patent 3,058,876; *Chem. Abstr.* **58**, 7975 (1963).
48. W. J. Middleton, U.S. Patent 3,067,233; *Chem. Abstr.* **58**, 11402 (1963).
49. F. Ramirez and O. P. Madan, *Abstr. Papers, 148th Meeting, Am. Chem. Soc.* p. 13S (1964); *J. Org. Chem.* **30**, 2284 (1965).
50. F. Ramirez, O. P. Madan and S. R. Heller, *J. Am. Chem. Soc.* **87**, 731 (1965).

6

IMINOPHOSPHORANES

The chemistry of iminophosphoranes (phosphinimines) will be discussed in this book because these substances (I) are isoelectronic with phosphorus ylids and their chemistry is very similar. Iminophosphoranes can be represented as resonance hybrids of the two contributing structures Ia and Ib, the nature of the multiple bonding between the nitrogen

$$\underset{(a)}{R_3\overset{\oplus}{P}-\overset{\ominus}{\ddot{\underset{..}{N}}}-R'} \longleftrightarrow \underset{(b)}{R_3P{=}\underset{..}{N}-R'} \qquad \text{I} \tag{6.1}$$

and phosphorus atoms depending on the degree of overlap of the filled nitrogen $2p$-orbital(s) with the vacant phosphorus $3d$-orbital(s).

Discussion of the chemistry of iminophosphoranes leads directly into a description of the chemistry of the phosphonitrilic compounds. However, since there have been recent reviews covering the relationship of these two classes of compounds (*1*, Paddock) this discussion will be limited to substances containing single phosphinimine units. Some qualitative information is available regarding the chemistry of iminophosphoranes but virtually nothing is known about the quantitative aspects of their chemistry.

I. Preparation of Iminophosphoranes

Three different methods commonly have been used for the preparation of iminophosphoranes. One of the earliest methods was exemplified by the reaction between triphenylphosphine and chloramine-T in ethanolic solution to afford *N-p*-toluenesulfonyliminotriphenylphosphorane (II) (*2*, Mann and Chaplin). This reaction afforded crystalline imide

$$(C_6H_5)_3P + \begin{matrix} H \\ Cl \end{matrix}\!\!>\!N-SO_2C_6H_4CH_3(p) \longrightarrow \underset{\text{II}}{(C_6H_5)_3P{=}N-SO_2C_6H_4CH_3(p)} \tag{6.2}$$

in good yield from a variety of substituted triphenylphosphines. The reaction probably proceeded via nucleophilic attack of the phosphine on nitrogen to form an amidophosphonium salt which then was deprotonated.

More recently phosphine displacements on mono-substituted amines have been used for the preparation of iminophosphoranes. Appel *et al.* (*3, 4*) found that triphenylphosphine and hydroxylamine-*O*-sulfonic acid afforded aminotriphenylphosphonium hydrogen sulfate. In the presence of methoxide in methanol the salt was hydrolyzed to triphenylphosphine oxide and no imine could be obtained. However, when the reaction was carried out in liquid ammonia the initially formed salt was deprotonated to the parent iminotriphenylphosphorane (III).

$$(C_6H_5)_3P + H_2NOSO_3H \longrightarrow (C_6H_5)_3\overset{\oplus}{P}{-}NH_2 \quad HSO_4^{\ominus} \xrightarrow{NH_3(l)} \underset{\text{III}}{(C_6H_5)_3\overset{\oplus}{P}{-}\overset{\ominus}{N}H} \qquad [6.3]$$

Reaction between chloramine and triphenylphosphine also afforded an aminophosphonium salt (*5*, Sisler *et al.; 6, 7*, Appel and Hauss). The latter group found that this salt could be deprotonated with sodamide to III [6.4]. Sisler *et al.* (*8*) effected the same conversion but in the

$$(C_6H_5)_3P + H_2NCl \longrightarrow (C_6H_5)_3\overset{\oplus}{P}{-}NH_2 \quad Cl^{\ominus} \xrightarrow{\text{base}} \underset{\text{III}}{(C_6H_5)_3P{=}NH} \qquad [6.4]$$

presence of magnesium hydride. Iminophosphoranes could be prepared from a variety of tertiary phosphines by this method. All attempts to effect the deprotonation of the aminophosphonium salts in aqueous media led to hydrolysis and the formation of tertiary phosphine oxides and ammonia.

Horner and Oediger (*9*) developed a second method for the preparation of iminophosphoranes, one that is, in general, the most widely applicable of the three methods to be discussed. Reaction of a wide range of primary arylamines with triphenylphosphine dibromide in the presence of two equivalents of triethylamine afforded the imines in 70%

$$(C_6H_5)_3PBr_2 + ArNH_2 \longrightarrow \left[Ar{-}\underset{H}{\overset{H}{N}}{-}P(C_6H_5)_3\right]^{2+} 2\,Br \xrightarrow{(C_2H_5)_3N} Ar{-}\overset{\ominus}{\ddot{N}}{-}\overset{\oplus}{P}(C_6H_5)_3 \qquad [6.5]$$

yield. The reaction probably involved the attack of the amine on phosphorus followed by two deprotonations [6.5].

Zimmer and Singh (*10*) recently have shown that this reaction also can be applied to the preparation of *N*-alkyliminotriphenylphosphoranes using sodamide as the base. Similarly, reaction of triphenylphosphine dibromide with hydrazine in the presence of sodamide afforded *N*-amino-

$$(C_6H_5)_3PBr_2 + N_2H_4 \xrightarrow{NaNH_2} (C_6H_5)_3P{=}N{-}NH_2$$

[6.6]

$$2\,(C_6H_5)_3PCl_2 + N_2H_4 \longrightarrow \underset{\text{IV}}{(C_6H_5)_3P{=}N{-}N{=}P(C_6H_5)_3}$$

iminotriphenylphosphorane but use of excess phosphine dihalide afforded the bis-imine (IV) (*11*, Appel and Schollhorn). In all examples of the reaction of phosphine dihalides with amines there have not developed any limitations on the nature of the amines that can be employed.

The reaction of organic azides with tertiary phosphines is the oldest known method for the preparation of iminophosphoranes. Staudinger and Meyer (*12*) found that warming phenyl azide with triphenylphosphine led to the evolution of nitrogen and the formation of crystalline *N*-phenyliminotriphenylphosphorane [6.7]. They applied this reaction to

$$(C_6H_5)_3P + C_6H_5N_3 \longrightarrow (C_6H_5)_3P{=}NC_6H_5 + N_2\uparrow \qquad [6.7]$$

several other organic azides and, for example, prepared *N*-benzoyliminotriphenylphosphorane from benzoyl azide. In recent years many other azides have been used including triphenylsilyl azide (*13, 14*), trimethylsilyl azide (*15*), 2,3,4,6-tetraacetoxy-β-D-glucosyl azide (*16*) and sulfonyl azides (*17–19*). The structure of the phosphine component of the reaction can be varied without altering the course of the reaction. Thus, Birkofer and Kim (*20*) obtained good yields of imines using several trialkylphosphines while others have used trialkylphosphites (*21*), tris-(dimethylamino)phosphine (*22*) and 1,2-bis(diphenylphosphonio)ethane (*23*). It appears that the scope of the reaction between phosphines and azides for the synthesis of iminophosphoranes is limited only by the availability of the requisite reactants.

The mechanism of the reaction between tertiary phosphines and azides probably involves nucleophilic attack by the phosphine on the azide nitrogen since Horner and Gross (*24*) found a ρ value of +1.36 for the reaction of triphenylphosphine with a series of substituted phenyl azides. The question of exactly where on the azide molecule the phosphine carries out this attack has not been resolved. There have been several successful attempts to isolate the initial adducts from the reaction. Bergmann and Wolff (*25*) reported that triphenylmethyl azide and triphenylphosphine reacted in ether solution to afford an adduct of m.p. 104–105°, which retained all three nitrogen atoms. Heating of this adduct was reported to result in the loss of nitrogen and the formation of *N*-triphenylmethyliminotriphenylphosphorane (VI). No characterization was reported for the imine but it was found to react with ketene to afford a Schiff base, a reaction characteristic of other iminophosphoranes (*11*). More recently, Leffler *et al.* (*27*) have confirmed the formation of an adduct from this reaction and have formulated it as the linear mole-

$$(C_6H_5)_3P + (C_6H_5)_3CN_3 \rightleftharpoons (C_6H_5)_3\overset{\oplus}{P}-\overset{\ominus}{\ddot{N}}-N{=}N-C(C_6H_5)_3 \quad \text{V}$$

$$\xrightarrow{\Delta} (C_6H_5)_3\overset{\oplus}{P}-\overset{\ominus}{\ddot{N}}-C(C_6H_5)_3 \quad \text{VI} \qquad [6.8]$$

cule (V) on the basis of the absence of the asymmetric azide absorption near 2100 cm^{-1}. The adduct appeared to dissociate when in solution since this absorption slowly appeared. However, these workers were unable to obtain the iminophosphorane (VI) upon warming in either neutral or acidic solution.

In a somewhat similar case Thayer and West (*14*) found that triphenylsilyl azide also formed an isolable 1:1 adduct with triphenylphosphine. However, since this adduct easily was decomposed to *N*-triphenylsilyliminotriphenylphosphorane (VIII) and since it showed the asym-

$$(C_6H_5)_3P + (C_6H_5)_3SiN_3 \longrightarrow (C_6H_5)_3\overset{\oplus}{P}-\underset{:N{=}\ddot{N}^{\ominus}}{\underset{|}{\ddot{N}}}-Si(C_6H_5)_3 \quad \text{VII}$$

$$\longrightarrow (C_6H_5)_3\overset{\oplus}{P}-\overset{\ominus}{\ddot{N}}-Si(C_6H_5)_3 \quad \text{VIII} \qquad [6.9]$$

metric azide absorption at 2018 cm^{-1}, both as a solid and in solution, it was formulated as the non-linear adduct (VII). These authors suggested that the σ bond between nitrogen and phosphorus was formed by electron donation from nitrogen to phosphorus. This seems unlikely in view of the kinetic data of Horner and Gross (*24*). The formation of the non-linear adduct with triphenylsilyl azide but not with triphenylmethyl azide may have been facilitated by delocalization of the lone pair of electrons on nitrogen to the *3d*-orbitals of both phosphorus and silicon in the former.

Two groups have claimed the isolation of linear adducts from sulfonyl azides and triphenylphosphine. Bock and Wiegrasse (*17*) found that the adduct from *p*-toluenesulfonyl azide and triphenylphosphine showed no azidic absorption in the infrared region. Franz and Osuch (*18*) found that the crystalline adduct from benzenesulfonyl azide and triphenylphosphine also did not show the azide asymmetric absorption at 2130 cm^{-1}, but a solution of the adduct did. In both of these cases the supposed linear adducts could be thermally converted to the corresponding iminophosphoranes. The question of the structure of all of these azide-phosphine adducts still seems very much open and in need of definitive investigation.

Staudinger and Hauser (*26*) reported that triphenylphosphine and hydrazoic acid afforded aminotriphenylphosphonium azide but this result was questioned by Leffler *et al.* (*27*) who in turn proposed phosphatetrazole structures. More recently Cookson and Hughes (*28*) have confirmed the original assignment by detection of the azide ion. An oxidation-reduction must take place since nitrogen is evolved in the reaction.

Bestmann and Seng (*29*) have reported an interesting reaction which resulted in the conversion of a methylenephosphorane into an iminophosphorane [6.10]. Benzylidenetriphenylphosphorane and benzylidene-

$$\begin{matrix} (C_6H_5)_3P{=}CHC_6H_5 \\ + \\ C_6H_5CH{=}NC_6H_5 \end{matrix} \longrightarrow \left[\begin{matrix} (C_6H_5)_3\overset{\oplus}{P}-CH-C_6H_5 \\ \ominus\quad | \\ C_6H_5-\ddot{N}-CH-C_6H_5 \end{matrix}\right] \longrightarrow \begin{matrix} (C_6H_5)_3P{=}NC_6H_5 \\ + \\ C_6H_5CH{=}CHC_6H_5 \end{matrix} \qquad [6.10]$$

aniline afforded *N*-phenyliminotriphenylphosphorane and stilbene, presumably by way of a betaine similar to those encountered in the Wittig reaction.

There have been four reports of the preparation of the bis-imine (IX). Appel *et al.* (*30*) found that triphenylphosphine and *N*-bromoiminotriphenylphosphorane afforded the bromide salt of IX. The same salt was obtained when triphenylphosphine dibromide was treated with iminotriphenylphosphorane (*31*, Appel and Hauss). Appel and Buchler (*32*) obtained the chloride of IX from two moles of triphenylphosphine and one mole of nitrogen trichloride. More recently, Cookson and Hughes (*28*) obtained the azide of IX by heating aminotriphenylphosphonium azide. The bis-imine is isoelectronic with the hexaphenylcarbo-

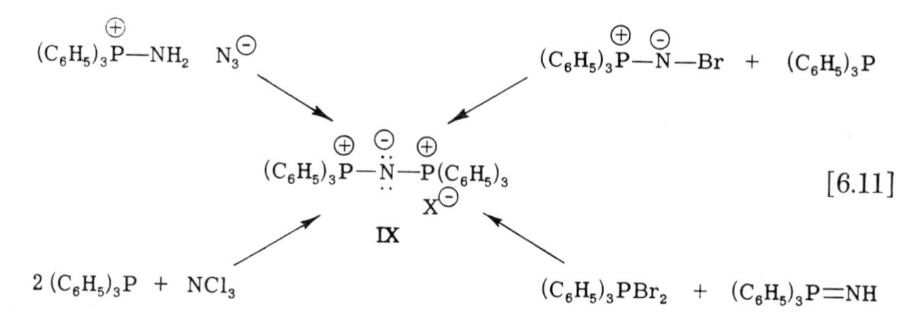

diphosphorane obtained by Ramirez *et al.* (*33*) but the former is considerably more stable.

The three general methods for the synthesis of iminophosphoranes, the phosphine displacement on a substituted amine, the amine displacement on a phosphine dihalide and the reaction between an azide and a phosphine all have proven sufficiently versatile to date. The latter method probably is the most straightforward but the second method perhaps is the most flexible.

II. Reactions of Iminophosphoranes

Iminophosphoranes are being discussed in this chapter because their chemistry is so similar to that of the isoelectronic methylenephosphoranes. The outstanding chemical property of the former is their nucleophilicity but the most interesting properties are those that depend on a combination of a nucleophilic reaction and an elimination of the phosphorus group in an oxidized state. In other words, those reactions which take advantage of the availability of the vacant 3*d*-orbitals of phosphorus.

Iminophosphoranes are readily protonated to the corresponding aminophosphonium salts in the presence of mineral acids (*24, 26, 31*). Numerous salts have been characterized (*5*). There have been no quantitative studies on the relationship of basicity to molecular structure but they would be expected to parallel the behavior of phosphonium ylids.

$$(C_6H_5)_3\overset{\oplus}{P}—\overset{\ominus}{N}—R \underset{-H^{\oplus}}{\overset{H^{\oplus}}{\rightleftharpoons}} [(C_6H_5)_3P—NHR]^{\oplus} \qquad [6.12]$$

A qualitative relationship is apparent from the observation that sodamide is required to form iminotriphenylphosphorane from its conjugate acid (*7*) while triethylamine will form *N*-phenyliminotriphenylphosphorane from its conjugate acid (*9*). Therefore, the *N*-phenyl imine appears to be the least basic, its conjugate acid the most acidic, probably due to stabilization afforded the imine by delocalization of the unshared electrons through the phenyl ring.

Iminophosphoranes are prone to hydrolyze. The ease with which the hydrolysis occurs seems to be correlated to the basicity of the imine. Simply upon exposure to the atmosphere iminotriphenylphosphorane is hydrolyzed to ammonia and triphenylphosphine oxide (*6,* Appel and Hauss; *8,* Sisler *et al.*). On the other hand, *N*-aryliminotriphenylphosphoranes are stable on exposure to the atmosphere and in aqueous solu-

$$(C_6H_5)_3P{=}NH \xrightarrow{H_2O} (C_6H_5)_3PO + NH_3$$

$$(C_6H_5)_3P{=}N{-}Ar \xrightarrow[H^+]{H_2O} (C_6H_5)_3PO + ArNH_2 \qquad [6.13]$$

tion but are hydrolyzed rapidly in dilute acid media [6.13] (*12,* Staudinger and Meyer). *N*-Triphenylsilyliminotriphenylphosphorane (VIII) is a weakly basic imine and is hydrolyzed only in the presence of strong acid. In this case overlap of the filled orbitals of nitrogen with the vacant 3*d*-orbitals of silicon undoubtedly adds to the delocalization of the electron density from nitrogen.

Birkofer and Kim (*20*) found that *N*-trimethylsilyliminotriphenylphosphorane was cleaved preferentially at the silicon-nitrogen bond under mild conditions. In the presence of methanolic sulfuric acid at —20° they were able to isolate methoxytrimethylsilane and iminotriphenylphosphorane. Apparently the intermediate salt underwent more

$$(C_6H_5)_3P{=}N{-}Si(CH_3)_3 \xrightarrow[CH_3OH]{H_2SO_4} (C_6H_5)_3P{=}NH + CH_3OSi(CH_3)_3$$

$$\underset{X}{(C_6H_5)_3P{=}\overset{\oplus}{N}H{-}Si(CH_3)_3} \longleftrightarrow \underset{XI}{(C_6H_5)_3\overset{\oplus}{P}{-}NH{-}Si(CH_3)_3} \qquad [6.14]$$

rapid attack by methanol on silicon than on phosphorus. This might indicate that the salt had a structure better represented by X than XI.

The mechanism of the hydrolysis of iminophosphoranes probably involves the initial protonation of the imine to form a salt followed by attack of water or hydroxide on phosphorus, via a pentavalent intermediate (XII), to form the phosphine oxide and amine. Horner and Winkler (*34*) have shown that the phosphorus atom in such hydrolyses mainly is inverted in agreement with this proposal. Optically active *N*-(*p*-nitrophenyl)iminomethylphenylpropylphosphorane (XIII) was hydrolyzed with dilute hydrochloric acid to afford mainly, but not exclusively, inverted phosphine oxide. When the imine was alkylated first and

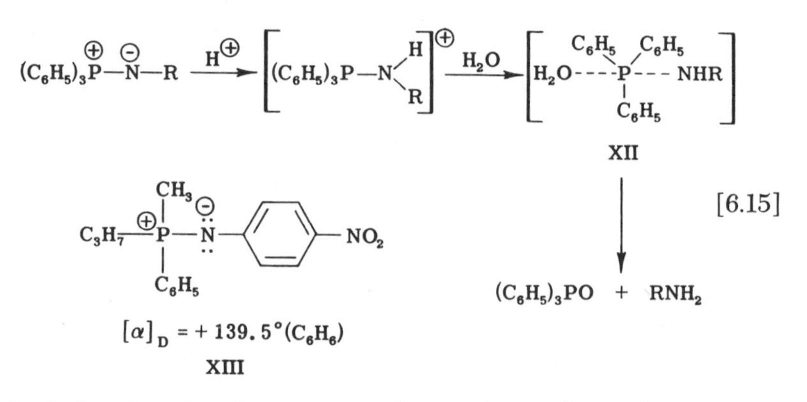

then hydrolyzed under the same conditions the oxide product was exclusively inverted.

Phosphinimines will undergo reaction with a variety of Lewis acids. Crystalline and stable adducts were formed with a variety of boron compounds such as boron trifluoride (*35, 36*), triphenylboron (*35*) and diborane (*37*). As a result of their nucleophilicity imines also have served as ligands in metal complexes such as $[(C_6H_5)_3P{=}NH]_2CoCl_2$ and the copper (II) analog (*38*, Appel and Schaaff). Iminotriphenylphosphorane (III) also reacted with triphenylphosphine dibromide to form *N*-triphenylphosphonioiminotriphenylphosphorane bromide (IX) (*31*, Appel and Hauss).

Zimmer and Singh (*39*) reported an interesting reaction between *N*-phenyliminotriphenylphosphorane and nitrosyl chloride which afforded triphenylphosphine oxide and phenyldiazonium chloride at −70°. The latter was identified by trapping it with 2-naphthol. Horner and Winkler (*34*) carried out this same reaction with their optically active imine (XIII) and found that the phosphine oxide was produced with net re-

$$(C_6H_5)_3P{=}NC_6H_5 + NOCl \longrightarrow \left[(C_6H_5)_3\overset{\oplus}{P}-\underset{\underset{O=N}{|}}{N}-C_6H_5 \longleftrightarrow (C_6H_5)_3\overset{\oplus}{P}-\underset{\underset{\overset{\ominus}{O}-N}{\|}}{\overset{\oplus}{N}}-C_6H_5 \right] \quad \textbf{XIV} \qquad [6.16]$$

$$\downarrow$$

$$(C_6H_5)_3PO + C_6H_5\overset{\oplus}{N}{\equiv}N\ \ Cl^{\ominus}$$

tention accompanied by some racemization. Zimmer had proposed the intermediacy of a *N*-nitrosoaminophosphonium salt (XIV) which eliminated triphenylphosphine oxide via a four-membered transition state similar to that involved in the Wittig reaction. This mechanism predicted the phosphorus atom to retain its configuration.

Iminophosphoranes can be readily alkylated with alkyl halides. Reaction of equimolar amounts of imine and alkyl halide leads to formation of the corresponding *N*-alkylaminophosphonium salt (*10, 11, 24*). However, in the case of the parent iminotriphenylphosphorane (III), only half of the imine was alkylated, the other half acting as a base to convert the initially formed *N*-alkyl salt into a *N*-alkylimine [6.17]. The

$$(C_6H_5)_3P{=}N{-}R \quad + \quad R'X \longrightarrow \left[(C_6H_5)_3P{-}N\begin{matrix} R \\ R' \end{matrix}\right]^{\oplus} X^{\ominus}$$

$$\underset{\text{III}}{2\,(C_6H_5)_3P{=}NH} \quad + \quad 2\,C_2H_5I \longrightarrow (C_6H_5)_3\overset{\oplus}{P}{-}N(C_2H_5)_2 I^{\ominus} + (C_6H_5)_3\overset{\oplus}{P}{-}NH_2 \quad I^{\ominus} \qquad [6.17]$$

$$\underset{\text{III}}{2\,(C_6H_5)_3P{=}NH} \quad + \quad (C_6H_5)_3C{-}Cl \longrightarrow (C_6H_5)_3P{=}N{-}C(C_6H_5)_3 + (C_6H_5)_3\overset{\oplus}{P}NH_2 \quad Cl^{\ominus}$$

latter could be alkylated again. For example, III with ethyl iodide afforded *N,N*-diethylaminotriphenylphosphonium iodide (*31*, Appel and Hauss). In the case of a bulky alkylating agent the second alkylation did not occur. Thus III with trityl chloride afforded *N*-tritylιminotriphenylphosphorane (*35*, Appel and Vogt).

Reaction of iminotriphenylphosphorane (III) with acid halides followed a course similar to that discussed for methylenetriphenylphosphoranes [6.18]. Benzoyl chloride converted III into *N*-benzoylimino-

$$\underset{\text{III}}{(C_6H_5)_3P{=}NH} \quad + \quad C_6H_5COCl \longrightarrow \left[(C_6H_5)_3\overset{\oplus}{P}{-}N\begin{matrix} COC_6H_5 \\ H \end{matrix}\right] \xrightarrow{\text{III}} (C_6H_5)_3P{=}NCOC_6H_5 \quad + \quad (C_6H_5)_3\overset{\oplus}{P}NH_2 \quad Cl^{\ominus} \qquad [6.18]$$

triphenylphosphorane but only in 50% maximum yield since the intermediate amidotriphenylphosphonium salt was deprotonated by unreacted III (*6, 7*, Appel and Hauss). Likewise, *p*-toluenesulfonyl chloride and III afforded the *N*-tosylimine (*7*). There have been no examples reported of the reaction between N-substituted imines and acyl halides but they would be expected to follow suit.

Halogenation of iminophosphoranes proceeded in a somewhat analogous manner [6.19]. Reaction of III with chlorine, bromine or iodine

$$(C_6H_5)_3P{=}NH + X_2 \longrightarrow \left[(C_6H_5)_3\overset{\oplus}{P}{-}NHX \quad X^{\ominus}\right] \xrightarrow{III} (C_6H_5)_3P{=}N{-}X$$

III [6.19]

$$2\,(C_6H_5)_3P{=}NH + I{-}Cl \longrightarrow (C_6H_5)_3P{=}NI + (C_6H_5)_3\overset{\oplus}{P}{-}NH_2 \quad Cl^{\ominus}$$

afforded one-half equivalent of the corresponding *N*-haloiminotriphenylphosphorane, all of which were crystalline compounds. That the reaction involved the initial nucleophilic attack of the imine on halogen was evident from the formation of the *N*-iodoimine from iodine monochloride and III (*30, 31,* Appel *et al.*)

Iminophosphoranes have been shown to react with carbonyl compounds to form Schiff bases and phosphine oxides in a manner reminiscent of the Wittig reaction. The first report of this reaction was by Staudinger and Meyer (*12*) in 1919 who found that *N*-phenyliminotriphenylphosphorane (XV) reacted with diphenylketene to afford triphenylketenimine (XVI), identical to that formed from benzhydrylidenetriphenylphosphorane and phenyl isocyanate. The scope of this reaction

$$(C_6H_5)_3P{=}NC_6H_5 \;(\text{XV}) + (C_6H_5)_2C{=}C{=}O$$

$$\downarrow$$

$$(C_6H_5)_2C{=}C{=}NC_6H_5 \;(\text{XVI}) + (C_6H_5)_3PO \qquad [6.20]$$

$$\uparrow$$

$$(C_6H_5)_3P{=}C(C_6H_5)_2 + C_6H_5N{=}C{=}O$$

was expanded to include reaction of XV with benzophenone, benzaldehyde and phenyl isocyanate, all of which involved the replacement of a carbonyl oxygen by a *N*-phenylimine group [6.21]. The same group (*12, 26*) found that XV reacted with carbon dioxide and carbon disulfide to form an isocyanate and an isothiocyanate, respectively. Reaction of *N*-phenyliminotriethylphosphorane with carbon dioxide, however, afforded *N,N*-diphenylcarbodiimide, presumably via the intermediacy of phenyl isocyanate which reacted with additional imine (*26*). An analogous conversion of *N*-glucosyliminotriphenylphosphorane to a carbodiimide was reported by Messmer *et al.* (*16*).

Appel *et al.* (*7, 31*) have shown that the parent iminotriphenylphos-

$$(C_6H_5)_3P{=}NC_6H_5 + (C_6H_5)_2CO \longrightarrow (C_6H_5)_2C{=}NC_6H_5 + (C_6H_5)_3PO$$
XV

$$\text{XV} + C_6H_5CHO \longrightarrow C_6H_5CH{=}NC_6H_5 + (C_6H_5)_3PO$$

$$\text{XV} + C_6H_5NCO \longrightarrow C_6H_5N{=}C{=}NC_6H_5 + (C_6H_5)_3PO \quad [6.21]$$

$$\text{XV} + CO_2 \longrightarrow C_6H_5N{=}C{=}O + (C_6H_5)_3PO$$

$$\text{XV} + CS_2 \longrightarrow C_6H_5N{=}C{=}S + (C_6H_5)_3PS$$

$$2\,(C_2H_5)_3P{=}NC_6H_5 + CO_2 \longrightarrow C_6H_5N{=}C{=}NC_6H_5 + 2\,(C_2H_5)_3PO$$

phorane (III) also reacted with a variety of carbonyl compounds to afford the corresponding imines and triphenylphosphine oxide. Appel (*11*), Bergmann (*25*) and Horner (*24*) have found the reaction of other N-substituted iminotriphenylphosphoranes with carbonyl compounds to produce analogous results.

There have been no mechanistic studies on the reaction between carbonyl compounds and iminophosphoranes. It is likely that the mechanism will parallel rather closely the mechanism of the Wittig reaction between methylenephosphoranes and carbonyl compounds as shown in [6.22]. The first step should be betaine formation via nucleophilic at-

$$R''_3P{=}N{-}R' + R_2CO \longrightarrow \left[\overset{\oplus}{R''_3P}{-}N{-}R' \;\; \overset{\ominus}{O}{-}CR_2 \right] \longrightarrow R''_3PO + R_2C{=}N{-}R' \quad [6.22]$$

tack of the nitrogen on the carbonyl carbon. This step could be reversible or irreversible. The second step should be the transfer of oxygen from carbon to phosphorus and it could be either reversible or irreversible.

Campbell *et al.* (*40*) found that isocyanates could be converted into carbodiimides in the presence of 3-methyl-1-(phenyl or ethyl)phospholene-1-oxide. For example, phenyl isocyanate afforded a 94% yield of diphenylcarbodiimide and *n*-butyl isocyanate afforded a 60% yield of di-*n*-butylcarbodiimide. The phospholene oxide served a catalytic role and carbon dioxide was evolved from the reaction mixture [6.23]. The carbodiimide synthesis also was effective with isothiocyanates.

The same group (*41*) proposed that the reaction proceeded via the initial formation of an iminophosphorane (equation 1 in [6.24]) and that this imine reacted with additional isocyanate to afford the carbodiimide and regenerated phosphine oxide (equation 2 in [6.24]). The

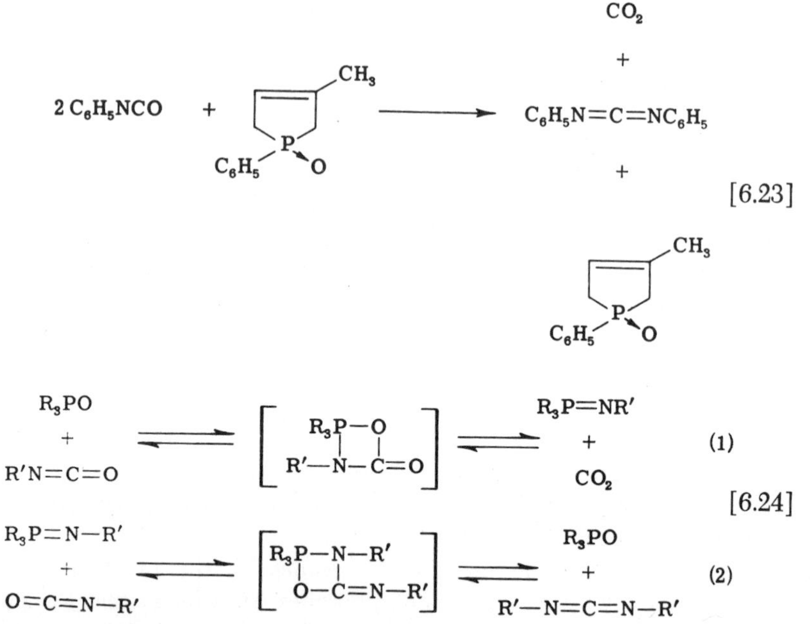

reaction was shown to be reversible since operation in a closed system led to the establishment of an equilibrium. The observation that a carbodiimide would react even with carbon dioxide to afford isocyanate made it seem likely that equation (2) was reversible while equation (1) had been shown previously to proceed in the reverse direction (*26*, Staudinger and Hauser). The formation of the iminophosphorane was thought to be the slow step for several reasons: (i) none ever was isolated from the reaction mixtures although no attempts were made to trap any with a more reactive carbonyl compound; (ii) even in the presence of large excesses of catalyst the carbodiimide was formed and no imine could be found (if the second step were slow all of the isocyanate should have been tied up as imine); (iii) the reaction followed pseudo-first-order kinetics and (iv) the use of substituted phenyl isocyanates led to a Hammett ρ-σ plot with a ρ value of +1.19 at 50°, a value consistent with those of reactions of other nucleophiles with isocyanates. The relatively high entropy of activation for the reaction, -37.9 ± 3.9 eu, indicated the likelihood of a highly ordered transition state, one such as that shown in equation (1). This value is very similar to that (−40 eu) found in the Wittig reaction (*42*, Speziale and Bissing). Additional evidence for a mechanism approximating that indicated above was provided by Monagle's observation that the O^{18} content of O^{18}-labeled catalysts (triphenylphosphine oxide) slowly decreased (*43*). In addition, Monagle (*44*) found that a wide variety of phosphine oxides, arsine oxides, sulfoxides and phosphates also would catalyze the reaction, albeit with differing efficiency.

The recently reported reaction between iminophosphoranes and dimethyl acetylenedicarboxylate (XVII) provides another illustration of the nucleophilicity of the iminophosphorane operating in conjunction with the valence shell-expanding ability of the phosphorus atom therein. Brown *et al.* (*45*) reported that XVII and *N*-phenyliminotriphenylphosphorane (XV) afforded a 1:1 adduct that had a dipole moment of 5.3 D and formed salts with hydrogen bromide and perchloric acid. Similar adducts were formed with the *p*-bromo isomer of XV (XVIII) and with the parent iminotriphenylphosphorane (III). However, iminophos-

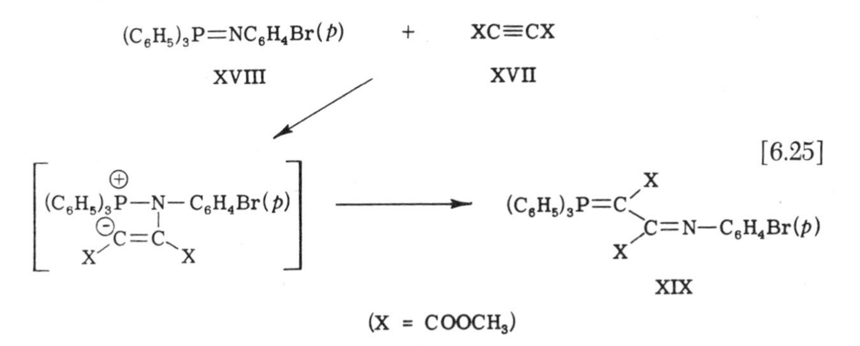

phoranes carrying electron-withdrawing groups on nitrogen, those such as 2,4-dinitrophenyl, carbethoxy, benzoyl and *p*-toluenesulfonyl, all failed to react with XVII, presumably because of their reduced nucleophilicity (*46*, Brown *et al.*). The structure of the *p*-bromo adduct (XIX) recently was determined by X-ray crystallographic analysis (*47*, Mak and Trotter). The mechanism of the formation of the adducts probably involved the nucleophilic addition of the phosphinimine to the triple bond followed by phosphorus transfer from nitrogen to carbon. This last step would involve the formation of a pentavalent phosphorus atom in the transition state but would be quite analogous to the betaine decomposition step of the Wittig reaction. The overall result of this reaction is to convert an iminophosphorane into a methylenephosphorane but it is not clear what the driving force for this reaction might be. The situation is especially confusing when it is recalled that Bestmann and Seng (*29*) reported the reverse transformation of a methylenephos-

$$C_6H_5CH{=}P(C_6H_5)_3 + C_6H_5CH{=}NC_6H_5 \longrightarrow C_6H_5N{=}P(C_6H_5)_3 + C_6H_5CH{=}CHC_6H_5 \qquad [6.26]$$

$$(C_6H_5)_3P{=}NC_6H_5 + XC{\equiv}CX \longrightarrow (C_6H_5)_3P{=}C(X){-}C(X){=}NC_6H_5$$

phorane into an iminophosphorane by reaction with benzylideneaniline [6.26].

Pyrolysis of *N*-acyliminotriphenylphosphoranes proceeds in a manner analogous to that for acylmethylenetriphenylphosphoranes which afforded alkynes and phosphine oxides. Staudinger and Hauser (*26*) reported the pyrolysis of *N*-benzoyliminotriphenylphosphorane (XX) to afford benzonitrile and triphenylphosphine oxide. The reaction probably involved a four-membered transition state for transfer of oxygen from carbon to phosphorus. Horner and Gross (*24*) reported the same be-

$$C_6H_5-\underset{\underset{O}{\|}}{C}-\overset{\ominus}{N}-P(C_6H_5)_3 \longleftrightarrow C_6H_5-\underset{\underset{O^{\ominus}}{|}}{C}=N-\overset{\oplus}{P}(C_6H_5)_3 \longrightarrow \left[\begin{matrix} C_6H_5-C\equiv N \\ \vdots \quad \vdots \\ O\cdots P(C_6H_5)_3 \end{matrix}\right] \longrightarrow C_6H_5C\equiv N + (C_6H_5)_3PO \qquad [6.27]$$

XX

havior for the thiono analog of XX, and Campbell *et al.* (*40*) detected benzonitrile in the phospholene-catalyzed dimerization of benzoyl isocyanate which, according to the proposed mechanism, should involve the intermediacy of XX. Later work by this group, however, indicated that the benzonitrile may arise from another source (*48*). Trippett and Walker (*49*) reported the conversion of *N*-bromobenzamide into benzonitrile and triphenylphosphine oxide in the presence of triphenylphosphine but claimed the reaction did not involve the intermediacy of a phosphinimine (XX) because of the mild conditions employed (mixing in benzene at room temperature).

This discussion of the reactions of iminophosphoranes indicates there are two basic types of reactions known for these substances: first, their reactions as ordinary nucleophiles in which the phosphorus atom plays no discrete role, and, second, their reactions as nucleophiles in which the phosphorus atom plays a direct role, usually involving its transitory pentavalency. In both of these cases the reactions of iminophosphoranes parallel those of the methylenephosphoranes rather closely although much remains to be done in elucidating the detailed mechanisms of most of the reactions.

III. Iminotrihalophosphoranes

In general, iminotrihalophosphoranes have been prepared by routes that are analogous to those employed for other iminophosphoranes.

Hoffmann *et al.* (*50*) found that chlorodiphenylphosphine and phenyl azide formed the expected imine but that the latter was extremely susceptible to hydrolysis [6.28].

$$(C_6H_5)_2PCl + C_6H_5N_3 \longrightarrow (C_6H_5)_2\overset{\oplus}{\underset{Cl}{\underset{|}{P}}}-\overset{\ominus}{N}C_6H_5 \xrightarrow[H_2O]{} (C_6H_5)_2P(\rightarrow O)NHC_6H_5 \qquad [6.28]$$

Gilpin (*51*), as early as 1897, found that aniline (as its hydrochloride) would react with phosphorus pentachloride to afford what appeared to be *N*-phenyliminotrichlorophosphorane but again the compound was unstable in the presence of moisture. However, this same type of reaction, similar to the general method of Horner described in Section I of this chapter (*9*), was utilized by Kirsanov *et al.* for the preparation of a variety of iminophosphoranes, usually those containing electron withdrawing groups on nitrogen. Thus, reaction of sulfonamides (*52*), acyl amides (*53*) or aryl amines (*54*) with phosphorus pentachloride afforded the corresponding N-substituted iminotrichlorophosphoranes [6.29]. Similarly Bock and Wiegrasse (*55*) showed that chlorodiphenyl-

$$\begin{array}{lllll} ArSO_2NH_2 & + PCl_5 & \longrightarrow & ArSO_2N{=}PCl_3 & + 2\,HCl \\ RCONH_2 & + PCl_5 & \longrightarrow & RCON{=}PCl_3 & + 2\,HCl \\ ArNH_2 & + PCl_5 & \longrightarrow & ArN{=}PCl_3 & + 2\,HCl \end{array} \qquad [6.29]$$

phosphine dichloride and *p*-toluenesulfonamide afforded *N*-*p*-toluenesulfonyliminodiphenylchlorophosphorane.

Some of the halophosphinimines have been shown to have dimeric rather than monomeric structures. Thus, *N*-2,4-dinitrophenyliminotrichlorophosphorane was monomeric whereas *N*-phenyliminotrichlorophosphorane was dimeric, both as determined cryoscopically (*54*, Zhmurova and Kirsanov). The authors proposed that the nitrogen was more basic in the latter imine and was therefore better able to complex with the phosphorus atom. The dimeric imine was stable in the atmosphere whereas the monomeric compound was quite unstable.

Chapman *et al.* (*56*) also found that *N*-methyliminotrichlorophosphorane, prepared by heating methylamine hydrochloride with phosphorus pentachloride, also was dimeric (XXI). Trippett (*57*) has con-

$$\begin{array}{ccc} Cl_3P-N-CH_3 \\ | \quad\ | \\ CH_3-N-PCl_3 \end{array} \qquad\qquad \begin{array}{c} \ominus\ \oplus \\ Cl_3P{=}N-CH_3 \\ | \quad\ | \\ CH_3-N{=}PCl_3 \\ \oplus\ \ominus \end{array}$$

XXI XXII

firmed this conclusion and assigned a cyclic structure to the dimer on the basis of the observation that its H^1 NMR spectrum showed a triplet for the methyl group, indicating coupling with two phosphorus atoms with $J_{PH} = 20$ cps. The fact that the resonance occurred at 7.0 τ indicates some deshielding of the methyl protons and perhaps a contribution of the delocalized structure (XXII). The bond dissociation energy of the phosphorus-nitrogen bond (74.3 kcal/mole) was considerably above that for $P[N(C_2H_5)_2]_3$ but was close to that (72.3 kcal/mole) found for the phosphonitrilic chlorides. This again indicated some additional stabilization for the phosphorus-nitrogen bond, perhaps a contribution represented by structure XXII (*58*, Fowell and Mortimer).

There appears to be a consistent correlation between the probable basicity of a given iminophosphorane and its tendency to dimerize. Only those imines which carry phosphorus substituents which are electron withdrawing and nitrogen substituents which are not electron withdrawing appear to be capable of dimerization. These are the ideal conditions for effective head-to-tail complexation between two molecules such as represented by structure XXI.

In some respects iminotrihalophosphoranes undergo reactions similar to those reported for other iminophosphoranes. For example, Ulrich and Sayigh (*59*) found that heating dimeric *N*-methyltrichlorophosphinimine (XXI) apparently afforded the monomeric form which then underwent reactions with a variety of carbonyl compounds, resulting in the formation of phosphorus oxychloride and the corresponding Schiff base

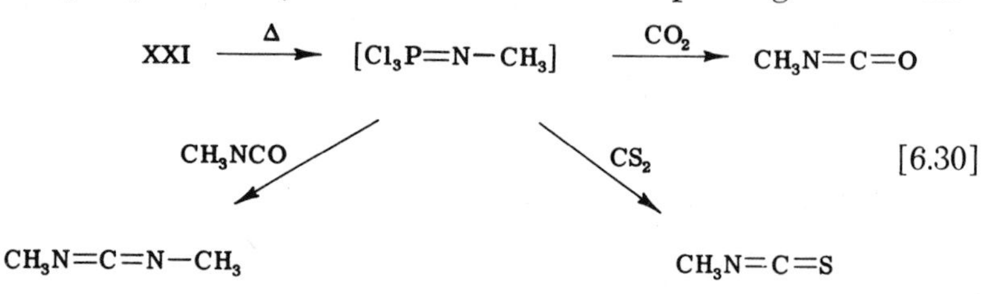

[6.30]

[6.30]. This reaction undoubtedly proceeded via a mechanism analogous to that proposed for the Wittig reaction.

The halogen atoms in *P*-halophosphinimines also are subject to displacement by appropriate nucleophiles. Thus *N*-*p*-toluenesulfonyliminodiphenylchlorophosphorane reacted with azide ion to afford the *P*-azido derivative (*55*, Bock and Wiegrasse). Similarly, *N*-benzenesulfonyliminotrichlorophosphorane was converted into a phosphiteimine upon treatment with sodium methoxide (*60*, Kirsanov and Shevchenko), and the hexachloroimine (XXIII) was converted into its hexaphenyl analog with phenylmagnesium bromide [6.31] (*61*, Moeller and Vandi). The halogens of *P*-halophosphinimines also are subject to hydrolyses but this

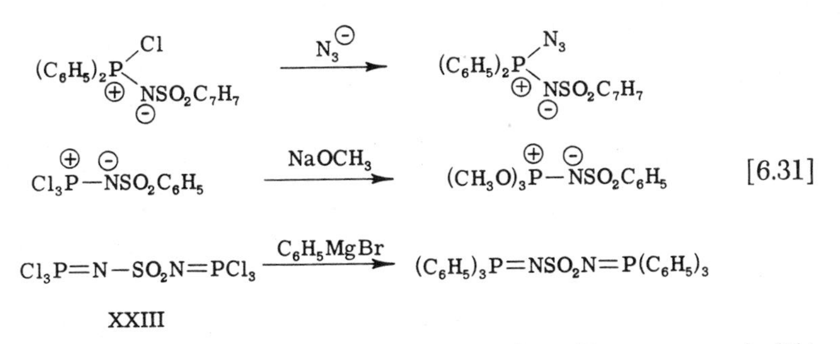

also results in the protonation of the imine. Thus, Kirsanov *et al.* (*54*, *62*) found that one of the three halogens in XXIV could be cleaved by water in the absence of a catalyst.

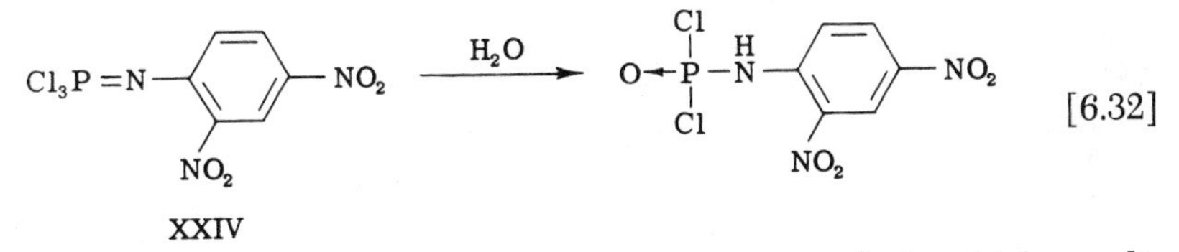

N-Acyliminohalophosphoranes are subject to pyrolysis which results in the elimination of phosphorus oxyhalide, probably via a four-membered transition state (XXV) analogous to that proposed for methylenephosphoranes and other phosphinimines (*62*, Kirsanov and Derkach).

$$Cl_3C-\overset{\overset{O}{\|}}{C}-\overset{\ominus}{N}-\overset{\oplus}{P}Cl_3 \xrightarrow{\Delta} \underset{\text{XXV}}{\begin{matrix} O\cdots PCl_3 \\ \vdots \quad\ \vdots \\ Cl_3C-C\equiv N \end{matrix}} \longrightarrow \begin{matrix} Cl_3C-C\equiv N \\ + \\ POCl_3 \end{matrix} \qquad [6.33]$$

The overall synthetic conversion of an acyl amide into a nitrile by heating with phosphorus pentachloride well may proceed via the intermediacy of *N*-acyliminotrichlorophosphoranes.

IV. Oxygenated Iminophosphoranes

The Russian workers have studied the chemistry of iminotrialkoxyphosphoranes(phosphiteimines). These compounds can be prepared by three routes: the coupling of an azide with a phosphite (*21*, Kabachnik and Gilyarov), the reaction between an amine and a phosphite dichloride (*63*, Zhmurova and Kirsanov) and the displacement of halogen from an iminotrihalophosphorane with alkoxide [6.34] (*60*, Kirsanov and Shevchenko).

The reaction of the iminotrialkoxyphosphoranes with carbonyl compounds has not been studied in detail. The only report was the reaction with carbon disulfide which afforded an isothiocyanate and a phosphite

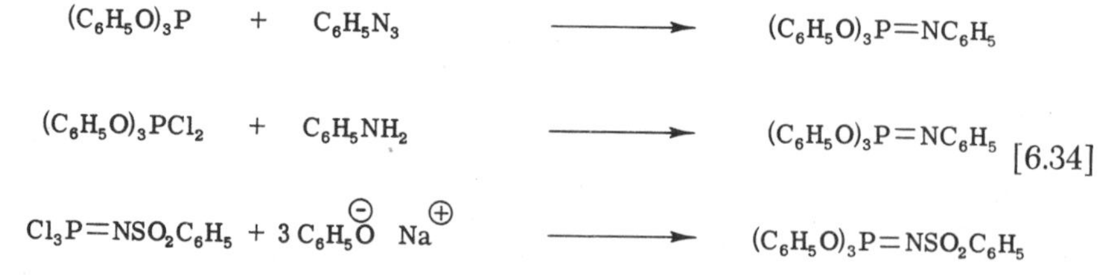

sulfide [6.35]; this is analogous to the reaction reported for iminotriarylphosphoranes (*21*).

$$(RO)_3P{=}NC_6H_5 + CS_2 \longrightarrow (RO)_3PS + C_6H_5NCS \qquad [6.35]$$

Hydrolysis of the phosphiteimines under both acidic and basic conditions resulted in the loss of one of the O-alkyl or O-aryl groups rather than cleavage of the phosphorus-nitrogen bond as for phosphinimines. Thus, *N*-phenyliminotriphenoxyphosphorane, in the presence of water only, was converted to *N*-phenyldiphenylphosphoramidate [6.36] (*63*).

$$(C_6H_5O)_3P{=}NC_6H_5 + H_2O \longrightarrow (C_6H_5O)_2P(O)NHC_6H_5 + [C_6H_5OH] \qquad [6.36]$$

The nature of the N-substituent did not affect the course of this reaction (*21, 60, 62*). The other product, presumably phenol in the case illustrated, was not isolated or identified in any way.

Reaction of the phosphiteimines with alkylating or acylating agents proceeded initially as for the phosphinimines with nucleophilic attack by the nitrogen atom on the reagent to form *N*-alkyl or *N*-acyl derivatives. However, in each instance the anion released by such an attack apparently removed one of the *O*-alkyl groups to again afford a phos-

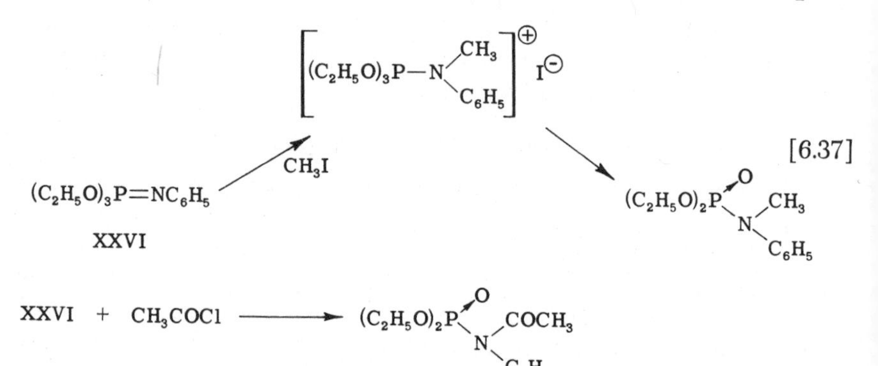

phoramidate. Thus, *N*-phenyliminotriethoxyphosphorane (XXVI) and methyl iodide afforded a *N*-methyl phosphoramidate and XXVI with acetyl chloride afforded a *N*-acetyl phosphoramidate (*21,* Kabachnik and Gilyarov).

Wadsworth and Emmons (*64*) prepared and studied the behavior of a series of phosphoramidate anions. These hold the same relationship to phosphinimines as phosphonate carbanions hold to phosphonium ylids (see Chapter 5). The phosphoramidates were prepared from diethylphosphorochloridates or from diethyl phosphite. The anions were generated in benzene or diglyme solvents by treatment with sodium hydride [6.38]. These workers found that the phosphoramidate anions

$$(C_2H_5O)_2P(\rightarrow O)Cl + RNH_2 \longrightarrow (C_2H_5O)_2P(\rightarrow O)NHR \xrightarrow{NaH} (C_2H_5O)_2P(\rightarrow O)\overset{\ominus}{N}-R \qquad [6.38]$$

$$(C_2H_5O)_2P(\rightarrow O)H + RNH_2 \xrightarrow{CCl_4} (C_2H_5O)_2P(\rightarrow O)NHR$$

reacted with a variety of carbonyl compounds including aldehydes, isocyanates, ketenes, carbon dioxide and carbon disulfide. In each case the C=O or C=S group was replaced by a C=NR group and diethyl phosphate was ejected. The latter usually precipitated from the reaction

$$(C_2H_5O)_2P(\rightarrow O)NHR + CO_2 \xrightarrow{NaH} RN{=}C{=}O + (C_2H_5O)_2PO_2^{\ominus}$$

$$(C_2H_5O)_2P(\rightarrow O)NHR + CS_2 \longrightarrow RN{=}C{=}S + (C_2H_5O)_2POS^{\ominus}$$

$$(C_2H_5O)_2P(\rightarrow O)NHR + R'NCO \longrightarrow RN{=}C{=}NR' + (C_2H_5O)_2PO_2^{\ominus} \qquad [6.39]$$

$$(C_2H_5O)_2P(\rightarrow O)NHR + R_2'C{=}C{=}O \longrightarrow RN{=}C{=}CR_2' + (C_2H_5O)_2PO_2^{\ominus}$$

$$(C_2H_5O)_2P(\rightarrow O)NHR + R'CHO \longrightarrow RN{=}CHR' + (C_2H_5O)_2PO_2^{\ominus}$$

mixture and was separated much easier than was triphenylphosphine oxide from the analogous phosphinimine reaction. The driving force for the reaction, which probably occurred by a mechanism analogous to that of the Wittig reaction, must be the formation of the diethyl phosphate [6.39]).

The carbonyl condensation reactions all proceeded in high yield under mild conditions. The phosphoramidate anion route to a C=N bond perhaps is somewhat more useful than the iminophosphorane route because the reagent is easier to obtain and perfectly stable whereas some iminophosphoranes are sensitive to moisture. The overall result of both routes is the same, however. The phosphoramidate anions would be expected to be more nucleophilic than the phosphinimines due to the lower net positive charge on the phosphorus atom. This relationship has not been tested, however.

V. The Structure of Iminophosphoranes

There has been no systematic investigation of the variation in the properties and structure of the iminophosphoranes with a change in substituents. In addition, there have been relatively little physical data uncovered. The most interesting feature of the iminophosphoranes, of course is the nature of the phosphorus-nitrogen bond. An X-ray analysis will be required to determine the geometry about this bond.

The hybridization about the phosphorus most likely is very similar to that in phosphonium ylids—tetrahedral hybridization with any multiple bonding occurring by overlap of the appropriate filled nitrogen orbitals with vacant $3d$-orbitals of the phosphorus atom. The nitrogen atom of an iminophosphorane could be trigonally hybridized, the P—N—R bond being near 120° with one lone pair in an sp^2-hybrid orbital and the other in a p-orbital (Figure 6.1). Tetrahedral hybridization of the nitrogen

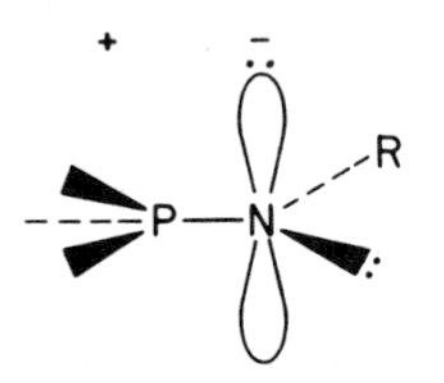

FIG. 6.1. Trigonal hybridization of phosphinimine.

would result in a P—N—R angle of 109.5° with both unshared pairs in sp^3-hybrid orbitals (Figure 6.2). Digonal hybridization would result in

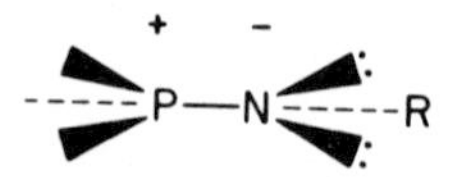

FIG. 6.2. Tetrahedral hybridization of phosphinimine.

a P—N—R angle of 180° with both lone pairs being in p-orbitals (Figure 6.3).

The dipole moment of N-phenyliminotriphenylphosphorane is 4.40 D, very close to that of triphenylphosphine oxide (4.3 D) (*65*, Phillips *et al.*). The polarity of the phosphorus-nitrogen bond must be somewhat greater than the phosphorus-oxygen bond since it has as high a dipole moment in spite of the lower electronegativity of nitrogen than

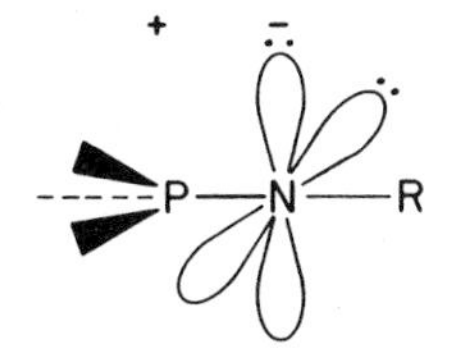

FIG. 6.3. Digonal hybridization of phosphinimine.

oxygen. The imines appear to be more basic and more nucleophilic than the phosphine oxides in agreement with this supposition.

The phosphorus-nitrogen stretching frequency occurs in the infrared region at 1160–1180 cm^{-1} (*9*, Horner and Oediger), very close to that for the phosphorus-oxygen bond. There have been no correlations of structure with frequency as yet and insufficient data are available to begin here. The ultraviolet spectrum of *N*-phenyliminotriphenylphosphorane has little character to it—much less than triphenylphosphine or its oxide. It does have a yellow color with absorption tailing off in the 350 mμ region. The *N*-(*p*-substitutedphenyl)-iminotriphenylphosphoranes do exhibit maxima in the ultraviolet region, extending from about 260 mμ for the *p*-chloro imine to about 430 mμ for the *p*-nitro imine (*9*, Horner and Oediger). These bands and the shifts they show reflect conjugation between the imine nitrogen and the phenyl substituent.

Mortimer *et al.* (*66*) have determined the bond dissociation energy of the phosphorus-nitrogen bond in $(C_6H_5)_3P{=}N{-}C_2H_5$ to be about 126 kcal/mole and in $(CH_3)_3P{=}NC_2H_5$ to be about 97 kcal/mole. In the phosphine oxides the trend is in the opposite direction, the dissociation energy of the phosphorus-oxygen bond being higher for the trialkylphosphine oxide (138 kcal/mole) than for the triphenylphosphine oxide (128 kcal/mole) (*67*, Mortimer). The reasons for this reversal are not at all clear. It might be argued that due to the greater electronegativity of the phenyl groups the P=N bond should be stronger in the triphenyl than in the trimethyl derivative, the result of better overlap with the 3*d*-orbitals of phosphorus. This explanation has been applied to the D(P=O) values for the phosphorus oxyhalides (*67*). The reversal of the D(P=O) values for the phosphine oxides remains to be explained, however.

Practically no P^{31} NMR data has been recorded for iminophosphoranes. The unique bis-imine (IX) and the bis-ylid (XXVII) showed almost identical chemical shifts (*31*, Appel and Hauss; *68*, Driscoll *et al.*). Both showed single resonance lines, which indicated that both

$$(C_6H_5)_3\overset{\oplus}{P}-\overset{\ominus}{\ddot{\underset{..}{N}}}-\overset{\oplus}{P}(C_6H_5)_3 \quad Br^{\ominus} \qquad (C_6H_5)_3\overset{\oplus}{P}-\overset{\ominus}{C}H-\overset{\oplus}{P}(C_6H_5)_3 \quad Br^{\ominus}$$

$\delta = -21$ ppm (IX) $\qquad \delta = -21.2$ ppm (XXVII)

phosphorus atoms were similarly shielded. Insufficient spectra have been obtained on either ylids or imines to permit correlations. It is somewhat surprising, however, that these chemical shifts are to the downfield side of the reference, 85% phosphoric acid, just as are those of the corresponding phosphonium salts.

All of the N-substituted and unsubstituted triaryl and trialkyl iminophosphoranes are monomeric in solution (*6*). The only phosphinimines that have been shown to be dimeric are those which have trihalophosphorus groups. Not even all of the N-substituted iminotrihalophosphoranes are dimeric, but only those which carry substituents on nitrogen which are not powerfully electron-withdrawing. Thus the *N*-phenyl and the *N*-methyliminotrichlorophosphoranes are the only ones which have been shown to be dimeric to date (*54, 56–58*). The *N*-(2,4-dinitrophenyl) and *N*-(*p*-toluenesulfonyl) derivatives are monomeric (*52, 54*). Evidence for the state of polymerization was discussed in Section III of this chapter.

The iminophosphoranes have shown no indication of thermal instability with but one exception. Appel and Schollhorn (*11*) recently have found that the bis-imine (IV) decomposed upon warming to nitrogen and triphenylphosphine. The driving force must be the formation of molecular nitrogen [6.40].

$$\underset{\textbf{IV}}{(C_6H_5)_3\overset{\oplus}{P}-\overset{\ominus}{\ddot{\underset{..}{N}}}-\overset{\ominus}{\ddot{\underset{..}{N}}}-\overset{\oplus}{P}(C_6H_5)_3} \longrightarrow 2\,(C_6H_5)_3P + N_2\uparrow \qquad [6.40]$$

The brevity of the preceding discussion is evidence of the paucity of data available regarding the relationship of molecular structure to the chemical and physical properties of iminophosphoranes. Information regarding the P^{31} NMR spectra, the X-ray analysis and the relative basicity of the imines is the most urgently needed.

VI. Phosphinazines

In some respects the chemistry of phosphinazines resembles that of phosphinimines but there also are many differences. The first report of the preparation of this class of compounds was by Staudinger and Meyer (*69*) in 1919 who found that a variety of stable diazo compounds such

$$(C_6H_5)_2C{=}\overset{\oplus}{N}{=}\overset{\ominus}{N}: \;+\; (C_6H_5)_3P \longrightarrow (C_6H_5)_2C{=}N{-}N{=}P(C_6H_5)_3 \qquad [6.41]$$

XXVIII

as diazofluorene, diphenyldiazomethane and diazoacetic ester, etc., reacted spontaneously with triethylphosphine and triphenylphosphine without nitrogen evolution to afford the corresponding phosphinazine (e.g. XXVIII). Since this original report the scope of the reaction has been broadened to include trialkylphosphites as the phosphorus component (*70, 71*) and diazomethane (*71, 72*), diazocyclopentadiene (*73, 74*), diazoketones (*75, 76*) and quinone diazides (*77, 78*) as the diazo components. Recently, Goetz and Juds (*79*) have shown that the reaction between diazofluorene and *p*-substituted phenyldiphenylphosphines was a second-order reaction, first order in each component [6.42]. A

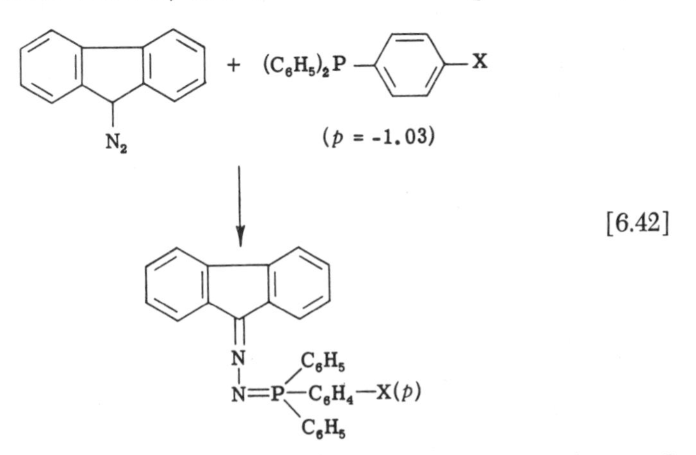

[6.42]

Hammett ρ-σ correlation of the reaction rates with various substituted phosphines gave a ρ value of -1.03, indicating that the more nucleophilic the phosphine, the faster was the rate of phosphinazine formation.

Bestmann and Fritzsche (*80, 81*) showed that Horner's method for the preparation of iminophosphoranes could be applied to the prep-

$$\begin{matrix}R^1\\ R^2\end{matrix}\!\!>C{=}N{-}NH_2 \;+\; (C_6H_5)_3PBr_2 \xrightarrow{(C_2H_5)_3N} (C_6H_5)_3P{=}N{-}N{=}C\!\!<\!\!\begin{matrix}R^1\\ R^2\end{matrix} \qquad [6.43]$$

aration of phosphinazines. Reaction of a hydrazone with triphenylphosphine dibromide in the presence of triethylamine afforded the corresponding phosphinazine [6.43].

Phosphinazines are stable compounds not subject to spontaneous decomposition. Upon pyrolysis, however, nitrogen is evolved (*69*) and triphenylphosphine occasionally could be isolated (*72*). In most cases, however, the phosphinazines did not afford any recognizable products (*69, 75, 82*) or the phosphinazine was recovered unchanged (*73*). Staudinger and Meyer (*12, 69*) did find that the azine (XXVIII) could be pyrolyzed to benzhydrylidenetriphenylphosphorane and nitrogen [6.44]. This single observation of the conversion of a phosphinazine to

$$\underset{\text{XXVIII}}{(C_6H_5)_2C{=}N{-}N{=}P(C_6H_5)_3} \xrightarrow{\Delta} (C_6H_5)_2C{=}P(C_6H_5)_3 + N_2\uparrow \qquad [6.44]$$

a phosphonium ylid led to innumerable unsuccessful attempts to find other examples of the reaction and develop a general synthesis of phosphonium ylids (*72, 73*). Poshkus and Herweh (*70*) found that benzhydrylidenetrisisopropylphosphiteazine (XXIX) underwent an intramo-

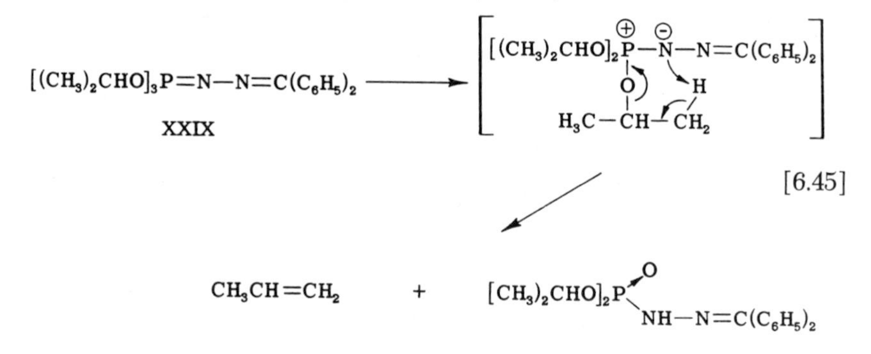

lecular elimination reaction upon pyrolysis to afford propylene and a phosphoramidate. A concerted cyclic mechanism may have been involved.

Wittig and Schlosser (*83*) recently have found that decomposition of phosphinazines in the presence of cuprous chloride does afford low yields of phosphonium ylids. These ylids have not been isolated but their presence has been detected by their conversion into phosphonium salts in the presence of acid or by the isolation of olefin when the decomposition was carried out in the presence of a carbonyl compound. Thus, the azine from phenyldiazomethane and triphenylphosphine, when heated in tetrahydrofuran solution with cuprous chloride, afforded a 10% yield of the ylid conjugate acid, 45% of triphenylphosphine oxide and 41% of benzal azine [6.46]. The azine from diazomethane and triphenylphosphine was decomposed in the presence of benzophenone to afford a 35% yield of olefin.

Phosphinazines are basic since they are insoluble in neutral aqueous media but will dissolve in dilute mineral acid, indicating salt formation. (*69*, Staudinger and Meyer). The phosphinazine can be reprecipitated upon basifying the solution.

$$C_6H_5CH{=}N{-}N{=}P(C_6H_5)_3 \xrightarrow[(2)\ HCl]{(1)\ CuCl,\ \Delta} \overset{\oplus}{C_6H_5CH_2P}(C_6H_5)_3Cl^{\ominus} + (C_6H_5CH{=}N{-})_2 + (C_6H_5)_3PO \qquad [6.46]$$

$$(C_6H_5)_3P{=}N{-}N{=}CH_2 + (C_6H_5)_2CO \xrightarrow[\Delta]{CuCl} (C_6H_5)_2C{=}CH_2 + (C_6H_5)_3PO$$

Phosphinazines might be regarded as derivatives of phosphinimines via resonance structure XXX or as azophosphonium ylids via structure XXXI. Horner and Schmelzer (*77*) found that *p*-benzoquinotriphenyl-

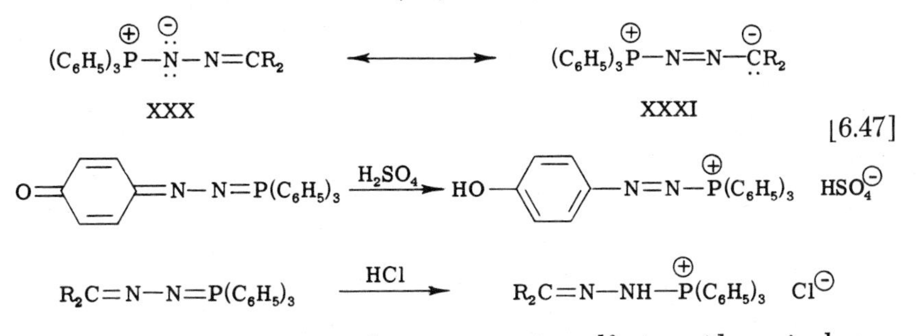

phosphinazine was protonated on oxygen in sulfuric acid, a vinylogous protonation. However, Bestmann and Gothlich (*84*) showed that a wide variety of phosphinazines were protonated on nitrogen (i.e. via structure XXX) as indicated by the appearance of a typical nitrogen-hydrogen absorption in the infrared region at 3.0 μ.

Bestmann and Gothlich (*84*) and Singh and Zimmer (*85*) have found that most phosphinazines were alkylated on the α-nitrogen and not on carbon. For example, methylenetriphenylphosphinazine and benzyl iodide afforded an iodide salt which was hydrolyzed to triphenyl-

$$CH_2{=}N{-}N{=}P(C_6H_5)_3 \xrightarrow{C_6H_5CH_2I} CH_2{=}N{-}\underset{\underset{C_6H_5CH_2}{|}}{N}{-}\overset{\oplus}{P}(C_6H_5)_3\ I^{\ominus}$$

$$\xrightarrow[(2)\ H^{\oplus}]{(1)\ \overset{\ominus}{O}H} C_6H_5CH_2NHNH_2 + [CH_2O] + (C_6H_5)_3PO \qquad [6.48]$$

phosphine oxide and benzylhydrazine [6.48]. The α-alkylation originally was proposed by Staudinger and Meyer (*69*).

Attempts to alkylate phosphinazines occasionally have led to cleavage of the phosphorus-nitrogen bond and formation of an alkylphosphonium salt. Bestmann *et al.* (*76*) found that acylidenetriphenylphosphinazines and methyl iodide afforded triphenylmethylphosphonium iodide and the diazoketone [6.49]. Similarly, cyclopentadienylidenetriphenylphosphina-

$$C_6H_5COCH{=}N{-}N{=}P(C_6H_5)_3 \xrightarrow{CH_3I} C_6H_5COCHN_2 + CH_3\overset{\oplus}{P}(C_6H_5)_3\ I^{\ominus} \quad [6.49]$$

$$C_5H_4{=}N{-}N{=}P(C_6H_5)_3 \xrightarrow{CH_3I} C_5H_4{-}N_2 + CH_3\overset{\oplus}{P}(C_6H_5)_3\ I^{\ominus}$$

zine and methyl iodide also resulted in phosphorus-nitrogen cleavage (*84*). The same group showed that phosphinazines from especially stable diazo compounds, such as diazocyclopentadiene and diazoketones, existed in equilibrium with the phosphine and the diazo compound. In these cases, the phosphine probably reacted directly with the methyl iodide while in other cases the methyl iodide reacted with the phosphinazine in a normal alkylation.

Phosphinazines are stable in the presence of moisture but are hydrolyzed in the presence of appropriate catalysts. The phosphorus-nitrogen bond is subject to cleavage in alkaline media, affording phosphine oxides and hydrazones (*69*). This reaction probably proceeds via hydroxide attack on the phosphorus atom as was the case for phosphonium

$$(C_6H_5)_2C{=}N{-}N{=}P(C_6H_5)_3 \xrightarrow{^{\ominus}OH/H_2O} (C_6H_5)_2C{=}N{-}NH_2 + (C_6H_5)_3PO \quad [6.50]$$

ylids and phosphinimines [6.50]. However, cyclopentadienylidenetriphenylphosphinazine is a particularly stable azine resisting all attempts at hydrolysis (*73*, Ramirez and Levy). Occasionally carbonyl azines can be isolated from the hydrolysis reaction mixtures of phosphinazines. These may be formed by a recondensation between carbonyl compound and hydrazone or between two molecules of hydrazone (*70, 86*).

There have been several attempts to induce reactions between phosphinazines and carbonyl compounds similar to a Wittig reaction. In this case the products should be carbonyl azines and phosphine oxides. Staudinger and Braunholtz (*87*) were unable to identify the products of the reaction between XXVIII and carbon disulfide or phenyl isocyanate. Later, Braunholtz (*86*) reported no reaction between XXVIII and carbon dioxide but obtained some triphenylphosphine oxide from the reaction with phenyl isocyanate. Triphenylphosphine sulfide and what was claimed to be the carbodiimide (XXXII) were obtained from the reaction with carbon disulfide (*86*). Kabachnik and Gilyarov (*71*)

$$\underset{\text{XXVIII}}{(C_6H_5)_3P{=}N{-}N{=}C(C_6H_5)_2} + CS_2 \longrightarrow \underset{\text{XXXII}}{(C_6H_5)_2C{=}N{-}N{=}C{=}N{-}N{=}C(C_6H_5)_2} + (C_6H_5)_3PS \qquad [6.51]$$

reported that methylenetriethylphosphiteazine and carbon disulfide also afforded triethyl thiophosphate but they made no mention of the other products.

Bestmann and Fritzsche (*80, 81*) have effected clear-cut Wittig-type reactions between a series of phosphinazines and carbonyl compounds. For example, XXVIII and benzaldehyde afforded a 70% yield of benzophenone benzylidenehydrazone (XXXIII). Similar condensations were

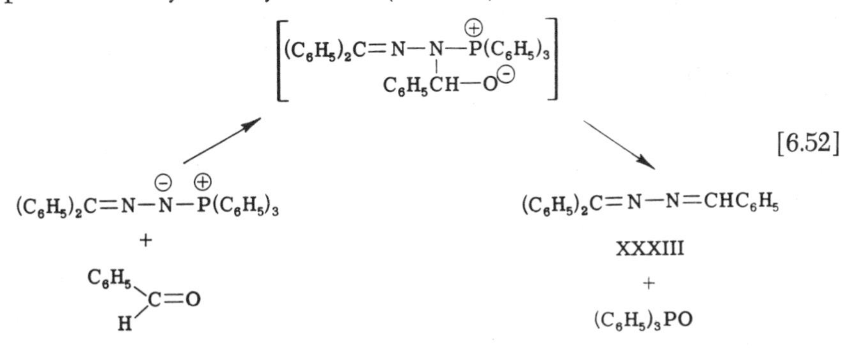

effected using other phosphinazines and carbonyl compounds. This reaction probably proceeded via a betaine analogous to that involved in the Wittig reaction between phosphonium ylids and carbonyl compounds or between phosphinimines and carbonyl compounds. The course of the reaction is consistent with the portrayal of XXX as the dominant ionic contributing structure for the phosphinazines with little or no nucleophilicity residing on the γ-position as represented in structure XXXI. If the latter structure were important the products of a reaction with a carbonyl compound should have been a phosphine oxide, an olefin and nitrogen formed via a betaine (XXXIV). The driving force for such a reaction should not be insignificant.

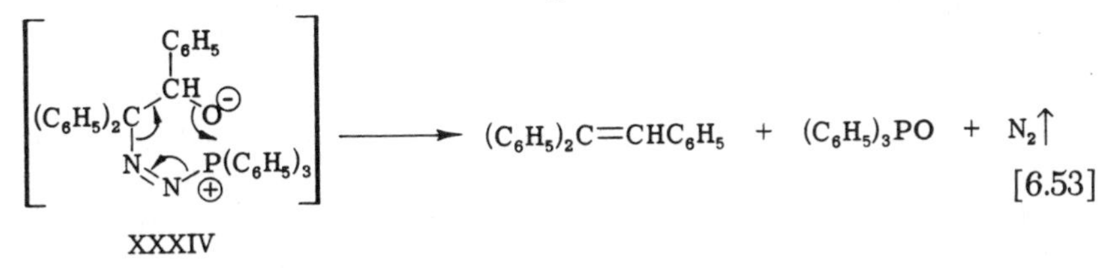

Interestingly, Schonberg and Brosowski (*88*) did find that XXVIII reacted with nitrosobenzene to afford triphenylphosphine oxide, nitrogen and benzophenone-anil [6.54]. This reaction may have occurred via

$$(C_6H_5)_3P{=}N{-}N{=}C(C_6H_5)_2 \;(\text{XXVIII}) + C_6H_5NO \longrightarrow (C_6H_5)_2C{=}NC_6H_5 + N_2 + (C_6H_5)_3PO \quad [6.54]$$

an intermediate similar to XXXIV or by prior conversion of XXVIII to the corresponding phosphonium ylid and nitrogen. Since the reaction was effected in refluxing benzene the phosphinazine should have been stable and not decomposed.

Brown *et al.* (*46*) found that fluorenylidenetriphenylphosphinazine reacted with dimethyl acetylenedicarboxylate by undergoing phosphorus-nitrogen cleavage and forming a phosphonium ylid in a manner analogous to that shown by phosphinimines. However, the monosubstituted

$$C_{13}H_8{=}N{-}N{=}P(C_6H_5)_3 + XC{\equiv}CX \longrightarrow C_{13}H_8{=}N{-}N{=}C(X){-}C(X){=}P(C_6H_5)_3$$

$$(C_6H_5)_3P{=}N{-}N{=}CHCO_2CH_3 \;(\text{XXXV}) + XC{\equiv}CX \longrightarrow (C_6H_5)_3P{=}N{-}N{=}C(X){-}C(X){=}CH(X) \quad [6.55]$$

$$\xrightarrow{\Delta} \text{pyrazole (N–H, tri-X substituted)} + (C_6H_5)_3PO$$

$(X = COOCH_3)$

phosphinazine (XXXV) underwent Michael addition to the diester as indicated by conversion of the adduct to a pyrazoline and triphenylphosphine upon warming in ethyl acetate solution. The reasons for the different courses of reaction for the two phosphinazines have not been elucidated. The fluorenylidenephosphinazine likely would be too sterically hindered to attack with carbon rather than with nitrogen. In addition, there may be a higher electron density on the γ-carbon in XXXV than in the fluorenylidene case.

In summary, the chemistry of phosphinazines is quite similar to that of phosphinimines both in the normal nucleophilic reactions of the α-nitrogen atom and in those reactions which also depend on the potential pentavalency of the phosphorus atom. Phosphinazine chemistry is complicated somewhat by the ambident nature of the nucleophilic portion of the molecule.

REFERENCES

1. N. L. Paddock, *Quart. Rev.* **18,** 168 (1964).
2. F. Mann and E. J. Chaplin, *J. Chem. Soc.* 527 (1937).
3. R. Appel, W. Buchner and E. Guth, *Ann.* **618,** 53 (1958).
4. R. Appel and E. Guth, *Z. Naturforschg.* **15b,** 57 (1960).
5. H. H. Sisler, A. Sarkis, H. S. Ahuja, R. J. Drago and N. L. Smith, *J. Am. Chem. Soc.* **81,** 2982 (1959).
6. R. Appel and A. Hauss, *Angew. Chem.* **71,** 626 (1959).
7. R. Appel and A. Hauss, *Chem. Ber.* **93,** 405 (1960).
8. H. H. Sisler, H. S. Ahuja and N. L. Smith, *J. Org. Chem.* **26,** 1819 (1961).
9. L. Horner and H. Oediger, *Ann.* **627,** 142 (1959).
10. H. Zimmer and G. Singh, *J. Org. Chem.* **28,** 483 (1963).
11. R. Appel and R. Schollhorn, *Angew. Chem.* **76,** 991 (1964).
12. H. Staudinger and J. Meyer, *Helv. Chim. Acta* **2,** 635 (1919).
13. N. Wiberg and F. Raschig, *Angew Chem.* **74,** 388 (1962).
14. J. S. Thayer and R. West, *Inorg. Chem.* **3,** 406 (1964).
15. L. Birkofer, A. Ritter and S. M. Kim, *Chem. Ber.* **96,** 3099 (1963).
16. A. Messmer, I. Pinter and F. Szego, *Angew. Chem.* **76,** 227 (1964).
17. H. Bock and W. Wiegrasse, *Angew. Chem.* **75,** 789 (1963).
18. J. E. Franz and C. Osuch, *Tetrahedron Letters* 841 (1963).
19. J. Goedler and H. Ullmann, *Chem. Ber.* **94,** 1067 (1961).
20. L. Birkofer and S. M. Kim, *Chem. Ber.* **97,** 2100 (1964).
21. M. I. Kabachnik and V. A. Gilyarov, *Izvest. Akad. Nauk. S.S.S.R.* 790 (1956); *Bull. Acad. Sci. U.S.S.R.* 809 (1956).
22. H. J. Vetter and H. Noth, *Chem. Ber.* **96,** 1308 (1963).
23. A. M. Aguiar and J. Beisler, *J. Org. Chem.* **29,** 1660 (1964).
24. L. Horner and A. Gross, *Ann.* **591,** 117 (1955).
25. E. Bergmann and H. A. Wolff, *Chem. Ber.* **63,** 1176 (1930).
26. H. Staudinger and E. Hauser, *Helv. Chim. Acta* **4,** 861 (1921).
27. J. E. Leffler, U. Honsberg, Y. Tsuno and I. Forsblad, *J. Org. Chem.* **26,** 4810 (1961).
28. R. C. Cookson and A. N. Hughes, *J. Chem. Soc.* 6061 (1963).
29. H. J. Bestmann and F. Seng, *Angew. Chem.* **75,** 475 (1963).
30. R. Appel, A. Hauss and G. Buchler, *Z. Naturforschg.* **16b,** 405 (1961).
31. R. Appel and A. Hauss, *Z. Anorg. u. Allg. Chem.* **311,** 290 (1961).
32. R. Appel and G. Buchler, *Z. Naturforschg.* **17b,** 422 (1962).
33. F. Ramirez, N. B. Desai, B. Hansen and N. McKelvie, *J. Am. Chem. Soc.* **83,** 3539 (1961).
34. L. Horner and H. Winkler, *Tetrahedron Letters* 175 (1964).
35. R. Appel and F. Vogt, *Chem. Ber.* **95,** 2225 (1962).
36. H. Zimmer and G. Singh, *J. Org. Chem.* **29,** 3412 (1964).
37. M. F. Hawthorne, *J. Am. Chem. Soc.* **83,** 367 (1961).
38. R. Appel and R. Schaaff, *Z. Naturforschg.* **16b,** 405 (1961).
39. H. Zimmer and G. Singh, *Angew. Chem.* **75,** 574 (1963).
40. T. W. Campbell, J. J. Monagle and V. S. Foldi, *J. Am. Chem. Soc.* **84,** 3673 (1962).
41. J. J. Monagle, T. W. Campbell and H. F. McShane, Jr., *J. Am. Chem. Soc.* **84,** 4288 (1962).
42. A. J. Speziale and D. E. Bissing, *J. Am. Chem. Soc.* **85,** 3878 (1963).

43. J. J. Monagle and J. V. Mengenhauser, *Abstr. Papers, 149th Meeting, Am. Chem. Soc.* p. 48P (1965).
44. J. J. Monagle, *J. Org. Chem.* **27**, 3851 (1962).
45. G. W. Brown, R. C. Cookson, I. D. R. Stevens, T. C. W. Mak and J. Trotter, *Proc. Chem Soc.* 87 (1964).
46. G. W. Brown, R. C. Cookson and I. D. R. Stevens, *Tetrahedron Letters* 1263 (1964).
47. T. C. W. Mak and J. Trotter, *Acta Cryst.* **18**, 81 (1965).
48. L. A. McGrew, W. Sweeney, T. W. Campbell and V. S. Foldi, *J. Org. Chem.* **29**, 3002 (1964).
49. S. Trippett and D. M. Walker, *J. Chem. Soc.* 2976 (1960).
50. H. Hoffmann, R. Grunewald and L. Horner, *Chem. Ber.* **93**, 861 (1960).
51. J. E. Gilpin, *Am. Chem. J.* **19**, 352 (1897).
52. A. V. Kirsanov, *Z. Obshchei Khim.* **22**, 269 (1952); *J. General Chem.* **22**, 329 (1952).
53. A. V. Kirsanov and G. I. Derkach, *Z. Obshchei Khim.* **26**, 2009 (1956); *J. General Chem.* **26**, 2245 (1956); *Chem. Abstr.* **51**, 1821 (1957).
54. I. N. Zhmurova and A. V. Kirsanov, *Z. Obshchei Khim.* **30**, 3044 (1960); *J. General Chem.* **30**, 3018 (1960).
55. H. Bock and W. Wiegrasse, *Angew. Chem.* **74**, 327 (1962).
56. A. C. Chapman, W. S. Holmes, N. L. Paddock and H. T. Searle, *J. Chem. Soc.* 1825 (1961).
57. S. Trippett, *J. Chem. Soc.* 4731 (1962).
58. P. A. Fowell and C. T. Mortimer, *Chem. & Ind. (London)* 444 (1960).
59. H. Ulrich and A. A. Sayigh, *Angew. Chem.* **74**, 900 (1962).
60. A. V. Kirsanov and V. I. Shevchenko, *Z. Obshchei Khim.* **24**, 474 (1954); *J. General Chem.* **24**, 483 (1954).
61. T. Moeller and A. Vandi, *J. Org. Chem.* **27**, 3511 (1962).
62. A. V. Kirsanov and G. I. Derkach, *Z. Obshchei Khim.* **26**, 2631 (1956); *Chem. Abstr.* **51**, 1821 (1957).
63. I. M. Zhmurova and A. V. Kirsanov, *Z. Obshchei Khim.* **29**, 1687 (1959).
64. W. S. Wadsworth and W. D. Emmons, *J. Am. Chem. Soc.* **84**, 1316 (1962); *J. Org. Chem.* **29**, 2816 (1964).
65. G. M. Phillips, J. S. Hunter and L. E. Sutton, *J. Chem. Soc.* 146 (1945).
66. A. P. Claydon, P. A. Fowell and C. T. Mortimer, *J. Chem. Soc.* 3284 (1960).
67. C. T. Mortimer, *Pure and Applied Chem.* **2**, 71 (1961).
68. J. S. Driscoll, D. W. Grisley, Jr., J. V. Pustinger, J. E. Harris and C. N. Matthews, *J. Org. Chem.* **29**, 2427 (1964).
69. H. Staudinger and J. Meyer, *Helv. Chim. Acta* **2**, 619 (1919).
70. A. C. Poshkus and J. E. Herweh, *J. Org. Chem.* **27**, 2700 (1962).
71. M. I. Kabachnik and V. A. Gilyarov, *Doklady Akad. Nauk. S.S.S.R.* **106**, 473 (1956).
72. G. Wittig and W. Haag, *Chem. Ber.* **88**, 1654 (1955).
73. F. Ramirez and S. Levy, *J. Org. Chem.* **23**, 2037 (1958).
74. T. Weil and M. Cais, *J. Org. Chem.* **28**, 2472 (1963).
75. H. Staudinger and G. Luscher, *Helv. Chim. Acta.* **5**, 75 (1922).
76. H. J. Bestmann, H. Buckschewski and H. Leube, *Chem. Ber.* **92**, 1345 (1959).
77. L. Horner and H. G. Schmelzer, *Chem. Ber.* **94**, 1326 (1961).
78. W. Ried and H. Appel, *Ann.* **646**, 82 (1961).
79. H. Goetz and H. Juds, *Ann.* **678**, 1 (1964).

80. H. J. Bestmann, *Angew. Chem.* **72**, 326 (1960).
81. H. J. Bestmann and H. Fritzsche, *Chem. Ber.* **94**, 2477 (1961).
82. L. Horner and E. Lingnau, *Ann.* **591**, 135 (1955).
83. G. Wittig and M. Schlosser, *Tetrahedron* **18**, 1023 (1962).
84. H. J. Bestmann and L. Gothlich, *Ann.* **655**, 1 (1962).
85. G. Singh and H. Zimmer, *J. Org. Chem.* **30**, 417 (1965).
86. W. Braunholtz, *J. Chem. Soc.* **121**, 300 (1922).
87. H. Staudinger and W. Braunholtz, *Helv. Chim. Acta* **4**, 897 (1921).
88. A. Schonberg and K. H. Brosowski, *Chem. Ber.* **92**, 2602 (1959).

PART 2

YLIDS OF OTHER HETEROATOMS

7

NITROGEN YLIDS

From the discussion presented in Part I and from that to follow in succeeding chapters it is apparent that phosphonium, arsonium, stibonium and sulfonium groups exert powerful acidifying effects on α-hydrogen atoms. In these cases the strength of this effect has been attributed to stabilization of the resulting carbanion by overlap with the vacant, low energy *d*-orbitals of the 'onium atom. On the basis of the deuteroxide-catalyzed exchange of deuterium in tetramethylammonium and tetramethylphosphonium salts, it is apparent that the difference between the almost complete lack of exchange in the former and the rapid exchange in the latter cannot be explained solely by coulombic stabilization of the carbanionic (ylid) intermediates or transition states (*1*, Doering and Hoffmann; see discussion in Chapter two). In a related series of experiments Doering and Schreiber (*2*) observed that vinyldimethylsulfonium salts would undergo rapid nucleophilic addition to the double bond whereas vinyltrimethylammonium salts were inert under the same conditions.

On the basis of these and many other observations of a similar nature it has been assumed generally that an ammonium group is capable of stabilizing an adjacent carbanion only by an electrostatic interaction between the opposite charges. Being a first row element with filled 2*s*- and 2*p*-orbitals, the next available empty orbital is the 3*s*-orbital which is at a much higher energy level. In contrast, phosphorus and sulfur, being second row elements with filled 3*s*- and 3*p*-orbitals, have as their next available empty orbitals the 3*d*-orbitals which are at only a slightly higher energy level and therefore more available for overlap with considerable stabilization for a carbanion, especially when the heteroatom carries a positive charge. It must be pointed out that the absence of other than electrostatic stabilization by positively charged nitrogen has not been rigorously proven. In spite of this, the larger acidifying ability of the lower row elements usually is interpreted as being due to valence shell expansion and assumes the absence of such stabilization for nitrogen. However invalid this assumption may be, it is still apparent that an ammonium group is much less effective than a phosphonium group for carbanion stabilization. Accordingly, nitrogen ylids would be expected to be more difficult to prepare and, perhaps, to be less stable once prepared. This expectation has been verified as will be indicated in this chapter. There are other significant differences between nitrogen ylids and other ylids that also will be explored.

The evolution of the chemistry of nitrogen ylids really began with the early attempts to demonstrate the existence of pentavalent nitrogen. Schlenk and Holtz (3) had isolated tetramethylammonium triphenylmethide as a highly colored conducting substance. It underwent typical

$$(C_6H_5)_3C^{\ominus}Na^{\oplus} + (CH_3)_4N^{\oplus}Cl^{\ominus} \longrightarrow (CH_3)_4N^{\oplus}\overset{\ominus}{C}(C_6H_5)_3$$

$$(CH_3)_4N^{\oplus}\overset{\ominus}{C}(C_6H_5)_3 \xrightarrow{H_2O} (C_6H_5)_3CH \qquad (CH_3)_4N^{\oplus}\overset{\ominus}{C}(C_6H_5)_3 \xrightarrow{CO_2} (C_6H_5)_3C-CO_2^{\ominus} \qquad [7.1]$$

carbanionic reactions with water and carbon dioxide [7.1]. The same workers later prepared tetramethylammonium benzylide (*4*). However subsequent attempts by Hager and Marvel (*5*) to prepare analogs in which all five groups were more nearly alike were unsuccessful. Thus, ethyllithium and triethyl-*n*-butylammonium bromide failed to yield any pentavalent nitrogen derivatives. Similarly, triethylbenzylammonium bromide with ethyllithium did not afford tetraethylammonium benzylide, ruling out the existence of any structures in which the five groups bound to nitrogen approached equivalency. Hager and Marvel (*5*) concluded that the existence of Schlenk's examples was due mainly to the stability of the triphenylmethyl and benzyl groups as carbanions and not due to the formation of pentavalent nitrogen derivatives.

In 1947 Wittig and Wetterling (*6*) re-examined the pentavalent nitrogen question and attempted to prepare tetramethylammonium phenyl by reaction of phenyllithium with tetramethylammonium bromide. None was obtained but benzene was produced along with an ether-insoluble substance that subsequently was shown to be the lithium

$$(CH_3)_3\overset{\oplus}{N}-CH_3 \quad Br^{\ominus} \xrightarrow{C_6H_5Li} C_6H_6 + \underset{\textbf{I}}{(CH_3)_3\overset{\oplus}{N}-\overset{\ominus}{C}H_2 \cdot LiBr} \qquad [7.2]$$

bromide complex of trimethylammoniummethylide (I). Wittig and others vigorously have explored the chemistry of ammonium ylids since this report. It should be pointed out, however, that the first report of a nitrogen ylid appears to be due to Krohnke (*7*) in 1935. Since his original report on proton removal from *N*-phenacylpyridinium bromide [7.3] he and his coworkers have continued actively to investigate the chemistry of pyridinium ylids and related substances.

$$C_5H_5\overset{\oplus}{N}-CH_2-\overset{O}{\overset{\|}{C}}-C_6H_5\ Br^{\ominus} \xrightarrow{K_2CO_3} C_5H_5\overset{\oplus}{N}-CH\!=\!\!=\overset{O^{\ominus}}{\overset{|}{C}}-C_6H_5 \qquad [7.3]$$

In the following sections of this chapter the chemistry of ammonium and pyridinium ylids and imines will be reviewed. In addition, several well-known reactions and rearrangements which appear to involve nitrogen ylids as intermediates will be discussed.

I. Ammonium Ylids

Wittig and Wetterling (*6*) found that treatment of an ethereal slurry of tetramethylammonium bromide with phenyllithium afforded a complex of trimethylammoniummethylide with lithium bromide which was insoluble in ether but soluble in tetrahydrofuran. This complex (I) exhibited typical nucleophilic properties in that it afforded tetramethylammonium hydroxide with water, iodomethyltrimethylammonium iodide with iodine, ethyltrimethylammonium iodide with methyl iodide and (2-hydroxy-2,2-diphenyl)ethyltrimethylammonium bromide with benzophenone after acidification with hydrobromic acid [7.4]. All of these reactions are to be expected of trimethylammonium methylide.

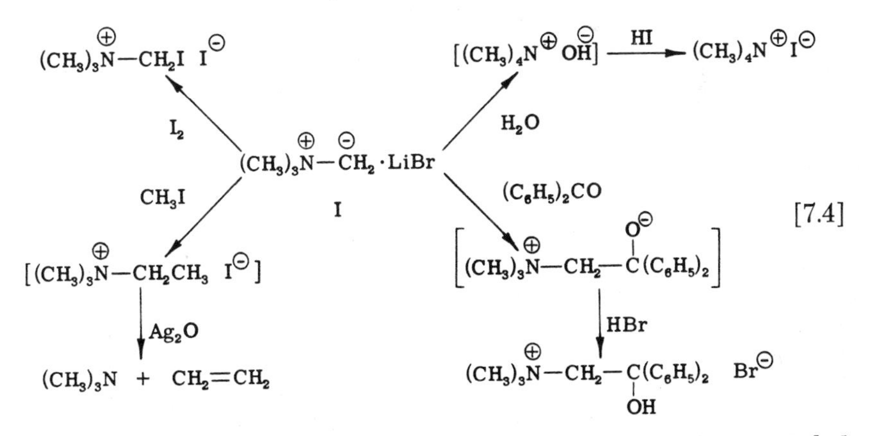

Some indication of the acidity of the α-hydrogen of tetramethyl ammonium bromide is afforded by the observation (*6*) that phenyllithium will afford the ylid whereas benzylsodium will not, affording instead tetramethylammonium benzyl (*4*). Thus, the acidity of the salt must be between that of benzene and toluene. This is considerably weaker than the corresponding phosphonium and sulfonium salts in agreement with the kinetic data of Doering and Hoffmann (*1*).

Evidence for the complexing of the ylid with lithium bromide was first provided by an analysis of the solid complex (*6*). Wittig and Polster (*8*) found that the characteristics of the ylid changed markedly if this complexing was prevented or destroyed. Treatment of tetramethylammonium bromide with phenyllithium in ether at room temperature for a period of 90 hours led to the formation of 20% of trimethylamine and 12% of polymethylene, presumably by decomposition of the ylid into the amine and methylene; the latter polymerized. However, repetition of this reaction but using the dimethyl ether of ethylene glycol

$$(CH_3)_3\overset{\oplus}{N}-\overset{\ominus}{C}H_2\cdot LiBr \xrightarrow{\text{solvent}} (CH_3)_3\overset{\oplus}{N}-\overset{\ominus}{C}H_2 + \text{solvent}\cdot LiBr$$

$$\text{I} \qquad\qquad \text{II}$$

$$\text{II} \longrightarrow (CH_3)_3N + [CH_2:] \longrightarrow (CH_2)_n \qquad [7.5]$$

as the solvent led to the formation of the polymethylene in 74% yield. This solvent was shown to form a strong complex with lithium bromide, apparently stronger than that formed by the ylid. These workers proposed that when trimethylammoniummethylide was complexed with lithium bromide the ylid was stable but that the free ylid (II) rapidly decomposed to trimethylamine and methylene. The same general results were obtained by the use of phenylsodium in place of phenyllithium as the base for ylid formation, resulting in the formation of 78% of trimethylamine and 53% of polymethylene. Sodium bromide would be expected to complex much less strongly with the ylid than would lithium bromide.

The decomposition of ammonium ylids to carbenes has been mentioned in two other reports. Franzen and Wittig (*9*) claimed that a solution of the methylide (II), prepared with phenylsodium and phenyl-

$$(CH_3)_4\overset{\oplus}{N}\ Br^{\ominus} \xrightarrow[C_6H_5Li]{C_6H_5Na} (CH_3)_3\overset{\oplus}{N}-\overset{\ominus}{C}H_2 \longrightarrow [CH_2:] + (CH_3)_3N$$

$$[CH_2:] + \text{cyclohexene} \longrightarrow \text{norcarane (bicyclo[4.1.0]heptane, }CH_2\text{ bridge)} \qquad [7.6]$$

lithium, converted cyclohexene to norcarane in 5–18% yield presumably via the intermediacy of methylene [7.6]. More recently, however, Wittig and Krauss (*10*) have been unable to repeat this work. They were able, however, to isolate 7-(*n*-butoxy)norcarane (15%) and 7-phenoxynorcarane (48%) from treatment of *n*-butoxymethyltrimethylammonium bromide and phenoxymethyltrimethylammonium bromide, respectively, with organolithium reagents in the presence of cyclohexene. These observations were interpreted in terms of *n*-butoxymethylene and phenoxymethylene intermediates.

Weygand *et al.* (*11*) found that a solution of I in tetrahydrofuran reacted with methyl bromide to afford a mixture of trimethylethylammonium bromide and tetramethylammonium bromide in a 1:5 ratio. Repetition of this reaction using completely C-^{14}-labeled ylid and unlabeled methyl bromide resulted in tetramethylammonium bromide containing only 79–85% of the original radioactivity and polymethylene formed in the reaction containing 19% of the radioactivity. These workers proposed that the ylid (II) decomposed to trimethylamine and methylene; the latter polymerized and the former reacted with methyl bromide to afford the quaternary salt. These observations of Wittig and Weygand do not prove the intermediacy of carbenes but certainly do indicate the facile cleavage of a carbon-nitrogen bond in an ammonium ylid.

Wittig and Rieber (*12*) found that treatment of tetramethylammonium bromide with two equivalents of phenyllithium afforded an ethereal solution and slurry that reacted with benzophenone to afford 18% of a monoadduct (III) and 20% of a bis-adduct (IV) after quenching with hydrobromic acid. They found that the ether-insoluble portion of the

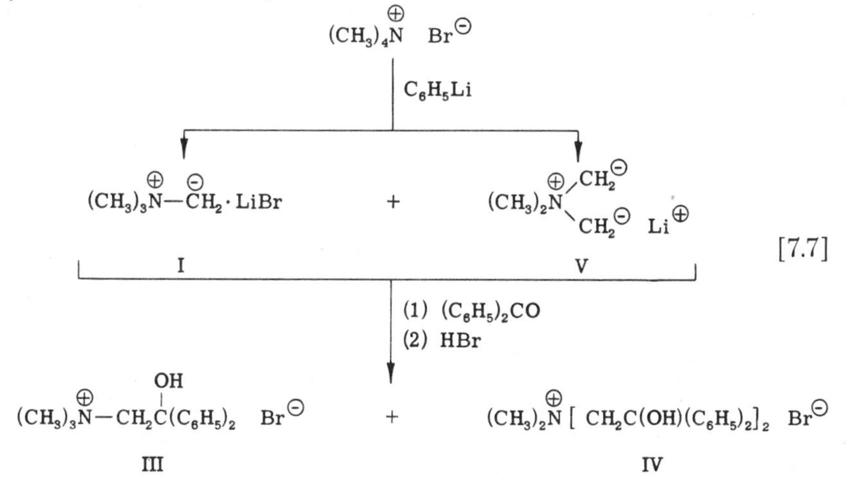

ylid solution contained the lithium bromide complex (I) and that what must have been a bis-ylid (V) was in the ether-soluble portion of the solution. Furthermore, increasing the time of contact of the phenyllithium with the ammonium salt before the addition of benzophenone led to a decrease in the amount of monoadduct (III) formed. This appeared to indicate the gradual conversion of the monoylid (I) into the bis-ylid (V) (*8*, Wittig and Polster).

Wittig and Tochtermann (*13*) more recently have observed another example of bis-ylid formation. *N*-Methyl-*N*-bromomethylpyrolidinium bromide with butyllithium and benzophenone afforded a bis-adduct when reacted at room temperature. The same reaction, but at —70°, afforded only a mono-adduct [7.8]. As expected, therefore, the formation of bis-

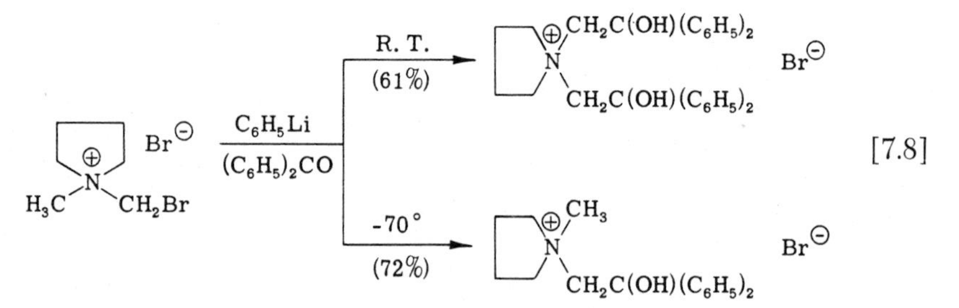

ylids seems to be a much higher energy process than the formation of mono-ylids.

Weygand *et al.* (*11, 14*) recently have explored other nucleophilic reactions of trimethylammoniummethylide (II) in tetrahydrofuran solution. The ylid was very difficult to alkylate because it tended both to decompose to trimethylamine and effect eliminations with the alkyl halides. For example, reaction with cyclohexyl bromide afforded a 92% yield of cyclohexene and reaction with benzyl bromide afforded a 58% yield of stilbenes, perhaps via an α-elimination, and only a small yield of 2-phenylethyltrimethylammonium bromide. The ylid (II) could be acylated by a variety of reagents. Carbon dioxide afforded a mono- or bis-carboxylic acid after acidifaction depending on which component was in excess. Benzonitrile and ethylbenzoate both afforded phenacyltrimethylammonium bromide (VI) after acidification. Acetyl chloride afforded acetonyltrimethylammonium salts and benzoyl chloride afforded the corresponding phenacyl derivative along with an enol ester (VII), the result of acylation of the anion of VI. In a separate experiment VI was converted to VII by benzoyl chloride in the presence of sodium carbonate.

All of the above acylations are typical of carbon nucleophiles in general and show no special characteristics. The facile conversion of

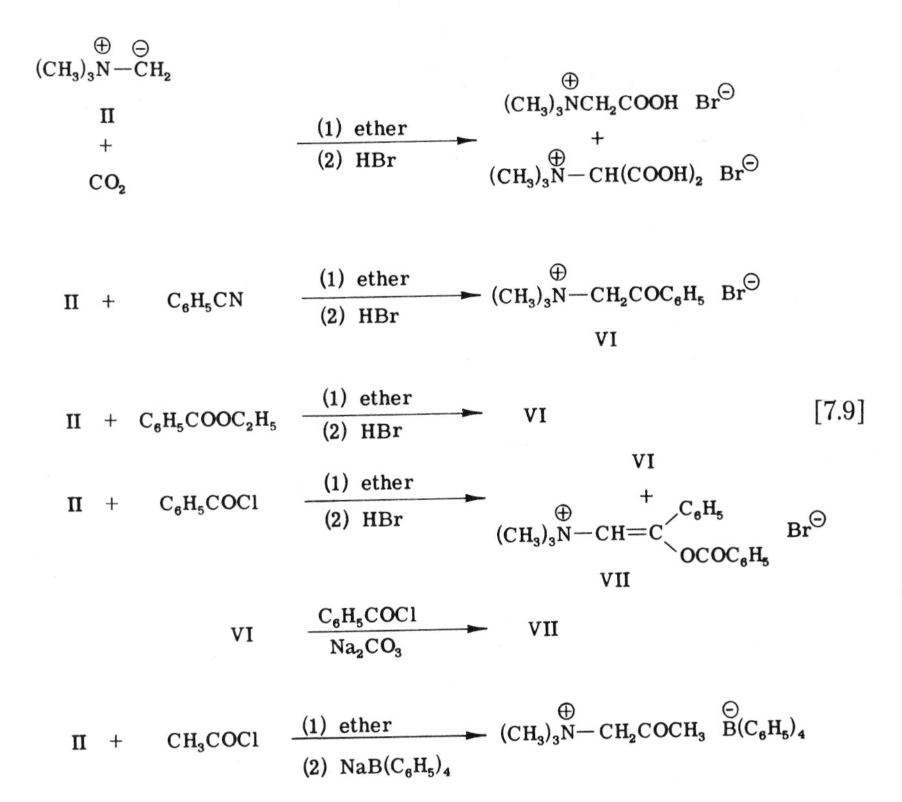

VI to VII is not unusual and indicates that the carbanionic intermediate in the conversion must be rather stable since such a weak base could be used. The stability of the intermediate must be due largely to its enolate characteristics in addition to its ylid characteristics. Much earlier, Krohnke and Heffe (*15*) observed an analogous acylation and claimed to have isolated the ylid intermediate (VIII). The same ylid also reacted with nitroso compounds to afford a nitrone, a reaction typical of

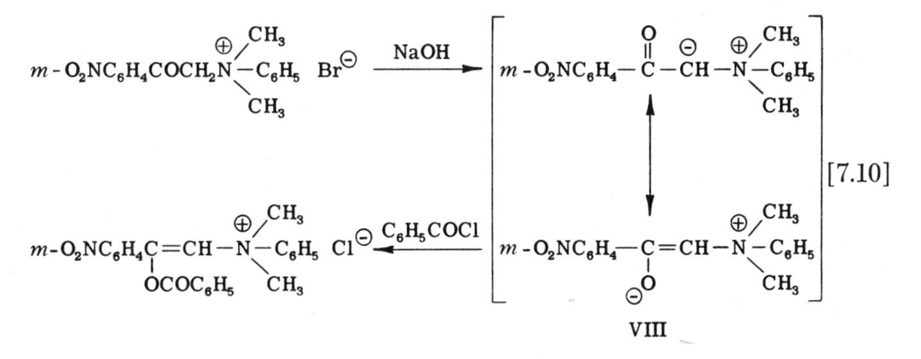

pyridinium and sulfonium ylids. Thus, in agreement with the observations with phosphonium ylids, it appears that nitrogen ylids also are subject to stabilization by appropriate substitution on the ylid carbanion.

In 1929 Ingold and Jessop (*16*) proposed the existence of trimethylammoniumfluorenylide (IX) and found, in accord with earlier discussion, that it was much less stable than the corresponding sulfonium ylid. Treatment of 9-fluorenyltrimethylammonium bromide with silver oxide afforded a deep purple solution whose color could be removed with dilute acid. Heating of the solution to 170° afforded trimethylamine

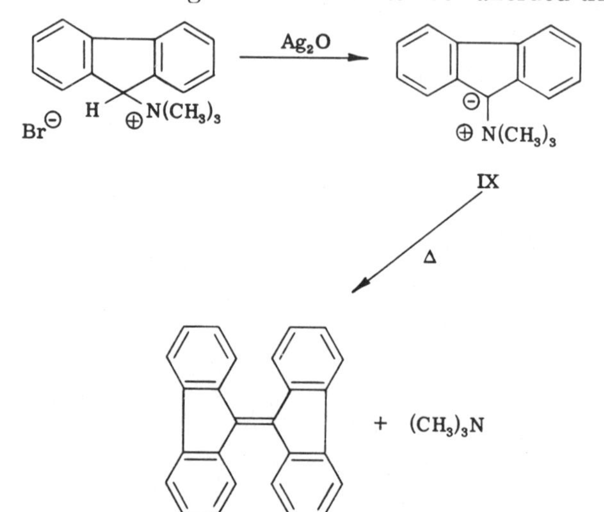

[7.11]

and difluorenylidene, a transformation presumed to have occurred via a carbene intermediate. The decomposition of IX into a carbene may have been verified by Franzen's observation (*17*) that heating the ylid with dimethylbenzylamine afforded 9-benzyl-9-dimethylaminofluorene, probably via coupling of fluorenylidene with the amine to form a new nitrogen ylid (X) which underwent a Steven's rearrangement. The latter proposal was substantiated by Wittig and Laib's observation (*18*) that diethylbenzylammoniumfluorenylide, generated from the corresponding conjugate acid with phenyllithium, rearranged in 90% yield to 9-benzyl-9-diethylaminofluorene.

Wittig and Felletschin (*19*) isolated the crude yellow ylid (IX) by treatment of the salt with phenyllithium in ether. Hydrolysis of the ylid afforded the corresponding quaternary ammonium hydroxide and reaction of the ylid with hydrogen bromide reformed the original ammonium salt. The ylid could be alkylated on the fluorenyl carbon with methyl iodide and benzyl iodide. However, the ylid did not react with benzaldehyde or benzophenone as had the methylide (I) and it would

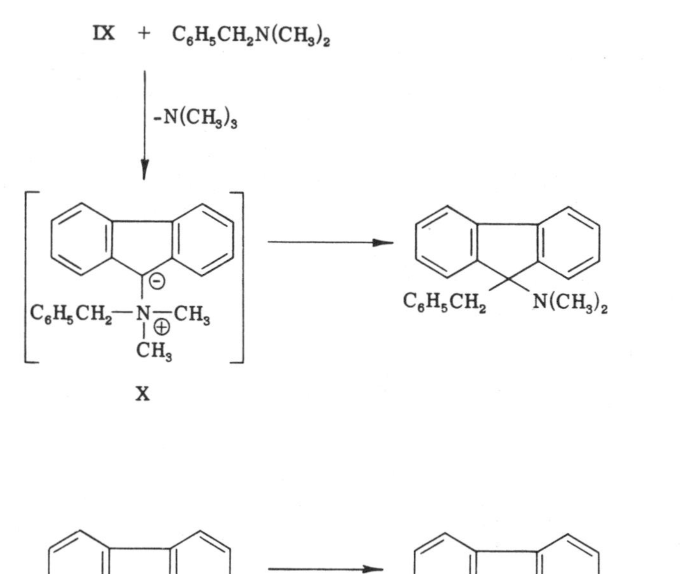

not undergo a Steven's rearrangement, even upon pyrolysis. In general, therefore, the fluorenylide (IX) reacted as a normal carbanion with the ammonium group playing no special role in its reactions other than providing some coulombic stabilization for the carbanion.

Dauben and Spooncer (*20*) reported the isolation of trimethylammoniumcyclopentadienylide (XI) as a pink solid which was stable for short periods of time under nitrogen. The ylid was stable in deoxy-

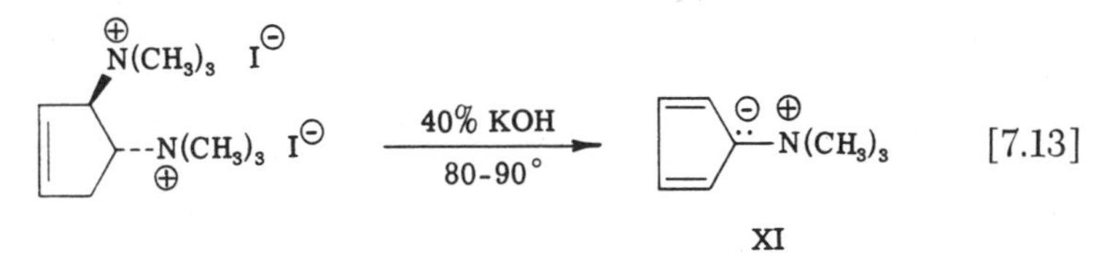

genated aqueous solution and the pK_a of its conjugate acid was determined to be 10.06. The structure of the ylid was determined only by its method of formation, its microanalysis and by hydrogenation to cyclopentane and trimethylamine. No trimethylcyclopentylammonium salts were found. The ylid appeared to react with a variety of electrophiles, methyl iodide, picric acid, mercuric chloride and bromine, but in no instance could any identifiable products be isolated. In acidic solution the ultraviolet spectrum was essentially that of a diene but at high pH long wavelength absorption appeared. Thus, while there is some evidence for the existence of the ylid (XI) its characterization as such has left much to be desired.

The most stable ammonium ylid reported to date appears to be trimethylammoniumdicyanomethylide (XII) obtained from its conjugate acid with aqueous hydroxide (*21*, Arnold). The ylid has a melting point of 153° and appeared to be stable in the presence of oxygen and water for indefinite periods. Unfortunately, the chemical characteristics of XII

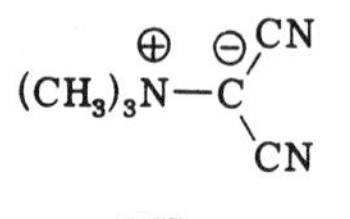

XII

have not been reported. The stability of the ylid can be attributed to delocalization of the lone pair electrons on the carbanion through the two cyano groups in addition to the electrostatic stabilization afforded by the ammonium group.

From this discussion it is quite apparent that ammonium ylids are quite different from phosphonium ylids in two respects. They are much less stable in both a thermodynamic sense and a reactivity sense. In addition, they appear to undergo only normal carbanionic reactions, not those that appear to be unique for ylidic carbanions. The nitrogen atom plays no role other than affording stabilization for the carbanion.

II. Pyridinium Ylids

Pyridinium ylids were discovered by Krohnke (*7*) in 1935. Treatment of *N*-phenacyl pyridinium bromide with potassium carbonate afforded the crystalline ylid (XIII) of m.p. 74°. Analogous ylids also were prepared using the corresponding quinolinium and isoquinolinium

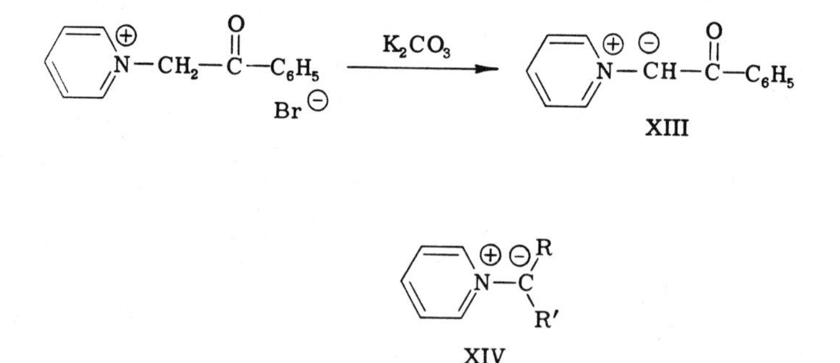

salts. These were but the first of a series of ylids of general structure XIV prepared by Krohnke and coworkers over a period of nearly thirty years. Some of the ylids were capable of isolation while others were observed only by their reactions. In order for the ylid to be capable of isolation at least one of the groups R or R′ in XIV must be capable of stabilizing the carbanion by resonance. Thus, *N*-(2-benzoylethyl)pyridinium bromide did not form an ylid but instead underwent an elimination reaction when treated with sodium hydroxide to afford phenylvinyl ketone (*22*, Krohnke). The following pyridinium ylids (XIV) have been isolated and characterized: R = H, R′ = C_6H_5CO— (*7*); R = R′ = C_6H_5CO— (*7*); R = R′ = —$COOC_2H_5$ (*22*); R = C_6H_5CO—, R′ = C_6H_5NH—CO— (*23*); R = C_6H_5CO—, R′ = —$CSSC_2H_5$ (*24*); R = H, R′ = —$CSSC_2H_5$ (*25*). Many other pyridinium ylids have been prepared in solution and used in subsequent reactions, for example those of the general type X—C_6H_4—CH^-—$N^+C_5H_5$ (*26*). From all these observations it is clear that the pyridinium group is a much more powerful activating group for α-hydrogens than is a simple ammonium group. This may be due to stabilization of the carbanion by both electrostatic attraction and resonance interaction of the carbanion with the pyridinium ring [7.15]. From the above examples it also is clear that considerable

$$\text{C}_5\text{H}_5\overset{\oplus}{\text{N}}-\overset{\ominus}{\text{C}}\!\!< \;\longleftrightarrow\; \overset{\ominus}{\text{(ring)}}\overset{\oplus}{\text{N}}=\text{C}< \;\longleftrightarrow\; \ominus\text{(ring)}\overset{\oplus}{\text{N}}=\text{C}< \quad \text{etc.} \qquad [7.15]$$

stabilization must be afforded by the substituents on the carbanion.

Pyridinium ylids were found to undergo a wide range of reactions typical of nucleophiles in general and carbanions in particular. As was

$$C_5H_5\overset{\oplus}{N}-\overset{\ominus}{C}HCSSCH_3 \xrightarrow{C_6H_5COCl} C_5H_5\overset{\oplus}{N}-CH=C(SCH_3)(SCOC_6H_5)\quad Cl^{\ominus}$$

$$C_5H_5\overset{\oplus}{N}-\overset{\ominus}{C}HCSSCH_3 \xrightarrow{(C_6H_5CO)_2O} C_5H_5\overset{\oplus}{N}-CH(CSSCH_3)(COC_6H_5)$$

$$C_5H_5\overset{\oplus}{N}-\overset{\ominus}{C}HCOC_6H_5 \xrightarrow{(C_6H_5CO)_2O} C_5H_5\overset{\oplus}{N}-\overset{\ominus}{C}(COC_6H_5)_2$$

$$C_5H_5\overset{\oplus}{N}-\overset{\ominus}{C}HCOC_6H_5 \xrightarrow{C_6H_5NCO} C_5H_5\overset{\oplus}{N}-\overset{\ominus}{C}(COC_6H_5)(CONHC_6H_5) \qquad [7.16]$$

$$C_5H_5\overset{\oplus}{N}-\overset{\ominus}{C}HCOC_6H_5 \xrightarrow{CS_2} C_5H_5\overset{\oplus}{N}-CH(COC_6H_5)(CSS^{\ominus}) \xrightarrow{CH_3I} C_5H_5\overset{\oplus}{N}-CH(CSSCH_3)(COC_6H_5)\quad I^{\ominus}$$

the case with ammonium ylids, however, most of these reactions were not peculiar to ylids.

Acylation of pyridinium ylids with benzoyl chloride led to O-acylation or S-acylation (*25*) whereas benzoic anhydride led to C-acylation (*7*, *25*), a situation reminiscent of analogous reactions with phosphonium ylids (*27*, Chopard *et al.*). Phenyl isocyanate likewise led to C-acylation and the formation of new ylids (*23*). Carbon disulfide led to the formation of new betaines which could be alkylated [7.16] (*24*).

The only alkylation reaction reported for pyridinium ylids resulted in the S-alkylation of an ylid with ethyl iodide [7.17] (*25*). Pyridinium ylids would partake in Michael additions to conjugated carbonyl sys-

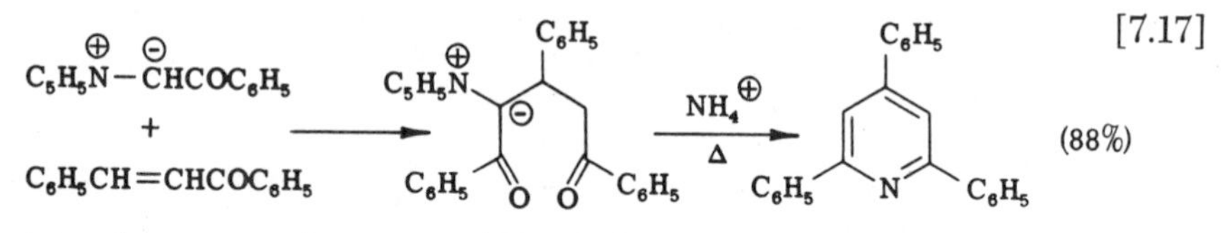

[7.17]

tems in a normal manner (*28*, Krohnke *et al.*). In some instances the resulting products could be converted into heterocyclic systems (*29*, Zecher and Krohnke).

Pyridinium ylids also have been found to react with aldehydes but in a manner different from that of ammonium ylids and phosphonium ylids. The reaction is analogous to that of any active methylene compound with an aldehyde, the Knoevenagel reaction. With ammonium ylids the reaction stopped after the initial stage to produce a betaine (*6*, *12*) whereas with phosphonium ylids the initial adduct (betaine) underwent an elimination of the phosphonium and oxygen groups to afford olefins (Wittig reaction). With pyridinium ylids, however, an olefin was formed but the pyridinium groups also were retained and

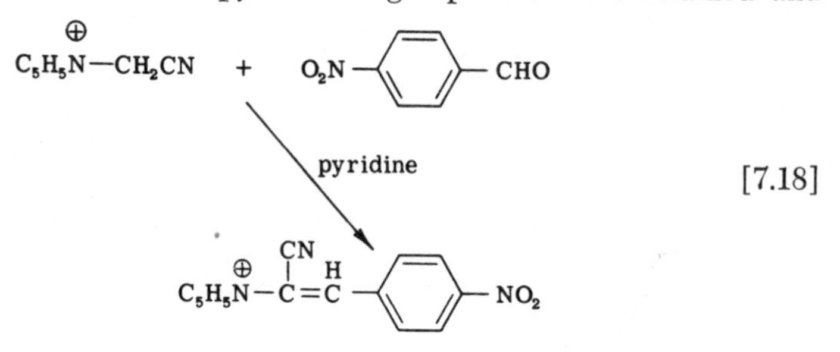

[7.18]

acted only as acidifying groups [7.18] (*24*, *30*, Krohnke *et al.*). The mechanism of the reaction presumably is similar to that of the aldol-type condensation.

One of the most interesting reactions of pyridinium ylids is with nitrosoaromatic compounds. Krohnke and Borner (*31*) first observed that the ylids generated from *N*-phenacylpyridinium bromide and *N*-phenacylisoquinolinium bromide both reacted with nitrosobenzene to afford the same nitrone (XV) in 76% and 70% yield, respectively. Nu-

$$C_5H_5\overset{\oplus}{N}-\overset{\ominus}{C}HCOC_6H_5 + C_6H_5NO \longrightarrow \underset{\text{XV}}{C_6H_5\overset{O}{\overset{\nearrow}{N}}=CHCOC_6H_5} + C_5H_5N \qquad [7.19]$$

merous other examples of this reaction with a variety of pyridinium ylids subsequently were reported (*26, 32*) and summarized in a review article by Krohnke (*33*). The reaction has developed into a useful synthetic tool since it permits the conversion of an alkyl group ultimately into a carbonyl group as shown in [7.20]. This sequence was

$$R-CH_3 \longrightarrow R-CH_2Br \longrightarrow R-CH_2-\overset{\oplus}{N}C_5H_5 \longrightarrow RCH=\overset{O}{\overset{\uparrow}{N}}-Ar \longrightarrow RCHO \qquad [7.20]$$

explored by Reid and Gross (*34*) where R was a variety of heterocyclic aromatic systems. The nitrone formation reaction does not depend on the availability of an isolable pyridinium ylid. For example, *N*-benzylpyridinium bromide in the presence of *p*-nitroso-*N,N*-dimethylaniline and aqueous sodium hydroxide afforded a good yield of nitrone (*26*). The mechanism of this reaction has not been elucidated. It seems likely, however, that it involves attack of the ylid carbanion on the electrophilic

$$C_5H_5\overset{\oplus}{N}-\overset{\ominus}{C}R_2 + ArNO \longrightarrow \left[\begin{array}{c} Ar-\ddot{N}-O^{\ominus} \\ | \\ C_5H_5\overset{\oplus}{N}-C-R \\ | \\ R \end{array}\right] \longrightarrow Ar-\overset{\oplus}{N}(O^{\ominus})=CR_2 + C_5H_5N \qquad [7.21]$$

nitrogen of the nitroso group to afford a betaine intermediate which then ejects pyridine [7.21] (*35*, Krohnke). The course of this reaction is identical to that followed by sulfonium ylids (see Chapter 9).

In recent years other workers have investigated the chemistry of pyridinium ylids containing other than acyl groups. Pinck and Hilbert (*36*) reported the preparation, in solution only, of 9-pyridiniumfluorenylide (XVI) by treating the conjugate acid with aqueous sodium hydroxide. Krohnke (*35*) reported the blue ylid (XVI) to react with *p*-nitroso-*N,N*-dimethylaniline to afford the corresponding *N*-(*p*-dimethylaminophenyl)fluorenone ketoxime. Allowing the ylid to stand in

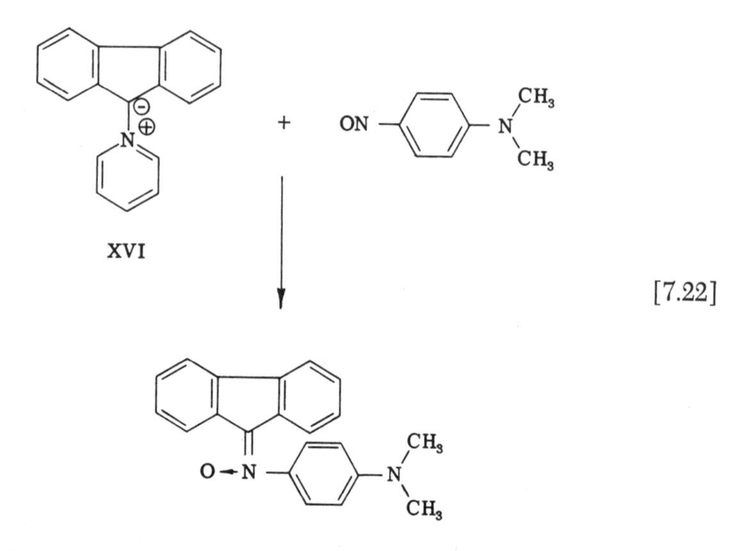

[7.22]

aqueous solution led to eventual hydrolysis and isolation of fluorenone. Hartmann and Gossel (*37*) reported the dipole moment of the crude ylid to be 4.13 ± 0.08 D. This value seems very low when it is noted that the triphenylphosphoniumfluorenylide had a dipole moment of 7.1 D (*38*, Johnson).

Lloyd and Sneezum (*39*) reported the preparation and isolation of pyridiniumcyclopentadienylide (XVII) in 1955. It was obtained as a red-brown, high-melting solid by treating 3,5-dibromocyclopentene with excess pyridine then sodium hydroxide. It was insoluble in water but

Br Br + N ⟶ salt(s) $\xrightarrow{NaOH}$ XVII [7.23]

dissolved to afford a colorless solution in dilute acid. The ylid could be reprecipitated upon the addition of base. It was sensitive to oxygen. The color of the ylid varied depending on the solvent used for dissolution, and Kosower and Ramsey (*40*) subsequently detected an intramolecular charge-transfer band near 5000 Å. The pK_a of the ylid conjugate acid was reported to be 10.0, almost identical to that reported for trimethylammoniumcyclopentadienylide (*20*, Dauben and Spooncer). Kursanov and Baranetskaia (*41*) reported a dipole moment of 13.5 D, far above the 7.0 D reported by Ramirez and Levy (*42*) for triphenylphosphoniumcyclopentadienylide. Evleth *et al.* (*43*) recently have reported calculations using the ω-technique for the dipole moment of XVII to give values between 12.6 and 13.6 D. Clearly the dipole moments of XVI and XVII are opposite in relative magnitude with those of the two analogous triphenylphosphonium ylids. The polarographic reduction and oxidation of XVII also has been studied (*44*).

Very little chemistry has been reported for XVII. Evidence for its structure was provided by catalytic reduction to *N*-cyclopentylpiperidine (*39*). Kursanov *et al.* (*45*) reported that the ylid absorbed two moles of bromine to afford a tetrabromoylid, the apparent result of electrophilic substitution on the cyclopentadienyl ring [7.24].

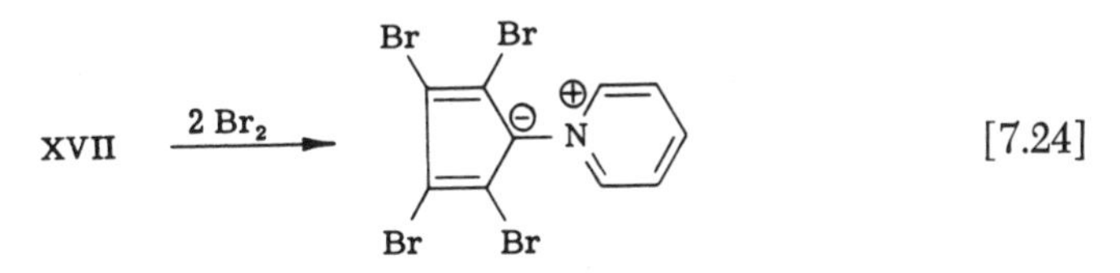

[7.24]

Linn *et al.* (*46*) recently reported that reaction of pyridine and tetracyanoethylene oxide at 0° afforded pyridiniumdicyanomethylide (XVIII) as a stable, high-melting (245–246°), momomeric solid. No chemistry has been reported for this interesting compound but Bugg

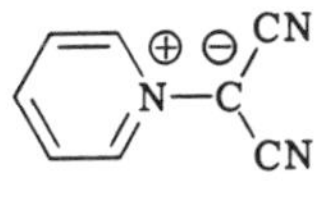

XVIII

et al. (*47*) have just completed an X-ray diffraction study. The carbanion nitrogen bond was 1.41 Å long, somewhat shorter than expected for a C_{sp^2}-N_{sp^2} bond. The ring carbon-nitrogen bonds were a normal 1.37 Å. The carbanion-carbon bonds were 1.42 Å, almost identical to that for the single carbon-carbon bond in acrylonitrile (1.426 Å, *sp*-sp^2) (*48*, Costain and Stoicheff), but considerably shorter than that for the single carbon-carbon bond of propionitrile (1.458 Å, *sp*-sp^3) (*49*, Costain). It appears as though the carbanion carbon is trigonally hybridized and that there is some interaction between the carbanion and the pyridinium ring resulting in the shortening of that bond. The bond angles about the carbanion all were 119–120°; this also indicates trigonal hybridization. The most interesting observation from the X-ray analysis indicated that the carbanion was not planar, the two cyano carbons were 0.08 Å and the two cyano nitrogens were 0.13 Å below a plane through the pyridinium ring and the carbanion.

In summary, pyridinium ylids appear to be much more stable than ammonium ylids, probably due to stabilization of the carbanion by resonance with the pyridinium ring. These ylids undergo normal carbanionic reactions such as alkylation and acylation. Reaction with nitroso compounds to afford nitrones appears to be a unique reaction but ammonium ylids also would be expected to undergo the reaction. Several pyridinium ylids have been isolated and characterized.

III. Nitrogen Imines

Following their success in preparing sulfidimines by reaction of sulfides with chloramine-T, Mann and Pope (*50*) attempted to follow an analogous route to ammonium imines in 1922. However, *N,N*,-dimethylaniline and chloramine-T failed to react under the conditions employed. In an isolated experiment, however, Wittig and Rieber (*12*) found that 1,1,1-trimethylhydrazinium iodide reacted with phenyllithium to afford a non-isolable intermediate which was alkylated by benzyl bromide to

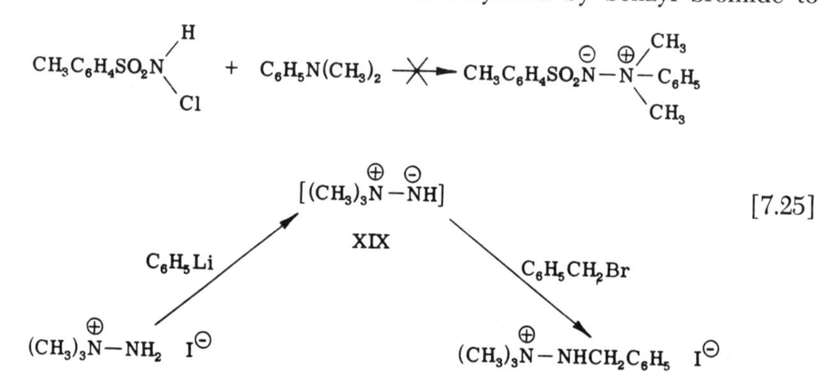

afford 1,1,1-trimethyl-2-benzylhydrazinium iodide. The intermediate probably was the ammonium imine (XIX).

Wawzonek and Meyer (*51*) developed an alternate approach to the type of compound envisaged by Mann and Pope (*50*). 1,1-Dimethylhydrazine and tosyl chloride afforded the expected sulfonamide which could be alkylated with methyl iodide. The alkylated amide, upon treatment with aqueous sodium hydroxide afforded trimethylammonium-*N*-

$$(CH_3)_2N-NH_2 \xrightarrow[(2)\ CH_3I]{(1)\ TsCl} (CH_3)_3\overset{\oplus}{N}-NH-Ts \ \ \overset{\ominus}{I} \xrightarrow{\ominus OH} (CH_3)_3\overset{\oplus}{N}-\overset{\ominus}{N}SO_2C_6H_4CH_3(p) \quad \text{XX} \qquad [7.26]$$

tosylimine (XX), m.p. 175–176°. The imine was stable in the presence of air and moisture. Reduction of XX with zinc and acetic acid afforded trimethylamine and *p*-toluenesulfonamide while oxidation with hydrogen peroxide also afforded *p*-toluenesulfonamide. Treatment of XX with hydriodic acid reformed the conjugate acid from which it had been prepared. The imine could be alkylated with benzyl chloride but not with methyl iodide. The stability of imines such as XX is not too surprising when it is recalled that ordinary sulfonamides are appreciably acidic.

Wawzonek and Yeakey (*52*) prepared the analogous *p*-nitrobenzyldimethylammonium-*N*-acetylimine (XXI) and found that it would undergo a thermal rearrangement which is analogous to the Steven's rearrangement. This reaction and the characteristics mentioned above

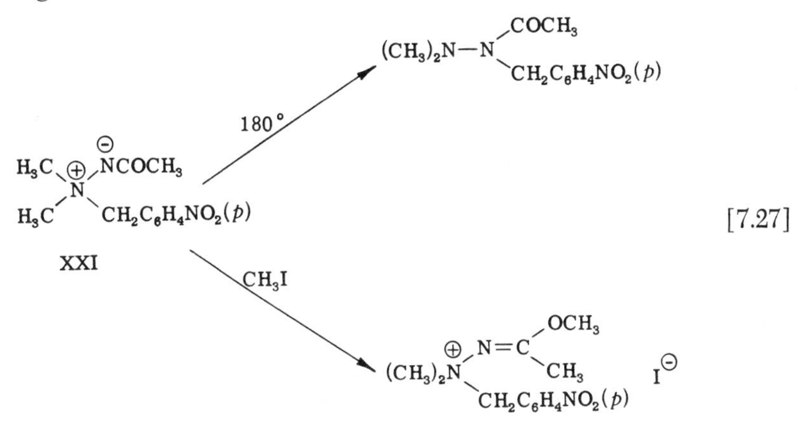

[7.27]

clearly establish the site of negative charge to be the nitrogen atom. However, the fact that the anion was stabilized in part by delocalization through the carbonyl group was evidenced by the *O*-alkylation undergone by XXI. Therefore, both the carbonyl group and the ammonium group play a key role in the stabilization of the imine.

Hinman and Flores (*53*) found that trimethylammonium-*N*-benzoylimine (XXII) also was a stable imine. Gibson and Murray (*54*) recently have shown that pyrolysis of this imine afforded trimethylamine

$$C_6H_5-\underset{\underset{O}{\|}}{C}-\overset{\ominus}{N}-\overset{\oplus}{N}(CH_3)_3 \xrightarrow{185°} [C_6H_5NCO] + (CH_3)_3N \qquad [7.28]$$

XXII

and phenyl isocyanate, the latter being isolated as its trimer or trapped as a urea. This result was expected since XXII is structurally analogous to the intermediate proposed for the Curtius, Schmidt and Hypobromite reactions, all of which are thought to involve the formation of an isocyanate by a carbon-to-nitrogen migration.

Schneider and Seebach (*55*) took advantage of the well-known conversion of pyrylium salts to pyridines by ammonolyses to carry out the conversion using phenylhydrazine as the base. The resulting pyridinium salt could be converted into a dark blue anhydro base, originally formulated as XXIII, by treatment with aqueous sodium hydroxide. The base could be reconverted to the salt with dilute acid. Several different anhydro bases were prepared and isolated. It subsequently was shown, however, that XXIII was an incorrect structure for the anhydro base

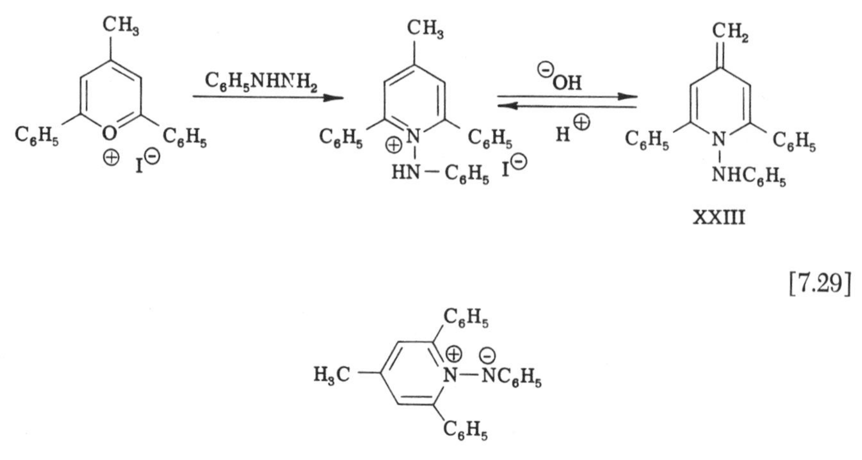

since it underwent methylation with methyl iodide on nitrogen rather than on carbon. The structure was accordingly revised to XXIV, a pyridiniumimine derivative (*56*, Schneider). The preparation of a 2,4,6-triphenyl derivative also ruled out a structure similar to XXIII.

Dimroth *et al.* (*57*) recently have reviewed Schneider's work and pursued it further. They have prepared additional examples of the pyridiniumimines, examined the influence of substituents on their stability and elaborated on a rearrangement of the imines to benzyl pyridines. The dark blue imines (XXV) were transformed into colorless pyridines

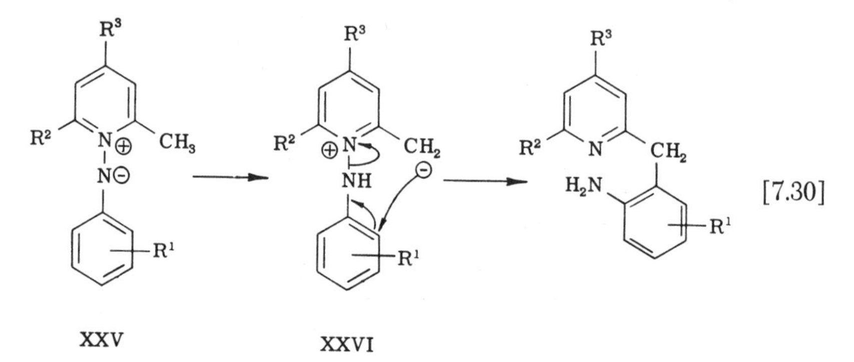

by warming to 40°. Since electron-donating substituents at R^1 and electron-withdrawing substituents at R^2 and R^3 favored the rearrangement, they proposed that the initial step involved a prototropic shift from carbon to nitrogen, the resulting methylide (XXVI) attacking the aniline ring at the *ortho* position in a Sommelet-type rearrangement.

Ashley *et al.* (*58*) prepared a pyridiniumimine by reaction of pyridine with *p*-acetamidobenzenesulfonyl azide, a reaction reminiscent of phosphine-imine preparations. The imine (XXVII) was a high-melting, stable, dark red substance which was insoluble in aqueous base but soluble in dilute acid. Acid hydrolysis afforded sulfanilic acid and

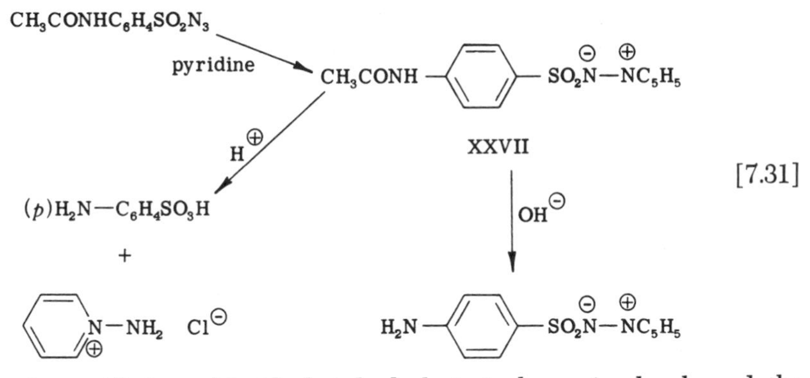

[7.31]

N-aminopyridinium chloride but hydrolysis in base simply cleaved the acetamido linkage and formed a new imine.

In the course of his investigation into the conversion of isocyanates into carbodiimides through the catalytic action of phosphine oxides, arsine oxides and sulfoxides, etc., Monagle (59) also found that pyridine-*N*-oxide would effect the conversion whereas trimethylamine-*N*-oxide would not. He and his coworkers previously had shown that the conversion involved the formation of a phosphinimine from a phosphine oxide and the isocyanate which then reacted with additional isocyanate in a Wittig-type olefin condensation. On this basis he claimed that neither pyridine-*N*-oxide nor trimethylamine-*N*-oxide should catalyze the conversion, at least not by a similar mechanism, since a pentavalent nitrogen and/or pyridinium imine would be involved. This should be the

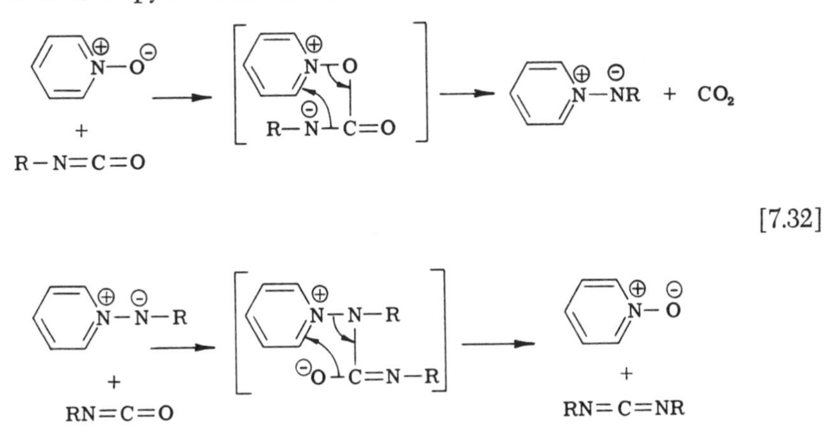

[7.32]

case for the trimethylamine-*N*-oxide but not for the pyridine-*N*-oxide since a pyridiniumimine could be involved as an intermediate as shown in [7.32]. The reaction of pyridiniumimines with carbonyl compounds has not been studied.

To summarize, nitrogen imines appear to bear the same relationship to nitrogen ylids as phosphinimines do to phosphonium ylids. They appear to be a little less basic than the ylids but their chemical and physical properties have not been explored adequately.

IV. Diazo Compounds

In many respects diazo compounds react as though they were nitrogen ylids, perhaps best represented by structure XXVIII. In those cases

$$>\overset{\ominus}{\ddot{C}}-\overset{\oplus}{N}\equiv N: \longleftrightarrow >C=\overset{\oplus}{N}=\overset{\ominus}{\ddot{N}}: \qquad [7.33]$$

XXVIII

where the diazo compound is stable enough to be isolated the substituents attached to carbon inevitably are those which could afford maximum stabilization for a carbanion. Thus, diazofluorene, diphenyldiazomethane, phenyldiazomethane and the diazoketones have been isolated, purified and characterized. These are the same structural features that often have led to the stabilization and, therefore, isolation of nitrogen, phosphorus and sulfur ylids, etc. No attempt will be made in this section to discuss all of the chemistry of diazo compounds. Instead, a few reactions will be mentioned to point out the similarity between the chemistry of these compounds and that of other nitrogen ylids.

One of the best known reactions of diazo compounds, and one that has been reviewed (*60*, Gutsche), is their reaction with aldehydes and ketones. Three products may be obtained from this reaction as shown in [7.34], the ratio of the three depending on the structure of both the

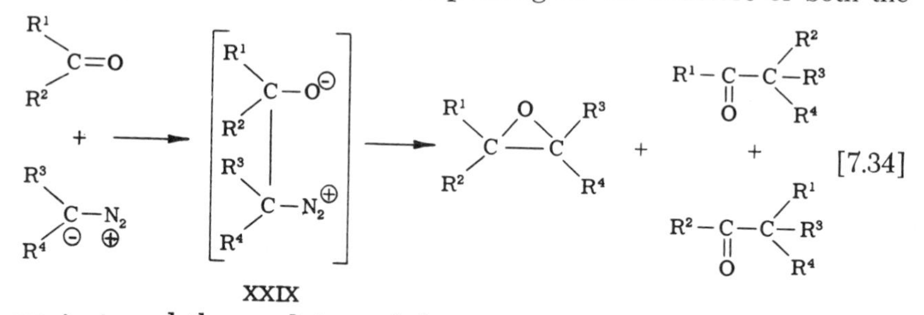

reactants and the conditions of the reaction. Carbonyl compounds with electron-withdrawing groups attached seem to form the oxide as the dominant product. The mechanism of this reaction probably proceeds via the initial attack of the diazo compound, through its carbanionic carbon, on the carbonyl carbon to afford a betaine intermediate (XXIX). In the case of oxide formation the oxyanion would displace nitrogen. This is the same mechanism that appears to operate in the conversion of carbonyl compounds to oxides via sulfonium ylids (*61*, Johnson *et al.*). In that case sulfides were ejected (see Chapter 9).

Schonberg and Junghans (*62*) have reported several examples of the reaction of diphenyldiazomethane with ketones resulting in the formation of epoxides. As early as 1920 Staudinger and Siegwart (*63*) reported the similar conversion of thiobenzophenone to tetraphenylthiirane with

$$(C_6H_5)_2C{=}S \; + \; (C_6H_5)_2CN_2 \longrightarrow \left[\begin{array}{l}(C_6H_5)_2C-S^{\ominus} \\ (C_6H_5)_2C-N_2^{\oplus}\end{array}\right] \longrightarrow (C_6H_5)_2C\overset{S}{\triangle}C(C_6H_5)_2 \qquad [7.35]$$

diphenyldiazomethane [7.35]. Schonberg *et al.* (*64, 65*) also have provided additional examples of this reaction.

The reaction of diazo compounds with nitrosobenzene probably follows a similar course although the ultimate products are different. Staudinger and Miescher (*66*) found that diazofluorene and diphenyldiazomethane reacted with nitrosobenzene rapidly at room temperature to afford the corresponding nitrones in high yield [7.36]. Johnson (*67*)

$$(C_6H_5)_2\overset{\ominus}{C}-\overset{\oplus}{N_2} \; + \; C_6H_5-N{=}O \longrightarrow \left[\begin{array}{l}(C_6H_5)_2C-N_2^{\oplus} \\ C_6H_5-N-O^{\ominus}\end{array}\right] \longrightarrow \left[\begin{array}{l}(C_6H_5)_2C \\ C_6H_5-N\end{array}\!\!>O\right] \longrightarrow (C_6H_5)_2C{=}N(\rightarrow O)C_6H_5 \qquad [7.36]$$

reinvestigated this reaction and concluded that it probably proceeded via an intermediate oxazirane which rearranged to the observed nitrone. This behavior also was typical of sulfonium and pyridinium ylids.

Reimlinger (*68*) has found that stable diazo compounds also will react with carbenes in a manner strictly analogous to that reported for the reaction between phosphonium ylids and the same carbenes (*69*, Oda *et al.*). Diazofluorene, in the presence of chloroform and potassium

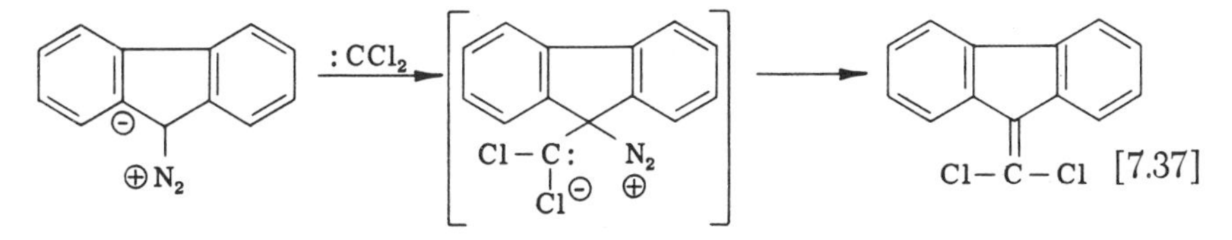

tert-butoxide, afforded an 80% yield of 9-(dichloromethylene)fluorene [7.37]. The reaction proceeded with a variety of both perhalo and haloaryl carbenes. Whether a free carbene was involved is beside the point in this reaction. It is obvious that the diazo compound must be acting as a nucleophile and, therefore, essentially as an ylid.

It is well known that diazo compounds also are able to fragment into nitrogen and carbenes under the appropriate conditions: usually thermal or photolytic excitation. For example, diazofluorene and the 2-butenes produced the expected cyclopropane derivatives (*70*, Horner and Lingnau; *71*, Doering and Jones). The report by Bamford and Stevens (*72*) that reaction between diazofluorene and *N,N*-dimethyl-

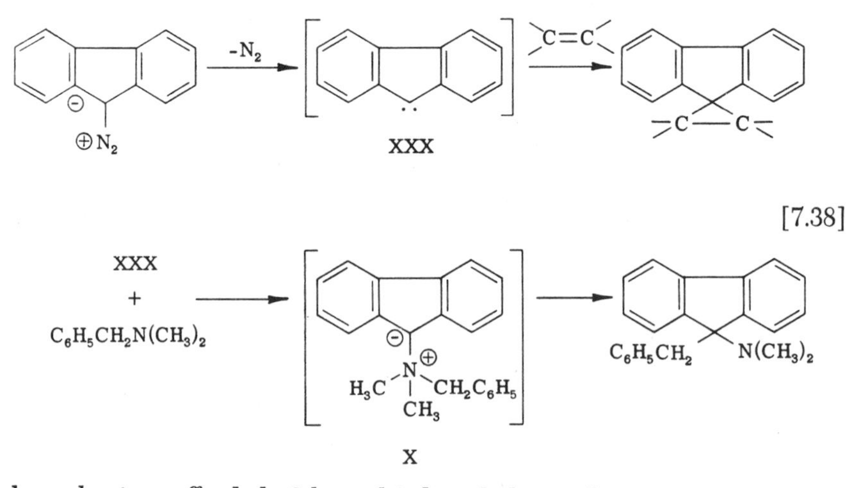

benzylamine afforded 9-benzyl-9-dimethylaminofluorene may be explained by proposing that the diazo compound decomposed to fluorenylidene (XXX) which was attacked by the amine to afford a new nitrogen ylid (X) as the key intermediate. This ylid would be expected to undergo a Steven's rearrangement to the observed product. As mentioned earlier in this chapter, the same product was obtained from reaction of 9-trimethylammoniumfluorenylide (IX) with *N,N*-dimethylbenzylamine and it probably occurred via the same overall route, the fragmentation of IX into fluorenylidene (XXX) and the formation of X (*17*, Franzen). The rearrangement step proposed for these mechanisms has been confirmed in a separate reaction (*19*, Wittig and Felletschin). In another example of the overall reaction Wittig and Schlosser (*73*) found that diazoacetic ester and *N,N*-dimethylbenzylamine afforded ethyl 2-(dimethylamino)dihydrocinnamate [7.39].

On the basis of the above few examples, it is apparent that diazo compounds can be considered simply a special form of nitrogen ylid in

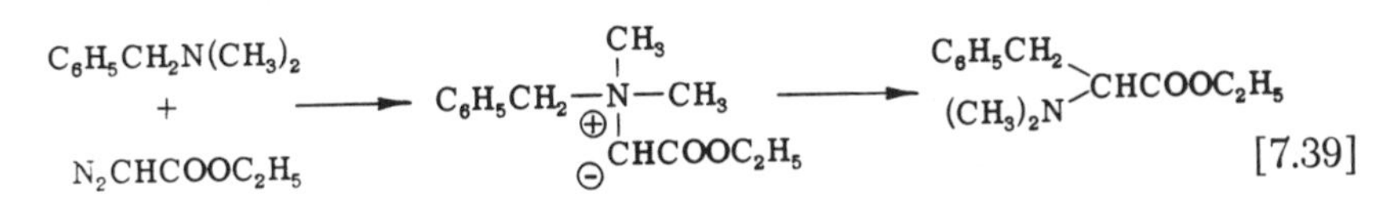

many of their reactions. They exhibit the nucleophilicity and thermal instability associated with such ylids and afford similar if not identical products from analogous reactions. If anything, they tend to be more stable than the corresponding ammonium and pyridinium ylids.

V. Rearrangements via Nitrogen Ylids

There are two closely related rearrangements of ammonium compounds that occur in the presence of strong base. From the evidence available it appears that both the Steven's rearrangement and the Sommelet (*ortho* substitution) rearrangement proceed via the intermediacy of nitrogen ylids.

Historically the Steven's rearrangement must be treated first. The initial observation of this rearrangement was by Stevens *et al.* (*74*) in 1928 when phenacylbenzyldimethylammonium bromide was treated with aqueous hydroxide and afforded (1-benzoyl-2-phenyl)ethyldimethylamine [7.40]. Later (*75*) the rearrangement was shown to be intramolecular by the absence of any cross over products from simultaneous reactions.

$$C_6H_5COCH_2\overset{\oplus}{N}(CH_3)_2-CH_2C_6H_5\quad Br^{\ominus} \xrightarrow{\ ^{\ominus}OH\ } C_6H_5COCH\begin{matrix} N(CH_3)_2 \\ CH_2C_6H_5 \end{matrix} \qquad [7.40]$$

The observation that electron-withdrawing substituents in the benzyl group, the apparent migrating group, favored the rearrangement led Thomson and Stevens (*76*) to propose that the reaction proceeded first by proton removal from the phenacyl group, formation of a carbon-nitrogen double bond as the benzyl group was ejected as an anion, and then attack of the benzyl group at the carbon end of the double bond [7.41].

$$C_6H_5-COCH_2\overset{\oplus}{N}(CH_3)_2-CH_2C_6H_5 \xrightarrow{\ ^{\ominus}OH\ } \left[C_6H_5CO\overset{\ominus}{C}H-\overset{\oplus}{N}(CH_3)_2-CH_2C_6H_5\right]$$

$$\downarrow \qquad [7.41]$$

$$C_6H_5COCH\begin{matrix} N(CH_3)_2 \\ CH_2C_6H_5 \end{matrix} \longleftarrow \left[C_6H_5COCH=\overset{\oplus}{N}(CH_3)_2 \quad {}^{\ominus}CH_2C_6H_5\right]$$

Subsequently it was shown that a variety of groups would undergo the migration including benzyl, phenacyl, allyl and propargyl (*77, 78,* Stevens *et al.*) and that a variety of bases including hydroxide, alkoxide, ammonia and organolithium would catalyze the rearrangement. The strength of base required depended on the acidity of the ammonium salt. For example, rearrangement of benzyltrimethylammonium bromide

$$C_6H_5-CH_2\overset{\oplus}{N}(CH_3)_3\ Br^{\ominus} \xrightarrow{C_6H_5Li} C_6H_5-CH(CH_3)N(CH_3)_2 \qquad [7.42]$$

to 1-phenylethyldimethylamine required the use of phenyllithium [7.42] (*79,* Wittig *et al.*).

Campbell *et al.* (*80*) showed that the migrating group retained its asymmetry during the course of the rearrangement. Optically active

$$C_6H_5COCH_2-\overset{\oplus}{N}(CH_3)_2-\overset{*}{C}H(CH_3)(C_6H_5)\ Br^{\ominus} \xrightarrow{\text{base}} C_6H_5COCH(N(CH_3)_2)-C(H)(CH_3)(C_6H_5) \qquad [7.43]$$

phenacyldimethyl(1-phenylethyl)ammonium bromide afforded optically active dimethyl(1-phenacyl-2-phenylpropyl)amine [7.43]. Therefore, the rearrangement was both intramolecular and concerted.

Wittig *et al.* (*79*) and Hauser and Kantor (*81*) proposed that the Stevens rearrangement proceeded via the formation of an ammonium ylid, the carbanion of which attacked the migrating group [7.44]. These

$$RCH_2-\overset{\oplus}{N}(CH_3)_2-CH_2R' \longrightarrow \left[R\overset{\ominus}{C}H-\overset{\oplus}{N}(CH_3)_2-CH_2R'\right] \longrightarrow RCH(N(CH_3)_2)-CH_2R' \qquad [7.44]$$

proposals accounted for the retention of optical activity in the migrating group and for the relative rate of rearrangement of substituted benzylammonium salts.

Many other examples of the Stevens rearrangement have been discovered including some mentioned earlier in this chapter resulting from the direct formation of an ammonium ylid (*17–19, 72*). The formation of *N*-methyl-*N*-ethylaniline from treatment of trimethylanilinium ion with phenyllithium (*82,* Weygand *et al.*) and the formation of the same compound from the reaction of trimethylamine with benzyne (*83,* Wittig and Benz) probably involved Stevens rearrangements. Wawzonek and Yeakey (*52*) found that dimethyl(*p*-nitrobenzyl)ammonium-*N*-acetylimine (XXI) also underwent a Stevens rearrangement, the migrating terminus being a nitrogen atom in this case [7.45].

The Sommelet rearrangement was discovered in 1937 (*84,* Sommelet)

$$\underset{\text{XXI}}{(CH_3)_2\overset{\oplus}{N}\begin{smallmatrix}\overset{\ominus}{N}COCH_3\\ CH_2C_6H_4NO_2(p)\end{smallmatrix}} \xrightarrow{180^{\circ}} (CH_3)_2N-N\begin{smallmatrix}COCH_3\\ CH_2C_6H_4NO_2(p)\end{smallmatrix} \qquad [7.45]$$

when it was found that benzhydryltrimethylammonium bromide, when heated in the presence of concentrated hydroxide, was converted into *o*-benzylbenzyldimethylamine. Wittig *et al.* (85) later found that treatment of dibenzyldimethylammonium bromide with phenyllithium afforded two amines, 1,2-diphenylethyldimethylamine and *o*-methylbenzhydryldimethylamine [7.46]. The former simply was a Stevens

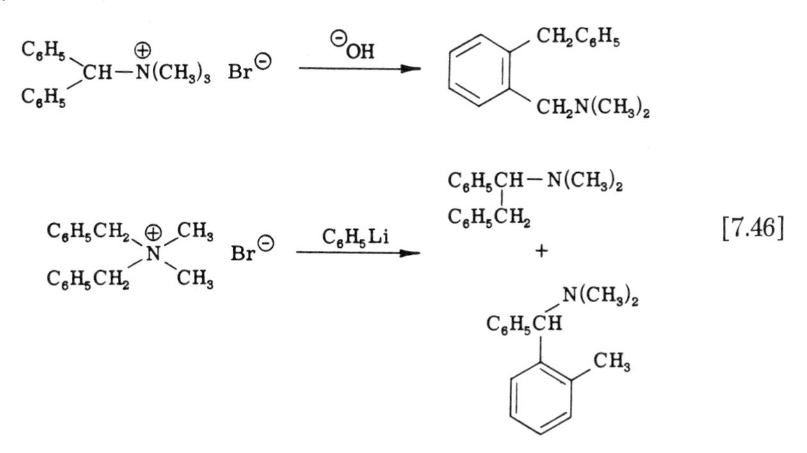

rearrangement product but the latter was the result of an apparent Sommelet rearrangement, a migration of a group into one of the aromatic rings. The mechanism of the Sommelet rearrangement probably involves the initial formation of an ammonium ylid but one which carries out a nucleophilic attack on the *ortho* position of an aromatic ring rather than on a carbon directly attached to the nitrogen atom as for the Stevens rearrangement [7.47].

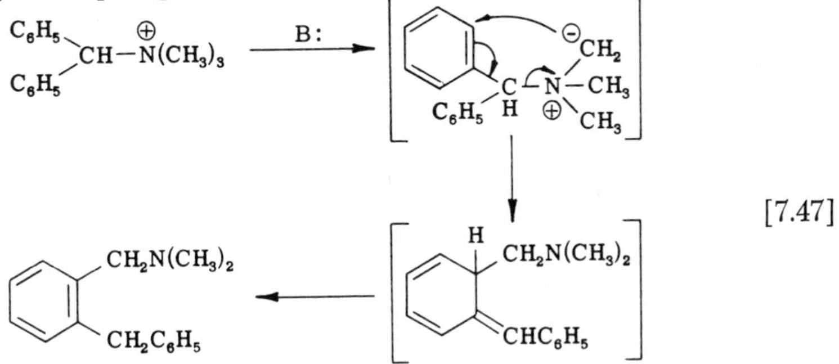

Kantor and Hauser (*86*) subsequently discovered that the Sommelet rearrangement could be effected in liquid ammonia with sodamide as the base. They called this the *ortho*-substitution rearrangement. Thus, benzyltrimethylammonium bromide under these conditions afforded *o*-methylbenzyldimethylamine. Subsequent work (*87*, Puterbaugh and Hauser) has shown that two carbanions are involved in the rearrange-

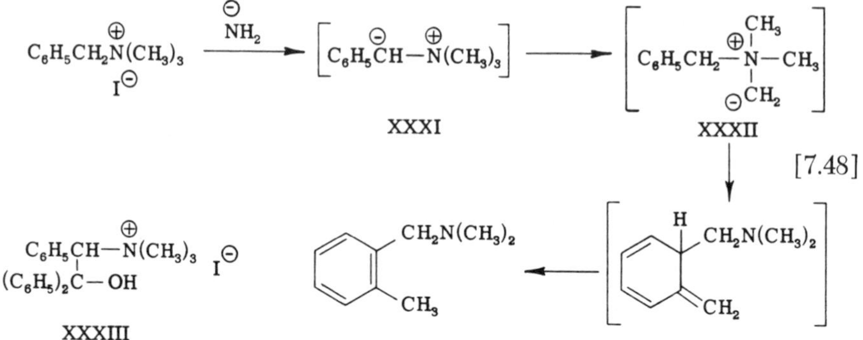

ment. Addition of benzophenone to the rearranging mixture at −80° afforded the ammonium carbinol (XXXIII) in 73% yield, the result of a reaction between the initially formed ylid (XXXI) and benzophenone. Repetition of the trapping experiment but at −33° resulted only in the isolation of the rearranged tertiary amine in 87% yield. Therefore, the slow step in the reaction appears to be a prototropic shift to convert what must be the most stable ylid (XXXI) into the least stable ylid (XXXII), the latter rearranging rapidly once formed.

Additional evidence for the involvement of two carbanions in the *ortho*-substitution rearrangement was obtained by the observation that ω-phenylphenacyltrimethylammonium bromide failed to undergo the rearrangement [7.49]. In this instance the initially formed carbanion

$$C_6H_5-\underset{\substack{\|\\ O}}{C}-\underset{\substack{|\\ C_6H_5}}{CH}-\overset{\oplus}{N}(CH_3)_3 \quad Br^{\ominus} \xrightarrow{KNH_2} C_6H_5-\underset{\substack{\|\\ O}}{C}-\underset{\substack{|\\ C_6H_5}}{\overset{\ominus}{C}}-\overset{\oplus}{N}(CH_3)_3 \qquad [7.49]$$

(no rearrangement)

probably was too weakly basic to abstract a proton from one of the methyl groups. The starting material was recovered unchanged from the reaction mixture (*88*, Puterbaugh and Hauser).

The extent to which nitrogen ylids undergo rearrangement is in striking contrast to the virtual absence of rearrangement of phosphorus ylids and the very few examples of sulfonium ylid rearrangements. This perhaps is a reflection of their inherent instability due to the absence of effective carbanion stabilization through valance shell expansion by the heteroatom.

VI. The Role of Nitrogen Ylids in the Hofmann Degradation

The normal mode of the Hofmann degradation of quaternary ammonium hydroxides is via a concerted *trans* E-2 elimination (*89*, Cope and Trumbull). One of the early conflicts with this general proposition was the observation by Arnold and Richardson (*90*) that both *cis*- and *trans*-2-phenylcyclohexyltrimethylammonium hydroxides afforded 1-

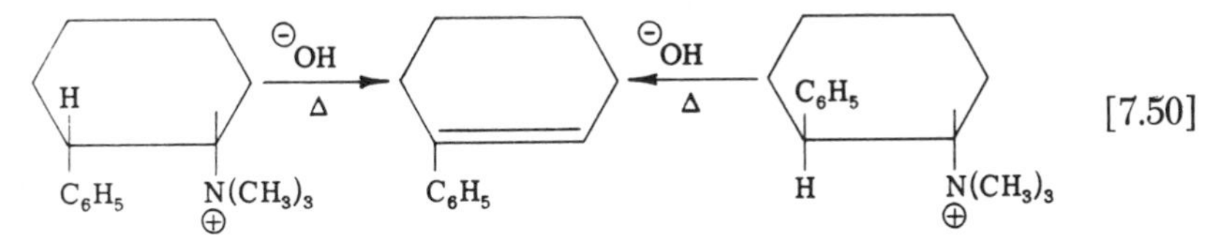

[7.50]

phenylcyclohexene [7.50]. The *trans* isomer does not have a proton on the 1-carbon which is *trans* to the ammonium group. Therefore, these workers proposed that the initial product of the elimination must have been 3-phenylcyclohexene which then underwent a prototropic shift to form the more stable 1-phenylcyclohexene. This proposition was ruled out by Weinstock and Bordwell's observations that the 3-phenylcyclohexene was stable under the conditions of the reaction (*91*). Cope *et al.* (*92*) confirmed this conclusion with their observation that *trans*-3,3,6,6-tetradeutero-2-phenylcyclohexyltrimethylammonium hydroxide underwent the elimination reaction in 86–91% yield without any appreciable loss of deuterium. Thus the conversion of the *trans* ammonium salt to 1-phenylcyclohexene must have been direct and may have been a *cis* elimination. One of the attractive mechanistic possibilities for such a direct *cis* elimination is the initial formation of an ammonium ylid

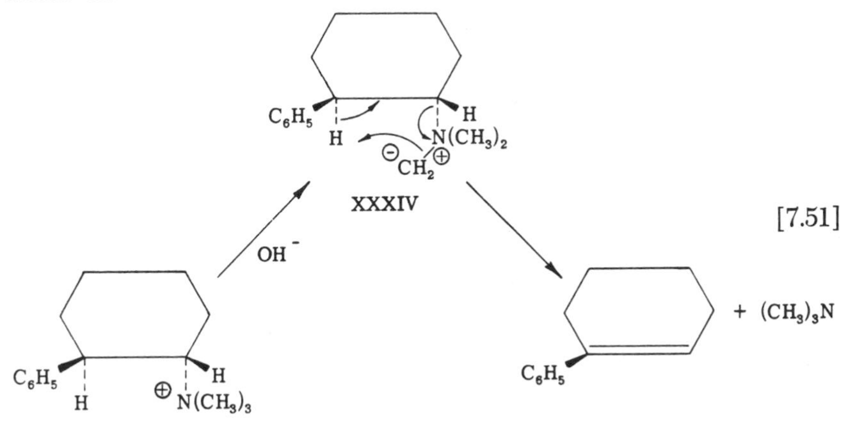

[7.51]

(XXXIV) which, in turn, removes the β-hydrogen. Such a process is termed an α',β-elimination and would be stereochemically analogous to the amine oxide pyrolytic elimination, a *cis* elimination (*89*). This proposal was eliminated in this specific example, however, by the observation that the trimethylamine formed from the elimination of *trans*-2-phenyl-2-deuterocyclohexyltrimethylammonium hydroxide did not contain deuterium (*93*, Bourns *et al.*).

Others also have proposed the operation of α',β-eliminations via ylid intermediates in certain Hofmann degradations, especially those in which there is a significant proportion of *cis* elimination. Grob *et al.* (*94*) invoked such a proposal to account for the high yield of norbornadiene from endo-2-(5-norbornenyl)trimethylammonium salts in the presence of phenyllithium as opposed to the lower yields in the presence of the weaker base, silver oxide. Cope *et al.* (*95*) also proposed an $\alpha'\beta$-elimination for the formation of norbornene from the exo-2-trimethylammonium

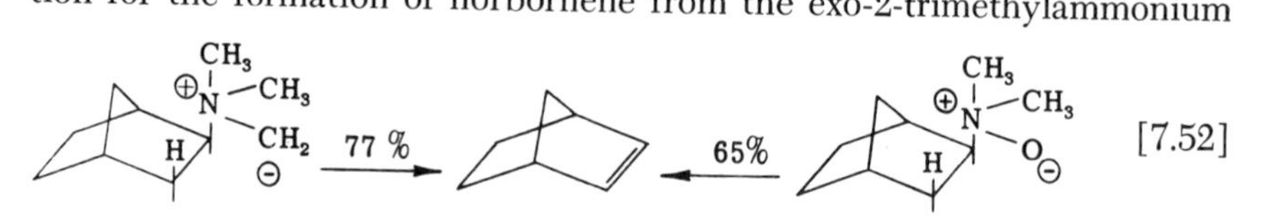

salt because of the similarity in yields to the amine oxide pyrolytic method; both mechanisms involve *cis* eliminations [7.52].

Wittig and Polster (*96*) found that the normal Hofmann degradation of trimethylammoniocyclooctane (via pyrolysis of the hydroxide) afforded cyclooctenes composed of 40% *cis* and 60% *trans* isomer. Potassium amide-catalyzed elimination of the same bromide gave a mixture composed of 15% *cis* and 85% *trans* isomer. Therefore, in both of these instances normal *trans* elimination dominated. However, elimination of the ammonium bromide with phenyllithium as the base afforded the same olefins but composed of 81% *cis* and 19% *trans* isomer. This change to dominant *cis* elimination was proposed to be due to the operation of α',β-elimination. Convincing evidence for this mode of elimination was provided by the observation that treatment of bromomethyldimethylcyclooctylammonium bromide with methyllithium afforded cyclooctenes composed of 90% *cis* and 10% *trans* isomer. The isolation of trimethylamine demanded the intermediacy of the ammonium ylid (XXXV), and the similarity of isomer composition indicated that the same intermediate probably was involved in the phenyllithium case. Replacement of one of the methyl groups of trimethylammoniocyclooctane with a benzyl group, the hydrogens of which are more acidic, and repetition of the amide-catalyzed elimination resulted in complete reversal of the stereochemical results, formation of essentially pure *cis*-cyclooctene. This

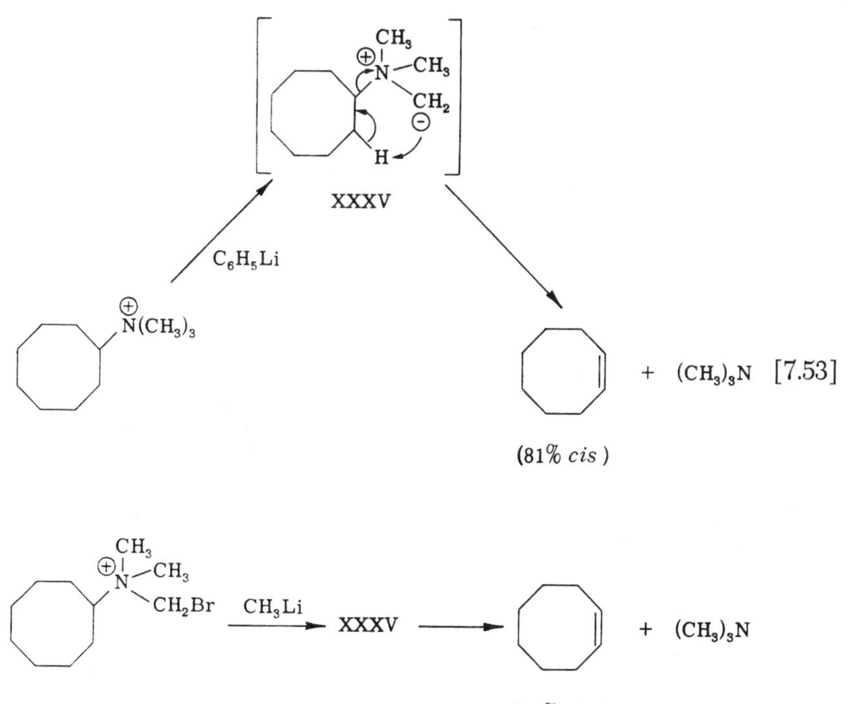

change probably was due to a change in mechanism from β-elimination to α′,β-elimination (*97*, Bumgardner).

The earliest evidence for the α′,β-elimination in open chain systems was provided by Wittig and Polster (*98*) using the phenyllithium modification of the Hofmann degradation on quaternary ammonium salts. Isopropyldimethyliodomethylammonium iodide with phenyllithium afforded 71% of iodobenzene along with trimethylamine and propylene

$$(CH_3)_2\overset{\oplus}{N}(CH_2I)-CH(CH_3)_2 \xrightarrow{C_6H_5Li} \left[(CH_3)_2\overset{\oplus}{N}(\overset{\ominus}{C}H_2)-CH(CH_3)-CH_2-H\right] \longrightarrow CH_3CH{=}CH_2 + (CH_3)_3N \qquad [7.54]$$

[7.54]. Therefore, the base must have attacked the iodomethyl group and formed a nitrogen ylid, the carbanion of which probably removed a β-hydrogen. This experiment proved the existence of α,β-eliminations in such systems but the observation of such a mechanism in the course of normal Hofmann degradations is quite another matter and required the use of isotopic labeling for its detection.

Weygand *et al.* (*99*) first undertook such isotopic studies [7.55]. 2-Tritioethyltrimethylammonium hydroxide produced trimethylamine containing 12% of the original tritium when heated to 150°. This 12% was

$$(CH_3)_3\overset{\oplus}{N}-CH_2CH_2T \xrightarrow[150^\circ]{^{\ominus}OH} (CH_3)_2N-CH_2T \quad (12\%\ \text{activity})$$

$$(CH_3)_3\overset{\oplus}{N}-CH_2CH_2T \xrightarrow[C_6H_5Li]{R.\,T.} (CH_3)_2N-CH_2T \quad (54\%\ \text{activity}) \qquad [7.55]$$

not due solely to the operation of the ylid elimination route, however, since it was shown that tritiated water and hydroxide exchanged with the starting material in the methyl groups upon extended contact at 150°. When the elimination was effected by treating the same tritiated bromide with phenyllithium at room temperature the trimethylamine contained 54% of the original activity indicating the probable important contribution of the ylid mechanism to the overall elimination.

Cope *et al.* (*100*) found that elimination of trimethylamine from 1-deuterocyclohexylmethyltrimethylammonium hydroxide by heating to 150° in steam with immediate collection of the amine led to the incorporation of less than 0.3% of the deuterium in the amine. Failure to collect the amine promptly led to exchange with the deuterated solvent and base. Therefore, there appeared to be insignificant participation of the ylid (α',β) mechanism in this normal Hofmann degradation. However, repetition of the elimination of the ammonium bromide using

$$C_6H_{10}(D)CH_2\overset{\oplus}{N}(CH_3)_3 \xrightarrow[\Delta]{^{\ominus}OH} C_6H_{10}{=}CH_2 + HOD + (CH_3)_3N$$

[7.56]

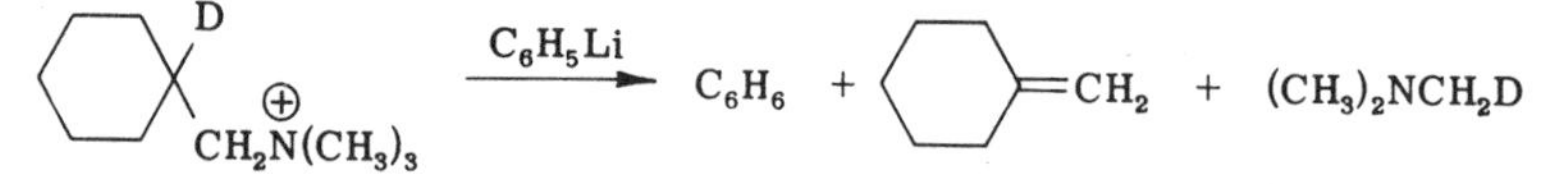

phenyllithium in ether as the base led to the incorporation of essentially 100% of the deuterium in the trimethylamine and indicated the dominant operation of the ylid route [7.56].

More recently, Cope and Mehta (*101*) found that in an open chain ammonium hydroxide where there was steric hindrance to the normal *trans*-β-elimination the *cis*-α',β-elimination can operate effectively. Thus, pyrolysis of XXXVI as the hydroxide led to the incorporation of about

70% of the deuterium into the trimethylamine evolved from the reaction. Models indicated that there was steric hindrance between the *tert*-butyl and ammonio groups in the conformation needed to effect a normal *trans* elimination.

$$\begin{matrix}(CH_3)_3C & & D \\ & C & \\ (CH_3)_3C & & CH_2\overset{\oplus}{N}(CH_3)_3\end{matrix} \xrightarrow[\Delta]{^{\ominus}OH} [(CH_3)_3C]_2C{=}CH_2 + (CH_3)_2NCH_2D \qquad [7.57]$$

XXXVI

The general conclusion that may be drawn from all of these results is that in the normal Hofmann elimination reaction, the pyrolysis of quaternary ammonium hydroxides, the *trans*-β-elimination usually dominates (disregarding for this discussion consideration of the E-1CB route). Such also is the case with sodium amide eliminations on the quarternary halides. Use of the Wittig modification, however, the treatment of the ammonium halide with an organolithium reagent, usually results in the dominant operation of the ylid route for elimination, the *cis*-$\alpha'\beta$-elimination. Structural modifications on the ammonium salt such as steric hindrance in the transition state or increased acidity of the methyl hydrogens will result in the ylid route and *cis* stereochemistry increasing in importance at the expense of *trans* elimination, even under the normal Hofmann elimination conditions.

REFERENCES

1. W. von E. Doering and A. K. Hoffmann, *J. Am. Chem. Soc.* **77,** 521 (1955).
2. W. von E. Doering and K. C. Schreiber, *J. Am. Chem. Soc.* **77,** 514 (1955).
3. W. Schlenk and J. Holtz, *Chem. Ber.* **49,** 603 (1916).
4. W. Schlenk and J. Holtz, *Chem. Ber.* **50,** 274 (1917).
5. F. D. Hager and C. S. Marvel, *J. Am. Chem. Soc.* **48,** 2689 (1926).
6. G. Wittig and M. Wetterling, *Ann.* **557,** 193 (1947).
7. F. Krohnke, *Chem. Ber.* **68,** 1177 (1935).
8. G. Wittig and R. Polster, *Ann.* **599,** 1 (1956).
9. V. Franzen and G. Wittig, *Angew. Chem.* **72,** 417 (1960).
10. G. Wittig and D. Krauss, *Ann.* **679,** 34 (1964).
11. F. Weygand, H. Daniel and A. Schroll, *Chem. Ber.* **97,** 1217 (1964).
12. G. Wittig and M. Rieber, *Ann.* **562,** 177 (1949).
13. G. Wittig and W. Tochtermann, *Chem. Ber.* **94,** 1692 (1961).
14. F. Weygand and H. Daniel, *Chem. Ber.* **94,** 3147 (1961).
15. F. Krohnke and W. Heffe, *Chem. Ber.* **70,** 1720 (1937).
16. C. K. Ingold and J. A. Jessop, *J. Chem. Soc.* 2357 (1929).
17. V. Franzen, *Chem. Ber.* **93,** 557 (1960).
18. G. Wittig and H. Laib, *Ann.* **580,** 57 (1953).
19. G. Wittig and G. Felletschin, *Ann.* **555,** 133 (1944).
20. H. J. Dauben, Jr. and W. W. Spooncer, *Abstr. Papers, 126th Meeting, Am. Chem. Soc.* p. 18-O; *Dissertation Abstr.* **16,** 458 (1956).
21. Z. Arnold, *Chem. & Ind.* London 1478 (1960); *Coll. Czech. Chem. Comms.* **26,** 1113 (1961).

22. F. Krohnke, *Chem. Ber.* **70,** 543 (1937).
23. F. Krohnke and H. Kubler, *Chem. Ber.* **70,** 538 (1937).
24. F. Krohnke and K. Gerlach, *Chem. Ber.* **95,** 1108 (1962).
25. F. Krohnke, K. Gerlach and K. E. Schnalke, *Chem. Ber.* **95,** 1118 (1962).
26. F. Krohnke, *Chem. Ber.* **71,** 2587 (1938).
27. P. A. Chopard, R. J. G. Searle and F. H. Devitt, *J. Org. Chem.* **30,** 1015 (1965).
28. F. Krohnke, W. Zecher, J. Curtze, D. Drechsler, K. Pfleghar, K. E. Schnalke and W. Weis, *Angew. Chem.* **74,** 811 (1962).
29. W. Zecher and F. Krohnke, *Chem. Ber.* **94,** 690 (1961).
30. F. Krohnke, *Angew. Chem.* **65,** 617 (1953).
31. F. Krohnke and E. Borner, *Chem. Ber.* **69,** 2006 (1936).
32. F. Krohnke, *Chem. Ber.* **72,** 527 (1939).
33. F. Krohnke, *Angew. Chem.* **75,** 317 (1963).
34. W. Reid and R. M. Gross, *Chem. Ber.* **90,** 2646 (1957).
35. F. Krohnke, *Chem. Ber.* **83,** 253 (1950).
36. L. A. Pinck and G. E. Hilbert, *J. Am. Chem. Soc.* **68,** 2011 (1946).
37. H. Hartmann and H. Gossel, *Z. Electrochem.* **61,** 337 (1957).
38. A. Wm. Johnson, *J. Org. Chem.* **24,** 282 (1959).
39. D. Lloyd and S. Sneezum, *Chem. & Ind.* (*London*) 1221 (1955); *Tetrahedron* **3,** 334 (1958).
40. E. M. Kosower and B. G. Ramsey, *J. Am. Chem. Soc.* **81,** 856 (1959).
41. D. N. Kursanov and N. K. Baranetskaia, *Izvest. Akad. Nauk. S.S.S.R.* 362 (1958); *Chem. Abstr.* **52,** 12864 (1958).
42. F. Ramirez and S. Levy, *J. Am. Chem. Soc.* **79,** 6167 (1957).
43. E. M. Evleth, J. A. Berson and S. L. Manatt, *Tetrahedron Letters* 3087 (1964).
44. S. I. Zhdanov and L. S. Mirkin, *Coll. Czech. Chem. Comms.* **26,** 370 (1961); *Chem. Abstr.* **55,** 10148 (1961).
45. D. N. Kursanov, N. K. Baranetskaia and Z. N. Parnes, *Izvest. Akad. Nauk. S.S.S.R.* 140 (1961); *Chem. Abstr.* **55,** 17632 (1961).
46. W. J. Linn, O. W. Webster and R. E. Benson, *J. Am. Chem. Soc.* **85,** 2032 (1963); **87,** 3651 (1965); A. Rieche and P. Dietrich, *Chem. Ber.* **96,** 3044 (1963).
47. C. Bugg, R. Desiderato and R. L. Sass, *J. Am. Chem. Soc.* **86,** 3157 (1964); C. Bugg and R. L. Sass, *Acta Cryst.* **18,** 591 (1965).
48. C. C. Costain and B. P. Stoicheff, *J. Chem. Phys.* **30,** 777 (1959).
49. C. C. Costain, *J. Chem. Phys.* **29,** 864 (1958).
50. F. G. Mann and W. J. Pope, *J. Chem. Soc.* **121,** 1052 (1922).
51. S. Wawzonek and D. Meyer, *J. Am. Chem. Soc.* **76,** 2918 (1954).
52. S. Wawzonek and E. Yeakey, *J. Am. Chem. Soc.* **82,** 5718 (1960).
53. R. L. Hinman and M. C. Flores, *J. Org. Chem.* **24,** 660 (1959).
54. M. S. Gibson and A. W. Murray, *J. Chem. Soc.* 880 (1965).
55. W. Schneider and F. Seebach, *Chem. Ber.* **54,** 2285 (1921).
56. W. Schneider, *Ann.* **438,** 115 (1924).
57. K. Dimroth, G. Arnoldy, S. von Eicken and G. Schiffler, *Ann.* **604,** 221 (1957).
58. J. N. Ashley, G. L. Buchanan and A. P. T. Easson, *J. Chem. Soc.* 60 (1947).
59. J. J. Monagle, *J. Org. Chem.* **27,** 3851 (1962).
60. C. D. Gutsche, *in* "Organic Reactions" (R. Adams, ed.), Vol. 8, Chapter 8, pp. 364–429, John Wiley and Sons, Inc., New York 1958.
61. A. Wm. Johnson, V. J. Hruby and J. L. Williams, *J. Am. Chem. Soc.* **86,** 918 (1964).

62. A. Schonberg and K. Junghans, *Chem. Ber.* **96**, 3328 (1963).
63. H. Staudinger and J. Siegwart, *Helv. Chim. Acta* **3**, 833 (1920).
64. A. Schonberg, K. H. Brosowski and E. Singer, *Chem. Ber.* **95**, 2144 (1962).
65. A. Schonberg and E. Frese, *Chem. Ber.* **95**, 2810 (1962).
66. H. Staudinger and K. Miescher, *Helv. Chim. Acta* **2**, 554 (1919).
67. A. Wm. Johnson, *J. Org. Chem.* **28**, 252 (1963).
68. H. Reimlinger, *Angew. Chem.* **74**, 153 (1962); *Chem. Ber.* **97**, 339 (1964).
69. R. Oda, Y. Ito and M. Okano, *Tetrahedron Letters* 7 (1964).
70. L. Horner and E. Lingnau, *Ann.* **591**, 21 (1955).
71. W. von E. Doering and M. Jones, Jr., *Tetrahedron Letters* 791 (1963).
72. W. R. Bamford and T. S. Stevens, *J. Chem. Soc.* 4675 (1952).
73. G. Wittig and M. Schlosser, *Tetrahedron* **18**, 1023 (1962).
74. T. S. Stevens, E. M. Creighton, A. B. Gordon and M. MacNicol, *J. Chem. Soc.* 3193 (1928).
75. T. S. Stevens, *J. Chem. Soc.* 2107 (1930).
76. T. Thomson and T. S. Stevens, *J. Chem. Soc.* 55 (1932).
77. T. Thomson and T. S. Stevens, *J. Chem. Soc.* 1932 (1932).
78. J. L. Dunn and T. S. Stevens, *J. Chem. Soc.* 279 (1934).
79. G. Wittig, R. Mangold and G. Felletschin, *Ann.* **560**, 116 (1948).
80. A. Campbell, A. H. J. Houston and J. Kenyon, *J. Chem. Soc.* 93 (1947).
81. C. R. Hauser and S. W. Kantor, *J. Am. Chem. Soc.* **73**, 1437 (1951).
82. F. Weygand, A. Schroll and H. Daniel, *Chem. Ber.* **97**, 857 (1964).
83. G. Wittig and E. Benz, *Chem. Ber.* **92**, 1999 (1959).
84. M. Sommelet, *Compt. rend.* **205**, 56 (1937).
85. G. Wittig, H. Tenhaeff, W. Schoch and G. Koenig, *Ann.* **572**, 1 (1951).
86. S. W. Kantor and C. R. Hauser, *J. Am. Chem. Soc.* **73**, 4122 (1951).
87. W. H. Puterbaugh and C. R. Hauser, *J. Am. Chem. Soc.* **86**, 1105 (1964).
88. W. H. Puterbaugh and C. R. Hauser, *J. Am. Chem. Soc.* **86**, 1108 (1964).
89. A. C. Cope and E. R. Trumbull, *in* "Organic Reactions" (R. Adams, ed.), Vol. 11, p. 328, John Wiley and Sons, Inc., New York 1960.
90. R. T. Arnold and P. N. Richardson, *J. Am. Chem. Soc.* **76**, 3649 (1954).
91. J. Weinstock and F. G. Bordwell, *J. Am. Chem. Soc.* **77**, 6706 (1955).
92. A. C. Cope, G. A. Berchtold and D. L. Ross, *J. Am. Chem. Soc.* **83**, 3859 (1961).
93. G. Ayrey, E. Bunal and R. N. Bourns, *Proc. Chem. Soc.* 458 (1961).
94. C. A. Grob, H. Kny and A. Gagneux, *Helv. Chim. Acta* **40**, 130 (1957).
95. A. C. Cope, E. Ciganek and N. A. LeBel, *J. Am. Chem. Soc.* **81**, 2799 (1959).
96. G. Wittig and R. Polster, *Ann.* **612**, 102 (1958).
97. C. L. Bumgardner, *J. Org. Chem.* **27**, 1035 (1962).
98. G. Wittig and R. Polster, *Ann.* **599**, 13 (1956).
99. F. Weygand, H. Daniel and H. Simon, *Chem. Ber.* **91**, 1691 (1958); *Ann.* **654**, 111 (1962).
100. A. C. Cope, N. A. LeBel, P. T. Moore and W. R. Moore, *J. Am. Chem. Soc.* **83**, 3861 (1961).
101. A. C. Cope and A. S. Mehta, *J. Am. Chem. Soc.* **85**, 1949 (1963).

8

ARSENIC AND ANTIMONY YLIDS

I. Introduction

With the investigations into the chemistry of phosphorus ylids having yielded so much useful and interesting synthetic and mechanistic information, the next logical step in the study of ylid chemistry was an investigation of the chemistry of ylids of other group V elements. As is typical of the periodic organization of the elements, the first element in a group shows properties rather different from those of the second element while the remaining elements in any group usually exhibit properties quite similar to those of the second element. Nitrogen ylids have been discussed as a separate group (see Chapter 7) in view of this fact and in view of the different atomic orbital description of nitrogen when compared to phosphorus. In this chapter the chemistry of arsenic and antimony ylids and imines will be discussed. Differences in degree but not in kind of chemical and physical properties may be expected from those of the analogous phosphorus ylids and imines.

The electronic configurations of atomic arsenic and antimony are similar in the outer shell, each having two paired electrons in an *s*-orbital and three unpaired electrons in *p*-orbitals. The configurations may be summarized as follows:

$$\text{As: } 1s^2 2s^2 2p^6 3s^2 3p^6 3d^{10} 4s^2 4p^3$$

$$\text{Sb: } 1s^2 2s^2 2p^6 3s^2 3p^6 3d^{10} 4s^2 4p^6 4d^{10} 5s^2 5p^3$$

As a result, both atoms form covalent σ bonds with three substituents, i.e. R_3As and R_3Sb, and the hybridization of the central atom appears to be between p^3, requiring bond angles of 90°, and sp^3, requiring angles near 109.5° (*1*, Wells). As expected from their electronic configurations and by analogy with phosphorus chemistry, trisubstituted arsenic and antimony derivatives act as Lewis bases and will form quaternary salts [8.1].

$$(C_6H_5)_3As + CH_3I \longrightarrow (C_6H_5)_3\overset{\oplus}{As}-CH_3 \quad I^{\ominus}$$

[8.1]

$$(C_6H_5)_3Sb + (C_2H_5)_3O^{\oplus}BF_4^{\ominus} \longrightarrow (C_6H_5)_3\overset{\oplus}{Sb}-C_2H_5 \quad BF_4^{\ominus}$$

If arsenic and antimony are to be involved in ylid formation they must be capable of stabilizing an adjacent carbanion by an electron-accepting mechanism. In order to do so, these atoms must be able to expand their outer shells. In other words, there must be vacant orbitals of rather low energy which effectively can overlap with a filled 2*p*-orbital of the carbanion. In the case of arsenic the 4*d*- and 5*s*-orbitals are vacant and of approximately the same energy while in the case of antimony the 4*f*-, 5*d*- and 6*s*-orbitals are vacant and of approximately the same energy. Whether overlap of a vacant 4*d*- or 5*s*-orbital, for example, with a filled 2*p*-orbital of carbon would provide effective stabilization for the carbanion is debatable, and certainly if based on the electronic configuration of the neutral atoms and even if based on the neutral trivalent arsine or stibine. However, conversion of the arsine or stibine into a quaternary salt, thereby forcing the central atom to carry a high degree of positive charge, probably would contract the normally diffuse *d*-orbitals such that effective overlap with a 2*p*-orbital of carbon could occur. This argument is exactly that used to account for the ability of sulfonium and phosphonium groups to play an electron-accepting role in ylids and transition metal complexes (*2*, Craig and Magnusson).

In an earlier chapter the evidence for the ability of phosphorus to expand its octet was discussed (Chapter 2). Considerable analogous evidence is available in the case of both arsenic and antimony. In general, an indication of similarity in the properties of arsenic and antimony compounds with those properties of phosphorus compounds attributable to valence shell expansion has been taken to mean that all three exhibit the properties for the same reason.

The ability of arsenic to expand its octet has been indicated clearly by the isolation of several pentavalent arsenic derivatives. Pentafluoroarsenic has been known for some time and Wittig *et al.* (*3, 4*) subsequently have reported the preparation of the novel compound, pentaphenylarsenane, $(C_6H_5)_5As$. Similarly, the pentachloro- and pentafluoro- derivatives of antimony have been known, and Wittig's group also have reported the preparation of pentaphenylantimony (*3, 5*). Electron diffraction data indicated that the pentahalides, for example, $SbCl_5$, have a trigonal bipyramid structure, consistent with a central atom hybridization of dsp^3 (*6*, Rouault). X-ray data indicated the same configuration for the solid state (*7*, Ohlberg). Sodium hexafluoroantimonate,

$NaSbF_6$, contains the octahedral SbF_6^- ion, consistent with the proposed sp^3d^2 hybridization for antimony (*8*, Teuffer). Preliminary X-ray examination of pentaphenylantimony has confirmed the gross structure and some of its general features. The five phenyl groups appear to be arranged in a square pyramid with the antimony atom just above the floor of the pyramid (*9, 10,* Wheatley and Wittig). From all of the above data it is clear that both arsenic and antimony can expand their outer valence shell to accommodate at least ten electrons. As indicated by the geometries involved this expansion most likely involves the use of 4*d*- and 5*d*-orbitals, respectively.

Some estimate has been obtained of the relative ease with which similarly substituted phosphines, arsines and stibines undergo valence shell expansion. Much of the information comes from the use of these substances as ligands in metal complexes. Chatt and Hart (*11*) examined the mono-substituted nickel tricarbonyls, $Ni(CO)_3L$, where L = $(C_6H_5)_3P$ or $(C_6H_5)_3As$. These were prepared by displacement of carbon monoxide from nickel tetracarbonyl, the reaction proceeding much faster with the phosphine than the arsine. However, the carbonyl stretching frequency for the phosphine case was 2063 cm^{-1} while that of the arsine case was 2072 cm^{-1}. This evidence was taken to indicate that the back donation of electrons from the filled *d*-orbitals of the nickel was in the direction of the carbon monoxide ligand more so in the case of the phosphine complex than in that of the arsine. In other words, the arsine, actually in the form of an arsonium group because of electron sharing with the nickel, accepted back donation into its vacant *d*-orbitals to a higher degree than did the phosphine (i.e. IIa contributed more than Ia to the actual electron distribution of the respective complexes).

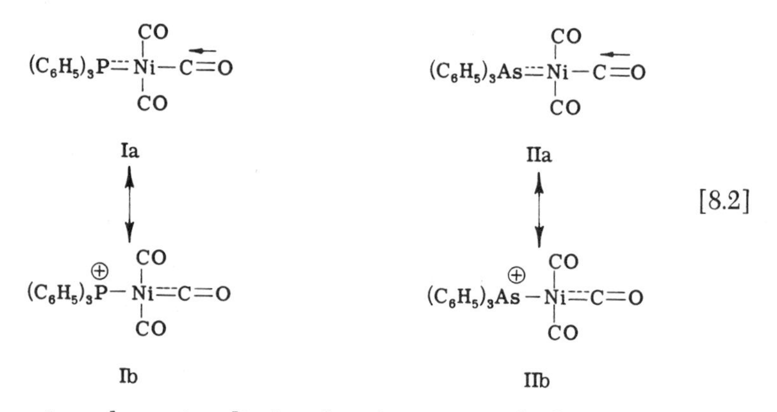

However, since the arsine displaced carbon monoxide from nickel tetracarbonyl more slowly than did the phosphine, Chatt and Hart compromised the conflicting data and concluded that the nickel-phosphorus and nickel-arsenic bonds were of about the same order, i.e. arsenic and phosphorus have about the same tendency to expand their octet.

From similar studies on the carbonyl frequencies in the infrared spectra of monosubstituted chromium pentacarbonyls Magee *et al.* (*12*) concluded that triphenylphosphine, triphenylarsine and triphenylstibine have nearly equivalent donor-acceptor properties. From a study of the "*trans* effect" (*13*) in a series of Pt^{II} complexes L_2PtCl_2, Chatt and Wilkins (*14*) esimated that the *d*-orbital contribution to bonding with Pt remained approximately constant for the groups $L = (C_2H_5)_3P$, $(C_2H_5)_3As$ and $(C_2H_5)_3Sb$, whether that contribution was viewed as double bonding or as rehybridization of single bonds. The conclusions of most studies on metal complexes of this type have been compromises between two or more pieces of experimental evidence that conflicted, some indicating the presence and others the lack of a difference in valence shell expansion between phosphorus, arsenic and antimony.

Doering and Hoffmann (*15*), in their studies of the exchange of deuterium from a solution of deuteroxide in D_2O with various permethyl 'onium salts, observed that under the same conditions tetramethylphosphonium iodide, tetramethylarsonium iodide and tetramethylstibonium iodide incorporated 73.9%, 7.44% and 0.78% deuterium, respectively. They concluded, however, on the basis of this one reaction (i.e. no heat of activation data and only one rate constant!) that the decrease in the rate of exchange in this series was due to the increased bond lengths between the heteroatom and the carbanion in the intermediate or transition state and the resultant decrease in coulombic stabilization. Furthermore, the contribution of valence shell expansion to the stabilization of the carbanion, i.e. the overlap of the filled 2*p*-orbital of the carbanion with the vacant *nd*-orbital of the heteroatom, was thought to remain approximately constant through the series P, As and Sb [8.3].

$$(CH_3)_3\overset{\oplus}{X}-\overset{\ominus}{C}H_2 \longleftrightarrow (CH_3)_3X{=}CH_2 \qquad [8.3]$$

From a very simple point of view it would be expected that as group VA of the periodic table was descended from phosphorus through arsenic to antimony less stabilization of an adjacent carbanion by an 'onium group (X) through electrostatic forces would occur since the C-X bond distances increase (1.87, 1.98 and 2.18 Å) (*15*) while the electronegativity of X also decreases (2.1, 2.0 and 1.9) (*16*, Pauling). In addition, however, and contrary to Doering's conclusion, stabilization of a carbanion by an adjacent 'onium atom through overlap of the filled 2*p*-orbital of the carbanion with a vacant *nd*-orbital of the heteroatom also would be expected to decrease down the group since the size of the two atoms involved (carbon-phosphorus, carbon-arsenic and carbon-antimony) becomes more dissimilar and the principal quantum number of the *d*-orbitals used increases, thereby decreasing the effectiveness of the $2p$-nd π overlap. Thus, it would seem to be somewhat premature, especially in view of the paucity of pertinent data, to conclude that valence shell expansion remains constant as group VA is descended from phosphorus to antimony.

From the above discussion it is clear that the elements below phosphorus in the periodic table can undergo valence shell expansion. Therefore, tetravalent derivatives of these elements should be convertible into ylids ($R_3X^+—C^-R'_2$) and these ylids should have appreciable stability and exhibit characteristic ylid reactions. The preparation and properties of such ylids will be described in this chapter. Attempts will be made to compare the properties of these ylids with the analogous phosphonium ylids.

The question might be raised as to why the chemistry of bismuth ylids has been avoided. There are very few examples of quaternary bismuth salts known and trivalent organobismuth compounds rarely have been used as ligands in transition metal complexes. One of the recent unsuccessful attempts to quaternize a trivalent bismuth compound was Henry and Wittig's (*17*) reaction of triphenylbismuth with trimethyloxonium tetrafluoborate. On the other hand, the pentacovalent compounds triphenylbismuth dichloride and pentaphenylbismuth have been reported although the latter was inadequately characterized (*18*, Wittig and Clauss). In addition, Wittig and Hellwinkel (*5*) claimed recently that triphenylbismuth reacted with chloramine-T to afford triphenylbismuthonium-*N*-tosylimine, $(C_6H_5)_3Bi^+—N^-—SO_2—C_6H_4CH_3(p)$. However, this substance was not isolated or characterized and upon standing the solution afforded *p*-toluenesulfonamide and triphenylbismuth. Thus, there is some, but meager, evidence that bismuth can expand its valence shell. Nonetheless there have been no bismuth ylids reported to date.

II. Arsonium Ylids

Very little work has been published on the chemistry of arsonium ylids. In spite of this significant information has been obtained on the physical and chemical properties of arsonium ylids.

A. Synthesis of Arsonium Ylids

The earliest report of an arsonium ylid seems to be due to Michaelis (*19*) who found that phenacyl bromide and triphenylarsine formed a salt which, upon addition of sodium hydroxide, afforded a new substance formulated as III. Krohnke (*20*) confirmed the experimental observa-

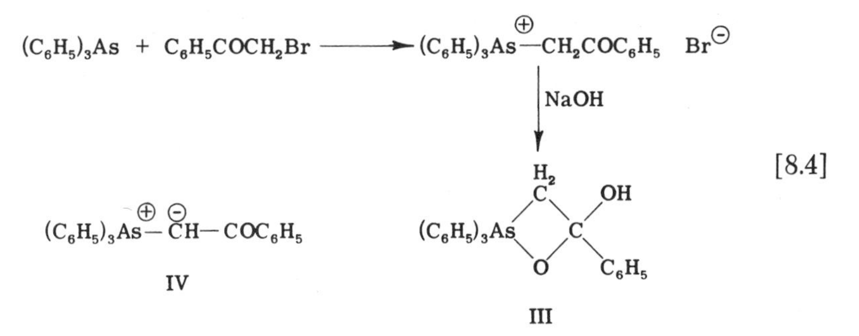

tions but formulated the substance as an arsonium ylid, phenacylidenetriphenylarsenane (IV). Nesmeyanov *et al.* (*21*) recently have reported the preparation of several analogous β-keto arsonium ylids by a similar route.

Most of the arsonium ylids synthesized to date have been prepared by the so-called "salt-method." Several groups, Henry and Wittig (*17*), Grim and Seyferth (*22*) and Seyferth and Cohen (*23*), have prepared triphenylmethylarsonium halides by direct alkylation. In all three cases the arsonium salt was converted into the ylid, methylenetriphenylarse-

$$(C_6H_5)_3\overset{\oplus}{As}-CH_3\ \ X^{\ominus}\ \xrightarrow[(C_2H_5)_2O]{C_6H_5Li}\ LiX + C_6H_6 + \underset{\mathbf{V}}{(C_6H_5)_3\overset{\oplus}{As}-\overset{\ominus}{C}H_2} \qquad [8.5]$$

nane (V), by treatment of an ethereal slurry with phenyllithium. There have been no attempts to isolate the ylid.

Johnson (*24*) and Wittig and Laib (*25*) both have prepared arsonium ylids from 9-bromofluorene. The latter group alkylated trimethylarsine with 9-bromofluorene and obtained the arsonium salt in 70% yield. Treatment of an ethereal slurry of the salt with phenyllithium afforded a yellow solution of fluorenylidenetrimethylarsenane. Again, however, the ylid was not isolated but was handled entirely in solution.

Johnson (*24*), however, found that treatment of an aqueous alcoholic solution of 9-fluorenyltriphenylarsonium bromide with dilute ammonia or

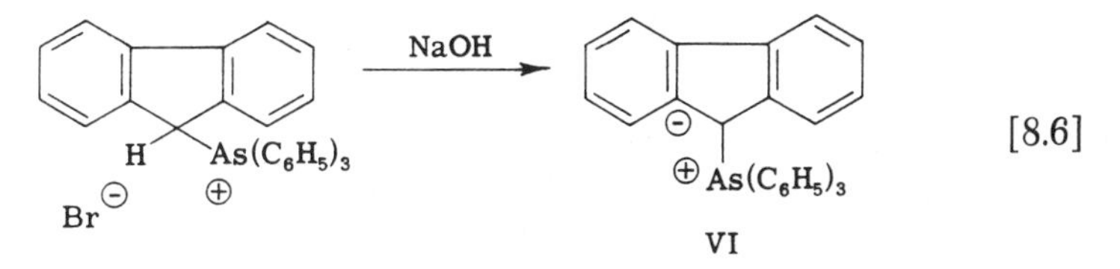

dilute sodium hydroxide afforded the yellow fluorenylidenetriphenylarsenane (VI) as a stable, isolable precipitate melting at 188–190°

Wittig and Laib (*25*) attempted to prepare an ylid from dimethyldibenzylarsonium salts but found, upon treatment of the salts with ethereal phenyllithium, that although the ylid apparently had formed as indicated by the appearance of a short-lived yellow color, it rapidly underwent a Stevens rearrangement and a subsequent elimination. The product of the reaction after quenching with water was stilbene but if methyl iodide were added first, trimethyl(1,2-diphenylethyl)arsonium iodide was isolated in 50% yield. These products were thought to have formed via the mechanism in [8.7]. The Stevens rearrangement seems straight-

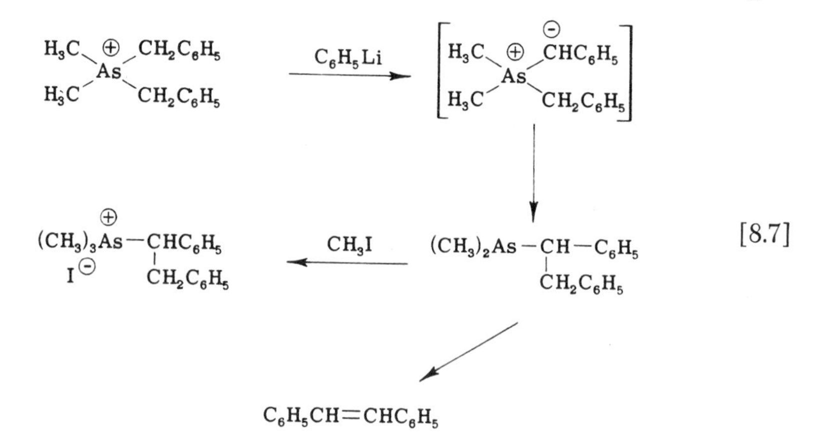

forward but there is no experimental evidence to substantiate the proposed route of stilbene formation.

It is apparent that, in principle at least, any arsonium ylid could be prepared via the "salt method" as demonstrated above. The restrictions applying to the nature of the groups attached to arsenic or to the carbanion should be the same as for the phosphorus cases (Chapter 3). In addition, however, there is no obvious reason why any method applicable to the synthesis of phosphonium ylids likewise could not be adapted to the synthesis of arsonium ylids. Such was, in fact, the case when Horner and Oediger (*26*) found that triphenylarsine dichloride reacted with active methylene compounds in the presence of triethylamine to afford stabilized arsonium ylids [8.8].

$$(C_6H_5)_3AsCl_2 + CH_2\langle^{X}_{Y} \xrightarrow{(C_2H_5)_3N} (C_6H_5)_3As{=}C\langle^{X}_{Y} \qquad [8.8]$$

X and Y = CN, $SO_2C_6H_5$, $COOCH_3$, NO_2, C_6H_5

B. Properties of Arsonium Ylids

As mentioned previously only one arsenic ylid, fluorenylidenetriphenylarsenane (VI), has been subjected to a detailed examination of its chemical and physical properties (*24*, Johnson). This ylid was a pale yellow, crystalline substance of melting point 188–190° It could be stored at room temperature exposed to the atmosphere for about one month after which time it began to lose its color as it slowly hydrolyzed. The ylid behaved as a typical covalent compound in that it was soluble in the usual organic solvents but insoluble in water and ethanol. The ylid dissolved in dilute mineral acid, however, presumably due to its conversion into the conjugate acid, an arsonium salt. In fact, Wittig and Laib (*25*) isolated the expected colorless arsonium iodide after treatment of fluorenylidenetrimethylarsenane with hydrobromic acid and potassium iodide. Thus, the salt-to-ylid conversion is completely reversible.

Horner and Oediger (*26*) reported the melting points and infrared spectra of a series of stabilized arsonium ylids, $(C_6H_5)_3As{=}CXY$, where X=Y=CN, $X{=}Y{=}SO_2C_6H_5$ and X=CN, $Y{=}COOCH_3$. The β-keto-ylids reported by Nesmeyanov *et al.* (*21*) also were stable to the atmosphere and were characterized by microanalysis and melting point.

Johnson and LaCount (*27*) determined the pK_a of triphenyl-9-fluorenylarsonium bromide to be 7.8 in 31.7% water-dioxane. This indicated that the arsonium ylid (VI) was more basic than the analogous triphenylphosphoniumfluorenylide ($pK_a = 7.5$), i.e. there was less driving force toward the formation of the arsonium carbanion than the phosphonium carbanion. It also indicates that there is less stabilization afforded an adjacent carbanion by valence shell expansion of an arsonium group than of a phosphonium group. In other words, VIa contributes less to the actual structure of the fluorenylidenetriphenylarsenane than does VIIa to the actual structure of the fluorenylidenetri-

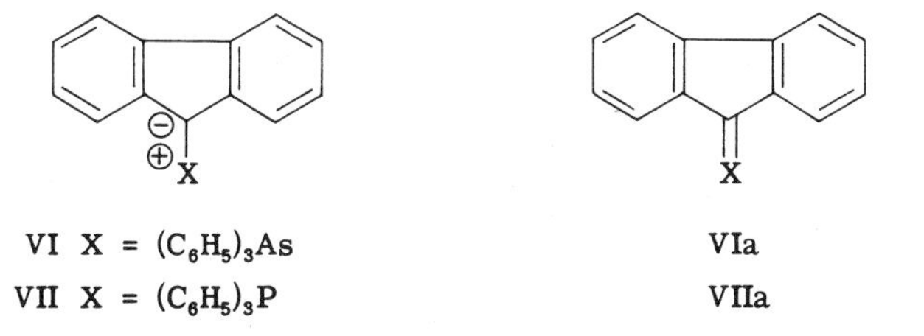

phenylphosphorane. This conclusion also has been reached on the basis of a comparative study of the reactivity of arsonium- and phosphonium-fluorenylides with carbonyl compounds as will be discussed in the following section of this chapter. The conclusion is at variance with that expressed by Doering and Hoffmann (*15*) and Chatt and Hart (*11*) (see Section I, this chapter).

The dipole moment of fluorenylidenetriphenylarsenane (VI) was 7.80 D (*28*, Johnson) compared to the moment of 7.09 D exhibited by the analogous phosphonium compound (VII). It is difficult to evaluate the significance of the dipole moment, however, since the needed bond distance and bond angle information is unavailable. The moment of the arsonium ylid would be expected to be larger than that of the phosphonium ylid since the moment of triphenylarsine oxide is 5.5 D while that of triphenylphosphine oxide is 4.3 D (*29*, Phillips *et al.*), presumably due to the increased bond lengths and decreased electronegativity of arsenic.

Only the more stable arsonium ylids remain unchanged in the presence of water. Methylenetriphenylarsenane (V) formed a triphenylmethylarsonium hydroxide by simple proton transfer. If picric acid were in the solution, the arsonium picrate precipitated (*23*, Seyferth and Cohen). This is in contrast to the phosphonium ylid which rapidly decomposed to a hydrocarbon and phosphine oxide. Fluorenylidenetriphenylarsenane (VI) was unchanged in the presence of water but it was hydrolyzed to fluorene and triphenylarsine oxide in the presence of aqueous base (*24*, Johnson). Likewise, phenacylidenetriphenylarsenane was hydrolyzed to acetophenone and triphenylarsine oxide (*21*, Nesmeyanov *et al.*). The latter two hydrolyses probably proceeded in a manner analogous to that envisioned for the phosphonium ylids and portrayed in [8.9]. The pentavalent intermediate would be expected to eject

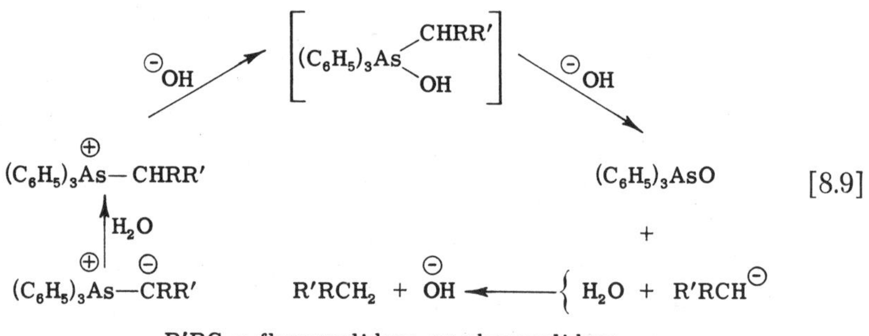

[8.9]

the most stable carbanion, in these cases the fluorenyl or the phenacyl anion. It should be emphasized that this mechanism is purely speculative in the case of arsonium ylids.

There have been no studies directed at a determination of the thermal stability of arsonium ylids. In all the work reported to date there is no indication, for example, that arsonium ylids will undergo alpha eliminations similar to those that have been reported for ammonium, sulfonium and some phosphonium ylids. Thermal instability has not proven to be a problem.

C. Reactions of Arsonium Ylids

Since arsonium ylids are but a special form of carbanion they should react at least as typical nucleophiles. Seyferth and Cohen (*23*) have verified this expectation by demonstrating that methylenetriphenylarsenane (V) could be alkylated with trimethylsilyl bromide in a typical displacement. Similarly, the same ylid reacted with boron trifluoride etherate to form a zwitterion, triphenylarsoniomethyltrifluoroborate [8.10]. Nesmeyanov *et al.* (*21*) found that phenacylidenetriphenylarsenane would react with both bromine and sulfur trioxide in typical nucleophilic fashion as had phosphonium ylids.

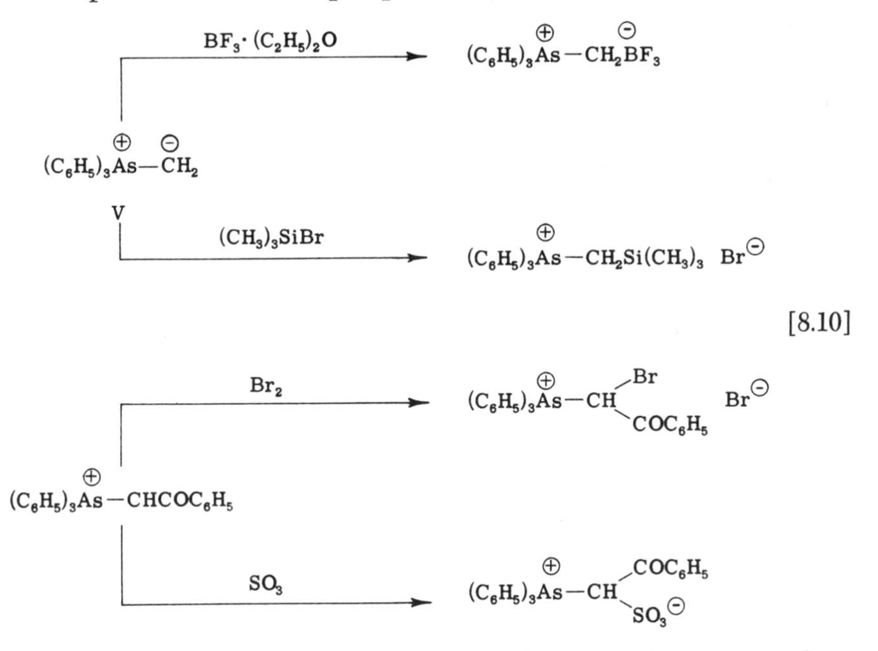

[8.10]

Although most have not been carried out, it is apparent from the information already discussed that arsonium ylids ought to partake in many of the reactions, such as alkylation, acylation, oxidation, Michael additions, etc., mentioned in Chapter 3 as being characteristic of phosphonium ylids. There is no *a priori* reason to suspect that the course of the reactions would be much different but this remains to be proven.

Perhaps the major initial justification for the work on arsonium ylids was the desire to explore the reaction of these ylids with carbonyl compounds and to compare the results with the behavior of phosphonium ylids. The first published report of a reaction between an arsonium ylid and a carbonyl compound was by Wittig and Laib (25). However, they did not investigate the reaction thoroughly at that time since they mainly were interested in detecting any Stevens rearrangement products. They did observe that a solution of fluorenylidenetrimethylarsenane, generated for some unknown reason with a molar excess of phenyllithium, reacted with benzophenone to afford phenyltrimethylarsonium iodide and 9-fluorenyldiphenylcarbinol [8.11], the latter being identified by dehydrating it with acetyl chloride to the characterizable benz-

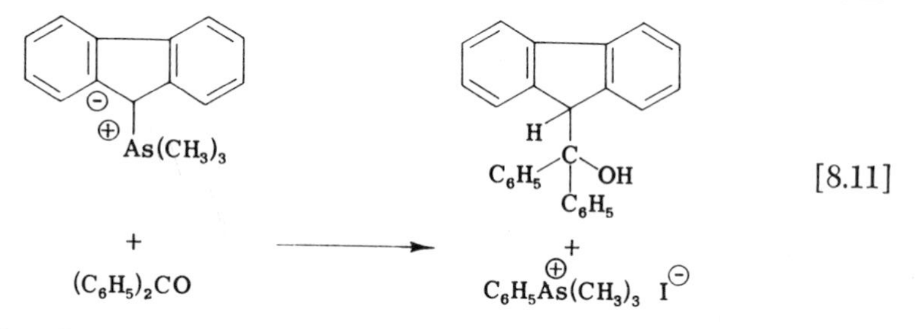

hydrylidenefluorene. There are several questions that need to be answered about this reaction. It is not clear why an excess of phenyllithium had to be used in generating the ylid, especially when a ketone was about to be added. In addition, the carbinol product was not identified directly. The latter point is extremely pertinent in this case since one might expect to observe the normal Wittig reaction product, benzhydrylidenefluorene, as a primary product. Instead it was isolated after a subsequent dehydration reaction. Could it have been the primary product?

Wittig and Laib (25) advanced the mechanism presented in [8.12]

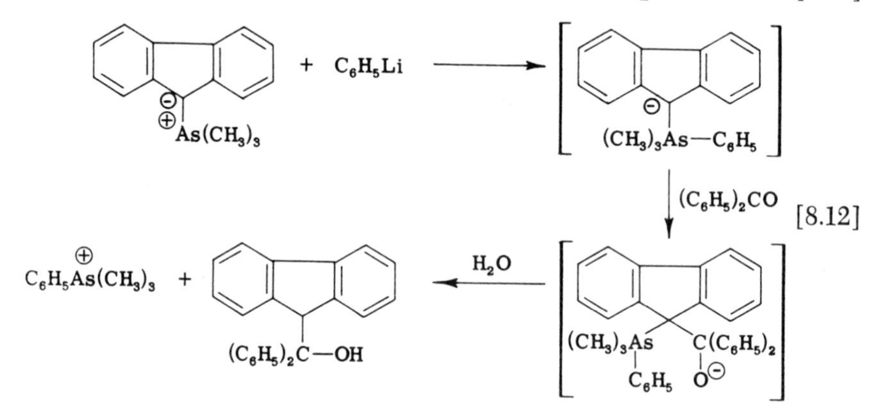

to account for the observed products. There is but little experimental data on which to base a realistic evaluation of this mechanistic proposal. Suffice it to mention that there is no indication that organolithium reagents will attack the 'onium atom of an ylid. In addition, the hydrolytic cleavage of the pentavalent arsenic has no precedent. The results of the reaction are somewhat surprising in view of the later data available on the reaction of similar arsonium ylids with carbonyl compounds. Clearly, this reaction needs to be reinvestigated.

Several years after the report of the above reaction Johnson (*24*) prepared and isolated fluorenylidenetriphenylarsenane (VI). He found that this ylid reacted with a variety of carbonyl compounds in a normal Wittig reaction to afford the expected olefin and triphenylarsine oxide. The ylid showed some selectivity in its reactions. For example, it re-

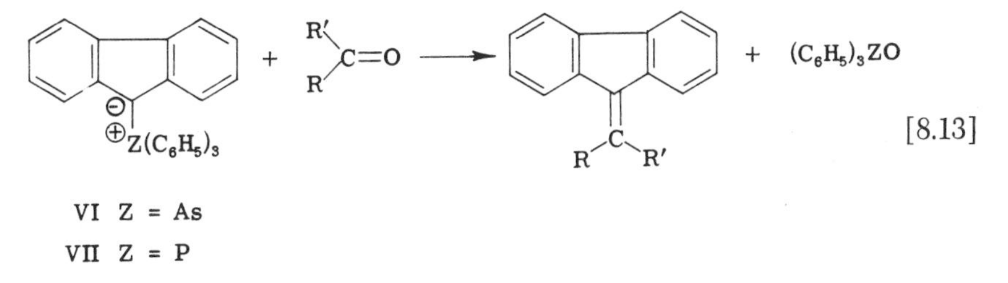

acted with a series of substituted benzaldehydes in 89–97% yields and reacted with acetaldehyde in 91% yield to afford ethylidenefluorene. However, VI would not react with acetone or benzophenone but it did react with *p*-nitroacetophenone. The direct olefin-forming ability of VI contrasts with the reported failure of fluorenylidenetrimethylarsenane to exhibit the same type of reaction (*25*). The behavior of the arsenane (VI) paralleled very closely that of the analogous phosphorane (VII). It was apparent, however, that the arsenane was somewhat more reactive than the phosphorane. For example, VI reacted with *p*-dimethylaminobenzaldehyde to form *p*-dimethylaminobenzalfluorene in 97% yield whereas the phosphorane (VII) failed to react with the same aldehyde under the same conditions.

W. Heffe, one of Krohnke's students at Berlin, apparently discovered the reaction between arsonium ylids and carbonyl compounds to form olefins and arsine oxides in 1937 but the work was not published except in his dissertation (*30*). He found that phenacyltriphenylarsonium bromide reacted with benzaldehyde in the presence of sodium hydroxide to afford benzalacetophenone, probably via the intermediacy of phenacylidenetriphenylarsenane (IV).

The mechanism of the reaction of arsonium ylids with carbonyl compounds was proposed to be the same as that for the Wittig reaction of phosphonium ylids, involving, as the slow step of the reaction with stabilized ylids, a nucleophilic attack of the ylid carbanion on the carbonyl carbon (*24*). The fact that the arsoniumfluorenylide (VI) was more reactive than the phosphoniumfluorenylide (VII) was attributed to a higher electron density on the carbanion of the former, due mainly to a decreased tendency for electron delocalization from the carbanion into the vacant *d*-orbitals of the arsonium atom. This conclusion was consistent with the observation, discussed in part B of this section, that the arsoniumfluorenylide also was more basic than the phosphoniumfluorenylide. In other words, both the reactivity data and the basicity data indicated that in the fluorenylide series at least, the tetravalent arsenic atom was less able or willing to undergo valence shell expansion than was an identically substituted tetravalent phosphorus atom (*24*, Johnson). This conclusion is at variance with that reached by Doering and Hoffmann (*15*) on the basis of deuterium exchange data and with that reached by Chatt *et al.* (*11, 14*) and Magee *et al.* (*12*) on the basis of infrared studies of metal carbonyls, all of whom believed that there was virtually no difference in the valence shell expanding abilities of arsenic and phosphorus.

At essentially the same time as the publication of Johnson's data on the fluorenylide (VI), Henry and Wittig (*17*) reported on the reaction of the simplest known arsonium ylid, methylenetriphenylarsenane (V), with benzophenone. Triphenylmethylarsonium bromide or iodide was treated with an equimolar amount of phenyllithium to generate the ylid (V) in ethereal solution. An equivalent amount of benzophenone was added, the reaction was carried out for twelve hours at 60° and the products were isolated by hydrolysis and chromatography. In a typical run using 10 mmoles of reactants the following products were obtained: 1.1 mmoles of unreacted benzophenone, 6.9 mmoles of triphenylarsine, 6.8 mmoles of diphenylacetaldehyde, 2.0 mmoles of triphenylarsine oxide and 2.1 mmoles of diphenylethylene. In all runs the triphenylarsine and diphenylacetaldehyde were formed in equimolar quantities and the triphenylarsine oxide and diphenylethylene were formed in equimolar quantities. It appeared, therefore, that two different reactions were occurring simultaneously. Henry and Wittig (*17*) advanced the mechanism in [8.14] to account for the observed products.

The reaction was thought to commence as for a normal Wittig reaction with the attack of the nucleophilic ylid on the carbonyl carbon to form a betaine intermediate (VIII). Transfer of oxygen from carbon to arsenic then would lead to the formation of triphenylarsine oxide and diphenylethylene in equimolar amounts as proposed by Johnson (*24*). The major portion of the betaine was thought to undergo an internal displacement reaction with the ejection of triphenylarsine and the forma-

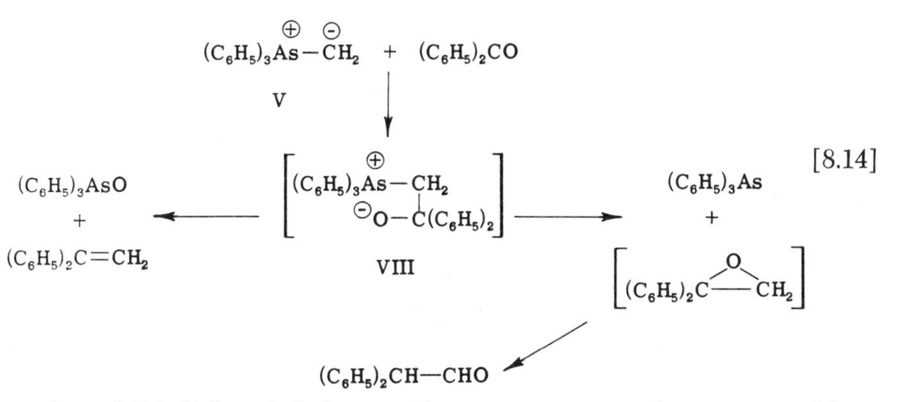

[8.14]

tion of 1,1-diphenylethylene oxide, again in equimolar amounts. There was adequate analogy for this latter proposal in the earlier report by Johnson and LaCount (*31*) that sulfonium ylids reacted with carbonyl compounds to form oxiranes with the ejection of a sulfide from a betaine intermediate (see Chapter 9). However, Wittig and Henry (*17*) did not isolate any oxirane, proposing instead that under the conditions of the reaction it rearranged to diphenylacetaldehyde. Johnson and LaCount (*31*) had shown that epoxides would not rearrange to aldehydes in the presence of triphenylarsine or triphenylarsine oxide under mild conditions and Cope *et al.* (*32*) had shown the rearrangement not to proceed under basic conditions. It would proceed under acidic conditions, however, and it should be noted that Henry and Wittig (*17*) worked up their reaction by subjecting it to hydrolysis with 6 *N* hydrochloric acid. Thus, it is likely that the oxirane was, in fact, the product of the reaction but not of the isolation.

The simultaneous formation of olefins and oxiranes in the reaction of arsonium ylids with carbonyl compounds subsequently has been verified by the isolation of *p*-nitrostilbene and *p*-nitrostilbene oxide from the reaction of benzylidenetriphenylarsenane with *p*-nitrobenzaldehyde

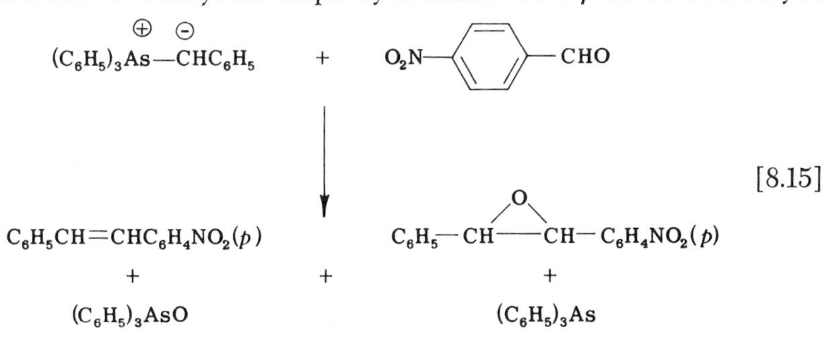

[8.15]

[8.15] (*28*, Johnson). The oxide and the olefin were formed in about equal amounts and each was accompanied by an equimolar quantity of the respective arsenic-containing by-product, the arsine or its oxide. These results tend to verify the supposition that the initially formed product in the reaction of methylenetriphenylarsenane (V) with benzophenone (*17*) was diphenylethylene oxide and that it rearranged, probably during workup, to diphenylacetaldehyde.

It appears that the arsonium ylids hold a position intermediate between the phosphonium ylids and sulfonium ylids in the course of their reaction with carbonyl compounds. All three ylids appear to react identically with the carbonyl compounds initially to form an intermediate betaine (IX, Z = 'onium group). At this point, however, the three betaines differ in their behavior. The phosphonium betaines($Z = R_3P$)

$$\overset{\oplus}{Z}-\overset{\ominus}{C}\langle \;+\; O{=}C\langle \;\longrightarrow\; \left[\begin{array}{c}\overset{\oplus}{Z}-C\langle \\ | \quad\;\; | \\ \overset{\ominus}{O}-C\langle\end{array}\right]_{\mathbf{IX}} \begin{array}{l}\xrightarrow{(a)} \overset{\oplus}{Z}-\overset{\ominus}{O} \;+\; \rangle C{=}C\langle \\ \xrightarrow{(b)} Z \;+\; \rangle C\overset{O}{-}C\langle \text{ (oxirane)}\end{array} \qquad [8.16]$$

follow path *a*, involving a transfer of oxygen to the 'onium group. The apparent driving force for this step is the formation of the phosphorus-oxygen bond, a rather high energy bond (*33*). On the other hand, the sulfonium betaines($Z = R_2S$) follow path *b*, involving displacement by the oxyanion on the carbon carrying the 'onium group (see Chapter 9). Apparently the potential formation of a sulfur-oxygen bond is not sufficient to alter the course of the reaction to path *a*.

Interestingly, the arsonium betaines($Z = R_3As$) apparently can follow either path *a* or *b* or both. In the triphenylarsonium series, the fluorenylide (*24*) led to exclusive olefin formation(path *a*), the benzylide (*28*) led to nearly equimolar formation of oxirane and olefin (paths *a* and *b*), and the methylide (*17*) led to dominant formation of the oxirane. Too little work has been reported to date to warrant a detailed mechanistic proposal to account for these observations. It appears, however, that potential formation of an arsenic-oxygen bond is not a sufficient driving force to control the course of the reaction. It should be noted that since the same arsine oxide or arsine was formed in the three examples quoted above, the nature of the heteroatom component cannot be affecting the ratio of path *a* product to path *b* product. It is conceivable that the potential conjugation of the carbanion substituents with the incipient double bond may control the course of the reaction.

Before the reaction of arsonium ylids with carbonyl compounds can be used in synthesis with any hope of reliability, the mechanistic intricacies of the reaction will have to be solved. This certainly is a fertile area for a concerted mechanistic study.

The course of the reaction of arsonium ylids with nitrosobenzene also indicates that there is less tendency to form an arsenic-oxygen bond than a phosphorous-oxygen bond. Johnson and Martin (*28*) found that whereas fluorenylidenetriphenylphosphorane (VII) and nitrosobenzene afforded fluorenone anil and triphenylphosphine oxide (i.e. normal Wittig-type reaction with oxygen transfer to phosphorus), fluorenylidene-triphenylarsenane (VI) and nitrosobenzene afforded *N*-phenyl fluorenone ketoxime and triphenylarsine. In the latter case oxygen was not transferred to arsenic, the oxyanion instead displaced triphenylarsine by attack on carbon. In a similar manner benzylidenetriphenylarsenane and nitrosobenzene afforded triphenylarsine and *N*-phenyl benzaldoxime. The reaction of arsonium ylids with nitrosobenzene to form nitrones followed the same course as the reaction of sulfonium ylids with nitrosobenzene (see Chapter 9).

III. Iminoarsenanes

As was the case for phosphonium systems, iminoarsenanes (arsonium imines, $R_3As^+—N^-—R'$) are isoelectronic with arsonium ylids. Accordingly, similarities should exist between the chemistry of the ylids and the imines. This expectation has not been tested adequately since there is even less known about arsonium imines than about arsonium ylids. However, what experimental information is available tends to bear out this expectation.

The simplest known example of this class of compounds is imino-triphenylarsenane (X) which was not prepared until 1960. Appel and Wagner (*34*) found that treatment of triphenylarsine with chloramine led to triphenylaminoarsonium chloride [8.17]. The imine was obtained

$$(C_6H_5)_3As + ClNH_2 \longrightarrow (C_6H_5)_3\overset{\oplus}{As}NH_2\ Cl^{\ominus} \xrightarrow[NH_{3(l)}]{NaNH_2} \underset{\textbf{X}}{(C_6H_5)_3\overset{\oplus}{As}—\overset{\ominus}{N}H} \qquad [8.17]$$

by treating the salt with sodamide in liquid ammonia. It was a colorless solid which could be handled in the atmosphere but was less stable than the corresponding phosphonium imine. Treatment with hydrogen chloride regenerated the conjugate acid. The imine could be acylated with acyl halides, a reaction typical of ylids and imines (*34*).

At a much earlier date (1937) Mann and Chaplin (*35*) observed that treatment of triphenylarsine with sodium chloramine-T led to the

$$(C_6H_5)_3As + CH_3C_6H_4SO_2\overset{\ominus}{N}(Cl)\ Na^{\oplus} \longrightarrow \underset{\textbf{XI}}{(C_6H_5)_3\overset{\oplus}{As}—\overset{\ominus}{N}SO_2C_6H_4CH_3(p)} \qquad [8.18]$$

formation of *N*-tosyliminotriphenylarsenane (XI). Wittig and Hellwinkel (*4*) recently have verified this report. Mann (*36*) earlier had provided some mechanistic information on a closely related reaction. He found that treatment of tri(*p*-tolyl)arsine with chloramine-T in aqueous medium led to the same product that could be obtained by treating the dihydroxide or oxide of the same arsine with *p*-toluenesulfonamide, the product being *N*-tosylaminotriphenylarsonium hydroxide rather than the expected arsine imine (formation of quaternary hydroxides appears to be the normal behavior of arsine imines in aqueous media). Mann found that the arsines rather than condensing directly with chloramine-T were being oxidized by the latter to the arsine oxide, the imine being formed by subsequent reaction with *p*-toluenesulfonamide [8.19].

$$(C_6H_5)_3As \xrightarrow{\text{chloramine-T}} (C_6H_5)_3AsO \xrightarrow{TsNH_2} XI \qquad [8.19]$$

Practically no chemistry has been reported for the arsonium imines. Mann (*36*) noted that they could be hydrolyzed to an arsine oxide and amine derivative in the cold with aqueous base. Appel and Wagner (*34*) reported their acylation, and Wittig and Hellwinkel (*4*) reported that reaction of XI with phenyllithium led to the very interesting pentavalent arsenic derivative, pentaphenylarsenane. There have been no attempts to study the alkylation of imines or their reaction with carbonyl compounds.

Monagle (*37*) recently has expanded the carbodiimide synthesis of Campbell *et al.* (*38*) by demonstrating that arsine oxides may be used. For example, reaction of a catalytic amount of triphenylarsine oxide with phenyl isocyanate led to the formation of diphenylcarbodiimide, carbon dioxide and recovery of the triphenylarsine oxide [8.20]. If the

$$(C_6H_5)_3AsO + C_6H_5N{=}C{=}O \longrightarrow \left[\begin{array}{c}(C_6H_5)_3\overset{\oplus}{As}-O \\ | \\ C_6H_5-\underset{\ominus}{N}-C{=}O\end{array}\right] \longrightarrow (C_6H_5)_3As{=}NC_6H_5 + CO_2$$

$$(C_6H_5)_3As{=}NC_6H_5 + C_6H_5N{=}C{=}O \longrightarrow \left[\begin{array}{c}(C_6H_5)_3\overset{\oplus}{As}-N-C_6H_5 \\ | \\ {}^{\ominus}O-C{=}NC_6H_5\end{array}\right] \longrightarrow (C_6H_5)_3AsO + C_6H_5N{=}C{=}NC_6H_5 \qquad [8.20]$$

proposed mechanism for the reaction is correct, *N*-phenyliminotriphenylarsenane must be a key intermediate in the reaction. It must undergo a normal carbonyl condensation reaction analogous to the Wittig reaction. The kinetic data available for the triphenylphosphine oxide case indicated that the formation of the phosphinimine was the slow step of the reaction. Therefore, it has been impossible to intercept the phosphinimine or arsinimine from these reactions.

IV. Stibonium Ylids and Imines

So little is known regarding the chemistry of stibonium ylids and imines that these two groups will be discussed in the same section.

Wittig and Laib (*25*) were the first to undertake a study of stibonium ylids. They were frustrated in their attempts to prepare trimethylstiboniumfluorenylide because they could not alkylate trimethylstibine with 9-bromofluorene. However, they did succeed in preparing dimethyldibenzylstibonium bromide. Treatment of this salt with an ethereal solution of phenyllithium afforded a yellow coloration which rapidly

$$(C_6H_5CH_2)_2\overset{\oplus}{Sb}(CH_3)_2 \xrightarrow{C_6H_5Li} \left[C_6H_5\overset{\ominus}{C}H-\overset{\oplus}{Sb}(CH_3)_2-CH_2C_6H_5 \right] \longrightarrow C_6H_5CH(CH_2C_6H_5)-Sb(CH_3)_2 \qquad [8.21]$$

faded [8.21]. Upon quenching the reaction mixture with water then adding methyl iodide they obtained trimethyl(1,2-diphenylethyl)stibonium iodide, the apparent result of a Stevens rearrangement.

In later work, Henry and Wittig (*17*) were able to prepare triphenylmethylstibonium salts by alkylating triphenylstibine with trimethyloxonium tetrafluoroborate. The corresponding ylid, triphenylstiboniummethylide (XII), was prepared in solution by treating the salt with ethereal phenyllithium. Reaction of the methylide (XII) with benzophenone led to the formation of very high yields of triphenylstibine and

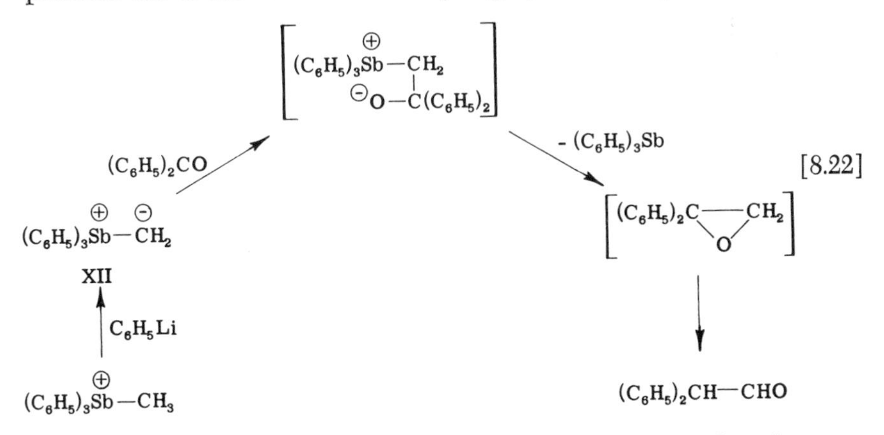

diphenylacetaldehyde. This reaction was carried out under the same conditions mentioned for the analogous arsonium methylide (V) and also was worked up with a hydrochloric acid hydrolysis. Therefore, in view of the discussion in the arsonium case (Section II,C of this chapter) it is likely that the original product of the reaction between the methylide (XII) and benzophenone was 1,1-diphenylethylene oxide. The mechanism of this reaction was assumed to be analogous to that proposed for the arsonium case.

No other work has been reported on the chemistry of stibonium ylids.

Equally little has been done with stibonium imines. Appel (39) verbally has reported the preparation of triphenylstiboniumimine by reaction of sodamide with triphenylstibine dichloride at elevated temperatures. This imine was reported to be the least thermally and hydrolytically stable of the various heteroatom imines. No experimental data have been published as yet.

Wittig and Hellwinkel (5) reported the preparation of triphenylstibonium-*N*-tosylimine by reaction of triphenylstibine with chloramine-T under conditions analogous to those used in the arsenic case. The imine was stable to atmospheric conditions but could be hydrolyzed to triphenylstibine oxide (or dihydroxide) in water. This same group reacted the imine with phenyllithium to produce the antimony analog (pentaphenylantimony) of the interesting series of pentaphenyl derivatives of the group V elements.

$$(C_6H_5)_3Sb + TsNHCl \longrightarrow (C_6H_5)_3\overset{\oplus}{Sb}-\overset{\ominus}{N}-Ts \xrightarrow{C_6H_5Li} (C_6H_5)_5Sb \quad [8.23]$$

$$(C_6H_5)_3\overset{\oplus}{Sb}-\overset{\ominus}{N}C_6H_5$$

XIII

Monagle (37) has examined the catalytic effect of triphenylstibine oxide on the conversion of phenyl isocyanate to diphenylcarbodiimide. The stibine oxide was found to be more effective than the phosphine oxide but less effective than the arsine oxide. The mechanism of the conversion was assumed to be identical to that using the arsine and phosphine oxide. If such is the case, it implies that triphenylstibonium-*N*-phenylimine (XIII) was an intermediate in the reaction and was capable of reacting with carbonyl groups in a Wittig-type condensation.

Obviously, there is room for considerable research into the chemistry of stibonium ylids and imines. A most pertinent question is why the stibonium ylids appear to afford oxiranes with carbonyl compounds while the imines afford unsaturated products.

REFERENCES

1. A. F. Wells, "Structural Inorganic Chemistry," p. 663, Oxford University Press, London and New York, 1962.
2. D. P. Craig and E. A. Magnusson, *J. Chem. Soc.* 4895 (1956).
3. G. Wittig and K. Clauss, *Ann.* **577**, 26 (1952).
4. G. Wittig and D. Hellwinkel, *Chem. Ber.* **97**, 769 (1964).
5. G. Wittig and D. Hellwinkel, *Chem. Ber.* **97**, 789 (1964).
6. M. Rouault, *Ann. Phys. Lpz.* **14**, 78 (1940).
7. S. Ohlberg, *J. Am. Chem. Soc.* **81**, 811 (1959).
8. G. Teuffer, *Acta Cryst.* **9**, 539 (1956).
9. P. J. Wheatley and G. Wittig, *Proc. Chem. Soc.* 251 (1962).
10. P. J. Wheatley, *J. Chem. Soc.* 3718 (1964).
11. J. Chatt and F. A. Hart, *J. Chem. Soc.* 1378 (1960).
12. T. A. Magee, C. N. Matthews, T. W. Wang and J. H. Wotiz, *J. Am. Chem. Soc.* **83**, 3200 (1961).
13. F. Basolo and R. G. Pearson, *in* "Progress in Inorganic Chemistry" (F. A. Cotton, ed.), Vol. 4, pp. 381–453, Interscience Publishers, New York, 1962.
14. J. Chatt and R. G. Wilkins, *J. Chem. Soc.* 4300 (1952).
15. W. von E. Doering and A. K. Hoffmann, *J. Am. Chem. Soc.* **77**, 521 (1955).
16. L. Pauling, "The Nature of the Chemical Bond," 3rd ed., p. 93, Cornell University Press, Ithaca, New York, 1960.
17. M. C. Henry and G. Wittig, *J. Am. Chem. Soc.* **82**, 563 (1960).
18. G. Wittig and K. Clauss, *Ann.* **578**, 136–146 (1952).
19. A. Michaelis, *Ann.* **321**, 174 (1902).
20. F. Krohnke, *Chem. Ber.* **83**, 291 (1950).
21. N. A. Nesmeyanov, V. V. Pravdina and O. A. Reutov, *Doklady. Akad. Nauk. S.S.S.R.* **155**, 1364 (1964); *Proc. Acad. Sci. U.S.S.R.* **155**, 424 (1964).
22. S. O. Grim and D. Seyferth, *Chem. & Ind. (London)* 849 (1959).
23. D. Seyferth and H. M. Cohen, *J. Inorg. and Nuclear Chem.* **20**, 73 (1961).
24. A. Wm. Johnson, *J. Org. Chem.* **25**, 183 (1960).
25. G. Wittig and H. Laib, *Ann.* **580**, 57 (1953).
26. L. Horner and H. Oediger, *Chem. Ber.* **91**, 437 (1958); *Ann.* **627**, 142 (1959).
27. A. Wm. Johnson and R. B. LaCount, *Tetrahedron* **9**, 130 (1960).
28. A. Wm. Johnson and J. O. Martin, *Chem. & Ind. (London)* 1726 (1965).
29. G. M. Phillips, J. S. Hunter and L. E. Sutton, *J. Chem. Soc.* 146 (1945).
30. W. Heffe, Dissertation, Berlin University (1937) [quoted by G. Wittig in *Pure and Applied Chem.* **9**, 249 (1964)].
31. A. Wm. Johnson and R. B. LaCount, *Chem. & Ind. (London)* 1440 (1958); *J. Am. Chem. Soc.* **83**, 417 (1961).
32. A. C. Cope, P. A. Trumbull and E. R. Trumbull, *J. Am. Chem. Soc.* **80**, 2844 (1958).
33. A. F. Bedford and C. T. Mortimer, *J. Chem. Soc.* 1622 (1960).
34. R. Appel and D. Wagner, *Angew. Chem.* **72**, 209 (1960).
35. F. G. Mann and E. J. Chaplin, *J. Chem. Soc.* 527 (1937).
36. F. G. Mann, *J. Chem. Soc.* 958 (1932).
37. J. J. Monagle, *J. Org. Chem.* **27**, 3851 (1962).
38. T. W. Campbell, J. J. Monagle and V. S. Foldi, *J. Am. Chem. Soc.* **84**, 3673 (1962).
39. R. Appel, *Abstr. Papers, 142nd Meeting, Am. Chem. Soc.*, p. 38N (1962).

9

SULFUR YLIDS

I. Introduction

In view of the evidence and discussion presented in Chapter 2 concerning the existence and stability of phosphorus ylids it would be expected that any other molecular system containing the same general structural features and a heteroatom group which was capable of providing adequate stabilization for a carbanion should form an ylid of some finite existence. Since delocalization of the electrons of an ylid carbanion into the vacant 3*d*-orbitals of phosphorus appeared to be a major factor contributing to the stability of phosphorus ylids, a search for new ylids logically might be expected to focus on those heteroatoms which, like phosphorus, have low energy vacant orbitals. Such reasoning was behind the recent re-awakening of interest in the chemistry of sulfonium ylids (*1*, Johnson and LaCount). It has led to interesting studies on many aspects of the chemistry of sulfur ylids which will be explored in the following pages.

A. Theoretical Arguments for *d*-Orbital Overlap

The general bonding characteristics of sulfur will not be discussed since several recent works have dealt with the subject in depth (*2*, Burg; *3*, Price and Oae). However, a brief discussion of the nature and role in bonding of the 3*d*-orbitals of sulfur must precede a study of the chemical and physical properties of sulfur ylids. Considerable experimental evidence and theoretical discussion has accumulated in recent years regarding the ability of sulfur to expand its outer shell to accommodate more than eight electrons (*4*, Cilento).

Sulfur belongs to the VIB subgroup and has the configuration $3s^2 3p_x^2 3p_y^1 3p_z^1$ in the outer shell of the free atom. In its divalent state the $3p_y$ and $3p_z$ orbitals are filled by electron sharing but there remain five empty 3*d*-orbitals. It is conceivable that overlap of a filled 2*p*-orbital of an ylid carbanion with a vacant 3*d*-orbital of sulfur would provide some measure of stability for that carbanion.

In their classic theoretical paper Craig *et al.* (5), using arguments based on overlap integrals, showed that *d*-orbitals in atoms such as phosphorus and sulfur could be used for *p-d* π bonding. It was apparent, however, that if the free sulfur atom were taken as the model for sulfur in its bonding state the size of the 3*d*-orbitals was such that they would be too diffuse to play an important role in bonding. For example, the distance from the nucleus to the radial maximum of the atomic orbitals of sulfur was calculated as 1.5 atomic units for the 3*p*-orbitals but 5.5 atomic units for the 3*d*-orbitals (6, Craig). The second value is far too large to permit effective bonding involving a 3*d*-orbital. In a later model Craig and Magnusson (7) described the polarization of such diffuse orbitals by electronegative substituents such that overlap of the *contracted* *d*-orbitals with an adjacent 2*p*-orbital would provide effective bonding. It is evident, however, that the energy of promotion of an electron into a *d*-orbital also would be an important factor in determining the effectiveness of the bonding. The promotional energy is related to the nuclear charge so that as the nuclear charge increases, the promotional energy decreases to the point where it requires less energy than is supplied by the resulting bond formation. Therefore, for maximum bonding using the *d*-orbitals of sulfur, the latter's substituents should be strongly electronegative and/or the sulfur atom should carry positive charge. Consistent with this picture is the observation that the tetrafluoride and hexafluoride of sulfur, both of which must use *d*-orbitals for σ bonding, are known but the corresponding chlorides and bromides are unknown. This description of the use of 3*d*-orbitals in bonding also is consistent with the more recent views of bonding in sulfinyl and sulfonyl systems (8, Gillespie and Robinson).

On the basis of the above theoretical picture of the use of sulfur 3*d*-orbitals in bonding, it could be predicted that maximum delocalization for a carbanion should occur if the α-sulfur atom carried a full unit positive charge. Such would be the case in an α-sulfonium carbanion,

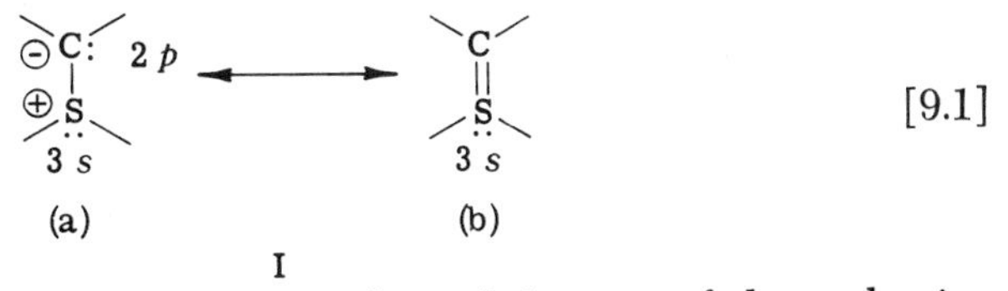

i.e. a sulfonium ylid (I). In that case, the stabilization of the carbanion probably would be due partly to electrostatic and partly to resonance (conjugative) interactions. The coulombic stabilization afforded the carbanion by the adjacent positively charged sulfur atom in Ia should be reinforced by delocalization of the electron pair into the outer shell of the sulfur atom. This delocalization should be permitted by overlap of the filled $2p$-orbital of carbon with the vacant $3d$-orbital of sulfur, the lone pair on sulfur remaining in a $3s$-orbital (Ib).

Cruickshank and Webster (*9*) and others recently have argued that contraction of the $3d$-orbitals need not be proposed to account for bonding in molecules such as sulfur hexafluoride. Using the Hartree-Fock SCF method they have calculated the $3d$-orbits to lie even nearer the nucleus in the ground state of sulfur than was predicted by Craig and Magnusson (*7*) after contraction in a ligand field. There seems to be little disagreement that $3d$-orbitals of sulfur can be used for bonding but considerable disagreement in accounting for it on theoretical grounds.

B. Experimental Evidence for *d*-Orbital Overlap

Considerable experimental evidence has accumulated which indicates that a sulfonium salt is extremely effective in stabilizing an adjacent carbanion. Some estimate of the degree to which this stabilization is due to conjugative interactions has been made available by comparing the physical and chemical properties of ternary sulfonium salts with those of quaternary ammonium salts. These comparisons all involve the assumption that the coulombic effects of a sulfonium and ammonium group are similar and that any difference in their behavior is due mainly to the conjugative electron-accepting ability of the sulfur atom. The validity of this assumption has been questioned though not too seriously (*10*, Doering and Hoffmann).

Bordwell and Boutan (*11*) have shown that while *m*-trimethylammoniophenol is more acidic than its *para* isomer, presumably due to inductive effects, *m*-dimethylsulfoniophenol is less acidic than its *para* isomer. The greater acidity of the *para* isomer is satisfactorily accounted for by proposing conjugative delocalization of the free electron pair of the phenoxide into the $3d$-orbitals of the sulfur atom (II). Such a conjugative effect could not operate in the *meta* isomer or in either of the ammoniophenols. The same authors found that the dimethylsulfonium group was much more effective than the methylthio group in increasing

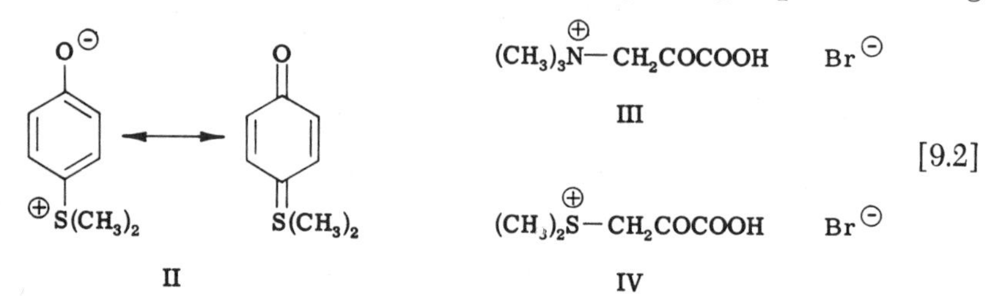

the acidity of phenol. This observation clearly is in line with the arguments of Craig and Magnusson (*7*).

In a similar vein Blau and Stuckwisch (*12*) found that while trimethylammoniopyruvic acid bromide (III) was a monobasic acid, the analogous dimethylsulfoniopyruvic acid bromide (IV) was a dibasic acid and the methyl ester of IV was a monobasic acid of pK_a 5.5. Apparently the methylene group of IV, but not of III, was acting as an acid. Once again the activating effect of the sulfonium group was attributed to stabilization of the conjugate base of IV and its ester by delocalization of the electron pair on the methylenic carbon into the 3*d*-orbitals of sulfur.

Since the ammonium or trimethylammonium group has little or no effect on the ultraviolet spectrum of benzene or other aromatic hydrocarbons (*13*, Doub and Vandenbelt; *14*, Fehnel and Carmack), the observation that the dimethylsulfonio group shifted the primary band of benzene from 203.5 to 220 mμ was striking (*11*). This indicates that the dimethylsulfonium group does conjugate with the aromatic π system in the photoexcited state. It is reasonable to expect this conjugation to be of an electron-accepting nature.

Doering and Schreiber (*15*) have shown that vinyldimethylsulfonium salts undergo typical Michael addition reactions whereas the corresponding vinyltrimethylammonium salts are relatively inert under the same conditions. For example, there is at least a 10^5 difference in the rate of addition of hydroxide ion to the two salts. The reaction is presumed to involve addition of the hydroxide anion to the β-carbon of the vinyl group to form an intermediate carbanion of low energy (V). The source

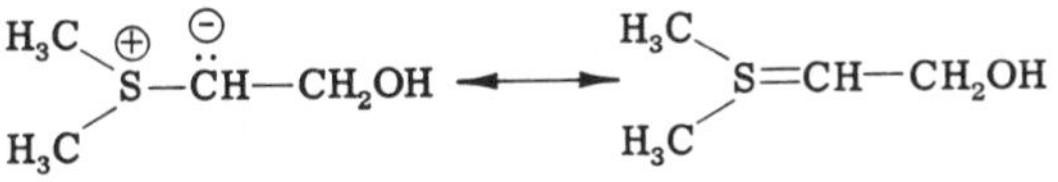

$$(H_3C)_2\overset{\oplus}{S}-\overset{\ominus}{\ddot{C}}H-CH_2OH \longleftrightarrow (H_3C)_2S{=}CH-CH_2OH \qquad [9.3]$$

V

of the stabilizing influence in the sulfonium intermediate, absent in the ammonium case, would appear to be overlap of the 2*p*-orbital of the carbanion in this intermediate (or a transition state similar in structure, i.e. with a high electron density on the α-carbon) with a vacant 3*d*-orbital of sulfur. The vinyl sulfonium salt was shown to react with a variety of nucleophiles, including ethoxide ion, sodiomalonic ester and methyl mercaptide ion.

Doering and Hoffmann (*10*) have presented some quantitative data on the extent to which a sulfonium group can provide stabilization for an adjacent carbanion. Trimethylsulfonium iodide was found to incorporate 98 atom per cent of deuterium from an equimolar solution of deuteroxide in deuterium oxide held at 62° for three hours. Under the same conditions, but for 504 hours, tetramethylammonium iodide underwent no detectable exchange. Using the data from the ammonium salt as their base line they calculated that the heat of activation for deuterium exchange in the sulfonium salt was lowered by 17.2 kcal/mole from the value expected if the only stabilization of the intermediate carbanion was due to coulombic interactions. However, it was apparent that the difference in the rates of exchange of the sulfonium and ammonium salts was due to both heat and entropy factors. The entropy factor was attributed to a release of solvent which accompanied a lowering of the dipole moment as the salt was converted into the less polar intermediate ylid, methylenedimethylsulfurane. The heat factor was attributed to the sulfonium group providing effective conjugative stabilization for the adjacent carbanion, most likely through orbital overlap of the $2p$-$3d$ π type. The same effects, perhaps differing only in degree, would be expected if the exchange mechanism proceeded by a concerted, one-step transition state since there still would be some localization of electron density on the carbon atom.

In summary it must be stated that in spite of the foregoing examples it is virtually impossible to *prove* that valence shell expansion of sulfur is involved in the stabilization of α-sulfonio carbanions. At best the accumulated weight of evidence is "accounted for" by invoking the use of vacant, low energy $3d$-orbitals. Until evidence is provided which is not accountable in a like manner, the use of such an explanation certainly is justifiable and in the best scientific tradition.

On the basis of the foregoing examples there appears to be considerable experimental evidence to back up the theoretical predictions of Craig *et al.* (5–7) that a sulfonium group should provide effective stabilization for an adjacent carbanion by an electron-accepting conjugative mechanism. It was logical, therefore, that the initial interest in the chemistry of sulfur ylids should have focused on sulfonium ylids, especially in view of the detailed information available regarding the analogous phosphonium ylids. However, before examining the chemistry of sulfonium ylids in detail it should be pointed out that other sulfur-containing compounds might be expected to afford stabilization sufficient for a carbanion to have appreciable stability and exhibit characteristic ylid properties. The only requirement should be for the sulfur atom to have a degree of positive charge sufficient to provide for contraction of the $3d$-orbitals and subsequent efficient overlap with a carbon $2p$-orbital. Several of these groups have been shown to exhibit characteristic conjugative electron-withdrawing properties and a few of them have been incorporated into ylids.

The sulfone group long has been known to exert strong electron attraction (*3*). That some of this effect is conjugative in nature is evident from the fact that *p*-methylsulfonylphenol is more acidic than the *meta* isomer (*16*, Bordwell and Cooper). Vinylsulfones also are known to undergo typical nucleophilic addition reactions, and bis(sulfonyl)alkanes are readily alkylated in the presence of hydroxide ion. As a result of this evidence both Corey (*17*) and Johnson (*18*) have initiated investigations into the properties of sulfonyl ylids (VI). On the basis of similar evidence regarding conjugative electron attraction by the sulfoxide group (*3*) both Corey (*17, 19*) and Walling (*20*) have studied the properties of sulfinyl ylids (VII). The observation (*21*, Smith and Winstein) that trimethylsulfoxonium nitrate underwent even faster exchange

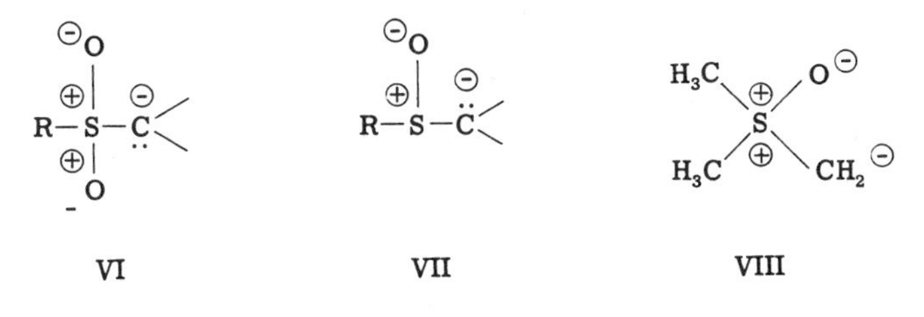

of deuterium than did trimethylsulfonium iodide probably led to Corey's study (*22*) of the unique ylid, methylenedimethyloxysulfurane (VIII).

Other sulfur-containing groups may be capable of supporting ylid formation. From studies of the electronic effects on the acidity of the anilinium ion and the observation that in each instance the σ values for the substituents were more positive in the *para* than in the *meta* position, Zollinger *et al.* (*23, 24*) concluded that both the amidosulfonyl ($-SO_2NH_2$) and the sulfonate ($-SO_3^-$) substituents were capable of affording conjugative stabilization to adjacent electron-rich groups. Others (*25*, Bredereck *et al.*) have claimed, on the basis of very meager and dubious evidence, that the methyl sulfinate group ($-SO_2CH_3$) acts as an electron-accepting group. The usefulness of these latter groups in ylid formation has not been investigated.

To summarize then, there are four types of sulfur ylids known at the present time. They are classified as the sulfonium ylids, oxysulfonium ylids, sulfonyl ylids and sulfinyl ylids. The chemistry of these ylids will be discussed in the following pages, and most of the emphasis will be on the first group. Coincidental with the development of sulfur ylid chemistry has been the slower and less spectacular evolution of sulfidimine (iminosulfurane) chemistry, that of sulfur-nitrogen systems isoelectronic with sulfur ylid systems. For example, the sulfidimines ($R_2S^+—N^-—R$) are isoelectronic with sulfonium ylids and the oxysulfidimines $[R_2S^+(O)—N^-—R]$ are isoelectronic with oxysulfonium ylids. The chemistry of these systems also will be discussed in this chapter.

II. Sulfonium Ylids

Sulfonium ylids, having a general structure IX, are the oldest known class of sulfur ylids. Interestingly enough the impetus for the synthesis of the first sulfonium ylid, fluorenylidenedimethylsulfurane (X), came from repeated failures to isolate the closely related nitrogen ylid, trimethylammoniumfluorenylide (XI) (*26*, Ingold and Jessop). In recent

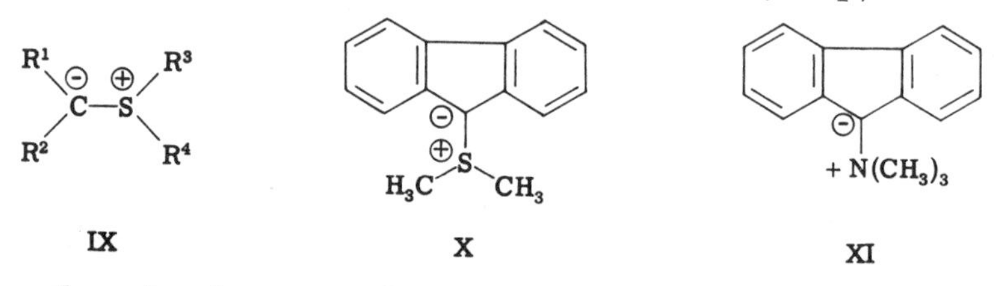

years there has been considerable emphasis on the chemistry of sulfonium ylids but now mainly due to their close structural and electronic relationship to the well-known phosphonium ylids.

As indicated in the introduction to this monograph it is proposed that sulfonium ylids be named as derivatives of a hypothetical sulfurane, SH_4. Thus, compound X is fluorenylidenedimethylsulfurane or 9-dimethylsulfuranylfluorene. The alternate system which has been used almost exclusively to date would name X as 9-dimethylsulfoniumfluorenylide. The former method is consistent with that recently adopted for phosphonium ylids and based on phosphorane (PH_5).

A. Methods of Synthesis

Only two general methods have been discovered to date for the synthesis of sulfonium ylids. The most general of these, the so-called "salt method," simply involves the removal of an α-hydrogen from a sulfonium salt by a base of the requisite strength. In principle any sulfonium salt carrying at least one α-hydrogen is convertible into an ylid [9.4]. For

$$(R^1)(R^2)CH-\overset{\oplus}{S}(R^3)(R^4)\ X^{\ominus} \xrightarrow{\text{base}} BH^{\oplus} + (R^1)(R^2)\overset{\ominus}{C}-\overset{\oplus}{S}(R^3)(R^4) \qquad [9.4]$$

$$(CH_3)_3S^{\oplus}\ Br^{\ominus} \xrightarrow{CH_3Li} (CH_3)_2\overset{\oplus}{S}-\overset{\ominus}{C}H_2 + CH_4 \qquad [9.5]$$

example, treatment of trimethylsulfonium bromide with an ethereal solution of methyllithium resulted in the evolution of methane and the formation of methylenedimethylsulfurane in solution (*27*, Wittig and Fritz).

In practice the salt method only is applicable in three structural situations. In the first instance, all three of the groups attached to sulfur must be identical so that it makes no difference which α-hydrogen is removed by the base. The method also is applicable in those cases in which one or two of the substituents have no α-hydrogen but those groups which do are identical. For example, it has been possible to prepare methylenephenylmethylsulfurane by treating dimethylphenylsulfonium salts with strong bases (*28*, Franzen and Driessen). The only other structural situation in which the sulfonium salt method is applicable necessitates there being an appreciable difference in the acidity of the various available α-hydrogens and the availability of a base of the proper strength. For example, dimethyl-9-fluorenylsulfonium salts have been converted to the corresponding fluorenylide (X) in good yield upon treatment with aqueous sodium hydroxide (*26*). In this instance the α-hydrogen on C-9 of the fluorenyl group is more acidic than the methyl hydrogens by at least 10 pK_a units. Likewise, it has been possible to convert dimethylbenzhydrylsulfonium bromide into benzhydrylidenedimethylsulfurane using butyllithium (*29*, Johnson).

There is insufficient data available to provide an accurate estimate of the difference in pK_a of the various α-hydrogens that is necessary before selective formation of one of several possible ylids from a given sulfonium salt could be achieved. However, it would be impractical to attempt to form ethylidenedimethylsulfurane from dimethylethylsulfonium salts since the system undoubtedly would contain some methylenemethylethylsulfurane [9.6].

$$CH_3CH_2\overset{\oplus}{S}(CH_3)_2 \xrightarrow{\text{base}} CH_3\overset{\ominus}{C}H-\overset{\oplus}{S}(CH_3)_2 + \overset{\ominus}{C}H_2-\overset{\oplus}{S}\begin{matrix} CH_3 \\ C_2H_5 \end{matrix} \quad [9.6]$$

A wide variety of bases has been used to effect the removal of the α-hydrogen from sulfonium salts. Probably the most common has been the alkyl- or aryllithiums. Corey (*30*) recently has introduced the methylsulfinyl carbanion ($CH_3SOCH_2^-$) as a versatile base for the generation of a variety of ylids from their corresponding salts. Other bases such as the triphenylmethyl anion, the *tert*-butoxide anion, other alkoxides and sodium hydride also have been used. It is difficult to obtain an accurate measure of the effectiveness of these various bases since few of the sulfonium ylids are isolable. However, most of the conversions from salt to ylid appear to proceed in high yield, such as the conversion of

9-fluorenyldimethylsulfonium bromide to ylid (X) in 82% yield (*31*). The strength of the base needed depends on the acidity of the sulfonium salt. These range from the fluorenyl salt requiring sodium carbonate to the trimethylsulfonium salt requiring the methylsulfinyl carbanion but being subjected to bases varying in strength up to methyllithium.

The applicability of the salt method for the preparation of sulfonium ylids depends on the availability of the appropriate sulfonium salts. The latter usually are prepared by direct alkylation of a sulfide with an alkyl halide. Dialkylsulfides are sufficiently powerful nucleophiles that the alkylation usually takes place under mild conditions, in the absence of catalysts to afford high yields of the sulfonium salts. For example, 9-bromofluorene and methylsulfide reacted with gentle warming in nitromethane solution to afford 9-fluorenyldimethylsulfonium bromide in quantitative yield (*26, 31*). Because of its availability at low cost from commercial sources methyl sulfide often is used as the source of the sulfur atom. However, the use of such a dialkylsulfide is not always practical since the resulting sulfonium salt then has a multiplicity of α-hydrogens and could form a mixture of two ylids upon treatment with base. Therefore, such a procedure is practical only when there is appreciable difference in the acidity of the α-hydrogens.

To avoid the above difficulty it seems preferable to use a sulfide that has no α-hydrogens. The simplest such sulfide would be phenylsulfide and the resulting salts, diphenylalkylsulfonium salts, would be structurally analogous to the common triphenylalkylphosphonium salts. Until just recently, however, it has been practically impossible to prepare such salts, at least by an alkylation method, since phenylsulfide is too weak a nucleophile to undergo alkylation by alkyl groups using common leaving groups, i.e. halides, tosylates, etc.

The synthesis of a diphenylalkylsulfonium salt could, in theory, be carried out via two substitution routes. On the one hand, a phenylalkylsulfide could be phenylated. Such a reaction has been developed by Makarova and Nesmeyanov (*32*) who found that phenylsulfide, when heated with a diphenyliodonium salt, was converted to a triphenylsulfonium salt but in rather low yield [9.7]. However, preliminary attempts

$$(C_6H_5)_2S + (C_6H_5)_2I^{\oplus}\ BF_4^{\ominus} \xrightarrow{\Delta} (C_6H_5)_3S^{\oplus}\ BF_4^{\ominus} \qquad [9.7]$$

to phenylate phenylmethylsulfide with diphenyliodonium tetrafluoroborate have been unsuccessful (*33*, Johnson).

The other possible route to a diphenylalkylsulfonium salt is alkylation of phenylsulfide. From previous statements it is clear that normal alkylation procedures are useless. However, Franzen *et al.* (*34*) and

Johnson *et al.* (*35*) discovered that phenylsulfide could be alkylated in good yield by treating it with a mixture of an alkyl bromide and silver tetrafluoroborate, the latter reagent probably facilitating the ionization of the carbon-halogen bond. In this manner, diphenylbenzylsulfonium tetrafluoroborate has been isolated in 73–84% yields and the diphenyl-*n*-butyl salt was obtained in 54–56% yields [9.8] (*34, 35*). The discovery

$$(C_6H_5)_2S + R'RCHBr \xrightarrow{AgBF_4} R'RCH-\overset{\oplus}{S}(C_6H_5)_2 \xrightarrow{base} R'R\overset{\ominus}{C}-\overset{\oplus}{S}(C_6H_5)_2 \quad [9.8]$$

of this procedure has cleared the way for the preparation virtually of any given diphenylalkylsulfonium salt. Such salts then can be converted to ylids of unambiguous structure since there is now only one type of α-hydrogen capable of being removed by base.

Franzen *et al.* (*36*) have reported a new reaction which may have potential in the synthesis of sulfonium salts. The addition of methylsulfide to a solution containing benzyne (dehydrobenzene) followed by acidification with perchloric acid afforded a 77% yield of dimethylphenylsulfonium perchlorate. They proposed that the nucleophilic sul-

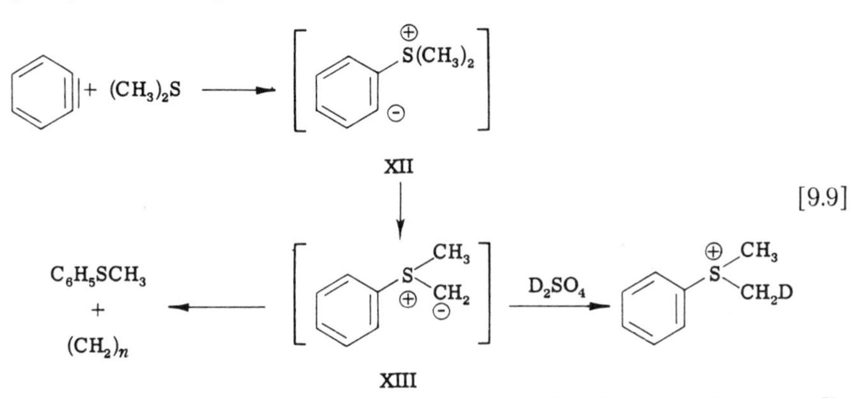

fide attacked benzyne to form a zwitterion (XII) which then rapidly underwent a prototropic shift to form the ylid, methylenemethylphenylsulfurane (XIII). Such a shift would be expected in view of the relative basicity of the two carbanions. Quenching of the reaction with deuterosulfuric acid afforded the methyl-d salt rather than the phenyl-d salt, indicating that the ylid (XIII) must have been protonated. Allowing a solution of the ylid to stand undisturbed afforded thioanisole and polymethylene, products to be expected from the decomposition of XIII.

It would appear that the above reaction could be applied readily to the synthesis of a wide variety of diphenylsulfoniumalkylides. Treatment of benzyne with an appropriate alkylphenylsulfide should afford, after the initial proton transfer, a solution of the diphenylsulfoniumalkylide

[9.10]. This would be a direct synthesis and would avoid having to iso-

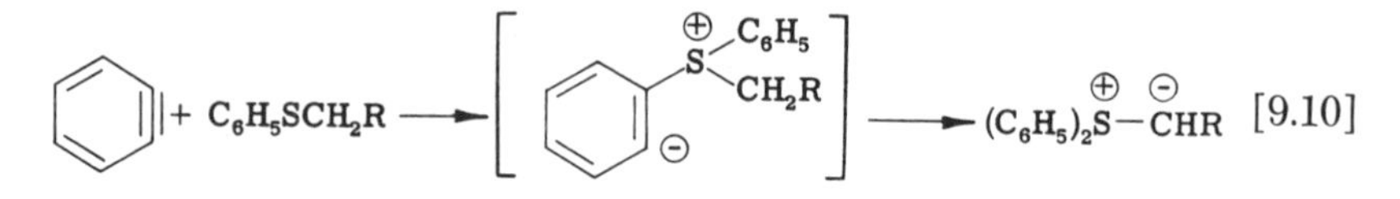

late a sulfonium salt and carry out ylid reactions in the presence of strong base. Furthermore, the alkylphenylsulfides needed for such a synthesis are readily available by alkylation of benzenethiol.

Franzen *et al.* (*36*) did apply the benzyne method to a long-chain sulfide, di-*n*-octylsulfide. They found that 1-octene (70%) and *n*-octylphenylsulfide (53%) were the major products, and they were unable to obtain di(*n*-octyl)-phenylsulfonium perchlorate upon treatment of what should have been a solution containing *n*-octylidenephenyl-*n*-octylsulfurane with perchloric acid. They surmised that the initially formed zwitterion (XIV) underwent an intramolecular elimination reaction.

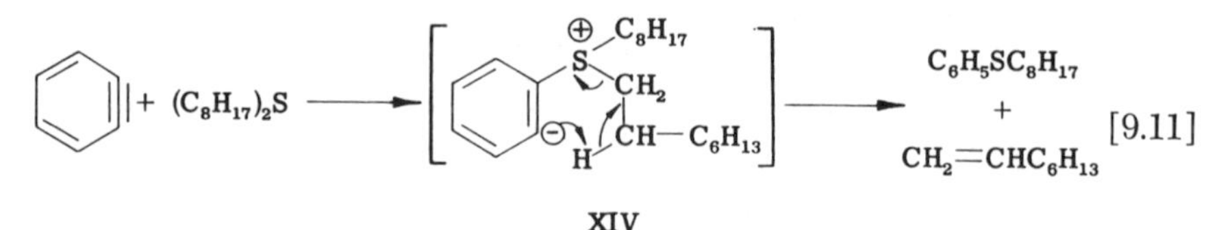

Hellmann and Eberle (*37*) also explored this reaction with several sulfides but were able to isolate only alkylphenylsulfides or Stevens rearrangement products.

While there are only the two methods available for the synthesis of sulfonium ylids at present, the salt method and the benzyne method, and only the former of any generality, they have been adequate to date. Further work in this area should be directed to making available a method which completely avoids basic media since the preparation of functionally substituted ylids may be difficult under presently available conditions.

In Table 9.1 is compiled a listing of all known sulfonium ylids together with an indication of the method employed for their preparation. Most of these ylids have been generated in solution and used without isolation. Those which have been isolated have been marked with an asterisk.

B. Physical Properties of Sulfonium Ylids

Relatively little is known about the physical properties of sulfonium ylids since only four have been isolated to date, fluorenylidenedimethylsulfurane (*26, 31*), 2-nitrofluorenylidenedimethylsulfurane (*38*), 2,7-

TABLE 9. 1

Known Sulfonium Ylids
($R^1R^2S{=}A$)

A[a]	Method of preparation	Base used	Ref.
Alkylenedimethylsulfuranes			
$=CH_2$	Salt	Alkyllithium	*27*
		Butoxide	*28, 66, 71*
		$CH_3SOCH_2^-$	*30*
$=CHCOOC_2H_5$	Salt	NaH	*72*
* $=CHCOC_6H_5$	Salt	NaOH	*33*
$=C(C_6H_5)_2$	Salt	C_4H_9Li	*29*
* $=C(o\text{-}C_6H_4)_2$	Salt	NH_3	*26*
		NaOH	*31*
* $=C(o\text{-}C_6H_4)_2\text{-}2\text{-}NO_2$	Salt	NH_3	*38*
		NaOH	*29, 33*
* $=C(o\text{-}C_6H_4)_2\text{-}2\text{-}NO_2\text{-}7\text{-}NO_2$	Salt	NaOH	*38*
Alkylenephenylmethylsulfuranes			
$=CH_2$	Salt	Butoxide	*28, 71*
$=CH_2$	Benzyne	–	*36*
* $=CHCOC_6H_5$	Salt	$(C_2H_5)_3N$	*39*
Alkylenediphenylsulfuranes			
$=CH{-}CH_3$	Salt	C_6H_5Li	*68*
$=CH{-}C_3H_7$	Salt	Trityl anion	*34*
		C_4H_9Li	*35, 68*
$=CH{-}CH(CH_3)_2$	Salt	Trityl anion	*34*
$=CH{-}C_6H_5$	Salt	Trityl anion	*34*
		C_4H_9Li	*35, 58*

[a] The ylids marked with an asterisk have been isolated and characterized.

dinitrofluorenylidenedimethylsulfurane (*38*) and phenacylidenemethylphenylsulfurane (*39*). All of the other sulfonium ylids have been prepared and used *in situ* without any attempt to isolate and characterize them.

Fluorenylidenedimethylsulfurane (X) exhibited solubility behavior typical of a covalent organic compound. It was soluble in the common organic solvents—ether, benzene, chloroform—and insoluble in water. The ylid dissolved in dilute hydrochloric acid solutions by virtue of the formation of the conjugate acid, a sulfonium salt, isolated as its picrate (*26*). The salt could be reconverted to the ylid upon treatment with base. The ylid could act as an indicator since it was yellow and its conjugate acid was colorless. An even sharper contrast in colors prevailed

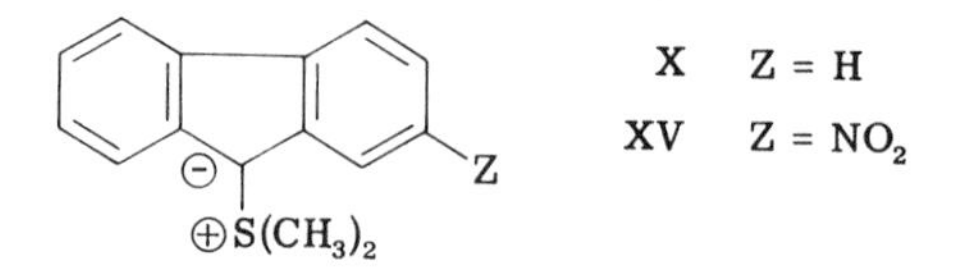

between 2-nitrofluorenylidenedimethylsulfurane (XV), a purple substance, and its colorless conjugate acid (*38*). The ylids can be titrated since Hughes and Kuriyan (*38*) used this method to determine the molecular weight of 2,7-dinitrofluorenylidenedimethylsulfurane.

The pK_a of 9-fluorenyldimethylsulfonium bromide in 31.7% water-dioxane, determined spectrophotometrically, was shown to be 7.3 (*31*, Johnson and LaCount). This indicates that the ylid (X) is a weak base and that the salt is an unusually strong carbon acid. By comparison the pK_a of phenol under the same conditions was 9.6. Therefore, the ylid was less basic than the phenoxide ion and only slightly more basic than the triacetylmethide anion. This very low basicity is a striking reflection of the ability of the dimethylsulfonium group to stabilize an adjacent carbanion and, when compared to the high basicity of trimethylammoniumfluorenylide (XI) (*26*, *40*), provides powerful evidence for the importance of an electron-accepting conjugative role for the sulfonium group.

The ultraviolet spectrum of fluorenylidenedimethylsulfurane (X) in chloroform solution was very similar to the spectra of other phosphonium and arsonium fluorenylides (*31*). The only significant difference was the hypsochromic shifting of the long wavelength band by 12 mμ to 368 mμ.

In view of the unique charge distribution that must exist in ylids a study of their dipole moments should be significant. At an early date, Phillips *et al.* (*41*) reported the dipole moment of the fluorenylide (X) to be 6.2 D. On the basis of this determination they concluded that the carbon-sulfur bond was single and the carbon electrons were unavailable to the sulfur atom because of their involvement with the aromatic system. Ingold (*42*) agreed "that the unshared carbon electrons are not absorbed into the sulfur atom to a great extent, even though the latter has unoccupied orbitals." On the other hand, Craig (*43*) felt that there was considerable interaction between the carbon electrons and the sulfur *d*-orbitals, otherwise the electron density would have been entirely diffused over the aromatic system resulting in a much higher dipole moment.

In our view it is difficult to envision the unshared electrons on the ylid carbon *not* being delocalized both into the vacant *d*-orbitals of the sulfur atom as well as into the π-electron system of the benzenoid rings of the fluorene nucleus, that is contributions of resonance forms such as X, XVI and XVII. If the major contributing resonance forms were X and XVII, that is overlap of the carbanion electrons into the vacant 3*d*-orbitals of the sulfur atom (XVI) were relatively unimportant, one would expect trimethylammoniumfluorenylide (XI) to be less basic and more stable than it is (*26, 40*). In other words, any stabilization due to the ammonium or sulfonium groups then would have to be due mainly

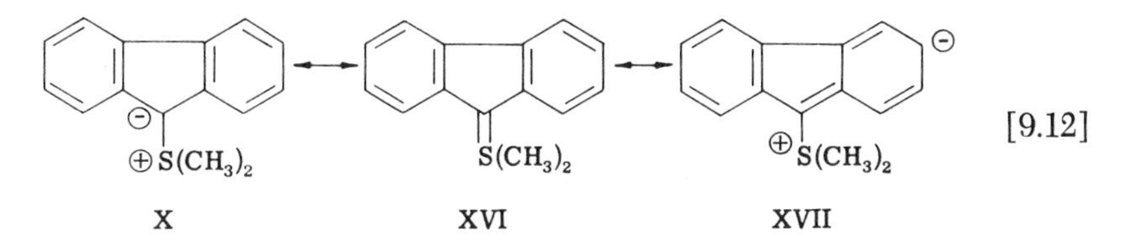

to coulombic forces. After comparing the normal carbon-nitrogen bond distance (1.47 Å) with the carbon-sulfur bond distance (1.81 Å) (*44*, Huggins) one would expect that coulombic stabilization would be more effective in the case of the ammonium ylid (XI) making it the least basic of the two. Just the opposite was found, indicating that there probably is conjugative stabilization by the sulfonium group and that contribution of resonance structures such as XVI is important.

Using a similar argument it is possible to demonstrate that resonance contributing forms such as XVII also are of major importance. In comparing the properties of fluorenylidenedimethylsulfurane (X) with methylenedimethylsulfurane it is apparent that the fluorenylide is both less basic and less nucleophilic as well as more stable (*28, 30, 31*). If the electron density on the ylid carbanion is delocalized throughout the fluorenyl system, this carbanion would be expected to have a lower electron density than the carbanion in the methylide which has no opportunity for delocalization of electron density other than through overlap with the *d*-orbitals of sulfur. From these arguments it is apparent that the attachment of any substituent to the ylid carbanion which can act as an electron acceptor should result in a lowering of the electron density on the ylid carbanion and be reflected by a decrease in the basicity and nucleophilicity of the sulfonium ylid. Incidentally, this often implies that it will be possible to isolate the ylid. Such has recently been the case for phenacylidenemethylphenylsulfurane (XVIII). The observation

$$(H_3C)(C_6H_5)\overset{\oplus}{S}-\overset{\ominus}{C}H-\overset{O}{\overset{\|}{C}}-C_6H_5 \longleftrightarrow (H_3C)(C_6H_5)\overset{\oplus}{S}-CH=\overset{O^{\ominus}}{\overset{|}{C}}-C_6H_5 \qquad [9.13]$$

XVIII

that the carbonyl stretching frequency was 1505–1470 cm^{-1} indicated the delocalization of the carbanion electrons through the carbonyl group (*39*, Nozaki *et al.*).

The argument on which we have expounded, that the resonance forms X, XVI and XVII, etc., all are important contributors to the electronic distribution in fluorenylidenedimethylsulfurane, also can be supported by a qualitative examination of the admittedly inadequate dipole moment data. Since there is no information available regarding the bond distances or bond angles in ylids it is impossible to carry out any quantitative calculations. However, if it is assumed that the ylid (X) is planar and using the known dipole moment of 1.4 D for methylsulfide (*45*, Smyth) the moment of the carbanion-sulfur bond could be estimated as $6.2 - 1.4 = 4.8$ D. Sutton *et al.* (*41*) estimated the carbon-sulfur moment to be 5.0 D in this ylid. This approach can be justified using the example of the sulfur-oxygen bond moment in sulfoxides. For example, methylsulfoxide has a moment of 3.9 D (*46*, Cotton and Francis) and simple subtraction of the methylsulfide moment gives an estimate of the sulfur-oxygen moment as 2.5 D. Considering the crudeness of the approach, this value is reasonably close to that of 2.8 D now accepted for the sulfur-oxygen moment in thionyl halides, sulfoxides, sulfones and sulfites (*47*, Pritchard and Lauterbur).

Using a carbon-sulfur single bond distance of 1.81 Å (*44*) the bond moment is calculated to be 8.6 D if the ylid is adequately represented by the dipolar structure X. Such obviously is not the case since the actual moment is about 5.0 D. If only structures X and XVII, etc., were used to describe the ylid, the apparent sulfur-carbon moment should be considerably larger than 8.6 D due to the greater distance separating the centers of charge density. However, if the contribution of structure XVI is increased relative to X and XVII a decrease in the sulfur-carbon moment would be expected. For example, if one third of the electron density on the carbanion were transferred to the sulfur atom with an appropriate shortening of the bond distance to about 1.7 Å, the bond moment would decrease to about 5.4 D, compared to the observed-estimated value of about 5.0 D. Therefore, it seems that the implications by Sutton *et al.* (*41*) and Ingold (*42*) that there is little delocalization from the carbanion to the sulfonium group are not valid.

It should be emphasized that the above calculated estimates are just exactly that, estimates. The electron distribution in the fluorenyl portion of the ylid has been completely neglected. It is apparent that information on the bond distances, bond angles and electron density about the ylid bond are sorely needed before any accurate description of the electronic distribution in the ylid can be produced. The problem of obtaining such information is complicated by the fact that in order to produce an ylid stable enough for study by X-ray crystallography, one must attach rather large groups to the ylid carbanion.

In view of the fact that fluorenylidenedimethylsulfurane (X) has a lower dipole moment than any of the other phosphonium or arsonium fluorenylides and has both a lower basicity and a lower nucleophilicity toward a carbonyl carbon than any of the other fluorenylides, it would seem logical to conclude that there is more double bond character to the sulfur-carbon bond than to a corresponding phosphorus-carbon or arsenic-carbon bond. In other words, the sulfonium group seems better able to provide stabilization for an adjacent carbanion through valence shell expansion than the phosphonium, arsonium and ammonium groups. This evidence, and the conclusion drawn from it, is in good agreement with the data provided by Doering and Hoffmann (*10*) on the rates of exchange of the permethyl 'onium salts with deuterium in deuteroxide which decreased in the order S > P > As > N. However, they attributed the observed decrease in the first three instances to the effect of increased bond distance with its attendant lowering of the coulombic stabilization of the ylid intermediate and therefore concluded that the contribution of *d*-orbital delocalization to the stability of the intermediate ylid was constant for sulfur, phosphorus and arsenic. While their calculations were made using meager kinetic data instead of heats of activation, they still indicated that the contribution of *d*-orbital stabilization to the exchange rates was in the order S (14.7 kcal) > P (13.8) > As (13.1) > N (0.0), in agreement with the evidence from the fluorenylides. Using more reliable data, the same authors (*10*) concluded that the heat of activation in the phosphonium case was lowered by 15.4 kcal while that of the sulfonium case was lowered by 17.2 kcal from the values expected if the stabilization of the respective ylids was due solely to coulombic forces and not to any contribution from valence shell expansion. Here again, sulfur seems better able to expand its outer shell than does phosphorus.

C. Chemical Stability of Sulfonium Ylids

In spite of the fact that few sulfonium ylids have been isolated and characterized, considerable information has been accumulated which bears on the chemical stability of these ylids.

Only a few sulfonium ylids are stable in the presence of water (those in Table 9.1 marked with an asterisk). All of the fluorenylides (*26, 31, 38*) were prepared in aqueous solution since their low basicity precluded reaction with water. However, the majority of sulfonium ylids have been prepared in non-aqueous media with rigorous precautions taken to prevent exposure to moisture. Most of these ylids are basic enough that reaction with water destroys the ylid, forming a sulfonium hydroxide. In many instances the reaction proceeded further with the hydroxide ion displacing methylsulfide from the salt and forming an alcohol [9.14]. Treatment of 9-fluorenyldimethylsulfonium bromide just

$$R_2\overset{\ominus}{C}-\overset{\oplus}{S}(CH_3)_2 \xrightarrow{H_2O} R_2CH-\overset{\oplus}{S}(CH_3)_2\ \overset{\ominus}{O}H \longrightarrow R_2CHOH + (CH_3)_2S \quad [9.14]$$

with boiling water resulted in the formation of fluorenol and methylsulfide (*26, 31*), and hydrolysis of benzyldimethylsulfonium tosylate with sodium hydroxide afforded benzyl alcohol (*48*, Swain and Thornton). It is apparent that sulfonium ylids usually must be handled in the absence of moisture.

An interesting example of the hydrolytic instability of a sulfonium ylid was discovered by Pinck and Hilbert (*49*). In spite of the fact that fluorenylidenedimethylsulfurane (X) was formed using an aqueous base, they found that further treatment of the ylid with alcoholic sodium hydroxide or with liquid ammonia resulted in the formation of 1-methylthiomethylfluorene (XXI). This rearrangement, discovered only shortly after Sommelet (*50*) reported the conversion of benzhydryltrimethylammonium hydroxide to *o*-benzylbenzyldimethylamine, became the second example of the Sommelet rearrangement. The structure of the rearranged thioether (XXI) was proven by degradation to fluorenone-1-carboxylic acid (*49*) and, much later, by synthesis from fluorene-1-car-

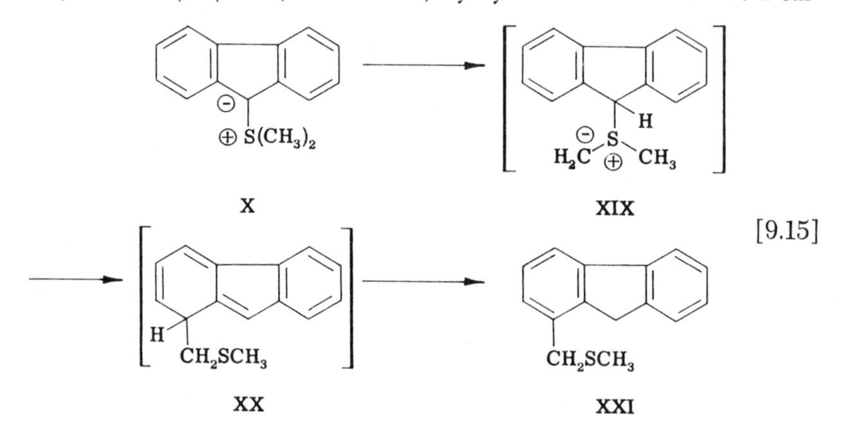

boxylic acid (*31*). Ingold proposed that the rearrangement proceeded by initial formation of a second, less stable sulfonium ylid (XIX) which then carried out a nucleophilic attack on the aromatic ring (*51*). A tautomeric shift would then result in the thioether (XXI). No proof has been offered for this mechanistic proposal although Johnson and LaCount (*31*) have discovered another example of the rearrangement but in a slightly different molecular environment.

The major problem faced in the handling of sulfonium ylids is the thermal instability of the ylid once formed. Some sulfonium ylids are stable only at low temperatures in the range of —40 to —70°C, while others are stable at room temperature but decompose on heating. At the other extreme, however, Nozaki *et al.* (*39*) claimed that phenacylidenemethylphenylsulfurane (XVIII) could not be thermally decomposed.

Hughes and Kuriyan (*38*) had shown that 2-nitrofluorenylidenedimethyl sulfurane (XV) was stable at room temperature for several days after which it slowly evolved methylsulfide. Upon heating a solution of the ylid in nitromethane, however, methylsulfide rapidly escaped and 2,2′-dinitrodifluorenylidene was formed [9.16]. Likewise, 2,7-dinitro-

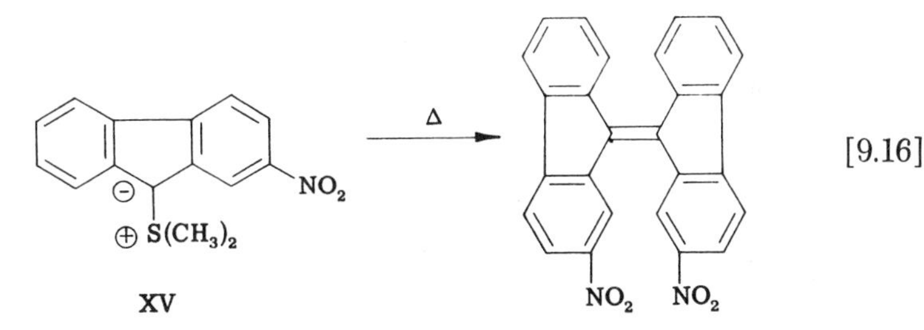

[9.16]

fluorenylidenedimethylsulfurane was thermally decomposed to 2,2′,7,7′-tetranitrodifluorenylidene. On the other hand, the unsubstituted fluorenylide (X) was stable, even in vacuo and under nitrogen, for but a few hours and then it too slowly evolved methylsulfide (*26, 31*). However, no difluorenylidene could be isolated; there was only an orange amorphous solid which defied characterization. Later, however, it was shown that difluorenylidene often was present as a minor by-product in a variety of ylid reactions. Therefore, it seems that even the so-called "stable" (isolable) sulfonium ylids can be thermally decomposed to a thioether and an olefin; the latter is a dimer of the hydrocarbon portion of the original ylid.

Several other investigators have observed analogous but spontaneous decomposition reactions with non-isolable sulfonium ylids. Both Corey and Chaykovsky (*30*) and Franzen and Driessen (*28*) noted that treatment of trimethylsulfonium salts with potassium *tert*-butoxide or the methylsulfinyl carbanion at room temperature led to the evolution of ethylene. The latter authors reported the same result with dimethylphenylsulfonium salts. Johnson *et al.* (*35*) found that solutions of benzylidenediphenylsulfurane were stable at —70° but slowly decolorized and formed a mixture of *cis*- and *trans*-stilbenes in 78% yield when the temperature was raised to —40°.

Several mechanisms have been advanced to account for the formation of olefin from the decomposition of a sulfonium ylid. Three of these are illustrated in [9.17] using the methylenedimethylsulfurane as an example.

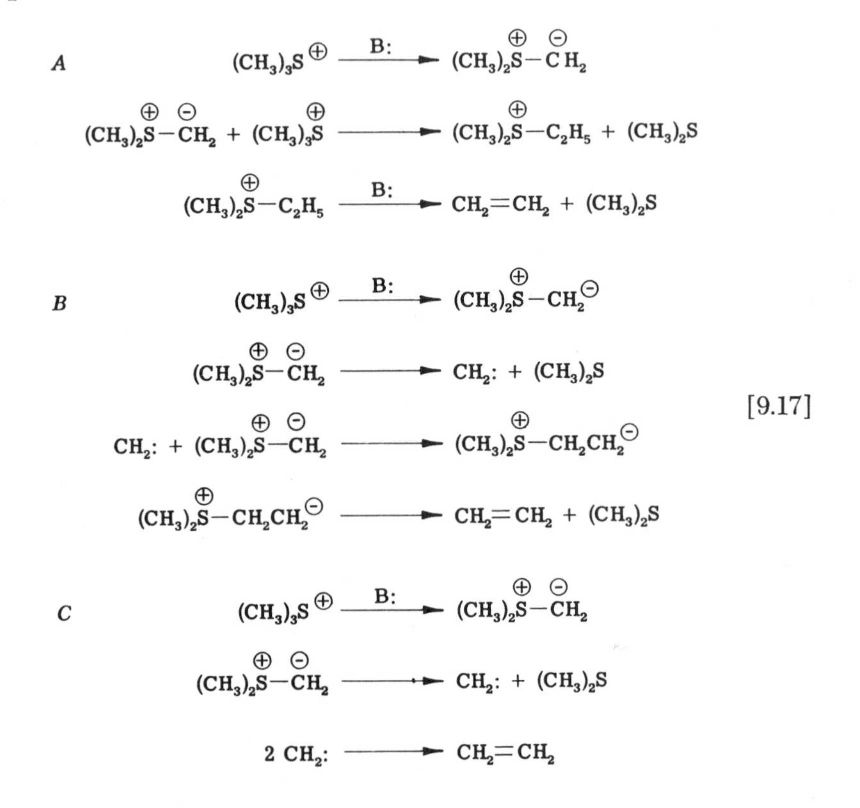

Mechanism *A* was proposed by Franzen and Driessen (*28*) for the specific case of the methylide. They eliminated any mechanism involving the formation of carbenes since they could find no products resulting from methylene insertion or addition reactions although no experimental details concerning their attempts were published. They noted that when the base (potassium *tert*-butoxide) was added to a solution of the sulfonium salt, olefin formation commenced almost immediately. On the other hand, if the sulfonium salt was added to a solution of the base, there appeared to be an induction period before ethylene was evolved. As a result they considered there had to be an excess of sulfonium salt present before the olefin could be formed and this demanded mechanism *A*. However, the results also could be accommodated by a mechanism similar to *B* in which the initially formed ylid decomposed to a carbene and then carried out an insertion reaction on trimethylsulfonium ion to form ethyldimethylsulfonium ion.

Franzen and Driessen (28) also observed that treatment of a solution of benzyl chloride and trimethylsulfonium salt in dimethylsulfoxide solution with three equivalents of potassium *tert*-butoxide afforded a 55% yield of styrene. This result was thought to support mechanism *A* in that the initially formed methylide was expected to displace chloride to form 2-phenylethyldimethylsulfonium chloride which then underwent Hofmann elimination to styrene [9.18]. However, these observations

$$C_6H_5CH_2Cl + (CH_3)_2\overset{\oplus}{S}-\overset{\ominus}{C}H_2 \longrightarrow (CH_3)_2\overset{\oplus}{S}-CH_2CH_2C_6H_5 \xrightarrow{B:} CH_2{=}CHC_6H_5 \quad [9.18]$$

also can be explained in several other ways. For example, benzyl chloride could undergo α-elimination to phenylcarbene which could effect an insertion on trimethylsulfonium ion or combine with methylenedimethylsulfurane. Alternatively, the methylide could fragment to methylene which could effect an insertion on benzyl chloride. All of these possibilities, in basic solution, would afford styrene as the end product.

That mechanism *B* is a feasible alternative to Franzen's proposals was demonstrated recently by Oda *et al.* (52). They showed that fluorenylidenedimethylsulfurane (X) reacted with chloroform in the presence of potassium *tert*-butoxide to afford 9-(dichloromethylene)-fluorene [9.19], probably via the initial formation of dichlorocarbene

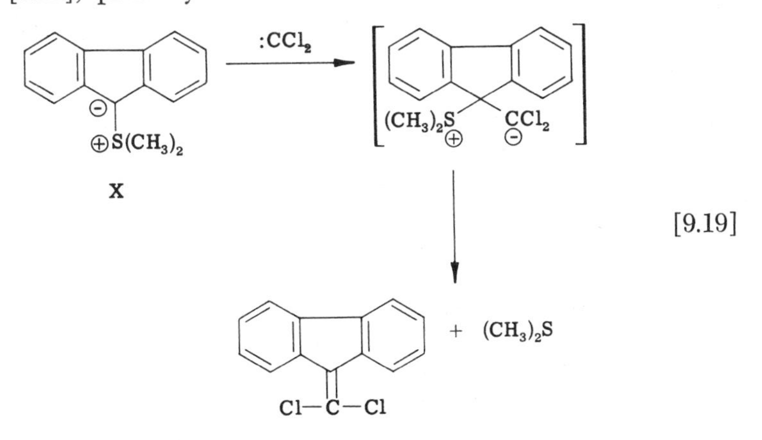

and its subsequent reaction with ylid as portrayed by mechanism *B*. The reaction clearly did not involve attack of the ylid on chloroform as required by mechanism *A* since there was no reaction in the absence of base (*31*). This is convincing evidence that a carbene will react with an ylid as required by mechanism *B*.

Swain and Thornton (*53*) recently have shown that in aqueous medium *p*-nitrobenzyldimethylsulfonium tosylate, when treated with sodium hydroxide at 60°, was converted quantitatively to a mixture of *cis*- and *trans*-4,4′-dinitrostilbene [9.20]. The reaction was shown to be

$$2\ (p)O_2NC_6H_4CH_2\overset{\oplus}{S}(CH_3)_2\ \overset{\ominus}{O}Ts \xrightarrow{\ominus OH} O_2NC_6H_4CH{=}CHC_6H_4NO_2 + 2\ (CH_3)_2S \qquad [9.20]$$

first order in each of the salt and hydroxide ions. When the reaction was carried out in deuterium oxide solution and the sulfonium salt isolated after only partial reaction, it was found that at least 97% of the methylene protons had been exchanged. Therefore, the conversion of salt to ylid was reversible and faster than any subsequent reaction (*54*, Rothberg and Thornton). Furthermore, since the reaction was second order and the first step fast, mechanism *A*, involving the attack of ylid on sulfonium salt, was eliminated since it would be a third-order reaction. The kinetic data were consistent with a rate-determining formation of a carbene from the sulfonium ylid, i.e. the α-elimination mechanisms *B* or *C*. Swain and Thornton (*53*) claimed that a mechanism such as *C*, involving dimerization of a carbene, was unlikely since the sulfur isotope effect in the hydrolysis was only 1.0066 ± 0.0008. However, it is conceivable that in the heterolysis of ylid to carbene the sulfur-carbon bond cleavage has not progressed very far in the transition state. Mechanism *B* perhaps is the most likely possibility in this case provided that the combining of carbene with ylid is fast.

The possibility that ylids could undergo α-elimination was recognized at an early date. Franzen and Wittig (*55*) claimed that treatment of tetramethylammonium salts with phenyllithium in the presence of cyclohexene led to the formation of norcarane, presumably via the fragmentation of trimethylammoniummethylide into trimethylamine and methylene, the latter adding to cyclohexene. This claim has been modified recently but there is still evidence for carbene formation from nitrogen ylids (*56*, Wittig and Krauss).

In 1961 Franzen *et al.* (*34*) attempted to demonstrate the heterolysis of sulfonium ylids into carbenes. They found that treatment of *n*-butyldiphenylsulfonium salts with trityl sodium afforded the following products: diphenylsulfide, triphenylmethane (42%), 1,1,1-triphenylpentane (15%) and, after bubbling the evolved gases through bromine in chloroform, 1,2-dibromobutane (7%), presumably from 1-butene. They claimed that the triphenylpentane was formed by an insertion reaction of

n-butylidene on triphenylmethane and that 1-butene was formed via the same carbene [9.21].

$$(C_6H_5)_2\overset{\oplus}{S}-\overset{\ominus}{C}HCH_2CH_2CH_3 \longrightarrow [C_3H_7CH:] \xrightarrow{H} CH_3CH_2CH=CH_2$$
$$[C_3H_7CH:] \xrightarrow{(C_6H_5)_3CH} (C_6H_5)_3C-CH_2C_3H_7 \qquad [9.21]$$

In another experiment, the same authors found that isobutyldiphenylsulfonium salts, in the presence of the same base, afforded a 35% yield of a 3:1 mixture of isobutylene and methylcyclopropane. Since that time Friedman and Berger (*57*) have shown that reaction of 1,1-dideuteroisobutyl chloride with phenylsodium afforded a mixture of isobutylene and

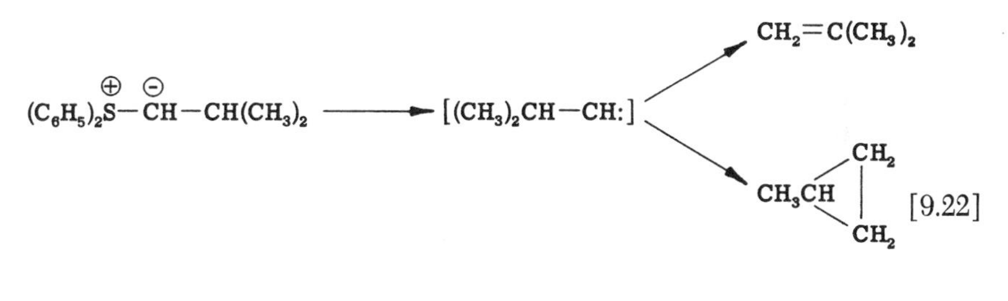

$$(CH_3)_2CHCD_2Cl \xrightarrow{C_6H_5Na} \underset{CDH}{\overset{CH_2}{|}}\!\!>CH-CH_3 + CDH=C(CH_3)_2$$

methylcyclopropane consisting of 89% monodeuterated material [9.22]. This indicated that the majority of the products were formed by an α-elimination, presumably involving a carbene intermediate. Therefore, the analogous products obtained by Franzen *et al.* (*34*) certainly provide adequate reason to suspect the intervention of a carbene mechanism in the decomposition of sulfonium ylids. However, the isolation of most of his products does not *demand* a carbene mechanism. In the case of n-butyldiphenylsulfonium salt, the triphenylpentane and 1-butene can be accounted for by simple displacement and elimination mechanisms, respectively. These workers (*34, 36*) also failed to trap any carbenes with cyclohexene as had been reported earlier for ammonium ylids (*55, 56*).

Johnson *et al.* (*35, 58*) may have been successful in their attempts to trap a carbene formed from a sulfonium ylid. They found that benzylidenediphenylsulfurane (XXII) was stable in solution at —70° under a nitrogen atmosphere. However, raising the temperature to —40° led to a loss of the yellow color of the ylid and the eventual isolation of a mixture of *cis*- and *trans*-stilbene (78%). Formation of the ylid but in the presence of an equivalent of acenaphthylene led to the isolation of a 43% yield of cyclopropane adduct (XXIV). The fact that no stilbenes

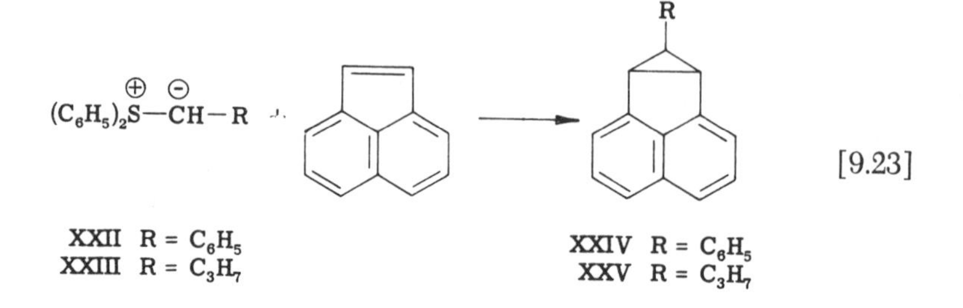

were found indicates that their precursor was trapped, an observation consistent with mechanism *B* and *C* but not mechanism *A*. The analogous reaction with *n*-butylidenediphenylsulfurane (XXIII) also led to the expected cyclopropane adduct (XXV). The initial conclusion was that the sulfonium ylids did dissociate to phenyl sulfide and a carbene, the latter carrying out a normal addition reaction with the olefin. However, it also is possible that the ylids themselves carried out a nucleophilic addition to the double bond [9.24]. In the case of a highly

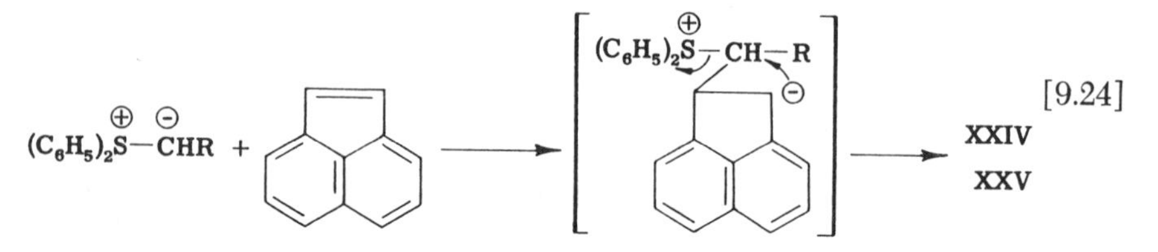

conjugated olefin such as acenaphthylene, the intermediate carbanion could be stabilized by resonance. In fact Corey and Chaykovsky (*30*) have reported the reaction of methylenedimethylsulfurane with diphenylethylene afforded 60% of 1,1-diphenylcyclopropane, and Mechoulam and Sondheimer (*59*) reported the addition of phosphonium ylids to dibenzofulvenes. On the other hand, there is no evidence that these latter two reactions did not proceed via a carbene addition to the olefinic group!

The likelihood of the ylid (XXII) attacking acenaphthylene directly is reduced when it is noted that fluorenylidenedimethylsulfurane (X) did not react with this olefin whereas 9-diazofluorene did so in high yield (*33*, Johnson). The latter is known to be a carbene precursor (*60*, Schonberg *et al.*). Therefore, cyclopropane formation in the reaction of XXII with acenaphthylene probably did involve carbene addition to the double bond.

It is abundantly clear that considerable work is required before the question of the intermediacy of carbenes in the decomposition of sulfonium ylids to olefins can be settled. It appears impossible to eliminate completely either of three possible mechanisms (*A*, *B* or *C*) with the information available at the present time. However, the weight of the evidence presented to date appears to favor a carbene mechanism, either *B* or *C*, the former being the more feasible.

Another type of thermal decomposition product has been formed from sulfonium ylids. Wittig and Fritz (*27*) found that treatment of trimethylsulfonium salts with either methyllithium or phenyllithium afforded high yields of amorphous polymethylene. Later, Franzen *et al.* (*36*) noted that methylenephenylmethylsulfurane, formed via the addition of methylsulfide to benzyne, also would decompose to polymethylene [9.25]. The latter group was unable to obtain any norcarane when

$$(CH_3)_2\overset{\oplus}{S}-\overset{\ominus}{C}H_2 \longrightarrow (CH_3)_2S + (CH_2)_n$$

$$(C_6H_5)(H_3C)\overset{\oplus}{S}-\overset{\ominus}{C}H_2 \longrightarrow C_6H_5SCH_3 + (CH_2)_n \qquad [9.25]$$

cyclohexene was added to the ylid solution so they concluded that the ylid was not heterolyzing to a carbene. As an alternate mechanism Franzen *et al.* proposed that polymethylene was formed by a series of methylene transfers between ylid and a growing zwitterion [9.26]. They

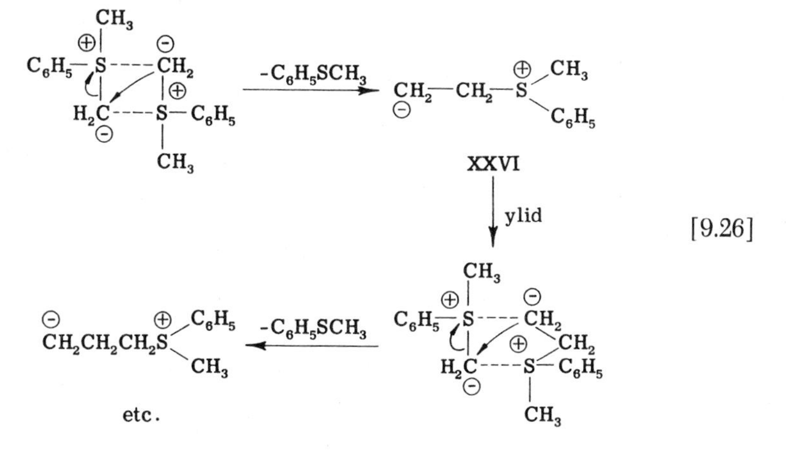

envisioned such transfers taking place via an initial ion-ion complex. However, this mechanism seems unlikely for two reasons. First, a site of high electron density is attacking another such site and, second, the resulting zwitterions, e.g. XXVI, are not ylids, do not have any stabilizing influences and most likely would undergo a spontaneous fragmentation to ethylene and thioanisole.

A more likely mechanism would involve successive attack of ylid on a homologating sulfonium salt [9.27]. Such a process would be an ionic

$$\begin{array}{ccc} \overset{\oplus}{R_2S}-CH_3 + \overset{\ominus}{CH_2}-\overset{\oplus}{SR_2} & \xrightarrow{-R_2S} & CH_3CH_2\overset{\oplus}{SR_2} \\ & & \downarrow \text{ylid} \\ CH_3(CH_2)_n-\overset{\oplus}{SR_2} & \xleftarrow{\text{etc.}} & \overset{\oplus}{R_2S}-CH_2CH_2CH_3 + R_2S \end{array} \qquad [9.27]$$

chain reaction and there would be as many growing chains as there were molecules of sulfonium salt to begin with. The chain easily could be terminated by, for example, an E-2 elimination with another ylid acting as the base. However, this would generate another molecule of sulfonium salt which would start a new chain.

In conclusion it is evident that in the course of utilizing sulfonium ylids for synthetic or other purposes, precautions must be taken in order to prevent the spontaneous decomposition of the ylid. This usually can be done by working rapidly in an inert atmosphere and at as low a temperature as possible. The exact conditions will depend on the stability of the individual ylids. It also is apparent that a thorough study of the mechanisms of the decomposition of sulfonium ylids is sorely needed.

D. Reactions of Sulfonium Ylids

Since sulfonium ylids are, in the simplest view, really nothing more than carbanions they would be expected to undergo normal carbanionic reactions. Therefore, they should be capable of being alkylated, acylated, combining with electrophiles directly and adding to conjugated unsaturated groups. In addition, however, the evolution of the chemistry of sulfonium ylids has been influenced by the chemistry of the better known phosphonium ylids. Therefore, considerable effort has been concentrated on studies of the reaction of sulfonium ylids with carbonyl groups.

There have been no reports of the alkylation or acylation of sulfonium ylids but these reactions would be expected to follow the same general course as found for the phosphonium ylids. There is only one reported reaction between a sulfonium ylid and a non-multiply bonded electrophile. Oda *et al.* (52) found that fluorenylidenedimethylsulfurane reacted with dichlorocarbene to form 9-(dichloromethylene)fluorene (see p. 323). Such a direct combination is analogous to that reported by Closs (*61*) for alkyllithium compounds with chlorocarbene and appears to be a typical reaction of carbanions.

Considerable effort has been expended on studies of the reactions of sulfonium ylids with carbonyl compounds. Much of this interest is a direct result of the potential analogy with the well-known Wittig reaction between phosphonium ylids and carbonyl compounds (see Chapter 4). What makes the addition of ylids to carbonyl compounds unique is that the reactions are not simple additions, but instead usually are addition-eliminations. Such is the case in the Wittig reaction and turns out to be so in the reaction of sulfonium ylids with aldehydes and ketones.

Johnson and LaCount (*1, 31*) were the first to study this reaction. They found that fluorenylidenedimethylsulfurane (X) reacted with substituted benzaldehydes to afford mainly two products, substituted benzalfluorene oxides (XXVII) and rearranged alcohols (XXVIII). The formation of these products was rationalized by the mechanism in [9.28]. The ylid reacted only with those benzaldehydes which carried powerful electron-withdrawing groups and thereby contained an especially electrophilic carbonyl group. Such an observation was in line with expectations based on the known correlation between ylid reactivity and ylid physical properties such as basicity and dipole moment (see Section II,B, this chapter) and indicated that the first step in the reaction probably was the nucleophilic attack of the ylid on the carbonyl group. The proposed second step, the displacement of methylsulfide by the oxyanion to form epoxide (XXVII) is analogous to many known displacements of sulfonium groups such as the hydrolysis of the conjugate acid of X to fluorenol and methylsulfide (*31*).

The rearranged alcohol (XXVIII) was accounted for by proposing an intramolecular prototropic shift in the original betaine (XXIX) to form a new sulfonium ylid (XXX). This ylid is structurally analogous to the intermediate (XIX) proposed for the Sommelet rearrangement of the original ylid (X) and would be expected to effect the same nucleophilic attack on the benzenoid ring. The structure of the alcohol was proven by degradation and that of the epoxide by total synthesis.

The total yield of products from the above reaction was good, usually 60–85%, but the maximum yields of epoxide were only about 40%, the remainder being rearranged alcohol. An interesting solvent effect on the ratio of alcohol (XXVIII) to epoxide (XXVII) was encountered (*31*). In the reaction between *p*-nitrobenzaldehyde and the ylid (X) the ratio of alcohol to epoxide was about 2 in benzene or carbon tetrachloride solution, near 0.6 in chloroform and methylene chloride, 0.07

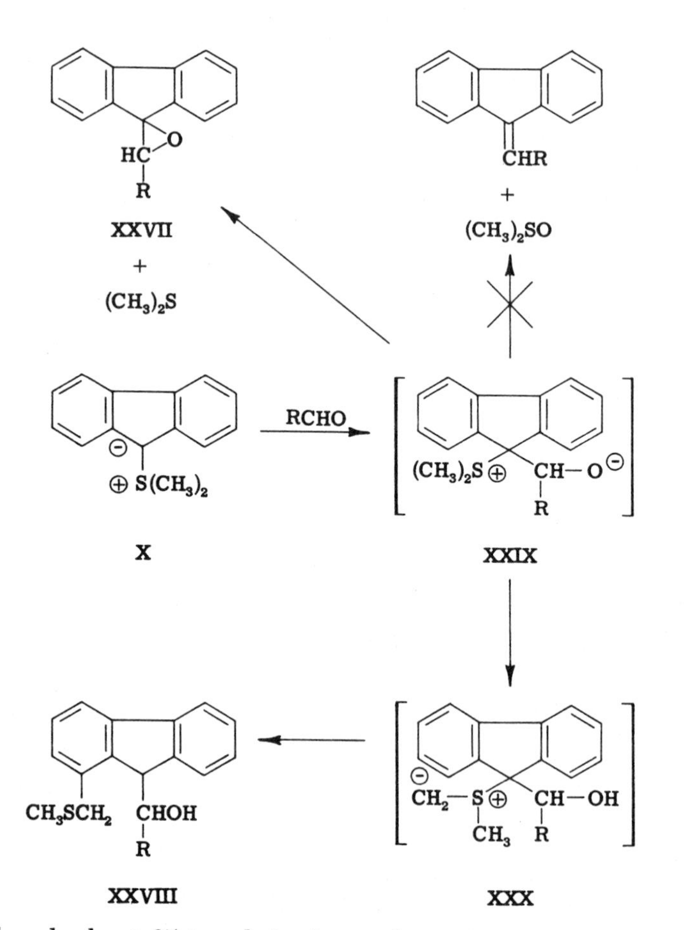

in ethanol, about 25 in ethyl ether and tetrahydrofuran, and there was no epoxide formed in diethylamine, acetonitrile or dimethylsulfoxide solutions. The exact nature of this solvent effect is unknown but it should be noted that the more basic solvents afford the least epoxide, perhaps by facilitating the conversion of the original betaine (XXIX) into the new sulfonium ylid (XXX).

If the mechanism proposed by Johnson and LaCount (*31*) is correct it is apparent that the reaction of sulfonium ylids with carbonyl compounds commenced in a manner analogous to that of the Wittig reaction of phosphonium ylids with carbonyls, i.e. nucleophilic attack of the ylid carbanion on the carbonyl carbon. The two reactions apparently diverged at the second step. Had the sulfur case followed the Wittig reaction route the oxyanion of the betaine (XXIX) would have attacked the sulfur atom, ultimately resulting in an oxygen transfer forming methylsulfoxide and an olefin. Instead, the oxyanion apparently displaced methylsulfide by attack at carbon rather than at sulfur.

At the present time it is possible only to speculate as to the reasons for the change in mechanism. In spite of the greater ability of sulfur to stabilize an adjacent carbanion by valence shell expansion (see Section II,B, this chapter), sulfur apparently is more reluctant than phosphorus to form a formally bonded intermediate of higher covalency. An additional example of this situation was discovered by Ingold *et al.* (*62*). They noted that phosphonium salts were converted by sodium hydroxide to a phosphine oxide and a hydrocarbon probably via direct attack of hydroxide on tetravalent phosphorus. This mechanism recently has been supported by the kinetic studies of VanderWerf *et al.* (*63*). On the other hand, sulfonium salts, when subjected to the same conditions, invariably afforded an alcohol or an olefin along with a thioether, the result of hydroxide attack on the α-carbon or on the β-hydrogen [9.29].

$$R_4P^{\oplus} \xrightarrow{^{\ominus}OH} [R_4P\text{—}OH] \xrightarrow{^{\ominus}OH} RH + R_3PO$$

$$R_3S^{\oplus} \xrightarrow{^{\ominus}OH} \begin{cases} R'CH{=}CH_2 + R_2S \\ ROH + R_2S \end{cases} \qquad [9.29]$$

$$(R = R'CH_2CH_2)$$

This reluctance of an oxyanion, either hydroxide as above or as alkoxide in XXIX, to attack tervalent sulfur compared to the ease of attack of oxyanion on tetravalent phosphorus may be due to the difference in the strengths of the sulfur-oxygen bond (D. E. 89 kcal/mole) (*64,* Mackle) and the phosphorus-oxygen bond (D. E. 128 kcal/mole) (*65,* Bedford and Mortimer). The formation of a phosphorus-oxygen bond is known to provide considerable driving force for a variety of reactions.

The methylsulfide group is known to be an excellent leaving group and this factor also may tend to favor epoxide formation over olefin formation from the betaine (XXIX). For example, 9-fluorenyldimethylsulfonium bromide was hydrolyzed in aqueous solution to 9-fluorenol (*26, 31*) but attempts to carry out analogous hydrolyses with phosphonium salts have been fruitless. Thus, it appears that a sulfide is a much better leaving group than a phosphine. Therefore, the combination of differences in the phosphorus-oxygen and sulfur-oxygen bond energies coupled with different leaving group abilities of sulfide and phosphine groups may explain why sulfonium ylids afford epoxides and phosphonium ylids afford olefins, both on reaction with carbonyl compounds. It should be pointed out that arsonium ylids exhibit intermediate behavior, affording both epoxides and olefins upon reaction with carbonyl compounds (see Chapter 8).

The reaction of the sulfoniumfluorenylide (X) with a few benzaldehydes as reported by Johnson and LaCount (*31*) was of mechanistic interest but of little practical use. However, the reaction did illustrate a potentially useful method for the synthesis of epoxides directly from carbonyl compounds rather than proceeding through an olefin and the Wittig reaction [9.30]. If the general reaction of sulfonium ylids with

$$R_2\overset{\oplus}{S}-\overset{\ominus}{C}R^1R^2 + R^3R^4C{=}O \longrightarrow R^1R^2C\overset{O}{—}CR^3R^4 + R_2S \qquad [9.30]$$

carbonyl compounds was to be of practical value several features of the original reaction would have to be modified. In particular, the rearrangement of the betaine initially formed would have to be prevented. If the Johnson mechanism was correct, this could be done by using groups on sulfur which contained no α-hydrogen, i.e. prevent the formation of new ylids analogous to XXX. In addition, the reactivity of the sulfonium ylids would have to be increased above that of the fluorenylide (X). It was expected that replacement of the fluorenyl group by any group which was less efficient at delocalizing the negative charge of the carbanion would lead to a more reactive ylid. By analogy with the phosphorus ylids, it was expected that a higher electron density on the ylid carbanion would lead to a more nucleophilic ylid, thereby increasing the range of carbonyl compounds susceptible to attack.

Within the space of a few months, three groups independently reported the successful application of the reaction between sulfonium ylids and carbonyl compounds to the general synthesis of epoxides. Franzen and Driessen (*66*) showed that methylenedimethylsulfurane reacted with carbonyl compounds such as benzaldehyde, cyclohexanone and benzophenone to afford 43–82% yields of mono- and disubstituted ethyl-

$$\begin{matrix}(CH_3)_2S{=}CH_2 \\ \text{or} \\ (C_6H_5)(H_3C)S{=}CH_2\end{matrix} + (C_6H_5)_2C{=}O \longrightarrow (C_6H_5)_2C\overset{O}{—}CH_2 + \begin{matrix}(CH_3)_2S \\ \text{or} \\ C_6H_5SCH_3\end{matrix} \qquad [9.31]$$

ene oxides [9.31]. Analogous behavior was shown by methylenephenylmethylsulfurane.

Corey and Chaykovsky (*30*) also studied the reactions of methylenedimethylsulfurane. They observed similar conversions but in somewhat higher yields. For example, treatment of trimethylsulfonium iodide with methylsulfinyl carbanion in tetrahydrofuran solution followed by addition of cycloheptanone gave a 97% yield of methylenecycloheptane oxide. Application either of Franzen's method (*66*) or Corey's method (*30*) permits the synthesis of any 1-substituted or 1,1-disubstituted epoxide, simply by the proper choice of carbonyl reactant. Their methods are limited, however, in that 1,2-disubstituted epoxides could not be prepared.

At about the same time Johnson *et al.* (*35, 67*) reported on their continuing attempts to broaden the scope of the reaction of sulfonium ylids with carbonyl compounds as a synthesis of epoxides. They found that alkylenediphenylsulfuranes with a series of carbonyl compounds afforded 46–72% yields of epoxides free from any rearrangement products. For example, benzylidenediphenylsulfurane (XXII) and benzaldehyde formed stilbene oxide [9.32]. The reaction was successful with

$$(C_6H_5)_2\overset{\oplus}{S}-\overset{\ominus}{C}HC_6H_5 + C_6H_5CHO \longrightarrow C_6H_5CH\overset{O}{\frown}CHC_6H_5 + (C_6H_5)_2S \qquad [9.32]$$

XXII

benzaldehydes carrying substituents ranging in electronic effect from nitro to methoxy, with aliphatic aldehydes and with acetophenones. This reaction was not successful with cyclohexanone. To illustrate further the scope of the reaction they reported that butylidenediphenylsulfurane (XXIII) also afforded epoxides, *p*-nitrobenzaldehyde forming 1-(*p*-nitrophenyl)-2-propyloxirane in 40% yield [9.33]. At a much later date

$$(C_6H_5)_2\overset{\oplus}{S}-\overset{\ominus}{C}HC_3H_7 + O_2NC_6H_4CHO \longrightarrow C_3H_7CH\overset{O}{\frown}CH-C_6H_4NO_2(p) \qquad [9.33]$$

XXIII

Corey and Oppolzer (*68*) reported further examples of Johnson's approach. They found that the butylide (XXIII) and corresponding ethylide reacted with a variety of carbonyl compounds to form the expected oxiranes in good yield. They reported the reaction to lack stereoselectivity whereas Johnson *et al.* (*67*) claimed stereoselective formation of *trans*-oxiranes.

The recently reported phenacylidenephenylmethylsulfurane (XVIII) was claimed not to react with *p*-nitrobenzaldehyde (*39*, Nozaki *et al.*). This is a somewhat surprising observation since it seems unlikely that the nucleophilicity of the ylid (XVIII) would be much lower than that of the fluorenylide (X) which did react with *p*-nitrobenzaldehyde (*31*).

In the phosphonium ylid series, the fluorenylide and phenacylide were of comparable nucleophilicity. Johnson (*33*) has found that the closely related phenacylidenedimethylsulfurane would form an oxirane with *p*-nitrobenzaldehyde, although in low yield.

From the above examples it is clear that any desired epoxide could, in principle, be formed by the reaction between an appropriately chosen alkylenediphenylsulfurane and a carbonyl compound. This reaction should assume an important position in the arsenal of available methods for the synthesis of epoxides. It has the distinct advantage of not proceeding through the intermediate formation of an olefin. In addition, it appears to be the most flexible method available, allowing virtually complete freedom in the choice of substituents. It is not complicated by the ketone formation that often accompanies the reaction of diazo compounds with carbonyl compounds.

In the course of their work with alkylenediphenylsulfuranes, Johnson *et al.* (*35*) noted that the reaction between the benzylide (XXII) and the benzaldehydes afforded only one stereoisomer, the *trans*-stilbene oxides. In no case was a mixture of stereoisomers detected. They showed that an authentic sample of *cis*-stilbene oxide was stable to the reaction and workup conditions, could be recovered in 88–97% yield and, therefore, was not absent in the ylid reaction mixture due to instability or isomerization. They concluded that the reaction between the ylid (XXII) and benzaldehydes was stereoselective.

On the basis of previous evidence they assumed that the first step in the reaction was attack of the ylid carbanion on the carbonyl group to form a betaine. This betaine was thought to be converted to the epoxide by a backside attack of the oxyanion on the carbon carrying the

$$\underset{\text{XXII}}{(C_6H_5)_2\overset{\oplus}{S}-\overset{\ominus}{C}HC_6H_5} + C_6H_5CHO \longrightarrow \left[C_6H_5\underset{}{\overset{O^{\ominus}}{\overset{|}{C}H}}-\underset{\oplus S(C_6H_5)_2}{\underset{|}{C}HC_6H_5} \right] \longrightarrow \underset{O}{C_6H_5CH-CHC_6H_5} + (C_6H_5)_2S \qquad [9.34]$$

sulfonium group [9.34]. This proposal is similar to that made for the conversion of the methohydroxide of 1,-2-diphenyl-2-aminoethanol into stilbene oxides (*69*, Read and Campbell) [9.35].

$$C_6H_5-\underset{\oplus N(CH_3)_3}{\underset{|}{C}H}-\overset{O-H}{\overset{|}{C}H}-C_6H_5 + OH^{\ominus} \longrightarrow \underset{O}{C_6H_5CH-CHC_6H_5} + (CH_3)_3N \qquad [9.35]$$

Johnson *et al.* (*35*) accounted for the stereochemistry of the reaction using this same mechanism and proposed that the first step of the reaction was reversible and could result in the formation of both the *erythro* (XXXI)- and the *threo* (XXXII)-betaines. If the second step were to be a backside attack by the oxyanion, the *threo* and *erythro* configurations would have to approach the conformations shown in the transition states for epoxide formation. In these conformations the *threo* form clearly would be the higher energy form since the three bulkiest

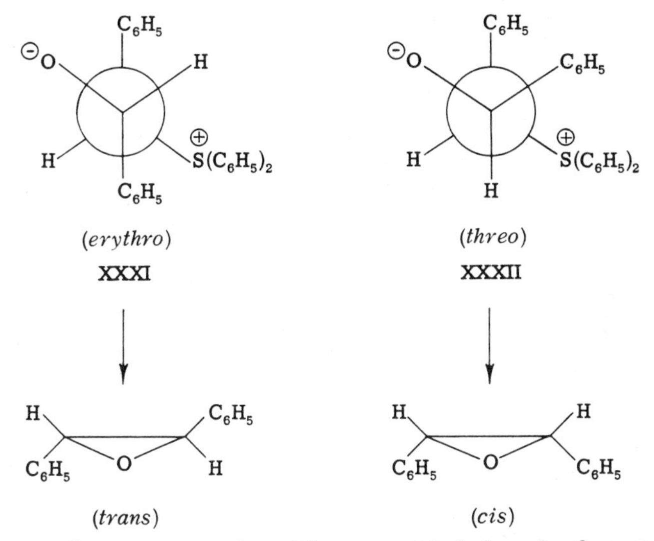

(*trans*) (*cis*)

groups are gauche to one another. Thus, provided that the first step was reversible, mainly the *erythro* form (XXXI) would be converted to epoxide (*trans*); the *threo* form would revert to ylid and carbonyl compound. As a result the reaction would be stereoselective. This stereochemistry is analogous to that encountered by Read and Campbell (*69*) for the decomposition of the *threo*- and *erythro*-methohydroxides of 1,2-diphenyl-2-aminoethanol. In addition, Speziale and Bissing (*70*) recently have shown that the stereochemistry of the Wittig reaction between stabilized phosphonium ylids and carbonyl compounds is best accounted for by reversible formation of the isomeric betaines.

An alternative explanation of the stereochemistry of the reaction of sulfonium ylids with carbonyl compounds would presume that betaine formation was not reversible and that the stereochemistry of the reaction therefore was determined in the first step, betaine formation. If the reaction truly is stereospecific this would imply that the two transition states for betaine formation would have quite different energies. If the transition states had structures approximating those of XXXI and XXXII, it is likely that XXXI would have considerably lower energy and therefore be converted to betaine and ultimately to epoxide more rapidly than the *threo* transition state, which would lead to the dominant formation of the *trans* epoxide as observed. However, it is unlikely that the transition states for the first step would have structures approximating XXXI and XXXII. More likely the transition states would have different rotational conformations as determined by ion-dipole interactions be-

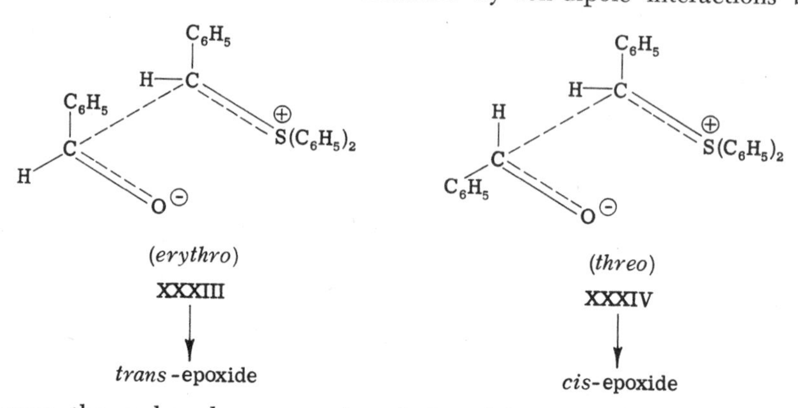

tween the carbonyl compound and the ylid grouping, structures such as XXXIII and XXXIV. If these are the structures leading to the betaines, it would be expected that the *threo* form would have the lowest energy since the *erythro* form (XXXIII) has the two phenyl groups eclipsed. However, this implies that the end product from the *threo* form, the *cis*-epoxide, should dominate the product mixture. This prediction is counter to the observed essentially exclusive formation of the *trans*-epoxide.

Little is known regarding the stereochemistry of sulfonium ylid reaction with cyclic ketones. Corey and Chaykovsky (*22*) found that methylenedimethylsulfurane underwent mainly, but not exclusively, axial addition to 4-*tert*-butylcyclohexanone to afford, after lithium aluminum hydride reduction, an 83/17 ratio of *cis-/trans*-4-*tert*-butyl-1-methylcyclohexanol. By contrast, methylenedimethyloxysulfurane (see Section III, A, this chapter) gave exclusive equatorial addition. The latter is the bulkier, less nucleophilic reagent. Steric approach control and product development control both would favor equatorial attack.

In the course of exploring the scope of the reactions of sulfonium ylids two research groups examined the course of the reaction between ylids and systems that are prone to conjugate addition, in particular, α, β-unsaturated carbonyl compounds. Corey and Chaykovsky (*30*) reported that methylenedimethylsulfurane transferred its methylene group to benzalacetophenone to afford the epoxide in 87% yield with no trace of the other possible product, 1-benzoyl-2-phenylcyclopropane [9.36].

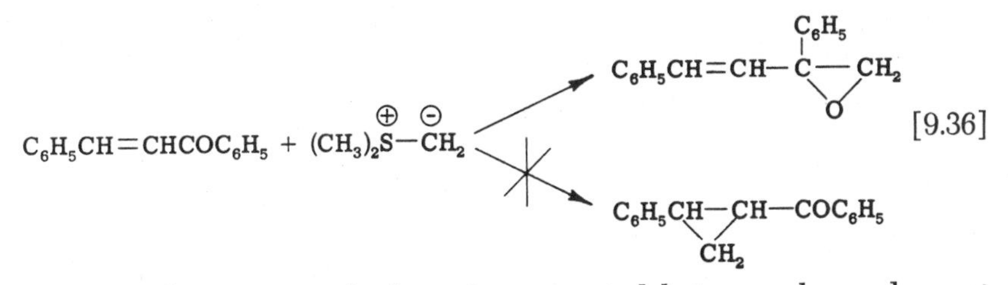

Carvone and eucarvone, both α, β-unsaturated ketones, also underwent addition at the carbonyl group to form the oxirane rather than addition at the olefinic group to form a cyclopropane.

Shortly after this report, Franzen and Driessen (*66*) reported that benzalacetophenone reacted with methylenedimethylsulfurane or methylenephenylmethylsulfurane to afford the conjugate addition product, 1-benzoyl-2-phenylcyclopropane. However, Corey and Chaykovsky (*71*) reaffirmed their observation of normal carbonyl addition, and in the full paper Franzen and Driessen (*28*) acknowledged their error.

Additional studies with conjugated carbonyl compounds by both Corey's group (*30, 71*) and Franzen's group (*28, 66*) have shown that methylenedimethylsulfurane will, upon occasion, transfer its methylene group to an olefin conjugated to a carbonyl group. For example, ethyl cinnamate was converted to 1-carbethoxy-2-phenylcyclopropane while cinnamaldehyde was converted to 1-styryloxirane [9.37] (*28*). It appears that Corey's generalization (*71*) that methylenedimethylsulfurane

$$C_6H_5CH{=}CHCOOC_2H_5 \xrightarrow{(CH_3)_2S{=}CH_2} C_6H_5\underset{\diagdown\ CH_2\ \diagup}{CH{-}CH}{-}COOC_2H_5$$

$$C_6H_5CH{=}CHCHO \xrightarrow{(CH_3)_2S{=}CH_2} C_6H_5CH{=}CH{-}\underset{\diagdown\ O\ \diagup}{CH{-}CH_2} \qquad [9.37]$$

is specific for the formation of oxiranes from conjugated carbonyls must be restricted to those instances where the carbonyl group is particularly susceptible to nucleophilic attack. Where the carbonyl group is less reactive, such as esters, conjugate addition to form cyclopropanes may even dominate the course of reaction.

Methylenedimethylsulfurane has been shown to react with a variety of other unsaturated systems to effect methylene transfer. Corey and Chaykovsky (*30*) reported the transfer of methylene to 1,1-diphenylethene to form 1,1-diphenylcyclopropane. The ylid also reacted with benzalaniline to afford the corresponding aziridine (*28, 30, 66*). Carbethoxymethylenedimethylsulfurane also reacted with benzalaniline but

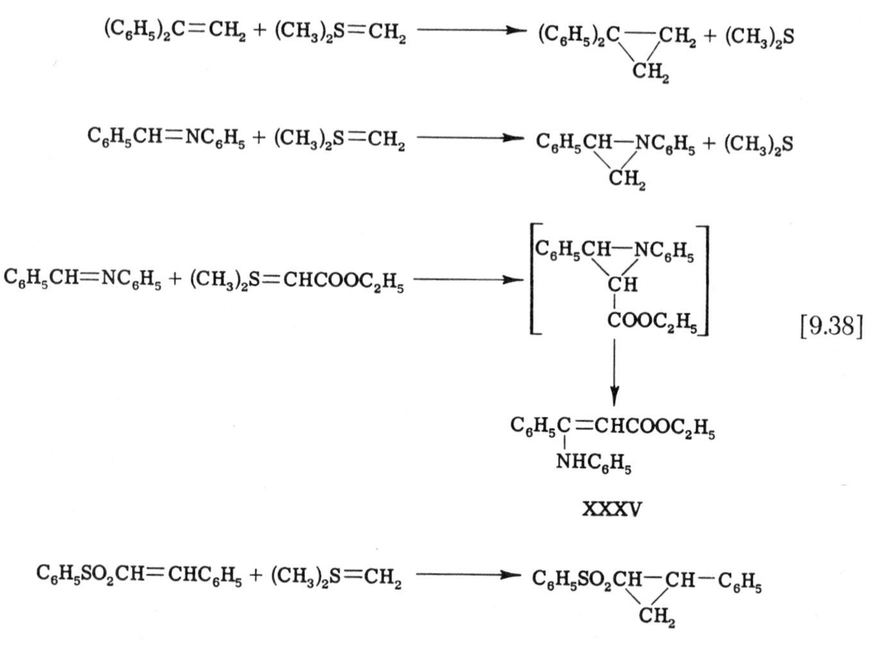

the initial product, undoubtedly an aziridine, rearranged to the conjugated vinyl amine (XXXV) (*72*, Speziale *et al.*). Truce and Badiger (*73*) found that the methylenesulfurane afforded a cyclopropane adduct (XXXVI) with both *cis*- and *trans*-styrylphenylsulfone.

Johnson (*29*) demonstrated what appears to be another fairly general addition-elimination, methylene transfer reaction of sulfonium ylids. Fluorenylidenedimethylsulfurane (X) and nitrosobenzene underwent an exothermic and very rapid reaction to afford the nitrone, *N*-phenylfluorenone ketoxime (XXXVII). Benzhydrylidenedimethylsulfurane and 2-nitrofluorenylidenedimethylsulfurane also afforded nitrones with nitrosobenzene. By analogy with the formation of epoxides from reaction of the fluorenylide (X) with benzaldehydes, it was initially expected that reaction of X with nitrosobenzene would afford the oxazirane (XXXVIII). It was concluded that the latter probably was the initial product of the reaction but that it rapidly isomerized to the nitrone (XXXVII). The facile isomerization of oxaziranes, especially *N*-phenyloxaziranes, to nitrones is well known (*74, 75*). In addition, all attempts to prepare an authentic sample of the oxazirane by standard procedures led to the isolation of the nitrone. Therefore, it appears that the reaction of sulfo-

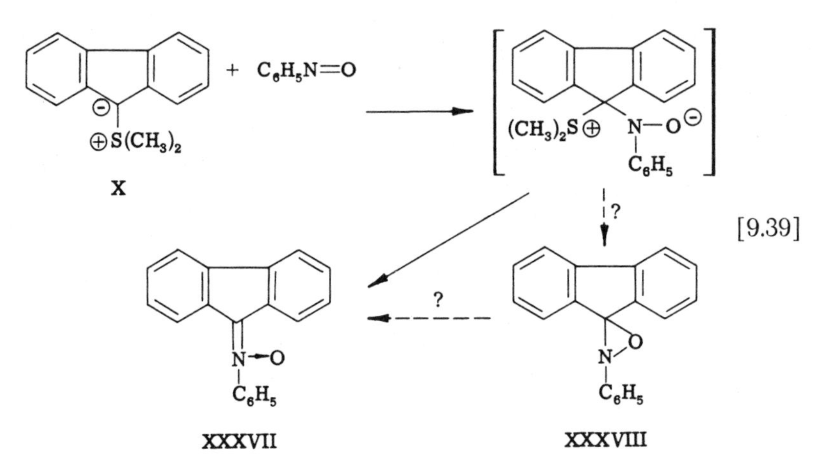

[9.39]

nium ylids with nitroso compounds is simply another example of methylene transfer.

It seems appropriate to envision all of these methylene transfer reactions by a common reaction mechanism as in [9.40]. The system

$$A{=}B + (CH_3)_2\overset{\oplus}{S}{-}\overset{\ominus}{C}H_2 \longrightarrow \left[(CH_3)_2\overset{\oplus}{S}{-}\underset{|}{C}H_2 \; A{-}B^{\ominus} \right] \longrightarrow (CH_3)_2S + \underset{CH_2}{A{-}B} \qquad [9.40]$$

A=B can be isolated olefin, conjugated olefin, carbonyl, imine and nitroso. This mechanism is especially consistent with the data available from the carbonyl reaction. However, there is considerable question about the actual mechanism in the other cases. The product results can be explained by the intervention of a carbene intermediate. More experimental work is needed.

It seems clear that one of the major practical uses of sulfonium ylids is in the synthesis of oxiranes. The method has a high degree of flexibility and, perhaps, a useful measure of stereoselectivity. It is equally apparent, however, that considerable work remains to be done in extending the scope of sulfonium ylid reactions and, most important, in elucidating the mechanistic and stereochemical fine points of the reactions in order that they may be applied to synthetic problems with a minimum of ambiguity.

E. Sulfonium Ylids as Reaction Intermediates

Long before a study of their chemistry came into prominence in its own right, sulfonium ylids had been proposed as intermediates in a variety of reactions. With what is now known of the chemistry of ylids it is possible to investigate such reactions thoroughly to determine whether or not ylids are, in fact, intermediates. The Hofmann-type elimination of sulfonium salts is one such reaction.

Sulfonium salts long have been known to undergo a Hofmann-type elimination in the presence of base to afford an olefin and a thioether (62, Ingold *et al.*). For any such elimination reaction four basic mechanisms are possible- a concerted *trans* E-2 elimination, carbanion formation on the β-carbon(E-1_{CB}), formation of an ylid and then α-elimination via a carbene, and α'-ylid formation followed by *cis* elimination (α', β). It seems fairly clear that where the stereochemistry of the system under study permits, the normal concerted, *trans* E-2 elimination prevails. Cope and Mehta (76) recently have demonstrated that the α', β-mechanism, via an ylid intermediate, does operate in an ammonium system where *cis* elimination is sterically favored.

In their studies of sulfonium elimination reactions Cristol and Stermitz (77) noted that both *cis*- and *trans*-1-dimethylsulfonio-2-phenylcyclohexane afforded 1-phenylcyclohexene as the dominant elimination

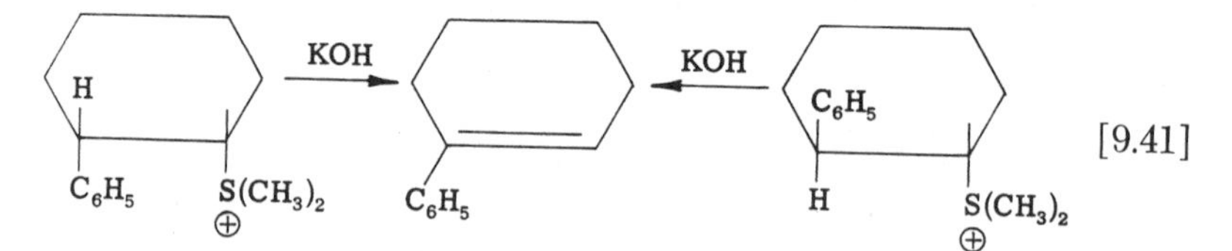

[9.41]

product when treated with potassium hydroxide in ethanol [9.41]. The *cis* isomer, which can undergo a *trans* E-2 elimination to form the observed product, did so in 70% yield. The other products were 7.5% of 1-methylthio-2-phenylcyclohexane and 1% of 3-phenylcyclohexene. The *trans* isomer, which cannot undergo a *trans* elimination to form its 20% yield of 1-phenylcyclohexene, also gave 61% of the substitution product and 2% of 3-phenylcyclohexene. The authors noted that the ratio of the elimination rates for the *cis/trans* isomers were smaller than normal, i.e. the *trans* isomer was more reactive than expected for a *cis* elimination.

Cristol and Stermitz (77) favored the α', β-elimination mechanism for the *cis* elimination undergone by the *trans* isomer or another mechanism involving the formation of a cyclohexylide which would permit isomerization of the *trans*-sulfonium salt to the *cis* isomer followed by normal E-2 elimination. The deuterium experiments which would distinguish these two mechanistic possibilities have not been reported to date.

Weygand and Daniel (78) have reported another elimination reaction of a sulfonium salt which has been proposed to proceed through a sulfonium ylid as an intermediate or transition state the net result being a fragmentation [9.42].

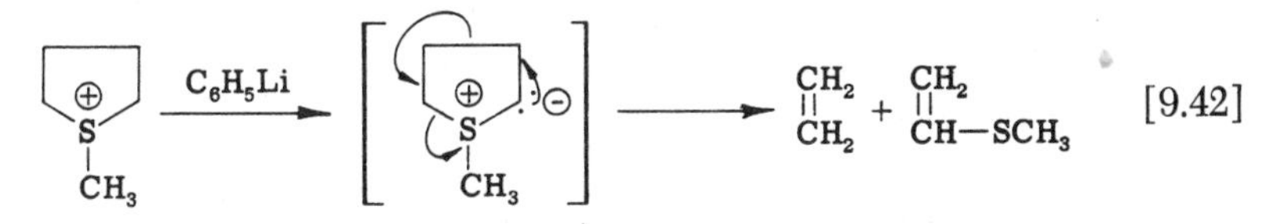

[9.42]

Franzen and Mertz (*79*) have provided the only concrete evidence for the intermediacy of an ylid in the elimination reactions of a sulfonium salt. They prepared triethylsulfonium bromide with the methylenic hydrogens all replaced by deuterium and then subjected it to elimination with the triphenylmethide anion. They found that 75% of the triphenylmethane contained deuterium, thereby indicating the operation

$$CH_3CD_2-\overset{\oplus}{S}(CD_2CH_3)-CD_2CH_3 \xrightarrow{(C_6H_5)_3C^{\ominus}} \left[CH_3CD_2-\overset{\oplus}{S}(-CD_2CH_3)-\overset{\ominus}{C}DCH_3\right] \quad \text{XXXIX}$$

$$\text{(a)} \xrightarrow{(C_6H_5)_3C^{\ominus}} CH_3CD_2-S-CD_2CH_3 \quad CH_2{=}CD_2 + (C_6H_5)_3CH \quad (E_2\text{-elimination})$$

$$\text{(b)} \quad CH_3CD_2-S-CD_2CH_3 \quad CH_2{=}CHD + (C_6H_5)_3CD \quad (\alpha\text{-elimination})$$

$$\text{(c)} \quad CH_3CD_2-S-CDH-CH_3 \quad CH_2{=}CD_2 + (C_6H_5)_3CD \quad (\alpha', \beta\text{-elimination}) \qquad [9.43]$$

of mechanistic pathway *b* and/or *c* [9.43]. Oxidation of the ethylsulfide obtained from the reaction afforded diethylsulfone, the NMR spectrum of which indicated the presence of one methylenic proton. This indicated that a portion of the elimination reaction probably occurred via an ylid intermediate (XXXIX) and an α', β-elimination (path *c*).

Franzen and Schmidt (*80*) concluded, on the basis of their isolation of *cis*-cyclooctene, that cyclooctyldimethylsulfonium salts undergo an α', β-elimination via an ylid intermediate when treated with triphenylmethide anion since the *trans*-olefin is the normal E-2 elimination product. It is likely, however, that use of a weaker base would result in a normal E-2 elimination as for the ammonium case. In the same article, Franzen and Schmidt (*80*) reported that 1,1-dideutero-2-phenylethyldimethylsulfonium salts reacted with triphenylmethide anion to afford styrene and triphenylmethane [9.44], the latter containing 79% deuterated material. This result indicated only that at some stage the anion abstracted an α-hydrogen from the salt but it does not mean that the resulting ylid necessarily was an intermediate in olefin formation. Saunders and Paulovic (*81*) showed that this system did not undergo

$$C_6H_5CH_2CD_2\overset{\oplus}{S}(CH_3)_2 \xrightarrow{(C_6H_5)_3C^{\ominus}} (C_6H_5)_3CD + (CH_3)_2S + C_6H_5CH{=}CH_2(CD_2)$$

$$C_6H_5CD_2CH_2\overset{\oplus}{S}(CH_3)_2 \xrightarrow{^{\ominus}OH} HOD + (CH_3)_2S + C_6H_5CD{=}CH_2 \qquad [9.44]$$

$$(\text{no } CH_3SCH_2D)$$

α', β-elimination since reaction of dimethyl(2-phenyl-2,2-dideutero)-ethylsulfonium bromide with aqueous hydroxide solution afforded dimethylsulfide containing essentially no deuterium.

It appears that Hofmann-type eliminations of sulfonium salts proceed via the normal E-2 *trans* elimination. If the geometry of the salt is such as to prevent a *trans* elimination, if the molecular structure is such as to provide no special stabilization for a β-carbanion(E-1_{CB} route) and the base is strong enough, sulfonium salts will undergo *cis* elimination via an ylid intermediate. However, such α', β-eliminations are not the normal course of events.

Sulfur ylids appear to be involved in several rearrangements that have been reported. Thomson and Stevens (*82*) reported a sulfur version of the Stevens rearrangement which probably proceeded via a sulfonium ylid [9.45]. The reaction of allylphenylsulfides with dichlorocarbene

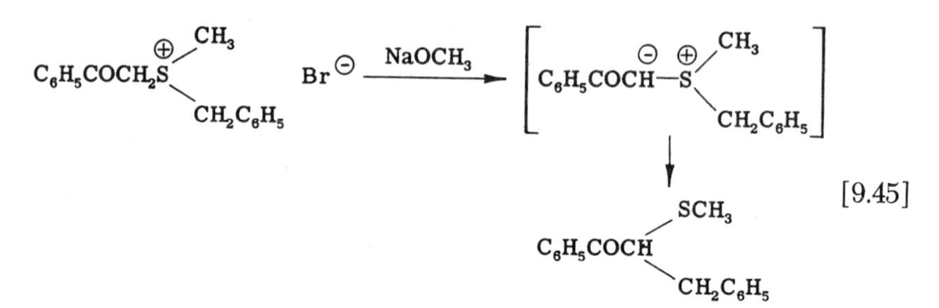

probably involved sulfonium ylid intermediates and, perhaps, Stevens rearrangement thereof (*83*, Parham and Groen). The rearrangement of benzyldimethylsulfonium ion to *o*-methylbenzylmethylsulfide in the presence of amide ion appears to be a normal Sommelet rearrangement involving intermediate ylid formation [9.46] (*84*, Hauser *et al.*). Finally, the rearrangement of fluorenylidenedimethylsulfurane (X) into 1-methylthiomethylfluorene (XXI) as discussed earlier is a Sommelet rearrangement (*31, 49*).

Two other instances should be mentioned where the intermediacy of sulfur ylids has been invoked. It has been observed that warming methyldodecyl(α-carboxymethyl)sulfonium tosylate with piperidine at 56° led to the evolution of carbon dioxide and the formation of dimethyldodecylsulfonium tosylate [9.47] (*85*, Burness). The authors proposed

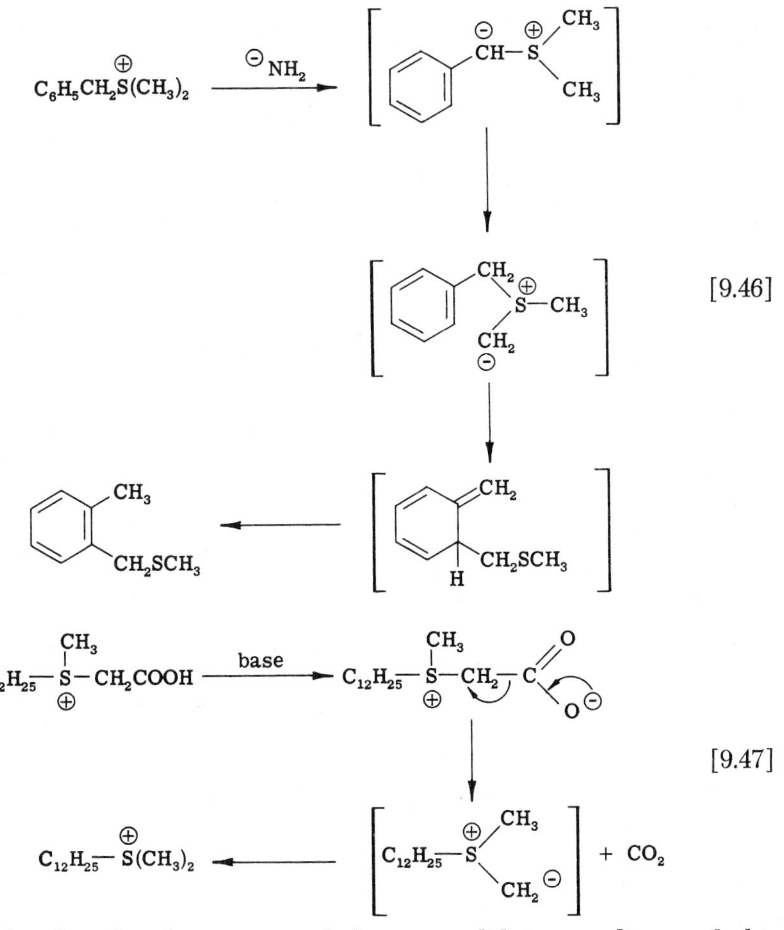

that the decarboxylation proceeded via an ylid intermediate and this lowered the activation energy for the process.

Considerable evidence has accumulated recently which indicates that the chemical and biological properties of thiamine may be due to the stability of the carbanion resulting from the removal of the proton on C-2 of the thiazolium portion of the molecule (*86,* White and Ingraham). Chemical studies have concentrated on simpler thiazolium ions. For example, Breslow (*87*) showed that 3,4-dimethylthiazolium bromide exchanged the C-2 hydrogen in neutral medium; this indicates the intermediacy of a carbanion. Several resonance structures may be written for the thiazolium ion (XL) but recent calculations and NMR spectra by Haake and Miller (*88*) have indicated that about 0.6 of the positive charge resides on the sulfur atom. It is tempting to explain the apparent stability of the thiazolium carbanion on the basis of the importance of

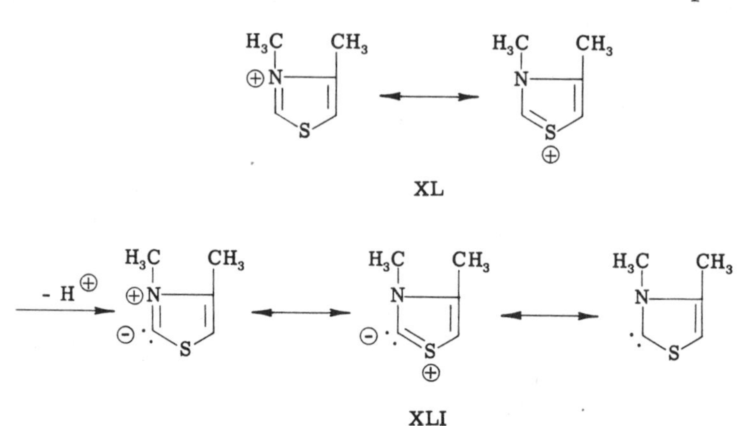

structure XLI, a sulfonium ylid. That valence shell expansion of the sulfur atom was not the sole factor was apparent by the observations (*87–89*) that imidazolium ions and oxazolium ions also exchanged hydrogen at C-2. However, the relative rates of such exchanges (*88*) indicated that overlap of the carbanion orbital with the *d*-orbitals of sulfur does afford some stabilization to the carbanion. Breslow (*87*) has doubted the extent of such stabilization due to the apparent geometric restrictions of overlap. However, structure XLI is isoelectronic with a benzyne system, and there probably is even better overlap in the former since *d*-orbitals would be used. Since so little is known about the specific geometric requirements for efficient overlap with *d*-orbitals, the question is difficult to settle.

III. Other Sulfur Ylids

As was pointed out during discussion of the theoretical basis for the existence of sulfur ylids, any system in which a sulfur atom carries a high degree of positive charge and which has a hydrogen on an adjacent carbon atom might be expected to be capable of ylid formation. In the following pages several types of sulfur ylids will be discussed in which the positive charge on sulfur is due to the formation of compounds other than sulfonium salts. In all of these cases, virtually nothing is known about the physical properties of the ylids themselves. Most have been prepared in solution and and used immediately in reactions. Few mechanistic studies have been reported.

It should be apparent that to design an ylid system which is to undergo reactions other than those of ordinary carbanions a heteratom group (Z) must be chosen which can easily be ejected (path *a*) to form cyclic compounds or can accept an oxyanion transfer (path *b*) to form unsaturated systems [9.49]. These two paths are most typical of sul-

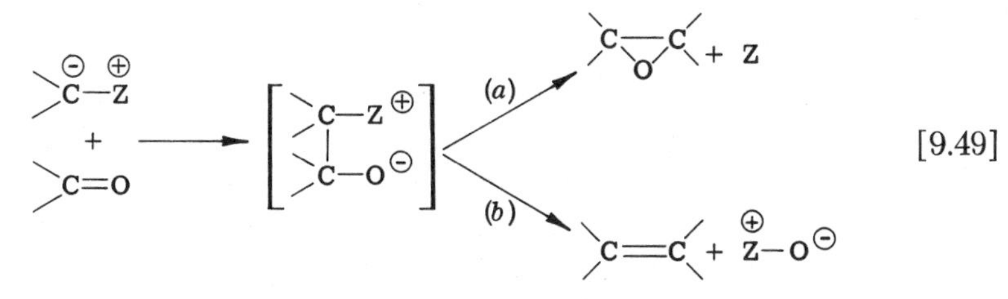

fonium and phosphonium ylids, respectively. Therefore, a study of a new ylid usually centers on its reaction with carbonyl compounds and must be concerned with the electronic nature of the group Z, i.e. can it be displaced and/or can it be oxidized.

A. Oxysulfonium Ylids

With sulfonium ylids as a starting point, one of the obvious means of increasing the positive charge on the sulfur atom is to somehow tie up the remaining unshared pair of electrons, the 3*s*-electrons. In a formal sense this could be done by oxidation of a sulfonium salt but this ap-

$$(CH_3)_3S^{\oplus}\ I^{\ominus} \xrightarrow{?} (CH_3)_3\overset{\oplus}{S}\rightarrow O\ I^{\ominus} \xleftarrow{CH_3I} (CH_3)_2S\rightarrow O \qquad [9.50]$$

pears not to have been done. In actual practice the desired substance, trimethyloxysulfonium iodide, has been prepared by reaction of dimethylsulfoxide with methyl iodide [9.50] (*90,* Kuhn and Trischmann). Smith and Winstein (*21*) also reported the preparation of this salt, sometimes called the S-methyl derivative of methylsulfoxide, and described conditions for obtaining the *O*-methyl derivative. The two isomers could be distinguished readily by their characteristic NMR spectra. The latter authors noted that trimethyloxysulfonium iodide underwent extremely rapid exchange with deuterium oxide in neutral solution. The exchange appeared to be about 10^2 faster than the analogous exchange of deuterium in trimethylsulfonium salts in the presence of deuteroxide catalysis (*10,* Doering and Hoffmann). Such a rapid exchange indicated that the carbanion (VIII) that must surely have been an intermediate had appreciable stability. Corey and Chaykovsky (*22*) undertook a study of the chemistry of the ylid, methylenedimethyloxysulfurane.

The ylid could be prepared in tetrahydrofuran or dimethylsulfoxide solution by the action of sodium hydride on trimethyloxysulfonium iodide [9.51]. It immediately became apparent that this ylid was considerably more stable than the closely related methylenedimethylsulfurane since the former could be prepared and stored in solution at room

$$(CH_3)_3\overset{\oplus}{S}{\rightarrow}O\ I^{\ominus} \xrightarrow{NaH} H_2 + NaI + (CH_3)_2\overset{\oplus}{S}(\rightarrow O)\overset{\ominus}{C}H_2 \qquad [9.51]$$

VIII

temperature or above whereas the sulfoniummethylide had to be prepared at low temperatures. In addition, there have not been any reports of the ylid (VIII) undergoing the decomposition reactions to olefins or polymethylene characteristic of some sulfonium ylids.

Methylenedimethyloxysulfurane (VIII) was found to react readily with a variety of aldehydes and ketones at room temperature or slightly above over a two-hour period (*22*). For example, reaction with benzophenone produced 1,1-diphenylethylene oxide in 90% yield. The reac-

$$(CH_3)_2\overset{\oplus}{S}(\rightarrow O)\overset{\ominus}{C}H_2 + (C_6H_5)_2CO \longrightarrow \left[(C_6H_5)_2C(O^{\ominus})-CH_2-\overset{\oplus}{S}(\rightarrow O)(CH_3)_2 \right] \longrightarrow (C_6H_5)_2C\overset{O}{—}CH_2 + (CH_3)_2SO \qquad [9.52]$$

XLII

tion is typical of those between sulfonium ylids and carbonyl compounds which also formed oxiranes, and it is likely that the mechanism is analogous as exemplified above. The ylid probably attacks the carbonyl carbon to form a betaine (XLII) which contains an excellent leaving group to be ejected by the oxyanion. There is no evidence in any of the reactions that the betaine intermediates will decompose to olefin and dimethylsulfone as for a Wittig reaction. The reaction appears to represent a rather general synthesis of oxiranes but so far is limited to varying substituents on the carbonyl carbon. It reacts similarly to methylenedimethylsulfurane but appears to be more stable and easier to handle.

Methylenedimethyloxysulfurane (VIII) reacted with conjugated carbonyl compounds but in a manner different from that exhibited by methylenedimethylsulfurane. The latter usually formed oxiranes with conjugated carbonyl systems (see p. 337) but VIII tended selectively to form cyclopropanes with the same compounds. For example, VIII with benzalacetophenone gave 1-benzoyl-2-phenylcyclopropane in 95% yield with only traces of the epoxide being detected [9.53] (*22*). With the same compound, methylenedimethylsulfurane afforded an 87% yield of epoxide (*30*, Corey and Chaykovsky). The cyclopropanes are easily accounted for by conjugate addition followed by ejection of dimethylsulfoxide. In a similar fashion VIII transferred a methylene group to

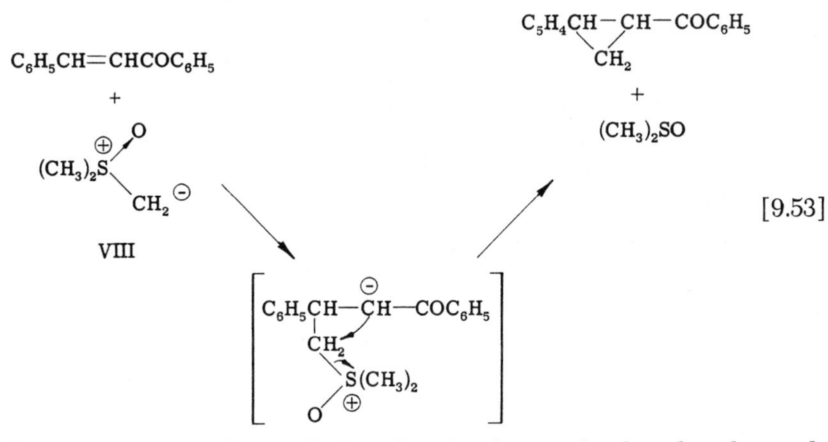

[9.53]

ethyl cinnamate to form ethyl 2-phenylcyclopropylcarboxylate but only in 32% yield (*71*). This reaction has been used by Izzo (*91*) for the transfer of a methylene group to the olefinic bond of a series of maleimides with the exclusive formation of the expected cyclopropanes.

Methylenedimethyloxysulfurane (VIII) appeared to be a less powerful methylene transfer agent than the sulfoniummethylide. Perhaps the best evidence for such a conclusion is the fact that the former would not convert 1.1-diphenylethene to 1,1-diphenylcyclopropane whereas the latter ylid would (*30*). However, the oxysulfonium ylid would react with an unsymmetrical double-bonded system such as benzalaniline to form a mixture of 1,2-diphenylaziridine and acetophenone anil [9.54]

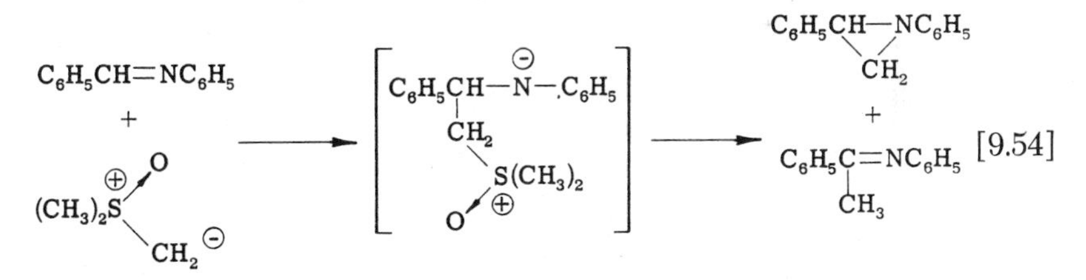

[9.54]

(*30*). The aziridine probably was formed by an addition-elimination sequence and the anil formation probably was due to a hydride shift either before or after the ejection of methylsulfoxide. At a later date

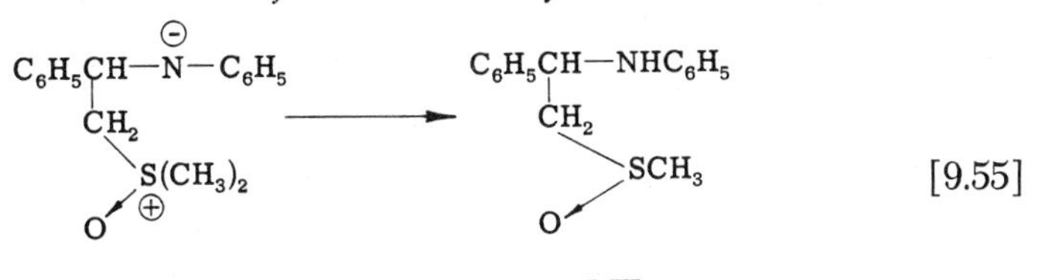

[9.55]

Metzger and Seelert (*92*) reported the isolation of the same two products but, in addition, reported the presence of a new substance (XLIII) in which only a methyl group had been lost from the proposed intermediate. Metzger and his group (*93a, b*) have reported the addition of VIII to a variety of other unsaturated centers.

Two groups have found that methylenedimethyloxysulfurane will serve as a methylating agent. Metzger *et al.* (*94*) have shown that treatment of VIII with active hydrogen compounds results in their methylation. For example, 2-naphthol was converted to its methyl ether, and *p*-nitrobenzoic acid afforded its methyl ester. These transformations presumably occurred by the ylid abstracting a proton to form a trimethyloxysulfonium salt and attack by the anion on this salt would lead to methylation. Traynelis and McSweeney (*95*) have found that VIII will methylate nitrobenzene in the *ortho* and *para* positions but no mechanism has been indicated.

Corey and Chaykovsky (*96*) have shown that the oxysulfonium ylid (VIII) can be acylated with benzoyl chloride to form a new ylid, phenacylidenedimethyloxysulfurane (XLIV) in 92% yield. Two equiva-

$$2\ (CH_3)_2\overset{\oplus}{S}(\to O)\text{—}CH_2^{\ominus} + C_6H_5COCl \longrightarrow C_6H_5CO\overset{\ominus}{C}H\text{—}\overset{\oplus}{S}(\to O)(CH_3)_2 + (CH_3)_3\overset{\oplus}{S}O\ \ Cl^{\ominus} \quad [9.56]$$

VIII XLIV

lents of ylid were required as for the phosphonium ylid cases. Some keto-ylids also could be prepared using phenyl esters as the acylating agents, and Konig and Metzger (*97*) affected acylations with anhydrides, ketenes and isocyanates to form a series of β-carbonyloxysulfonium ylids. The use of these substituted ylids in reactions or their characterization has not been reported to date. The ylid XLIV was isolated, m.p. 119–120°, and showed infrared absorption at 6.60 μ. This indicated considerable enolate character to the carbonyl group (*96*). Corey and Chaykovsky (*96*) did report that the keto-ylids could be cleaved photochemically to carbenes which rearranged to ketenes and then reacted

$$RCO\overset{\ominus}{C}H\text{—}\overset{\oplus}{S}(\to O)(CH_3)_2 \xrightarrow{h\nu} [RCOCH:] \longrightarrow [RCH{=}C{=}O] \xrightarrow{R'OH} RCH_2C(=O)OR' \quad [9.57]$$

with solvent [9.57]. This overall sequence may be a valuable counterpart to the homologation of acids via the Arndt-Eistert procedure.

In summary, methylenedimethyloxysulfurane (VIII) appears to be useful for the conversion of aldehydes and ketones to oxiranes by methylene transfer. It also is applicable to the transformation of α, β-unsaturated carbonyl compounds into cyclopropanes without the concurrent formation of epoxides. In this regard it seems to complement rather than duplicate methylenedimethylsulfurane which will convert the same unsaturated carbonyl compounds selectively into oxiranes.

B. Sulfinyl Ylids

Corey and Chaykovsky (*19*) found that dimethylsulfoxide was metallated readily by sodium hydride. The resulting methylsulfinyl carbanion (XLV) would add to carbonyl compounds such as benzophenone under mild conditions to form the simple hydroxy-sulfoxide addition

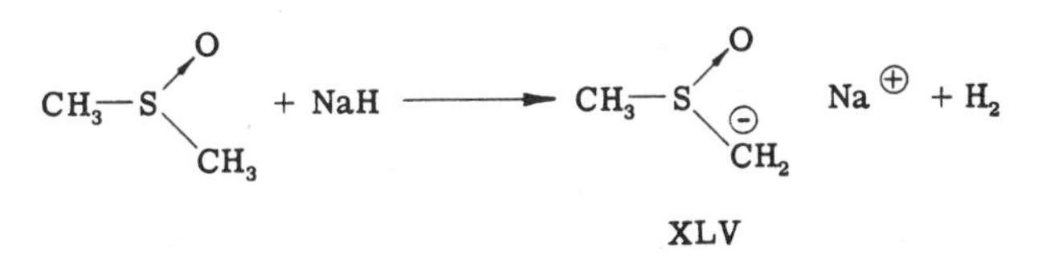

[9.58]

$$\text{XLV} + (C_6H_5)_2CO \longrightarrow (C_6H_5)_2\underset{\displaystyle OH}{C}-CH_2\overset{O}{\underset{\displaystyle CH_3}{S}}$$

product. Later, the same workers, as well as Walling and Bollyky (*20*), found that under more severe conditions the reaction would follow a different course although it may have commenced via the same initial step, the addition of the carbanion to the carbonyl group. A mixture of hydrocarbons of the composition in [9.59] was obtained. The total yield

$$CH_3SO\overset{\ominus}{C}H_2 + (C_6H_5)_2CO \xrightarrow{\Delta} \underset{(47\%)}{(C_6H_5)_2C{=}CH_2} + \underset{(30\%)}{(C_6H_5)_2CH_2} + \underset{(3\%)}{(C_6H_5)_2\overset{}{C}\!\!\underset{CH_2}{\diagdown\!\diagup}\!\!CH_2} + \underset{(20\%)}{(C_6H_5)_2CHCHO} \quad [9.59]$$

of hydrocarbons was only about 45% and it is obvious that the reaction is of no synthetic value. However, it is of mechanistic interest and the following steps have been proposed to account for the observed products [9.60].

The proposed formation of diphenylethylene as in steps 1 and 2 is analogous to a Wittig reaction with the betaine transferring its oxygen

$$CH_3SOCH_2^{\ominus} + (C_6H_5)_2CO \longrightarrow CH_3SOCH_2C(C_6H_5)_2O^{\ominus} \quad (1)$$

$$CH_3SOCH_2C(C_6H_5)_2O^{\ominus} \longrightarrow CH_2{=}C(C_6H_5)_2 + CH_3SO_2^{\ominus} \quad (2)$$

$$CH_3SOCH_2C(C_6H_5)_2O^{\ominus} \longrightarrow \left[\begin{matrix} (C_6H_5)_2C{-}CH_2 \text{ (oxirane, bridged by O)} \\ + \\ CH_3SO^{\ominus} \end{matrix}\right] \longrightarrow (C_6H_5)_2CHCHO \quad (3)$$

[9.60]

$$CH_3SOCH_2^{\ominus} + CH_2{=}C(C_6H_5)_2 \longrightarrow CH_3SOCH_2CH_2{-}\overset{\ominus}{C}(C_6H_5)_2 \quad (4)$$

$$CH_3SOCH_2CH_2{-}\overset{\ominus}{C}(C_6H_5)_2 \longrightarrow H_2C{-}C(C_6H_5)_2 \text{ (cyclopropane, bridged by } CH_2) + CH_3SO^{\ominus} \quad (5)$$

$$CH_3SOCH_2CH_2{-}\overset{\ominus}{C}(C_6H_5)_2 \longrightarrow CH_3SO\overset{\ominus}{C}HCH_2CH(C_6H_5)_2 \quad (6)$$

$$CH_3SO\overset{\ominus}{C}H{-}CH_2CH(C_6H_5)_2 \longrightarrow \begin{matrix} CH_3SOCH{=}CH_2 \\ + \\ (C_6H_5)_2CH^{\ominus} \end{matrix} \longrightarrow (C_6H_5)_2CH_2 \quad (7)$$

to the sulfur atom. If this is in fact the mechanism, it is the only reported instance where a sulfur-containing betaine produces an olefin rather than an oxirane as with sulfonium and oxysulfonium ylids. Its validity must be doubted in lieu of convincing evidence. The diphenylacetaldehyde has been accounted for by normal oxirane formation expected of sulfur ylids (step 3) followed by rearrangement to the aldehyde as proposed for arsonium ylids (*98*, Henry and Wittig). The rearrangement has not been demonstrated under these conditions, however.

The remaining hydrocarbon products have been accounted for by proposing attack of the methylsulfinyl carbanion on the product diphenylethylene (step 4). The feasibility of this step has been demonstrated by the observation that reaction of the carbanion with the olefin in a separate experiment afforded diphenylmethane and 1,1-diphenylcyclopropane. This indicates that the methylsulfinyl carbanion (XLV) is more reactive than methylenedimethyloxysulfurane (VIII) which would not react with this olefin (*30*).

Only the volatile hydrocarbon products from this reaction have been investigated and it is evident that there are several volatile sulfur-con taining compounds as well as non-volatile material formed. It may be seen, however, that the sulfinyl ylid appears to be capable of undergoing reactions that are typical of other sulfur ylids.

Russell *et al.* (*99, 100*) found that the carbanion (XLV) could be acylated by reaction with esters in dimethylsulfoxide solution. Thus, ethyl benzoate afforded phenacylmethylsulfoxide. Corey and Chaykovsky (*101*) found that reduction of the keto-sulfoxides with aluminum amalgam led to cleavage of the carbon-sulfur bond and formation of, for example, acetophenone [9.61]. This represents a method for conver-

$$C_6H_5COOC_2H_5 + \overset{\ominus}{C}H_2SOCH_3 \longrightarrow C_6H_5\overset{O}{\overset{\|}{C}}-CH_2SOCH_3 \xrightarrow{Al(Hg)} C_6H_5COCH_3 \quad [9.61]$$

sion of carboxylic acids into ketones and when applied to substituted sulfinyl ylids would permit the preparation of other than methyl ketones.

C. Sulfonyl Ylids

There have been only two reports on the chemistry of sulfone ylids. Chaykovsky and Corey (*17*) reported that addition of methylsulfonyl carbanion, prepared from methylsulfone and sodium hydride, to benzophenone at 100° afforded a mixture of hydrocarbons which were identified as diphenylmethane (38%), 1,1-diphenylethene (46%) and diphenylacetaldehyde (15%). These products could be accounted for by mechanisms analogous to those proposed for the reaction of the methylsulfinyl carbanion with benzophenone. If such proposals are correct, it appears that the sulfone ylid can form oxirane and methylsulfinate ion,

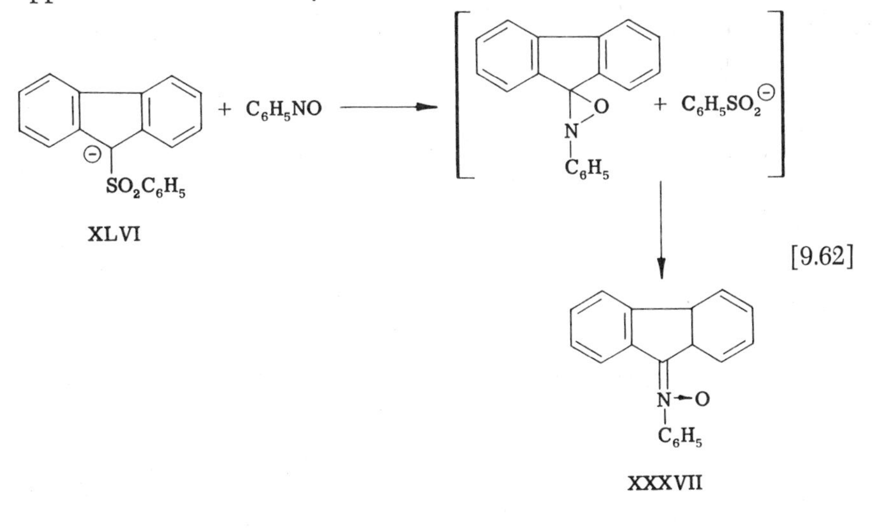

[9.62]

add to olefins or carry out a Wittig-type reaction with a carbonyl compound to form an olefin and methylsulfonate ion. No evidence has been provided to substantiate these proposals, especially the latter.

Johnson (*18*) reported that 9-phenylsulfonylfluorenylide (XLVI) reacted with nitrosobenzene to afford *N*-phenylfluorenone ketoxime (XXXVII) [9.62]. This behavior is exactly analogous to that shown by the sulfonium ylid, fluorenylidenedimethylsulfurane (X) (*29*), and is thought to indicate that the reactions probably proceeded via the same mechanism, in this case with the ejection of the phenylsulfinate ion from the betaine. The reaction of any of the sulfonyl ylids with carbonyl compounds under oxirane-forming conditions has not been investigated.

Hine and Porter (*102*) recently have proposed another instance of the phenylsulfinate ion acting as a leaving group. They found that difluoromethylphenylsulfone transferred its difluoromethyl group, probably as difluorocarbene, to thiophenoxide ion in the presence of sodium methoxide [9.63]. The original sulfone was shown to form a carbanion

$$C_6H_5SO_2CHF_2 \xrightarrow{^{\ominus}OCH_3} [C_6H_5SO_2\overset{\ominus}{C}F_2] \longrightarrow C_6H_5SO_2^{\ominus} + :CF_2$$

$$:CF_2 + C_6H_5S^{\ominus} \longrightarrow C_6H_5S\overset{\ominus}{C}F_2 \longrightarrow C_6H_5SCHF_2 \qquad [9.63]$$

which may have fragmented into phenylsulfinate ion and the carbene. Meyers *et al.* (*103*) also have proposed the dissociation of sulfonyl carbanions into carbenes.

There are many examples available of the normal carbanionic reactions of sulfonyl ylids such as alkylation by alkyl halides, acylation by esters (*104, 105*) and addition to conjugated carbonyl systems (*3*). Interestingly, the latter reaction afforded normal Michael addition products rather than cyclopropanes as did the oxysulfonium ylids.

D. Miscellaneous Sulfur Ylids

As early as 1908 the transitory existence of the sulfur analog of a ketene, formulated as $R_2C{=}SO_2$, was proposed (*106*, Zincke and Brune). Three years later an analogous proposal was advanced and the intermediate was called a "sulfene" (*107*, Wedekind and Schenk). To date there have been no authenticated reports of the isolation of a sulfene but evidence is accumulating which indicates their existence.

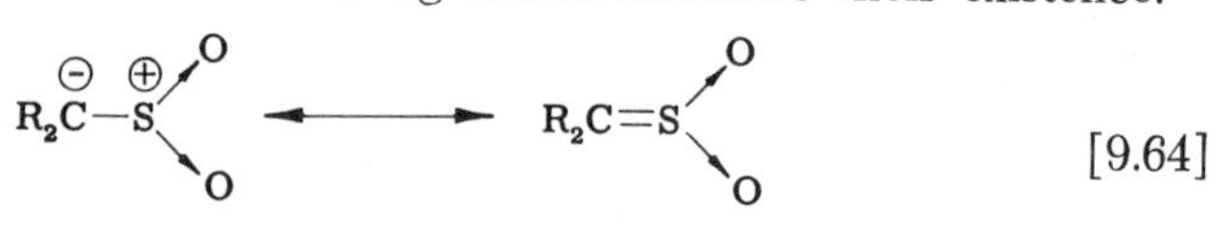

[9.64]

Sulfenes may well be considered as ylids [9.64]. Several groups (*108, 109*) have proposed such structures but there has been no attempt to test these proposals. The fact that ring formation (reaction 1 below) takes place only with highly polarized olefins indicates that the sulfene certainly is electrically unsymmetrical. However, there have been no attempts to react sulfenes with carbonyl compounds, a typical ylid substrate.

The intermediacy of sulfenes has been proposed to account for the products from three distinct types of reactions as shown in the following paragraphs.

(1) *The reaction of sulfonyl halides with tertiary amines* [9.65].

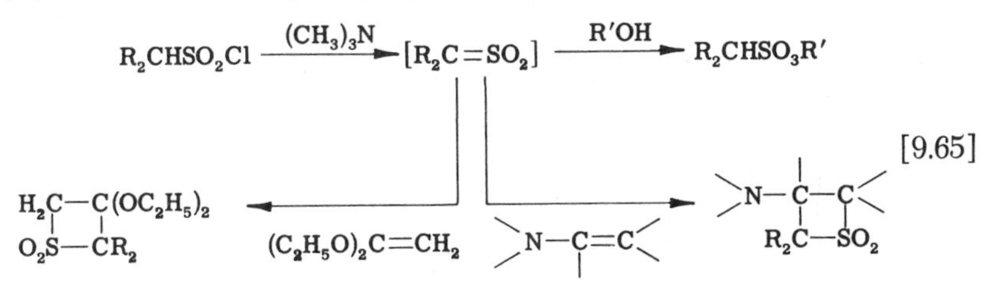

The sulfene has been shown to add to alcohols to form sulfonates (*108*, King and Durst; *109*, Truce *et al.*), to add to ketene diethyl acetal (*110*, Truce and Norell) and to add to enamines (*111*, Stork and Borowitz; *112, 113*, Opitz *et al.*).

(2) *The reaction of diazoalkanes with sulfur dioxide* [9.66]. Staud-

$$R_2CN_2 \xrightarrow{SO_2} R_2C{=}SO_2 \xrightarrow{R_2CN_2} \underset{\diagdown SO_2 \diagup}{R_2C-CR_2} + N_2 \qquad [9.66]$$

inger and Pfenninger (*114*) reported that the reaction of diphenyldiazomethane with sulfur dioxide afforded tetraphenylethylene sulfone. More recently, Hesse *et al.* (*115*) reported the formation of the heretofore unknown ethylenesulfone. They proposed that the initially formed sulfene possessed an ylid structure and that reaction with diazomethane proceeded via attack of the latter on the sulfur atom followed by expulsion of nitrogen [9.67].

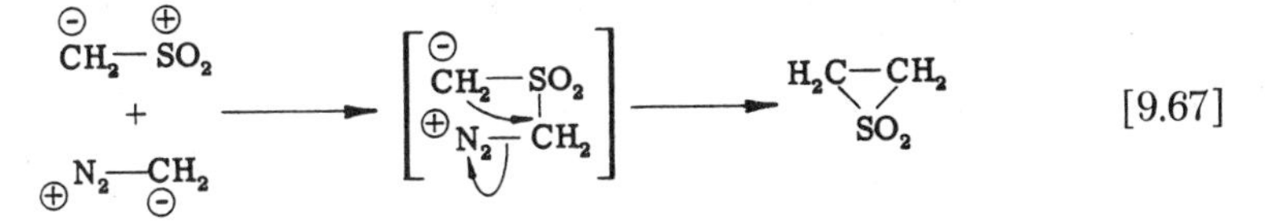

(3) *The photolysis of unsaturated sultones* [9.68]. King *et al.* (*116*) observed the formation of esters from a variety of sultones upon irradia-

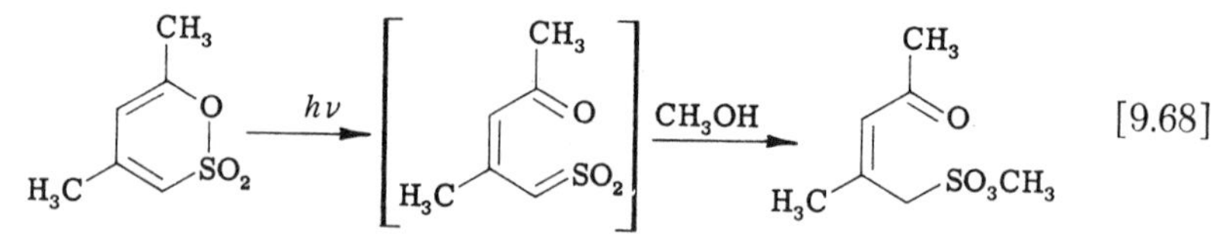

tion in methanolic solution. They proposed the intermediacy of conjugated sulfenes which resulted in the net migration of one double bond to a new position.

The only concrete evidence for the existence of sulfenes was provided in two simultaneous communications by the King group (*108*) and the Truce group (*109*). They observed that treatment of an alkyl sulfonyl chloride with triethylamine in a solution of an *O*-deutero-alcohol afforded the corresponding sulfonate esters. The esters were shown to be monodeuterated with little or no dideuterated material being detected. The lack of dideuterated material indicated that the exchange could not have involved a carbanion of the sulfonyl halide or the sulfonate and

$$R_2CHSO_2Cl \xrightarrow{(C_2H_5)_3N} [R_2C{=}SO_2] \xrightarrow{R'OD} R_2CD{-}SO_3R' \qquad [9.69]$$

indicates that the single deuterium atom probably was incorporated via an elimination-addition sequence involving a sulfene intermediate [9.69].

Suld and Price (*117*) recently have described an interesting series of aromatic sulfonium ylids. They found that treatment of 2,4,6-triphenylthiopyrylium perchlorate with phenyllithium afforded a violet, amorphous substance which was formulated as 1,2,4,6-tetraphenylthiabenzene (XLVII). The thiabenzene could be kept under nitrogen for a week but

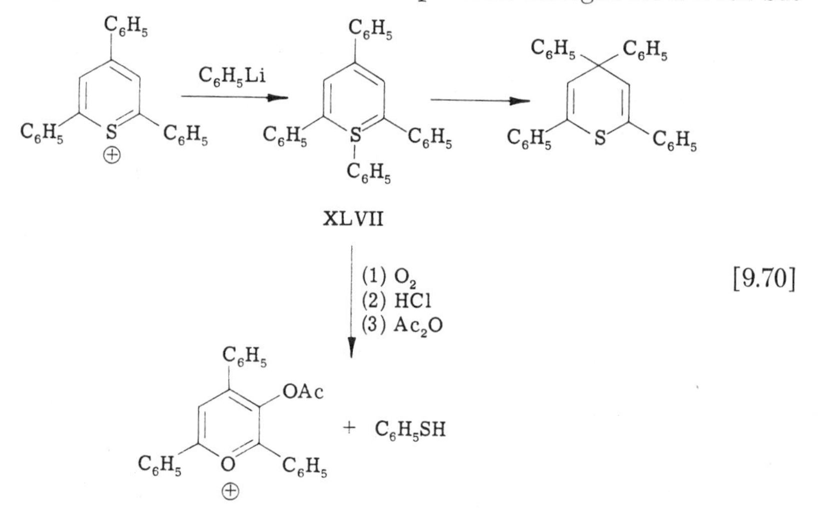

the color slowly faded and 2,4,4,6-tetraphenylthiapyran could be isolated. When the thiabenzene (XLVII) was treated with oxygen a peroxide was formed but addition of hydrogen chloride liberated phenylmercaptan and a zwitterion, the latter being converted to 3-acetoxy-2,4,6-triphenylpyrylium perchlorate by acetylation. The isolation of phenylmercaptan indicated that one of the phenyl groups in the thiabenzene, presumably that from the phenyllithium, must have been attached to sulfur or must have migrated to it during the oxidation procedure.

More recently Price *et al.* (*118*) have reported the preparation of four new thiabenzene derivatives, 1-phenyl-1-thianaphthalene, 2-phenyl-2-thianaphthalene, 10-phenyl-10-thiaanthracene and 9,10-diphenyl-10-thiaanthracene, all of which were prepared from the corresponding thiapyrylium perchlorates by treatment with phenyllithium. These systems all were more stable than the initially reported thiabenzene (XLVII). All were red-brown solids, capable of isolation and purification, and were stable to oxygen and even to boiling acetic acid solution.

The dipole moments of all four of the thiabenzene derivatives have been measured and were between 1.5 and 1.9 D. This is far less than the values of near 7 D reported for other ylids (*41*) and indicates a high degree of covalent character for these compounds. The NMR spectra of the thiabenzene derivatives showed only aromatic-type hydrogens with absorption centered near 3.0 τ. In view of the chemical stability and the above physical properties, it is apparent that while these compounds might be represented as resonance hybrids of at least two important contributing structures, the covalent form certainly must be the most important [9.71]. Therefore, while these compounds are, in a formal sense,

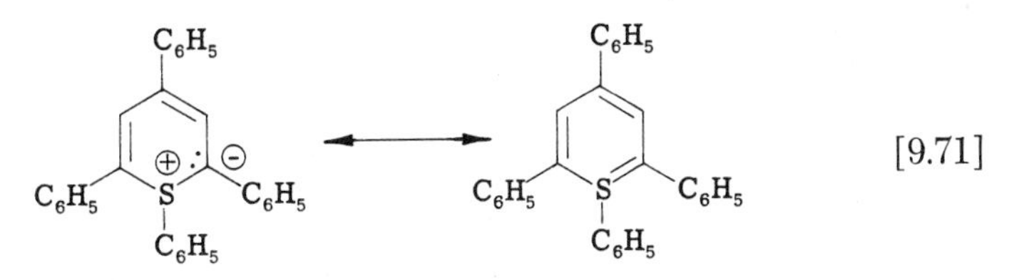

[9.71]

sulfonium ylids their properties are considerably different from the usual sulfonium ylid and they are best considered as a separate and unique class of compounds. These cyclic sulfonium ylids appear to be somewhat more stable than the similar cyclic phosphonium ylids reported by Markl (*119*).

Price (*118*) has accounted for the remarkable stability of the bicyclic and tricyclic derivatives of thiabenzene by proposing that the sulfur atom uses a $3p_z$-orbital for overlap with the $2p$-orbitals of the adjacent carbon atoms in the formation of the cyclic conjugated system. The unshared pair on sulfur would be promoted to a $3d$-orbital. The σ bond skeleton was assumed to be the result of trigonal hybridization. It also was proposed that the monocyclic derivative (XLVII) used p^3-orbitals for the σ skeleton in order to reduce the steric hindrance between the 1,2- and 6-phenyl groups and that this then required the use of $3d$-orbitals for overlap with the carbon $2p$-orbitals to form the cyclic conjugated system, the result being a less stable aromatic system.

IV. Iminosulfuranes

For every sulfur ylid that is known, an isoelectronic sulfur-nitrogen compound (iminosulfurane) could theoretically be prepared. Several compounds containing sulfur bonded to nitrogen have been prepared but very little has been done in the way of exploring the scope of their reactions.

One of the earliest preparations of such compounds was reported by Mann and Pope in 1922 (*120*). They found that ethylsulfide would react with chloramine-T to afford *N*-tosyliminodiethylsulfurane(*N*-tosyldiethylsulfidimide, XLVIII). This substance, a crystalline solid of m.p. 144°,

$$(C_2H_5)_2S + (p)CH_3C_6H_4SO_2NHCl \longrightarrow (C_2H_5)_2\overset{\oplus}{S}-\overset{\ominus}{N}-SO_2C_6H_4CH_3(p) \quad [9.72]$$

XLVIII

was stable in the atmosphere but could be hydrolyzed to ethylsulfoxide and p-toluenesulfonamide. This type of hydrolysis is typical of phosphonium ylids, affording phosphine oxides and a hydrocarbon, rather than of sulfonium ylids which normally form sulfides and alcohols.

Kenyon *et al.* (*121*) confirmed Mann's observations and noted that sulfidimides could be resolved into optical antipodes in a manner reminiscent of the resolution of sulfoxides. This led them to propose that the sulfidimides (and sulfidimines) were isoelectronic with sulfoxides, and all of their known properties are in accord with this suggestion.

Kenyon *et al.* (*121*) also proposed that the sulfidimides were formed by the transformation of chloramine-T into a nitrene followed by electrophilic attack on the sulfide [9.73]. Support for this proposal has been

$$Ts-N\begin{matrix} H \\ Cl \end{matrix} \longrightarrow [Ts-\ddot{\underset{..}{N}}] \xrightarrow{R_2S} R_2\overset{\oplus}{S}-\overset{\ominus}{N}-Ts \longleftrightarrow R_2S{=}N-Ts \quad [9.73]$$

provided by the recent work of Horner and Christmann (*122*) who found that reaction of sulfides with sulfonyl azides in the presence of ultraviolet light or under thermal conditions led to the evolution of nitrogen and the formation of sulfidimides in 48–55% yields.

Two other methods for the preparation of sulfidimides are available. Tarbell and Weaver (*123*) found that treatment of sulfoxides with sulfonyl amides in the presence of a dehydrating agent such a phosphorus pentoxide or acetic anhydride led to the formation of sulfidimides [9.74].

$$(C_6H_5)_2SO + TsNH_2 \xrightarrow{P_2O_5} (C_6H_5)_2\overset{\oplus}{S}-\overset{\ominus}{N}-Ts \qquad [9.74]$$

The generality of the reaction was demonstrated by the observation that even trichloroacetamide reacted with thiophane oxide to form the corresponding sulfidimide by an apparent dehydration. The sulfidimides were found to be resistant to alkaline hydrolysis but were readily hydrolyzed in acidic medium back to the sulfoxide and amide.

King (*124*) recently has discovered a most interesting condensation reaction which afforded sulfidimides in high yield in an exothermic reaction. *p*-Toluenesulfonyl isocyanate reacted with methylsulfoxide at room temperature and in the absence of a catalyst to afford an 87% yield of *N*-tosyliminodimethylsulfurane. The reaction may have taken place via a cyclic intermediate as shown in [9.75]. The evolution of carbon

$$\begin{matrix} Ts-N=C=O \\ + \\ (CH_3)_2SO \end{matrix} \longrightarrow \begin{matrix} Ts-N-C=O \\ | \quad\ | \\ (CH_3)_2S-O \end{matrix} \longrightarrow \begin{matrix} Ts-\overset{\ominus}{N}-\overset{\oplus}{S}(CH_3)_2 \\ + \\ CO_2 \end{matrix} \qquad [9.75]$$

dioxide certainly is a driving force for the reaction. This condensation is reminiscent of a similar reaction between phosphine oxides and isocyanates which formed phosphinimines and carbon dioxide in the first reaction (*125*, Campbell *et al.*). In that instance, however, the imine could not be isolated since it reacted rapidly with additional isocyanate. The failure to do so in the sulfur case may be due to a decreased nucleophilicity of the tosylimine. Monagle (*126*) and Appel and Rittersbacher (*127*) have reported additional examples of the formation of sulfidimides from sulfoxides and isocyanates. Kresze and his group (*128*) have reported similar condensations between the sulfur analogs of isocyanates and sulfoxides which resulted in the evolution of sulfur dioxide and the formation of sulfidimides [9.76].

$$\begin{matrix} CH_3SO_2N=S=O \\ + \\ (C_6H_5)_2SO \end{matrix} \longrightarrow \left[\begin{matrix} CH_3SO_2N-S=O \\ | \quad\ | \\ (C_6H_5)_2S-O \end{matrix}\right] \longrightarrow \begin{matrix} CH_3SO_2\overset{\ominus}{N}-\overset{\oplus}{S}(C_6H_5)_2 \\ + \\ SO_2 \end{matrix} \qquad [9.76]$$

Sulfidimides have been found to undergo the Stevens rearrangement under conditions varying from standing at room temperature to heating to 200° for sixteen hours. *N*-Tosyliminodibenzylsulfurane, under the lat-

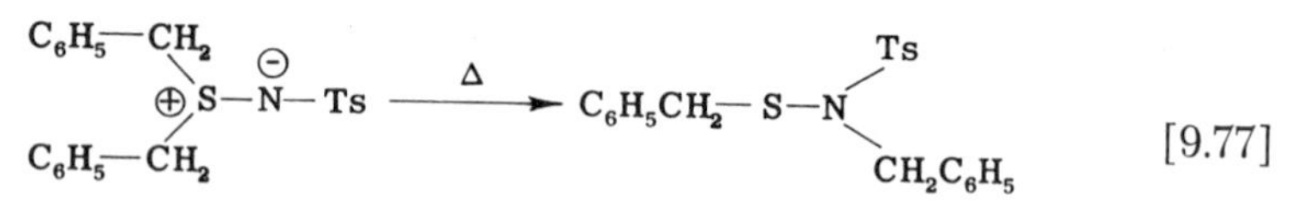

ter conditions, afforded a tertiary amine (XLIX) whereas *N*-tosyliminodiallylsulfurane underwent the same rearrangement upon standing at room temperature (*129*, Ash and Challenger).

The interesting material, thionylaniline, has been shown to behave essentially as a sulfidimine. The most striking example of the resemblance was the observation of its condensation with a variety of carbonyl compounds to eliminate sulfur dioxide and form imines [9.78] (*130, 131,*

$$C_6H_5-\overset{\ominus}{N}-\overset{\oplus}{S}{=}O + C_6H_5CHO \longrightarrow \left[\begin{array}{c} C_6H_5N-\overset{\oplus}{S}O \\ | \quad\quad | \\ C_6H_5CH-\underset{\ominus}{O} \end{array}\right] \longrightarrow C_6H_5N{=}CHC_6H_5 + SO_2 \quad [9.78]$$

Kresze *et al.; 132,* Senning). This reaction is strictly analogous to that of phosphinimines with carbonyl compounds and the mechanism probably is analogous.

In recent years and particularly through the efforts of Appel *et al.* (*133, 134*) methods have been developed for the preparation of the free sulfidimines. Several of these methods are presented in [9.79]. In some

$$R_2S + ClNH_2 \xrightarrow{NH_3} R_2\overset{\oplus}{S}-NH_2\ Cl^{\ominus} \xrightarrow{NH_2^{\ominus}} R_2\overset{\oplus}{S}-\overset{\ominus}{N}H$$

$$R_2S + H_2NOSO_3H \xrightarrow{^{\ominus}OCH_3} R_2\overset{\oplus}{S}-\overset{\ominus}{N}H \quad [9.79]$$

$$R_2SCl_2 + NH_3 \longrightarrow R_2\overset{\oplus}{S}-NH_2\ Cl^{\ominus} \xrightarrow{NH_2^{\ominus}} R_2\overset{\oplus}{S}-\overset{\ominus}{N}H$$

instances the sulfidiminium salt was isolated and characterized before treatment with a strong base (sodamide) to form the sulfidimine. These methods all are analogous to those employed in the synthesis of phosphinimines. The structures of the sulfidimines most often have been proven by tosylation to form the sulfidimide, an example of the acylation of the imines.

Most of the imines were low melting solids or oils. They behaved as expected for ylid-type compounds being insoluble in water but soluble in dilute acid. Upon hydrolysis in acidic or basic media they were converted to ammonia and a sulfoxide. Iminodiethylsulfurane reacted with carbon dioxide to form what appeared to be a carboxylate zwitterion

$[(C_2H_5)_2S^+—NH—COO^-]$. The imine also reacted with benzaldehyde but formed an as yet unidentified substance which apparently incorporated all of the elements of the two reactants into its structure.

A final series of sulfur-nitrogen compounds that should be mentioned are the iminooxysulfuranes (oxysulfidimines) and their sulfonated derivatives (oxysulfidimides). They were first reported by Bentley and Whitehead (*135*) from the oxidation of the corresponding sulfidimides as in [9.80]. The stability of the imide linkage to oxidation conditions

$$(CH_3)_2\overset{\oplus}{S}—\overset{\ominus}{N}—Ts \xrightarrow{KMnO_4} (CH_3)_2\overset{\oplus}{S}(\to O)—\overset{\ominus}{N}—Ts \xrightarrow{H^{\oplus}} (CH_3)_2\overset{\oplus}{S}(\to O)—\overset{\ominus}{N}H \qquad [9.80]$$

certainly is surprising, especially in view of the fact that phosphonium ylids are cleaved readily by oxidizing agents and in view of the report by Appel and Buchner (*134*) that oxidation of a sulfidimide with hydrogen peroxide led to cleavage of the imide bond and the formation of a sulfone. Whitehead and Bentley (*136*) have noted, however, that the oxysulfidimines were cleaved to sulfones by more vigorous oxidation conditions. They also noted that iminodiethyloxysulfurane was practically inert to hydrolysis in basic or acidic media. In contrast, Misani *et al.* (*137*) claimed that iminodiphenyloxysulfurane could be hydrolyzed in acidic or basic media to diphenylsulfone.

Several groups (*136, 137*) have found that the oxysulfidimines could be prepared directly by treating a sulfoxide with hyrazoic acid in sulfuric acid medium [9.81], essentially under the conditions of the Schmidt

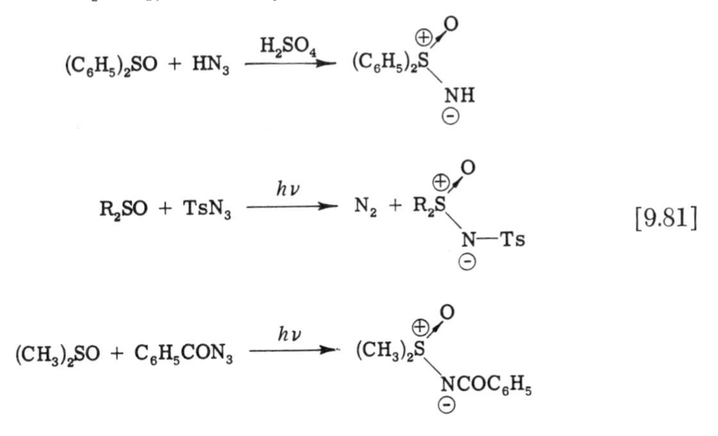

reaction. Horner and Christmann (*122*) found that a nitrene, generated photochemically or thermally from *p*-toluenesulfonyl azide, would react with a sulfoxide to form an oxysulfidimide but in relatively low yields.

They also noted that benzoyl azide and methylsulfoxide would react under photochemical conditions to afford *N*-benzoyliminodimethyloxysulfurane.

The ultraviolet spectra of a series of oxysulfidimines have been reported (*138*, Barash) and claimed to provide evidence for use of 3*d*-orbitals of sulfur. The oxysulfidimines exhibit typical nucleophilicity in that they could be acylated with acid chlorides (*139*, Wehr) and with acid anhydrides (*137*, *139*). There has been no attempt to determine whether they will react with carbonyl compounds to form aziridines as might be expected from their isoelectronic relationship to oxysulfuranes.

It is apparent that apart from the discovery of several methods for the preparation of the various sulfur-nitrogen compounds there is very little known about their chemistry. There has been no attempt to explore the variety of reactions that may be undergone by these interesting compounds nor has there been any serious attempt to study their physical properties. This obviously is a fertile area of potentially interesting research.

Appendix

Since the original manuscript was completed, activity in sulfur ylid research has increased leading to the isolation and characterization of several new sulfur ylids. Hochrainer and Wessely (*140*) isolated the cyclic bis-ester ylid (L) from reaction of methyl sulfoxide with the corresponding bromide. Linn *et al.* (*141*) and Middleton *et al.* (*142*) both have reported obtaining dicyanomethylenedimethylsulfurane (LI) from the reaction of methyl sulfide with tetracyanoethylene oxide or with bromomalononitrile. Use of the former procedure with a variety of different sulfides afforded a series of dicyanomethylenesulfuranes, all of which were isolated and characterized. These sulfuranes had dipole moments in the range 7.0–8.1 D. They would not react with carbonyl

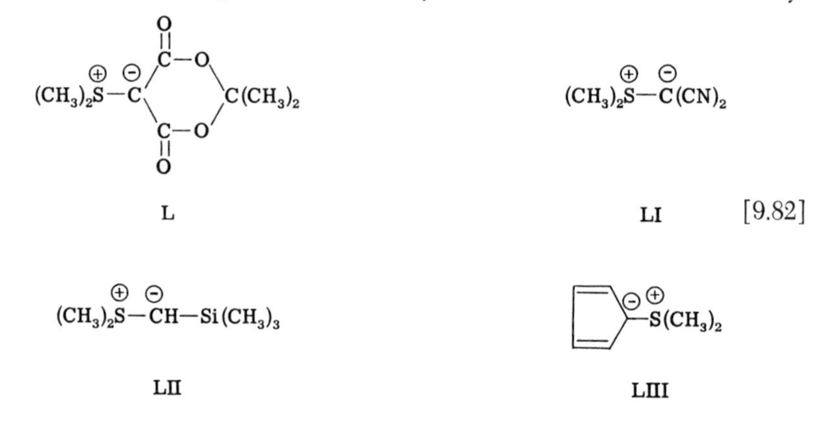

compounds and were stable to hydrogen peroxide, water, weak acids, and weak bases. Miller (*143*) has reported the preparation of the liquid trimethylsilylmethylenedimethylsulfurane (LII) by the salt method. Behringer and Scheidl (*144*) have isolated cyclopentadienylidenedimethylsulfurane (LIII). This ylid underwent electrophilic substitution reactions and had a dipole moment of 5.7 D, considerably smaller than the 6.99 D moment reported for cyclopentadienylidenetriphenylphosphorane (*145*).

Johnson and Amel (*146*) have prepared and characterized phenacylidenedimethylsulfurane (LIV). This ylid reacted with either its conjugate acid or with phenacyl bromide to afford tribenzoylcyclopropane, apparently by an alkylation-elimination-addition mechanism (see equation [3.104], p. 101). Thermal decomposition of LIV afforded the same cyclopropane, probably via a carbenoid mechanism. Benzoylation of LIV with benzoic anhydride or with benzoyl chloride resulted in C-benzoylation or O-benzoylation, respectively [9.83]. The ylid would form an oxirane

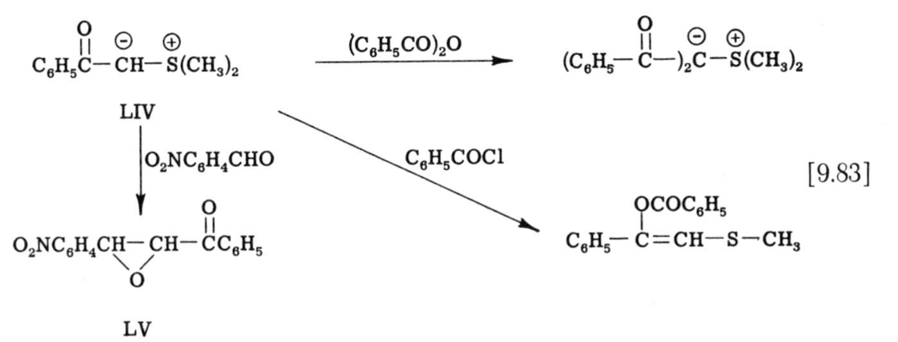

(LV), although in low yield, upon reaction with *p*-nitrobenzaldehyde. This is contrary to the behavior of the analogous phenacylidenesulfurane (XVIII) as reported by Nozaki *et al.* (*39*).

Cook *et al.* (*147*) have provided additional examples verifying Corey and Chaykovsky's (*22*) conclusions that methylenedimethylsulfurane usually attacks a cycloalkanone at an axial position while methylenedimethyloxysulfurane (VIII) attacks at an equatorial position. Konig, Metzger, and Seelert (*148–150*) have published full details of their study of the reactions of VIII with conjugated olefins, nitriles, and acylating agents.

Additional information pertaining to the chemistry of iminosulfuranes also has been reported. Appel and Buchler (*151*) have prepared the bisimine (LVI) from phenyl sulfide and nitrogen trichloride. Clemens *et al.* (*152, 153*) have shown that the dialkyl sulfurdiimines (LVII) have ylidic character since they were found to react with benzaldehyde to afford Schiff bases, presumably via a betaine intermediate as for the Wittig reaction [9.84]. The diimines also reacted with isocyanates and

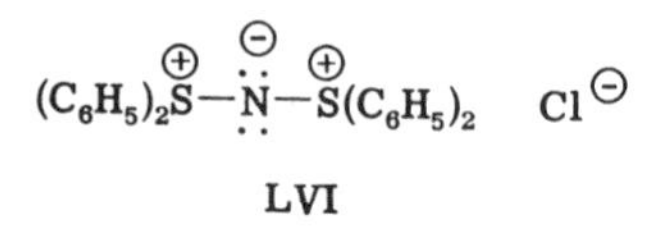

LVI

$$\underset{\text{LVII}}{R{-}N{=}S{=}N{-}R} + C_6H_5CHO \longrightarrow C_6H_5CH{=}N{-}R + RNSO \qquad [9.84]$$

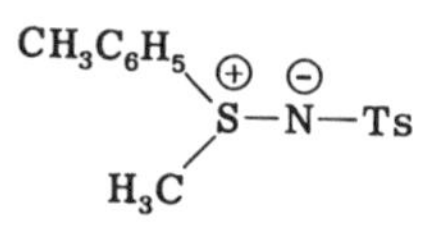

LVIII

isothiocyanates, again by nitrogen attack on carbon. The ylidic character of *N*-sulfinylaniline was demonstrated again (*154*, Bott) by its reaction with nitrosyl perchlorate to afford benzenediazonium perchlorate, a reaction very similar to that between iminiphosphoranes and nitrosyl chloride as reported by Zimmer and Singh (*155*).

Day and Cram (*156*) have prepared optically active *N*-tosyliminomethyl-*p*-tolylsulfurane (LVIII) either from optically active *p*-tolylmethylsulfoxide and *p*-toluenesulfonamide or from the same sulfoxide and *N*-sulfinyl-*p*-toluenesulfonamide. Both preparative reactions took place with inversion of the sulfur configuration. Hydrolysis of LVIII with methanolic potassium hydroxide to regenerate the sulfoxide and *p*-toluenesulfonamide also took place with inversion of configuration.

REFERENCES

1. A. Wm. Johnson and R. B. LaCount, *Chem. & Ind.* (*London*) 1440 (1958).
2. A. B. Burg, *in* "Organic Sulfur Compounds" (N. Kharasch, ed.), Vol. 1, pp. 30–40, Pergamon Press, New York, 1961.
3. C. C. Price and S. Oae, "Sulfur Bonding." Ronald Press, New York, 1962.
4. G. Cilento, *Chem. Revs.* **60,** 147 (1960).
5. D. P. Craig, A. Maccoll, R. S. Nyholm, L. E. Orgel and L. E. Sutton, *J. Chem. Soc.* 332 (1954).
6. D. P. Craig, *Chem. Soc. Spec. Publ. No.* **12,** 343 (1958).
7. D. P. Craig and E. A. Magnusson, *J. Chem. Soc.* 4895 (1956).
8. R. J. Gillespie and E. A. Robinson, *Can. J. Chem.* **41,** 2074 (1963).
9. D. W. J. Cruickshank and B. C. Webster, *J. Chem. Phys.* **40,** 3733 (1964).
10. W. von E. Doering and A. K. Hoffmann, *J. Am. Chem. Soc.* **77,** 521 (1955).
11. F. G. Bordwell and P. S. Boutan, *J. Am. Chem. Soc.* **78,** 87, 854 (1956).
12. N. F. Blau and C. G. Stuckwisch, *J. Org. Chem.* **22,** 82 (1957).
13. L. Doub and J. M. Vandenbelt, *J. Am. Chem. Soc.* **69,** 2714 (1947).

14. E. A. Fehnel and M. Carmack, *J. Am. Chem. Soc.* **71**, 84 (1949).
15. W. von E. Doering and K. C. Schreiber, *J. Am. Chem. Soc.* **77**, 514 (1955).
16. F. G. Bordwell and G. D. Cooper, *J. Am. Chem. Soc.* **74**, 1058 (1952).
17. E. J. Corey and M. Chaykovsky, *J. Org. Chem.* **28**, 254 (1963).
18. A. Wm. Johnson, *Chem. & Ind. (London)* 1119 (1963).
19. E. J. Corey and M. Chaykovsky, *J. Am. Chem. Soc.* **84**, 866 (1962), **87**, 1345 (1965).
20. C. Walling and L. Bollyky, *J. Org. Chem.* **28**, 256 (1963), **29**, 2699 (1964).
21. S. G. Smith and S. Winstein, *Tetrahedron* **3**, 317 (1958).
22. E. J. Corey and M. Chaykovsky, *J. Am. Chem. Soc.* **84**, 867 (1962), **87**, 1353 (1965).
23. H. Zollinger, W. Buchler and C. Wittwer, *Helv. Chim. Acta* **36**, 1711 (1953).
24. H. Zollinger and C. Wittwer, *Helv. Chim. Acta* **39**, 350 (1956).
25. H. Bredereck, E. Buder and G. Hoschele, *Chem. Ber.* **87**, 784 (1954).
26. C. K. Ingold and J. A. Jessop, *J. Chem. Soc.* 713 (1930).
27. G. Wittig and H. Fritz, *Ann.* **517**, 39 (1952).
28. V. Franzen and H. E. Driessen, *Chem. Ber.* **96**, 1881 (1963).
29. A. Wm. Johnson, *J. Org. Chem.* **28**, 252 (1963).
30. E. J. Corey and M. Chaykovsky, *J. Am. Chem. Soc.* **84, 3782 (1962), 87, 1353** (1965).
31. A. Wm. Johnson and R. B. LaCount, *J. Am. Chem. Soc.* **83**, 417 (1961).
32. L. G. Makarova and A. N. Nesmeyanov, *Izvest. Akad. Nauk. S.S.S.R.* 617 (1945).
33. A. Wm. Johnson, unpublished results.
34. V. Franzen, H. J. Schmidt and C. Mertz, *Chem. Ber.* **94**, 2942 (1961).
35. A. Wm. Johnson, V. J. Hruby and J. L. Williams, *J. Am. Chem. Soc.* **86**, 918 (1964).
36. V. Franzen, H. I. Joschek, and C. Mertz, *Ann.* **654**, 82 (1962).
37. H. Hellmann and D. Eberle, *Ann.* **662**, 188 (1963).
38. E. D. Hughes and K. I. Kuriyan, *J. Chem. Soc.* 1609 (1935).
39. H. Nozaki, K. Kondo and M. Takaku, *Tetrahedron Letters* 251 (1965).
40. G. Wittig and G. Felletschin, *Ann.* **555**, 133 (1944).
41. G. M. Phillips, J. S. Hunter and L. E. Sutton, *J. Chem. Soc.* 146 (1945).
42. C. K. Ingold, "Structure and Mechanism in Organic Chemistry," p. 177. Cornell University Press, Ithaca, New York, 1953.
43. D. P. Craig, personal communication as cited in ref. 42, p. 98, footnote 8.
44. M. L. Huggins, *J. Am. Chem. Soc.* **75**, 4126 (1953).
45. C. P. Smyth, "Dielectric Behaviour and Structure," p. 303. McGraw-Hill Book Co., Inc., New York, 1955.
46. F. A. Cotton and R. Francis, *J. Am. Chem. Soc.* **82**, 2986 (1960).
47. J. G. Pritchard and P. C. Lauterbur, *J. Am. Chem. Soc.* **83**, 2105 (1961).
48. C. G. Swain and E. R. Thornton, *J. Org. Chem.* **26**, 4808 (1961).
49. L. A. Pinck and G. E. Hilbert, *J. Am. Chem. Soc.* **60**, 494 (1938), **68**, 751 (1946).
50. M. Sommelet, *Compt. rend.* **205**, 56 (1937).
51. C. K. Ingold, ref. 42, p. 643.
52. R. Oda, Y. Ito and M. Okano, *Tetrahedron Letters* No. 1, 7 (1964).
53. C. G. Swain and E. R. Thornton, *J. Am. Chem. Soc.* **83**, 4033 (1961).
54. I. Rothberg and E. R. Thornton, *J. Am. Chem. Soc.* **85**, 1704 (1964), **86**, 3296, 3302 (1964).

55. V. Franzen and G. Wittig, *Angew. Chem.* **72**, 417 (1960).
56. G. Wittig and D. Krauss, *Ann.* **679**, 34 (1964).
57. L. Friedman and J. G. Berger, *J. Am. Chem. Soc.* **83**, 492 (1961).
58. A. Wm. Johnson and V. J. Hruby, *J. Am. Chem. Soc.* **84**, 3586 (1962).
59. R. Mechoulam and F. Sondheimer, *J. Am. Chem. Soc.* **80**, 4386 (1958).
60. A. Schonberg, A. Mustafa and N. Latif, *J. Am. Chem. Soc.* **75**, 2267 (1953).
61. G. L. Closs, *J. Am. Chem. Soc.* **84**, 809 (1962).
62. C. K. Ingold, J. A. Jessop, K. I. Kuriyan and A. M. M. Mandour, *J. Chem. Soc.* 533 (1933).
63. C. A. VanderWerf, W. E. McEwen and M. Zanger, *J. Am. Chem. Soc.* **81**, 3806 (1959).
64. H. Mackle, *Tetrahedron* **19**, 1159 (1963).
65. A. F. Bedford and C. T. Mortimer, *J. Chem. Soc.* 1622 (1960).
66. V. Franzen and H. E. Driessen, *Tetrahedron Letters* 661 (1962).
67. A. Wm. Johnson and V. J. Hruby, *Abstr. Papers, 142nd Meeting, Am. Chem. Soc.*, p. 31Q (1962).
68. E. J. Corey and W. Oppolzer, *J. Am. Chem. Soc.* **86**, 1899 (1964).
69. J. Read and I. G. M. Campbell, *J. Chem. Soc.* 2377 (1930).
70. A. J. Speziale and D. E. Bissing, *J. Am. Chem. Soc.* **85**, 3878 (1963).
71. E. J. Corey and M. Chaykovsky, *Tetrahedron Letters* 169 (1963).
72. A. J. Speziale, C. C. Tung, K. W. Ratts and A. Yao, *Abstr. Papers, 149th Meeting, Am. Chem. Soc.*, p. 44P (1965); *J. Am. Chem. Soc.* **87**, 3460 (1965).
73. W. E. Truce and V. V. Badiger, *J. Org. Chem.* **29**, 3277 (1964).
74. W. D. Emmons, *J. Am. Chem. Soc.* **79**, 5739 (1957).
75. J. S. Splitter and M. Calvin, *J. Org. Chem.* **23**, 651 (1958).
76. A. C. Cope and A. S. Mehta, *J. Am. Chem. Soc.* **85**, 1949 (1963).
77. S. J. Cristol and F. R. Stermitz, *J. Am. Chem. Soc.* **82**, 4692 (1960).
78. F. Weygand and H. Daniel, *Chem. Ber.* **94**, 3145 (1961).
79. V. Franzen and C. Mertz, *Chem. Ber.* **93**, 2819 (1960).
80. V. Franzen and H. J. Schmidt, *Chem. Ber.* **94**, 2937 (1961).
81. W. H. Saunders and D. Paulovic, *Chem. & Ind. (London)* 180 (1962).
82. T. Thomson and T. S. Stevens, *J. Chem. Soc.* 69 (1932).
83. W. E. Parham and S. H. Groen, *J. Org. Chem.* **30**, 728 (1965).
84. C. R. Hauser, S. W. Kantor and W. R. Brasen, *J. Am. Chem. Soc.* **75**, 2660 (1953).
85. D. M. Burness, *J. Org. Chem.* **24**, 849 (1959).
86. F. G. White and L. L. Ingraham, *J. Am. Chem. Soc.* **84**, 3109 (1962).
87. R. Breslow, *J. Am. Chem. Soc.* **80**, 3719 (1958).
88. P. Haake and W. B. Miller, *J. Am. Chem. Soc.* **85**, 4044 (1963).
89. T. J. Curphey, *J. Am. Chem. Soc.* **87**, 2063 (1965).
90. R. Kuhn and H. Trischmann, *Ann.* **611**, 117 (1958).
91. P. T. Izzo, *J. Org. Chem.* **28**, 1713 (1963).
92. H. Metzger and K. Seelert, *Z. Naturforschg.* **18b**, 335 (1963).
93a. H. Metzger and K. Seelert, *Z. Naturforschg.* **18b**, 336 (1963), *Angew. Chem.* **75**, 919 (1963).
93b. H. Konig and H. Metzger, *Z. Naturforschg.* **18b**, 976, 987 (1963).
94. H. Metzger, H. Konig and K. Seelert, *Tetrahedron Letters* 867 (1964).
95. V. J. Traynelis and J. V. McSweeney, *Abstr. Papers, 148th Meeting, Am. Chem. Soc.* p. 12S (1964).
96. E. J. Corey and M. Chaykovsky, *J. Am. Chem. Soc.* **86**, 1640 (1964).
97. H. Konig and H. Metzger, *Tetrahedron Letters* 3003 (1964).

98. M. C. Henry and G. Wittig, *J. Am. Chem. Soc.* **82**, 563 (1960).
99. H. D. Becker and G. A. Russell, *J. Org. Chem.* **28**, 1896 (1963).
100. H. D. Becker, G. J. Mikol and G. A. Russell, *J. Am. Chem. Soc.* **85**, 3410 (1963).
101. E. J. Corey and M. Chaykovsky, *J. Am. Chem. Soc.* **86**, 1639 (1964).
102. J. Hine and J. J. Porter, *J. Am. Chem. Soc.* **82**, 6178 (1960).
103. C. Y. Meyers, C. Rimaldi and L. Bonoli, *Abstr. Papers, 144th Meeting, Am. Chem. Soc.*, p. 4M (1963).
104. W. E. Truce and R. H. Knospe, *J. Am. Chem. Soc.* **77**, 5063 (1955).
105. H. D. Becker and G. A. Russell, *J. Org. Chem.* **28**, 1896 (1963).
106. T. Zincke and R. Brune, *Chem. Ber.* **41**, 902 (1908).
107. E. Wedekind and D. Schenk, *Chem. Ber.* **44**, 198 (1911).
108. J. F. King and T. Durst, *J. Am. Chem. Soc.* **86**, 287 (1964); **87**, 5684 (1965).
109. W. E. Truce, R. W. Campbell and J. R. Norell, *J. Am. Chem. Soc.* **86**, 288 (1964).
110. W. E. Truce and J. R. Norell, *J. Am. Chem. Soc.* **85**, 3231 (1963).
111. G. Stork and I. J. Borowitz, *J. Am. Chem. Soc.* **84**, 313 (1962), **86**, 1146 (1964).
112. G. Opitz and H. Adolph, *Angew. Chem.* **74**, 77 (1962).
113. G. Opitz and K. Fischer, *Z. Naturforschg.* **18b**, 775 (1963).
114. H. Staudinger and F. Pfenninger, *Chem. Ber.* **49**, 1941 (1916).
115. G. Hesse, E. Reichold and S. Majmudar, *Chem. Ber.* **90**, 2106 (1957).
116. J. F. King, P. deMayo, E. Markved, A. B. M. A. Sattar and A. Stoessl, *Can. J. Chem.* **41**, 100 (1963).
117. G. Suld and C. C. Price, *J. Am. Chem. Soc.* **83**, 1770 (1961), **84**, 2094 (1962).
118. C. C. Price, M. Hori, T. Parasaron and M. Polk, *J. Am. Chem. Soc.* **85**, 2278 (1963).
119. G. Markl, *Angew. Chem.* **75**, 168 (1963), *International Edn.* **2**, 153 (1963).
120. F. G. Mann and W. J. Pope, *J. Chem. Soc.* **121**, 1052 (1922).
121. S. G. Clarke, J. Kenyon and H. Phillips, *J. Chem. Soc.* 188 (1927).
122. L. Horner and A. Christmann, *Chem. Ber.* **96**, 388 (1963).
123. D. S. Tarbell and C. Weaver, *J. Am. Chem. Soc.* **63**, 2939 (1941).
124. C. King, *J. Org. Chem.* **25**, 352 (1960).
125. T. W. Campbell, J. J. Monagle and V. S. Foldi, *J. Am. Chem. Soc.* **84**, 3673 (1962).
126. J. J. Monagle, *J. Org. Chem.* **27**, 3851 (1962).
127. R. Appel and H. Rittersbacher, *Chem. Ber.* **97**, 852 (1964).
128. G. Schulz and G. Kresze, *Angew. Chem.* **75**, 1022 (1963).
129. A. S. F. Ash and F. Challenger, *J. Chem. Soc.* 1877 (1951), 2792 (1952).
130. R. Albrecht, G. Kresze and B. Mlakar, *Chem. Ber.* **97**, 483 (1964).
131. G. Kresze, D. Sommerfeld and R. Albrecht, *Chem. Ber.* **98**, 601 (1965).
132. A. Senning, *Acta Chem. Scand.* **18**, 1958 (1964).
133. R. Appel, W. Buchner and E. Guth, *Ann.* **618**, 53 (1958).
134. R. Appel and W. Buchner, *Angew. Chem.* **71**, 701 (1959), *Chem. Ber.* **95**, 849, 855, 2220 (1962).
135. H. R. Bentley and J. K. Whitehead, *J. Chem. Soc.* 2081 (1950).
136. J. K. Whitehead and H. R. Bentley, *J. Chem. Soc.* 1572 (1952).
137. F. Misani, T. W. Fair and L. Reimer, *J. Am. Chem. Soc.* **73**, 459 (1951).
138. M. Barash, *Chem. & Ind. (London)* 1261 (1964).
139. R. Wehr, *J. Chem. Soc.* 3004 (1965).
140. A. Hochrainer and F. Wessely, *Tetrahedron Letters* 721 (1965).

141. W. J. Linn, O. W. Webster and R. E. Benson, *J. Am. Chem. Soc.* **87**, 3651 (1965).
142. W. J. Middleton, E. L. Ruhle, J. G. McNally and M. Zanger, *J. Org. Chem.* **30**, 2384 (1965).
143. N. E. Mller, *Inorg. Chem.* **4**, 1458 (1965).
144. H. Behringer and F. Scheidl, *Tetrahedron Letters* 1757 (1965).
145. F. Ramirez and S. Levy, *J. Am. Chem. Soc.* **79**, 6167 (1957).
146. A. Wm. Johnson and R. T. Amel, *Tetrahedron Letters* 819 (1966).
147. C. E. Cook, R. C. Corley and M. E. Wall, *Tetrahedron Letters* 891 (1965).
148. H. Konig, H. Metzger and K. Seelert, *Chem. Ber.* **98**, 3712 (1965).
149. H. Konig, H. Metzger and K. Seelert, *Chem. Ber.* **98**, 3724 (1965).
150. H. Konig and H. Metzger, *Chem. Ber.* **98**, 3733 (1965).
151. R. Appel and G. Buchler, *Ann.* **684**, 112 (1965).
152. D. H. Clemens, A. J. Bell and J. L. O'Brien, *Tetrahedron Letters* 1487 (1965).
153. D. H. Clemens, A. J. Bell and J. L. O'Brien, *Tetrahedron Letters* 1491 (1965).
154. K. Bott, *Angew. Chem.* **77**, 132 (1965); *International Edn.* **4**, 148 (1965).
155. H. Zimmer and G. Singh, *Angew. Chem.* **75**, 574 (1963).
156. J. Day and D. J. Cram, *J. Am. Chem. Soc.* **87**, 4398 (1965).

AUTHOR INDEX

Numbers in parentheses are reference numbers and indicate that an author's work is referred to although his name is not cited in the text. Numbers in italic show the page on which the complete reference is listed.

A

Adolph, H., 353(112), *365*
Aguiar, A. M., 219(23), *245*
Ahuja, H. S., 218(5, 8), 222(5), 223(8), *245*
Aksnes, G., 7, *14*, 32(58), 40, 68(58), 91, *126, 129*
Albrecht, R., 358(130, 131), *365*
Alm, J. C., 30(112), 55(112), *128*
Amel, R. T., 361, *366*
Andree, F., 28(252), *131*, 147(58), *191*
Andrews, E. R., *79*
Appel, H., 239(78), *246*
Appel, R., 218, 219, 220(11), 221, 222 (7, 31), 223, 224(35), 225(11), 226, 227, 237, 238(6), *245*, 299, 300, 302, *303*, 357, 358, 359, 361, *365, 366*
Arbuzov, A. E., 203, *216*
Arnason, B., 46, 92, 103(81, 85), 104 (81), *127*
Arnold, R. T., 277, *283*
Arnold, Z., 260, *281*
Arnoldy, G., 268(57), *282*
Ash, A. S. F., 358, *365*
Ashley, J. N., 268, *282*
Ayrey, G., 278(93), *283*

B

Babad, H., 25(88), 32(88), 47, 103, *127*
Bach, R. D., 11, *15*
Badiger, V. V., 338, *364*
Bailey, W. J., 93, *129*
Bamford, W. R., 272, 274(72), *283*
Banitt, E. H., 29(38), 34, *126*
Banks, C. V., 195, *216*
Baranetskaia, N. K., 264, 265(45), *282*
Barash, M., 360, *365*
Barsukov, L. I., 23(28), 32(28), 47(28), 64(135), 103(28), *126, 128*, 136 (23), 185(93), *190, 191*
Bartlett, P. D., 109, *130*
Basolo, F., 287(13), *303*
Battiste, M. A., 145, *191*
Becker, H. D, 351(99, 100), 352(105), *365*
Bedford, A. F., 298(33), *303*, 331, *364*
Behringer, H., 361, *366*
Beisler, J., 219(23), *245*
Bell, A. J., 361(152, 153), *366*
Benson, R. E., 265(46), *282*, 360(141), *366*
Bentley, H. R., 359, *365*
Benz, E., 274, *283*
Berchtold, G. A., 277(92), *283*
Bercz, J. P., 206, *216*
Bergelson, L. D., 23, 24(134), 25(123), 32(28), 47, 58, 64, 103, *126, 128*, 136, 168, 169, 185, 186, 187(95), *190, 191*, 207, *216*
Berger, J. G., 325, *364*
Bergmann, E., 220, 227, *245*
Berrigan, P. J., 52(100), *127*
Berson, J. A., 264(43), *282*
Bestmann, H. J., 18(10), 24(10, 181, 184), 25(10), 26(10), 27(67, 71, 72, 74, 83, 198), 28(71), 30(182, 185), 31(77), 41, 43, 44, 45, 46, 57(73), 65, 66, 71, 72, 76, 92, 94, 95, 97, 98, 99, 100, 101, 102, 103, 104, 105, 106, 114, 119, 121, 123, *125, 127, 129, 130, 131*, 148, 149, 154(75), 166, 175, 181, *191*, 202, 216, 221, 229, 239(76), 241, 242 (84), 243, *245, 246, 247*
Bieber, T. I., 25(230), *131*, 151, *191*
Birkofer, L., 219(15), 223, *245*
Birum, G. H., 213, *216*
Bissing, D. E., 25(254), *131*, 144, 154 (49), 155, 156, 157, 160, 161, 165, 169, 170, 171, 173, 176, 177, 180,

189, *190, 192,* 197(12), *216,* 228, *245,* 315(70), 335, *364*
Blade-Font, A., 83(164), 84(164), 92 (164), 111(212), 112(212), 115 (217), *129, 130,* 173(89), 174(89), *191*
Blau, N. F., 307, *362*
Blomquist, A. T., 30(227), 32(227), *130,* 145, 147(53), 171, *191*
Bock, H., 219(17), 221, 231, 232, *245, 246*
Bohlmann, F., 24(106, 238, 243), 25 (247), 53, 72, 86, 117, *127, 129, 131,* 141, 142, 147, *190, 191*
Boll, W., 22, 26(23, 242), 28(23), *126, 131*
Bollyky, L., 309(20), 349, *363*
Bonoli, L., 352(103), *365*
Bordwell, F. G., 277, *283,* 306, 307(11), 309, *362, 363*
Borner, E., 263, *282*
Borowitz, I. J., 55, *128,* 353, *365*
Bose, A. K., 205, *216*
Boskin, M. J., 25(210), 111(210), *130,* 161, 183, *191,* 202(19), 210(19), *216*
Bott, K., 362, *366*
Bourns, R. N., 278(93), *283*
Boutan, P. S., 306, 307(11), *362*
Bradburg, A., *79*
Brandle, K. A., 108(204), *130*
Brasen, W. R., 342(84), *364*
Brauman, J. I., 64, *128*
Braunholtz, W., 242(86), 243(86), *247*
Bredereck, H., 309, *363*
Breslow, R., 343, 344(87), *364*
Brooks, T. W., 134(14), 136(14), 137 (14), 140(14), 144(14), 184(14), *190*
Brosowski, K. H., 31(206), 94(180), 109 (206), 114, 115(218), *129, 130,* 244, *247,* 271(64), *283*
Brown, G. W., 27(70), 43, 121, *127,* 229, 244, *246*
Bruckner, K., 24(107), 26(107), 54 (107), 117(107), *128*
Brune, R., 352(106), *365*
Buchanan, G. L., 268(58), *282*
Buchler, G., 221(30), 226(30), *245,* 361, *366*
Buchler, W., 309(23), *363*
Buchner, W., 218(3), *245,* 358(133, 134), 359, *365*
Buchta, E., 28(252), *131,* 147(58), *191*
Buckler, S. A., 93, *129*
Buckschewski, H., 239(76), 242(76), *246*
Buder, E., 309(25), *363*
Buegy, R., 24(238), *131*
Bugg, C., 265, *282*
Bumgardner, C. L., 279, *283*
Bunal, E., 278(93), *283*
Burchmann, G., 26(177), 92, *129*
Burg, A. B., 304, *362*
Burger, H., 21, *125*
Burkhardt, H., 24(246), *131*
Burlitch, J. M., 24(54), 38, 123(54), *126*
Burness, D. M., 342, *364*
Burske, M. W., 67(142), 68(142), *128*
Butler, G. B., 134(14), 136(14), 137 (14), 140(14), 144(14), 184(14), *190*

C

Cais, M., 239(74), *246*
Calvin, M., 338(75), *364*
Campbell, A., 274, *283*
Campbell, I. G. M., 334, 335, *364*
Campbell, R. W., 353(109), 354(109), *365*
Campbell, T. W., 26(97, 232), 30(97, 232), 51, 71, *127, 128, 131,* 141, 147, 149, *190, 191,* 227(41), 230 (48), *245, 246,* 300, *303,* 357, *365*
Campos, M. de M., 27(76), 45, 46, 103 (76), 108, *127*
Carmack, M., 307, *363*
Carretto, J., 26(241), *131*
Cava, M. P., 145, 147(52), *191*
Challenger, F., 358, *365*
Chaplin, E. J., 217, *245,* 299, *303*
Chapman, A. C., 231, 238(56), *246*
Chatt, J., 10, *14,* 286, 287, 292, 296, *303*
Chaykovsky, M., 57(118), *128,* 136(21), *190,* 309(17, 19, 22), 311(30), 315 (30), 317(30), 321, 326, 332, 333 (30), 336, 337(30), 345, 346(22), 347(30, 71), 348, 349, 350(30), 351, 361, *363, 364, 365*

Chopard, P. A., 25(249), 26(237, 249), 28(90), 48, 55(108), 68(143), 104, *127, 128, 131,* 262, *282*
Christmann, A., 356, 359, *365*
Christmann, K. F., 24(253), *131,* 135, 136, 189, *190, 192,* 196, *216*
Ciganek, E., 278(95), *283*
Cilento, G., 304, *362*
Clark, D., 9, *14*
Clarke, S. G., 356(121), *365*
Clauss, K., 285(3), 288, *303*
Claydon, A. P., 237(66), *246*
Clemens, D. H., 361, *366*
Closs, G. L., 329, *364*
Coffmann, D. D., 3(10), *4,* 24(117), 28 (117), 57, 89, *128*
Cohen, H. M., 289, 292, 293, *303*
Collins, C. H., 136, *190*
Cook, C. E., 361, *366*
Cookson, E. A., 86, *129*
Cookson, R. C., 27(70), 43(70), 121 (70), *127,* 221, 229(45, 46), 244 (46), *245, 246*
Cooper, G. D., 309, *363*
Cope, A. C., 277, 278(89), 280, *283,* 297, *303,* 340, *364*
Corey, E. J., 57(118), *128,* 136(21), *190,* 309, 311, 315(30, 68), 317 (30), 321, 326, 332, 333, 336, 337, 345, 346(22), 347(30, 71), 348, 349, 350(30), 351, 361, *363, 364, 365*
Corey, H. S., 19, 25(14), 57(14), 77, *125*
Corley, R. C., 361(147), *366*
Costain, C. C., 265, *282*
Cotton, F. A., 211, *216,* 318, *363*
Craig, D. P., 13, *15,* 75, *129,* 285, *303,* 305, 306, 307, 308, 316, *362, 363*
Cram, D. J., 81, *129,* 362, *366*
Creighton, E. M., 273(74), *283*
Cristol, S. J., 340, *364*
Crofts, P. C., 86, *129*
Cruickshank, D. W. J., 13, *15,* 306, *362*
Curphey, T. J., 344(89), *364*
Curtze, J., 262(28), *282*

D

Dahill, R. T., 205, *216*
Dallacker, F., 24(244), *131*
Daly, J. J., 80, *129*
Daniel, H., 255(11), 256(11, 14), 274 (82), 280(99), *281, 283,* 340, *364*
Dauben, H. J., Jr., 259, 264, *281*
Day, J., 362, *366*
Day, N. E., 57, *128*
Degani, C., 10, *14*
deMayo, P., 353(116), *365*
Denney, D. B., 19, 25(210), 27(93), 49, 77, 96, 97, 98, 111, *125, 127, 129, 130,* 161, 162, 183, *191,* 202, 210, *216*
Depoorter, H., 70, *128*
Derkach, G. I., 231(53), 233(62), 234 (62), *246*
Dershowitz, S., 7, *14,* 25(129), 26(129), 32(56), 39, 50, 51(56), 62, 68 (129), 69, 89, 100, *126, 128,* 138, *190,* 213, *216*
Desai, N. B., 26(91), 29(91, 115), 30 (91), 49(91), 55(91), 56(115), 66 (91), 96(186), *127, 128, 129,* 222 (33), *245*
Desiderato, R., 265(47), *282*
Devitt, F. H., 28(90), 48(90), 104(90), *127,* 262(27), *282*
Dewar, M. J. S., 87, *129*
Dicker, D. W., 20(17), *125*
Dietrich, P., 265(46), *282*
Dimroth, K., 30(233), *131,* 149, *191,* 268, *282*
Dmitriev, B. A., 141, *190*
Doering, W. von E., 10, 11, *14, 15,* 33, 67, 81, 107(30), *126, 128, 129,* 251, 254, 272, *281, 283,* 287, 292, 296, *303,* 306, 307, 319, 345, *362, 363*
Domagk, G. F., 24(107), 26(107), 54 (107), 117(107), *128*
Doub, L., 307, *362*
Douglas, B. E., 30(235), *131*
Drago, R. J., 218(5), 222(5), *245*
Drechsler, D., 262(28), *282*
Drefahl, G., 28(226), *130,* 141, 184, 185, 186, *190, 191,* 205, *216*
Driessen, H. E., 311, 315(28, 66), 317 (28), 321, 322, 323, 332, 333(66), 337, *363, 364*
Driscoll, J. S., 63(132), 76, 78, *128, 129,* 237, *246*
Duffey, D. C., 33, *126*

Duffner, P., 20(15), 30(15), 53(15), 57 (15), *125,* 140(37), 149(66), *190, 191,* 195(6), *215*
Duncan, W. G., 33(35), *126*
Dunin, A. A., 203(22), *216*
Dunn, J. L., 274(78), *283*
Durst, T., 353(108), 354(108), *365*

E

Eades, R. G., *79*
Easson, A. P. T., 268(58), *282*
Eberle, D., 314, *363*
Eggers, H., 20(15), 30(15), 53(15), 57 (15), *125,* 149(66), *191,* 195(6), *215*
Eichler, S., 134(18), 157(18), *190*
Eisert, M. A., 53(104), 61(104), *127*
Eisman, E. H., 25(230), *131,* 151, *191*
Eiter, K., 24(250), 25(251), 26(251), *131*
Emmons, W. D., 205, 206, 210, *216,* 235, *246,* 338(74), *364*
Engelhard, H., 26(248), *131,* 136, *190*
Erdmann, H. M., 24(107), 26(107), 54 (107), 117(107), *128*
Ertel, H., 194(5), 202(15), 206(5), 212 (5), *215, 216*
Evleth, E. M., 264, *282*

F

Fair, T. W., 359(137), 360(137), *365*
Fehnel, E. A., 307, *363*
Felletschin, G., 2(1), *4,* 9, *14,* 258, 272, 274(19, 79), *281, 283,* 316(40), 317 (40), *363*
Fenton, G. W., 60, 89, 93(128), *128*
Fiene, M. L., 10, *14*
Fischer, K., 353(113), *365*
Fletcher, T. L., 32(255), *131*
Fliszar, S., 26(139), 66, 67, *128,* 145, 154(50), 155, 156, 157, 158, 164, 170, *190*
Flores, M. C., 267, *282*
Fodor, G., 134, *190*
Fogel, J. S., 24(50), 37(50), 38(50), 61 (50), *126*
Foldi, V. S., 227(40), 230(40, 48), *245, 246,* 300(38), *303,* 357(125), *365*
Ford, J. A., Jr., 21, 30(20), *125,* 207 (38), *216*
Forsblad, I., 220(27), 221(27), *245*
Forster, H., 28(111), 55, 56, *128*
Fowell, P. A., 232, 237(66), 238(58), *246*
Francis, R., 318, *363*
Frank, G. A., 136(24), 185(24), 186 (24), 187(24), *190*
Franz, J. E., 40(59), *126,* 219(18), 221, *245*
Franzen, V., 29(34), 33, 34, 38, 123, *126,* 254, 258, 272, 274(17), *281,* 311, 312, 313(34), 314, 315(28, 34, 36, 66), 317(28), 321, 322, 323, 324, 325(36, 55), 327, 332, 333, 337, 341, *363, 364*
Freeman, J. P., 117, 118, *130,* 141, *190*
Frese, E., 115(218), *130,* 271(65), *283*
Freyschlag, H., 188(97), *192*
Friedman, L., 325, *364*
Fritz, H., 311, 315(27), 327, *363*
Fritzsche, H., 239, 243, *247*
Fujiwara, K., 205(30), *216*
Fuqua, S. A., 33, *126*

G

Gagneux, A., 278(94), *283*
Geissler, G., 8, *14,* 17, 24(4), 64, 74(4), *125,* 133, 152, 167, *190*
Gerlach, K., 261(24, 25), 262(24, 25), *282*
Gibson, M. S., 267, *282*
Gillespie, R. J., 305, *362*
Gilman, H., 26(234), 31(234), 60, *128, 131,* 139, *190*
Gilpin, J. E., 231, *246*
Gilyarov, V. A., 219(21), 233, 234(21), 239(71), 243, *245, 246*
Gimborn, H. V., 3, *4,* 7, *14,* 16, 25(2), 52, 57, *125*
Goedler, J., 219(19), *245*
Goetz, H., 144, 157, *190,* 239, *246*
Gordon, A. B., 273(74), *283*
Gossel, H., 264, *282*
Gothlich, L., 241, 242(84), *247*
Gough, S. T. D., 27(82), 28(82), 46 (82), 93, 103(82), 105, *127, 129,* 202, *216*
Grassner, H., 188(97), *192*
Grayson, M., 11, *15,* 26(119), 57(119), 76, 91, *128*

Greenwald, R., 57(118), *128,* 136(21), *190*
Griffin, C. E., 23, 24(29, 236), 25(236), 30(235), *126, 131,* 137, 144, 149, 151(69), *190, 191*
Grim, S. O., 26(32, 124), 33(32), 35 (32), 59(124), 106, 108, *126, 128, 130,* 289, *303*
Grisley, D. W., Jr., 30(112), 55(112), 76(154), 78(154), *128, 129,* 237 (68), *246*
Grob, C. A., 278, *283*
Groen, S. H., 342, *364*
Gross, A., 219, 221, 222(24), 225(24), 227(24), 230, *245*
Gross, R. M., 263, *282*
Grunewald, R., 231(50), *246*
Guth, E., 218(3, 4), *245,* 358(133), *365*
Gutmann, G., 24(238), *131*
Gutsche, C. D., 270, *282*

H

Haag, A., 140(38), 143, 161, 168, *190*
Haag, W., 36, 57(46), *126,* 133, 136, 161, 165, 171, *190,* 239(72), 240 (72), *246*
Haake, P., 343, 344(88), *364*
Haake, P. C., 84, *129,* 173, *191*
Haberlein, H., 30(185), 31(77), 45, 95 (185), 99, 100(192), *127, 129, 130*
Hager, F. D., 252, *281*
Hall, C., 124(257), *131*
Hall, C. D., 96(188), 97(188), *130*
Halmann, M., 10(20), *14*
Hamilton, W. C., 84, *129*
Hammond, G. S., 136, *190*
Hansen, B., 26(91), 29(91), 30(91), 49 (91), 55(91), 66(91), *127,* 222 (33), *245*
Harris, J. E., 63(132), 76(154), 78 (154), *128, 129,* 237(68), *246*
Harrison, I. T., 25(150), 72, *129*
Hart, F. A., 286, 292, 296(11), *303*
Hartley, S. B., 166, *191*
Hartmann, H., 264, *282*
Hartung, H., 105, *130*
Harvey, R. G., 212, *216*
Hassel, G., 32(96, 98), 50(96), 51(98), 69(98), *127*
Hauser, C. F., 134, 136, 137, 140, 144, 184, *190*
Hauser, C. R., 274, 276, *283,* 342, *364*
Hauser, E., 221, 222(26), 226(26), 228, 230, *245*
Hauss, A., 218, 221(30), 222(7, 31), 223 224, 225, 226(7, 30, 31), 237, 238 (6), *245*
Hawthorne, M. F., 106, *130,* 224(37), *245*
Heeren, J. K., 24(50, 127), 26(107), 37 (50), 38(50), 53(104), 59(124), 60(127), 61(50, 104, 127), *126, 127, 128,* 194(2), *215*
Heffe, W., 257, *281,* 295, *303*
Heitman, H., 18(9), 21, 30(9, 19), 76, *125*
Heller, S. R., 215(50), *216*
Hellmann, H., 53, *127,* 314, *363*
Hellwinkel, D., 10(18), *14,* 285(4, 5), 288, 300, 302, *303*
Hendrickson, J. B., 28(222), 42, 105, 121, 124(222), *127, 130, 131*
Henry, M. C., 288, 289, 296, 297, 298 (17), 301, *303,* 350, *365*
Herweh, J. E., 239(70), 240, 242(70), *246*
Hesse, G., 353, *365*
Hilbert, G. E., 8, *14,* 22, 31(116), 57, 64(116), 65(116), *126, 128,* 263, *282,* 320, 342(49), *363*
Hine, J., 33, 34, 67, 68(142), 107(40), *126, 128,* 352, *365*
Hine, M., 67(142), 68(142), *128*
Hinman, R. L., 267, *282*
Hochrainer, A., 360, *365*
Hoffman, A. K., 11, *15,* 33, 67, 81(141), 107(30), *126, 128,* 251, 254, *281,* 287, 292, 296, *303,* 306, 307, 319, 345, *362*
Hoffman, H., 27(60), 28(111), 32(96, 98), 40, 50, 51(98), 55, 56, 57 (120), 69(98), 109, *126, 127, 128,* 136(20), *190,* 193(1), 194(1, 5), 195(7), 199(14), 200(14), 202(7, 15, 18), 205(1), 206(5), 207(14), 212(5), *215, 216,* 231, *246*
Holmes, W. S., 166(84), *191,* 231(56), 238(56), *246*
Holtz, J., 252(4), 253(4), *281*

Honsberg, U., 220(27), 221(27), *245*
Hori, M., 355(118), *365*
Horner, L., 21, 27(22), 28(22), 32(96, 98), 40, 50(96), 51, 52, 53, 69, 124, *126, 127, 131,* 136(20), 173, *190, 191,* 193, 194, 195, 196, 197, 198 (11), 199, 200(14), 202, 205, 206, 207, 212, *215, 216,* 218, 219, 221, 222(9, 24), 223, 224, 225(24), 227, 230, 231(9, 50), 237, 239(77), 240 (82), 241, *245, 246, 247,* 272, *283,* 290, 291, *303,* 356, 359, *365*
Hoschele, G., 309(25), *363*
House, H. O., 25(88), 32(88), 47, 103, *127,* 136, 141, 166, 174, 185, 186, 187, 189, *190*
Houston, A. H. J., 274(80), *283*
Hruby, V. J., 30(227), 32(227), *130,* 145, 147(53), 171, *191,* 271(61), *282,* 313(35), 315(35, 58), 321 (35), 325(35, 58), 333(35, 67), 334(35), *363, 364*
Hudson, R. F., 25(249), 26(139, 249), 55(108), 66(139), 67(139), *128, 131,* 145(50), 154(50), 155(50), 156(50), 157(50), 158(50), 164 (50), 170(50), *190*
Huggins, M. L., 11, *15,* 317, 318(44), *363*
Hughes, A. N., 221, *245*
Hughes, E. D., 314(38), 315(38), 316 (38), 319(38), 321, *363*
Hughes, W. B., Jr., 24(127), 60(127), 61(127), *128*
Huisman, H. O., 18(9), 21(9, 19), 30(9, 19), 76(9), *125*
Hunter, J. S., 236(65), *246,* 292(29), *303,* 316(41), 318(41), 355(41), *363*

I

Imaev, M. C., 11, *14*
Ingold, C. K., 3, *4,* 60, 89, 93(128), *128,* 258, *281,* 310, 311(26), 312(26), 314(26), 315(26), 316(26), 317 (26), 318, 319(26), 320(26, 51), 321(26), 331(26), 340, *363, 364*
Ingraham, L. L., 343, *364*
Inhoffen, E., 117, *131,* 141, *190*
Inhoffen, H. H., 24(107, 238, 245, 246), 26(107), 54, *128, 131*
Inouye, Y., 112(214), *130,* 210(41), *216*
Irmacher, K., 24(245), *131*
Isler, O., 24(238), *131*
Ismail, A. F. A., 8, *14,* 40, *126*
Ito, Y., 107(202), *130,* 271(69), *283,* 323(52), 328(52), *363*
Izzo, P. T., 347, *364*

J

Jacques, J. K., 166(84), *191*
Jaffe, H. H., 13, *15,* 66, 75, *128, 129,* 157, *191*
Jambatkar, D., 26(239), *131,* 166(83), 175(83), *191*
Janiak, P. St., 81(158), *129*
Janzso, G., 206, *216*
Jenkes, G. J., *79*
Jensen, E. V., 212, *216*
Jessop, J. A., 3(9), *4,* 258, *281,* 310, 311 (26), 312(26), 314(26), 315(26), 316(26), 317(26), 319(26), 320 (26), 321(26), 331(26, 62), 340 (62), *363, 364*
Johnson, A. Wm., 3(8), *4,* 9, *14,* 18(7), 31(6, 7), 35, 63(44, 131), 64(44), 66, 67, 74, 75(7, 131), 89, *125, 126, 128,* 134, 135(17), 138, 139, 140(30), 144(33), 154(17), 156, 157, 158(30), 160, 162, 163(17), 167(17), 169, 188, *190, 192,* 264, 271, *282, 283,* 289, 291, 292, 295, 296(24), 297, 298(24, 28), 299, *303,* 304, 309, 311, 312(31), 313, 314(31), 315(29, 31, 33, 35, 58), 316, 317(31), 319(31), 320(31), 321, 324(31), 325, 326, 329, 330, 331(31), 332, 333(31), 334, 338, 342(31), 352(29), 361, *362, 363, 364, 366*
Jones, M., Jr., 272, *283*
Jones, R. G., 60, *128*
Jones, V. K., 136(24), 185(24), 186 (24), 187(24), *190*
Joschek, H. I., 38(53), 123(53), *126,* 313(36), 314(36), 315(36), 325 (36), 327(36), *363*
Juds, H., 239, *246*
Junghans, K., 271, *283*

K

Kabachnik, M. I., 11, *14,* 219(21), 233, 234(21), 239(71), 243, *245, 246*
Kantor, S. W., 274, 276, *283,* 342(84), *364*
Kawabata, T., 25(66), 41(66), *127*
Kenyon, J., 274(80), *283,* 356(121), *365*
Keough, P. T., 11, *15,* 26(119), 57(119), 76, 91, *128*
Ketcham, R., 26(239), *131,* 166, 175, *191*
Kim, S. M., 219(15), 223, *245*
Kimball, G. E., 81, *129*
King, C., 357, *365*
King, J. F., 353(108), 354, *365*
Kirmse, W., 34, 107(41), *126*
Kirsanov, A. V., 231(52, 53, 54), 232, 233(54), 234(60, 62, 63), 238(52, 54), *246*
Klahre, G., 136(20), *190,* 195(7), 202 (7, 15), *215, 216*
Klink, W., 194(5), 197, 198(11), 199 (14), 200(14), 206(5), 207(14), 212(5), *215, 216*
Klopman, G., 55(108), *128*
Klupfel, K., 40, *127*
Knauss, E., 148, *191*
Knospe, R. H., 352(104), *365*
Kny, H., 278(94), *283*
Kobrich, G., 57(121), 59, *128*
Koch, H. P., 81, *129*
Kochetkov, N. K., 141, *190*
Koenig, G., 275(85), *283*
Kohler, E., 7, *14,* 25(110), 26(110), 55, 57(110), 62, *128*
Kondo, K., 315(39), 318(39), 321(39), 333(39), 361(39), *363*
Konig, H., 348(93b, 94), 361, *364, 366*
Kornblum, N., 52, *127*
Kosolapoff, G. M., 204, *216*
Kosower, E. M., 264, *282*
Kratzer, O., 18(10), 24(10, 184), 26 (10), 30(182, 185), 72(149), 76, 94(182, 183), 95(185), 97(183), *125, 129,* 148(63), 148(63), 149 (63), 154(75), 166(75), 175, 181, *191,* 202, *216*
Kovalev, B. G., 205, *216*
Krauss, D., 255, *281,* 324, 325(56), *364*
Kresze, G., 357(128), 358(130), *365*
Krohnke, F., 3, *4,* 253, 257, 260, 261(7, 23, 24, 25), 262(7, 23, 24, 25), 263 (26, 32), *281, 282,* 289, *303*
Kruck, K. H., 26(242), *131*
Kubler, H., 261(23), 262(23), *282*
Kucherov, V. F., 205(29), *216*
Kuhn, R., 345, *364*
Kumli, K. F., 83(164), 84(164), 92 (164), *129,* 173(89), 174(89), *191*
Kuriyan, K. I., 314(38), 315(38), 316 (38), 319(38), 321, 331(62), 340 (62), *363, 364*
Kursanov, D. M., 264, 265, *282*
Kyllingstad, V. L., 188, *192*

L

LaCount, R. B., 3(8), *4,* 9, *14,* 18(7), 31(6, 7), 66, 67, 74, 75(7), *125,* 134, 135(17), 138, 144, 154(17), 156(17), 157(17), 158(17), 160, 162, 163(17), 167(17), 169(17), *190,* 291, 297, *303,* 304, 312(31), 314(31), 315(31), 316, 317(31), 319(31), 320(31), 321, 324(31), 329, 330, 331(31), 332, 333(31), 342(31), *362, 363*
Laib, H., 3(8), *4,* 22, 31(26), 89, 99 (26), *126,* 258, 274(18), *281,* 289, 290, 291, 294, 295(25), 301, *303*
LaLancette, E. A., 143, *190*
Langford, P. B., 67(142), 68(142), *128*
LaPlaca, S. J., 84(165), *129*
Latif, N., 326(60), *364*
Laulicht, I., 10(20), *14*
Lauterbur, P. C., *79,* 318, *363*
Lebedeva, N. M., 204, 208, *216*
LeBel, N. A., 278(95), 280(100), *283*
Lee, S. Y., 63(131), 75(131), *128*
Leffler, J. E., 220, 221, *245*
Lemal, D. M., 29(38), 34, *126*
LeNoble, W. J., 52(100), *127*
Leube, H., 239(76), 242(76), *246*
Levine, S. G., 26(225), *130,* 148, *191*
Levisalles, J., 18, *125,* 132, 145(2), 171 (2), *189*
Levy, L. K., 81(161), *129*
Levy, S., 32(130), 35, 63(129, 130), 64 (130), 70, 89, 110, *126, 128,* 138,

158(29), *190,* 239(73), 240(73), 242, *246,* 264, *282,* 361(145), *366*
Lichtenthaler, F. W., 55(108), *128*
Light, K. K., 24(52), 37(52), *126,* 152, *191*
Lindlar, H., 24(238), *131*
Lingnau, E., 240(82), *247,* 272, *283*
Linn, W. J., 265, *282,* 360, *366*
Lipp, M., 24(244), *131*
Lloyd, D., 264, 265(39), *282*
Lorenz, D., 184(92), 185(92), 186(92), *191*
Lucken, E. A. C., 87(172), *129*
Lunt, R. S., 23(27), *126*
Luscher, G., 132, 140(9), *190,* 239(75), 240(75), *246*
Luttke, W., 125, *131*
Lythgoe, B., 25(150), 72, *129*

M

Maccoll, A., 13(31), *15,* 75(152), *129,* 305(5), 308(5), *362*
McCormick, J. R. D., 19(14), 25(14), 57(14), 77(14), *125*
McCoubrey, J. C., 166(84), *191*
McDonald, R. N., 26(97, 232), 30(97, 232), 51, 71, *127, 128, 131,* 141, 147, 149, *190, 191*
McEwen, W. E., 26(211), 83, 84, 89 (174), 90(174), 92, 93(174), 111, 112, 115(217), *129, 130,* 167, 173, 174(89), *191,* 202, 210, *216,* 331 (63), *364*
McEwen, W. K., 9, *14*
McGrew, L. A., 230, *246*
McKelvie, N., 26(91), 29(91, 115), 30 (91), 49(91), 55(91), 56(115), 66 (91), *127, 128,* 222(33), *245*
Mackle, H., 331, *364*
McNally, J. G., 360(142), *366*
MacNicol, M., 273(74), *283*
McShane, H. F., Jr., 227(41), *245*
McSweeney, J. V., 348, *364*
Madan, O. P., 28(61, 62, 63), 40(61, 62, 63), *126,* 162(82), *191,* 212(44), 214, 215(50), *216*
Maercker, A., 188, *192*
Magee, T. A., 287, 296, *303*
Magnusson, E. A., 13(31), *15,* 72(152), *129,* 285, *303,* 305, 306, 307, 308 (7), *362*
Majmudar, S., 353(115), *365*
Mak, T. C. W., 80, 82, *129,* 229(45), *246*
Makarova, L. G., 312, *363*
Malbec, F., 26(241), *131*
Manatt, S. L., 264(43), *282*
Mandour, A. M. M., 331(62), 340(62), *364*
Mangold, R., 274(79), *283*
Mann, F. G., 217, *245,* 266, *282,* 299, 300, *303,* 356, *365*
Mannhardt, H. J., 24(106), 53, 86, *127*
Marcus, R., 28(114), 55, *128*
Mark, V., 29(37), 33, 78, 79, *126,* 139, *190*
Markl, G., 27(75, 92, 208), 28(75, 228), 36, 45, 49(95), 57(78), 84, 85, 86, 87, 97, 99, 103, 105, 109, 110, 115, *126, 127, 129, 130,* 138, *190,* 355, *365*
Markved, E., 353(116), *365*
Martin, K. R., 30(235), *131*
Martin, J. O., 292(28), 297(28), 298 (28), 299, *303*
Martinelli, L., 26(239), *131,* 166(83), 175(83), *191*
Marvel, C. S., 3, *4,* 24(117), 28(117), 57, 89, *128,* 252, *281*
Matsuo, H., 134, 138, *190*
Matthews, C. N., 30(112), 55(112), 63 (132), 76(154), 78(154), *128, 129,* 237(68), *246,* 287(12), 296(12), *303*
Mechoulam, R., 24(191), 100, 118, 119, *130,* 141, *190,* 326, *364*
Medved, T. Y., 11(25), *14*
Meguerian, G., 109, *130*
Mehta, A. S., 280, *283,* 340, *364*
Mengenhauser, J. V., 228(43), *246*
Mentrup, A., 53, *127*
Meriwether, L. S., 10, *14*
Mertz, C., 38(53), 123(53), *126,* 312 (34), 313(34, 36), 314(36), 315 (34, 36), 324(34), 325(34), 327 (36), 341, *363, 364*
Messmer, A., 219(16), 226, *245*
Metzger, H., 348(93b), 361, *364, 366*
Meyer, A. H., 9, *14*

Meyer, D., 266, *282*
Meyer, J., 3(4), *4*, 8(5), 9, *14*, 28(42), 35, 48, 63(42), 109, 115, *126*, *127*, 132, 140(8), *190*, 219, 223, 226, 238, 239(69), 240(69), 241, 242 (69), *245*, *246*
Meyers, C. Y., 352, *365*
Michaelis, A., 3, *4*, 7, *14*, 16, 52, 57, *125*, *128*, 288, *303*
Michaelis, H., 55, 57(110), 62, *128*, 144 (48), 157(48), *190*
Michaelis, R., 25(2, 110), 26(110), 39, 50, 51(55), *126*
Michalski, J., 204, 206, *216*
Middleton, W. J., 213, *216*, 360, *366*
Miescher, K., 271, *283*
Mikol, G. J., 351(99, 100), *365*
Miles, M. L., 134(14), 136(14), 137 (14), 140(14), 144(14), 184(14), *190*
Miller, N. E., 26(136), 64, 68, *128*, 361, *366*
Miller, W. B., 343, 344(88), *364*
Mirkin, L. S., 265(44), *282*
Misani, F., 359, 360(137), *365*
Misumi, S., 24(147), 72, *128*
Mitra, R. B., 96(186), *129*
Mlakar, B., 358(130), *365*
Moeller, T., 232, *246*
Moffitt, W. E., 81, *129*
Mogat, M., 9(15), *14*
Mole, M. F., 166(84), *191*
Monagle, J. J., 227(40, 41), 228(43), 230(40), *245*, *246*, 269, *282*, 300 (38), 302, *303*, 357(125), *365*
Mondon, A., 23(27), 24(27), 30(27), *126*
Montavon, M., 24(238), *131*
Moore, P. T., 280(100), *283*
Moore, W. R., 280(100), *283*
Mortimer, C. T., 232, 237(66), 238(58), *246*, 298(33), *303*, 331, *364*
Moureu, H., 9, *14*
Murray, A. W., 267, *282*
Musierowicz, S., 204, 206, *216*
Mustafa, A., 326(60), *364*

N

Nakagawa, M., 24(147), 72, *128*
Namkung, M. J., 31(255), *131*
Nerdel, F., 144(48), 157(48), *190*
Nesmeyanov, N. A., 27(201), 28(205), 107, 108, *130*, 289, 291, 292, 293, *303*, 312, *363*
Norell, J. R., 353(109), 354(109), *365*
Normant, H., 205, *216*
Noth, H., 219(22), *245*
Novikov, V. M., 108(205), *130*
Nozaki, H., 315(39), 318, 321, 333, 361, *363*
Nurrenbach, A., 188(97), *192*
Nyholm, R. S., 13(31), *15*, 75(152), *129*, 305(5), 308(5), *362*
Nys, J., 70(145), *128*

O

Oae, S., 304, 309(3), 352(3), *362*
O'Brien, J. L., 361(152, 153), *366*
Oda, R., 25(66), 41, 107, *127*, *130*, 271, *283*, 323, 328, *363*
Oediger, H., 21, 24(250), 25(251), 26 (251), 27(22), 28(22), 50, *126*, *131*, 218, 222(9), 231(9), 237, *245*, 290, 291, *303*
Ofner, A., 30(220), 116, *130*, 147, 171, *191*
Ohlberg, S., 285, *303*
Ohta, M., 205(30), *216*
Okano, M., 107(202), *130*, 271(69), *283*, 323(52), 328(52), *363*
Opitz, G., 353, *365*
Oppolzer, W., 315(68), 333, *364*
Orgel, L. E., 13(31), *15*, 75(152), *129*, 305(5), 308(5), *362*
Osuch, C., 40, *126*, 219(18), 221, *245*

P

Paddock, N. L., 217, 231(56), 238(56), *245*, *246*
Pandit, U. K., 21(19), 30(19), *125*
Panse, P., 134(18), 157(18), *190*
Parasaron, T., 355(118), *365*
Parham, W. E., 342, *364*
Parnes, Z. N., 265(45), *282*
Parrick, J., 145, *191*
Patai, S., 208, 209(40), *216*
Pauling, L., 80, *129*, 287, *303*
Paulovic, D., 341, *364*
Pearson, R. G., 287(13), *303*
Peters, J. A., 149, 151, *191*

Petragnini, N., 27(76), 45, 46, 103, 108, *127*
Pfenninger, F., 353, *365*
Pfleghar, K., 262(28), *282*
Phillips, H., 356(121), *365*
Phillips, G. M., 236, *246*, 292, *303*, 316, 318(41), 355(41), *363*
Pils, I., 100(192), *130*
Pinchas, S., 10(20), *14*
Pinck, L. A., 8, *14*, 22, 31(116), 57, 64 (116), 65(116), *126*, *128*, 263, *282*, 320, 342(49), *363*
Pinter, I., 219(16), 226(16), *245*
Plotner, G., 28(226), *130*, 141, *190*
Pohl, G., 30(233), *131*, 149, *191*
Pohl, R. J., 145, 147(52), *191*
Polikarpov, Y. M., 11(25), *14*
Polk, M., 355(118), *365*
Polster, R., 254, 256, 278, 279, *281*, *283*
Pommer, H., 53, *127*, 147, 188(97), *191*, *192*
Ponsold, K., 205(33), *216*
Pope, W. J., 266, *282*, 356, *365*
Porter, J. J., 352, *365*
Poshkus, A. C., 239(70), 240, 242(70), *246*
Powell, H. M., 9(17), *14*
Pravdina, V. V., 289(21), 291(21), 292 (21), 293(21), *303*
Prevost, C., 26(209), 110(209), *130*
Price, C. C., 40, 87, *127*, *129*, 304, 309 (3), 352(3), 354, 355, *362*, *365*
Price, J. R., 31(255), *131*
Pritchard, J. G., 318, *363*
Pudovik, A. N., 11, *14*, 204, 208, *216*
Pustinger, J. V., 76(154), 78(154), *129*, 237(68), *246*
Puterbaugh, W. H., 276, *283*

Q

Quinkert, G., 24(246), *131*

R

Rabinowitz, R., 28(114), 55, *128*
Ramirez, F., 7, *14*, 25(129), 26(91, 129), 28(61, 62, 63), 29(91, 115), 30(91), 32(56, 130), 35, 39, 40, 49, 50, 51(56), 55, 56, 62, 63(129, 130), 64(130), 66, 68(129), 69, 70, 84(165), 89, 96, 100, 110, *126*, *127*, *128*, *129*, 138, 158(29), 162, *190*, *191*, 212, 213, 214, 215, *216*, 222, 239(73), 240(73), 242, *245*, *246*, 264, *282*, 361(145), *366*
Ramsey, B. G., 264, *282*
Raschig, F., 219(13), *245*
Rasmusson, G. H., 141, 166(42), 174, 189, *190*
Ratts, K. W., 25(94), 26(140), 27(94), 28(31, 94, 140), 29(31), 33(31), 49(94), 67, 68, 76, 77, 78, 79, 100, 103(94), 124, *126*, *127*, *128*, *131*, 157, *191*, 315(72), 338(72), *364*
Raymond, R. A., 134(14), 136(14), 137 (14), 140(14), 144(14), 184(14), *190*
Razumov, A. I., 203(23), *216*
Read, J., 334, 335, *364*
Read, T. O., 26(32), 33(32), 35(32), *126*
Rees, R., 28(222), 121(222), 124(222, 257), *130*, *131*
Reichold, E., 353(115), *365*
Reid, W., 263, *282*
Reif, W., 188(97), *192*
Reimer, L., 359(137), 360(137), *365*
Reimlinger, H., 271, *283*
Relles, H. M., 162, *191*
Reutov, O. A., 27(201), 28(205), 107 (201), 108(205), *130*, 289(21), 291(21), 292(21), 293(21), *303*
Richards, J. J., 195, *216*
Richardson, P. N., 277, *283*
Rieber, M., 3(7), *4*, 8, 10(18), *14*, 17, 24(3), 57(3), 60, 74(3), 98, 99(3), *125*, *128*, 132, 153, 158, 160(10), 167, *190*, 255, 262(12), 266, *281*
Rieche, A., 265(46), *282*
Ried, W., 239(78), *246*
Rimaldi, C., 352(103), *365*
Ritter, A., 219(15), *245*
Rittersbacher, H., 357, *365*
Robinson, E. A., 305, *362*
Ross, D. L., 277(92), *283*
Ross, S. T., 27(93), 28(93), 49, 93, *127*
Rossi, C. J., 96(188), 97(188), *130*
Rothberg, I., 324, *363*
Rothe, O., 27(71), 28(71), 43, 121, *127*
Rouault, M., 9, *14*, 285, *303*
Royer, L. D., 63(131), 75(131), *128*

Ruchardt, Ch., 134, 157, *190*
Ruhle, E. L., 360(142), *366*
Russell, G. A., 351(99, 100), 352(105), *365*
Ryser, C., 24(238), *131*

S

Salvadori, G., 25(249), 26(139, 249), 66(139), 67(139), 68(143), *128, 131,* 145(50), 154(50), 155(50), 156(50), 157(50), 158(50), 164 (50), 170(50), *190*
Sarkis, A., 218(5), 222(5), *245*
Sarneck, W., 188(97), *192*
Sass, R. L., 265(47), *282*
Sattar, A. B. M. A., 353(116), *365*
Saunders, M., 26(177), 92, *129*
Saunders, W. H., 341, *364*
Sayigh, A. A., 232, *246*
Schaaff, R., 224, *245*
Schaefer, S. K., 31(255), *131*
Scheidl, F., 361, *366*
Schenk, D., 352, *365*
Schepers, R., 25(240), *131,* 151, *191*
Scherer, K. V., 23(27), *126*
Schick, H., 205(33), *216*
Schiemenz, G. P., 26(248), *131,* 136, *190*
Schiffler, G., 268(57), *282*
Schlenk, W., 252(4), 253(4), *281*
Schlosser, M., 18(11), 24(11, 48, 253), 26(33), 33, 36, 59, 90, 110, *125, 126, 129, 131,* 135, 136, 137, 139 (35), 144(35), 145(35), 148, 149 (64), 154(35), 156(35), 159(35), 160(35), 165(35), 167(35), 189, *190, 191, 192,* 196, *216,* 240, *247,* 272, *283*
Schmelzer, H. G., 52, *127,* 239(77), 241, *246*
Schmidt, H. J., 312(34), 313(34), 315 (34), 324(34), 325(34), 341, *363, 364*
Schnalke, K. E., 261(25), 262(25, 28), *282*
Schneider, W., 267, 268, *282*
Schnitt, G., 184(92), 185(92), 186(92), *191*
Schoch, W., 275(85), *283*
Schollhorn, R., 219, 220(11), 225(11), 227(11), 238, *245*
Schollkopf, U., 16(1), 17, 24(1), 26(1), 46, 62, 71, 88, 89(1), 103(1), *125,* 132, 133, 134(12), 135, 140(12), 145(3), 155(12), 158(12), 160, 164, 171(3), *189*
Scholz, R., 28(226), *130*
Schonberg, A., 8, *14,* 31(206), 39, 40, 50, 51, 94, 109, 114, 115, *126, 129, 130,* 244, *247,* 271, *283,* 326, *364*
Schreiber, K. C., 10, *14,* 251, *281,* 307, *363*
Schroll, A., 255(11), 256(14), 274(82), *281, 283*
Schulz, G., 357(128), *365*
Schulz, H., 27(74, 83), 44, 45, 46, 71, 98(72, 74), 99, 101(193, 194), 102, 103(83), 104, *127, 130*
Schumacher, O., 53, *127*
Schunn, R. A., 211, *216*
Schupp, O. E., 207(38), *216*
Schwartz, A., 208, 209(40), *216*
Schwarzenbach, K., 36(47), 108(47), *126*
Schweizer, E. E., 11, *15,* 24(52), 25(51, 240), 37, *126, 131,* 151(73), 152, *191*
Searle, H. T., 231(56), 238(56), *246*
Searle, R. J. G., 28(90), 48(90), 104 (90), *127,* 262(27), *282*
Seebach, F., 267, *282*
Seelert, K., 348(93a, 94), 361, *364, 366*
Seng, F., 27(67), 41, 100, 101(194), 114, 119, *127, 130,* 221, 229, *245*
Senning, A., 358, *365*
Seus, E. J., 207(38), *216*
Seyferth, D., 24(50, 54, 127), 26(32, 124), 33, 35, 37, 38, 53, 59, 60, 61, 106, 108(204), 123, *126, 127, 128, 130,* 194, *215,* 289, 292, 293, *303*
Shemyakin, M. M., 23(28), 24(134), 25 (123), 32(28), 47(28), 58(123), 64(134, 135), 103(28), *126, 128,* 136(23), 168, 169, 185, 186(86, 88), 187(88, 94, 95), *190, 191,* 207, *216*
Shevchenko, V. I., 232, 233, 234(60), *246*
Siegwart, J., 271, *283*
Siemiatycki, M. S., 26(209, 241), 100, 110(209), *130, 131*

Silverstein, R. M., 33(35), *126*
Simon, H., 72(149), *129*, 148(63), 149 (63), *191*, 280(99), *283*
Singer, E., 31(206), 94(180), 109(206), *129*, *130*, 271(64), *283*
Singh, G., 219, 224(36), 225(10), 241, *245*, *247*, 362, *366*
Sisler, H. H., 218, 222(5), 223, *245*
Smith, C. P., 40(62, 63), *126*, 162(82), *191*, 212(44), *216*
Smith, L. C., 19, 77, *125*
Smith, L. S., 96(187, 188), 97(188), *129*, *130*
Smith, N. L., 218(5, 8), 222(5), 223 (8), *245*
Smith, S. G., 309, 345, *363*
Smyth, C. P., 318, *363*
Sneezum, S., 264, 265(39), *282*
Sommelet, M., 274, *283*, 320, *363*
Sommer, N., 25(87), 47, 103(87, 196), *127*, *130*
Sommerfeld, D., 358(131), *365*
Sondheimer, F., 24(191), 100, 118, 119 (191), *130*, 141, *190*, 326, *364*
Song, J., 96(188), 97(188), *130*
Songstad, J., 91, *129*
Speziale, A. J., 25(94), 26(140), 27(94), 28(31, 94, 140), 29(31), 33, 49 (94), 67, 68, 76, 77, 78, 79, 100, 103, 124, *126*, *127*, *128*, *131*, 144, 154(49), 155, 156, 157, 160, 161, 165, 169, 170, 171, 173, 176, 177, 180, 189, *190*, *191*, *192*, 197(12), *216*, 228, *245*, 315(70, 72), 335, 338, *364*
Splitter, J. S., 338(75), *364*
Spooncer, W. W., 259, 264, *281*
Staab, H. A., 25(87), 47, 103(96), *127*, *130*
Staudinger, H., 3, *4*, 8, 9, *14*, 28(42), 35, 48, 63(42), 109, 115, *126*, *127*, 132, 140(8), *190*, 219, 221, 222 (26), 223, 226(26), 228, 230, 238, 239(69, 75), 240(69, 75), 241, 242 (69), 245, 246, 247, 271, *283*, 353, *365*
Stephens, F. S., 124, *131*
Stermitz, F. R., 340, *364*
Stevens, I. D. R., 27(70), 43(70), 121 (70), *127*, 229(45, 46), 244(46), *246*
Stevens, T. S., 272, 273(75), 274(72), *283*, 342, *364*
Stewart, R., 169, *191*
Stoessl, A., 353(116), *365*
Stoicheff, B. P., 265, *282*
Stork, G., 353, *365*
Streitwieser, A., 64, 65, *128*
Strzelecka, H., 26(209), 100, 110, *130*
Stuckwisch, C. G., 307, *362*
Sturtz, G., 205, *216*
Sugasawa, S., 134, 138, *190*
Sugita, T., 112(214), *130*, 210(41), *216*
Suld, G., 354, *365*
Surmatis, J. D., 30(220), 116, *130*, 147, 171, *191*
Sutton, L. E., 13(31), *15*, 75(152), *129*, 236(65), *246*, 292(29), *303*, 305 (5), 308(5), 316(41), 318(41), 355(41), *362*, *363*
Swain, C. G., 320, 324, *363*
Sweeney, W., 230(48), *246*
Swensen, W. E., 19(14), 25(14), 57 (14), 77(14), *125*
Swor, R. A., 63(131), 75(131), *128*
Szego, F., 219(16), 226(16), *245*

T

Takahashi, H., 205, *216*
Takaku, M., 315(39), 318(39), 321(39), 333(39), 361(39), *363*
Takashina, N., 40, *127*
Tanimoto, S., 25(66), 41(66), *127*
Tarbell, D. S., 357, *365*
Templeton, J. F., 28(222), 121(222), 124(222, 257), *130*, *131*
Tenhaeff, H., 275(85), *283*
Teuffer, G., 286, *303*
Thayer, J. S., 219(14), 220, *245*
Theilacker, W., 81, *129*
Thomson, T., 273, 274(77), *283*, 342, *364*
Thornton, E. R., 320, 324, *363*
Tochtermann, W., 256, *281*
Tomasi, R. A., 26(234), 31(234), *131*, 139, *190*
Tomoskozi, I., 27(198), 106, *130*, 134, *190*, 202, 206, 210, *216*

Toscano, V. G., 194(5), 202(15, 18), 206(5), 212(5), *215, 216*
Traynelis, V. J., 348, *364*
Trepka, R. D., 81(158), *129*
Trippett, S., 18, 20(16), 21, 22, 25(8, 12, 86), 26(8, 12), 27(21, 82), 28 (8, 21, 68, 82, 85, 86), 41, 42, 46, 47, 48, 53, 55, 76, 93, 94, 103, 105, 111, 112(102), 113, 115, 119, 120 (68), *125, 126, 127, 129,* 132, 137, 145(4, 6), 147, 159, 161, 163(80), 164, 171(4, 6), 172, 173, 181, 185, 188, 189, *190, 191, 192,* 197(13), 202, 206, *216,* 230, 231, 238(57), *246*
Trischmann, H., 345, *364*
Trotter, J., 80, 82, *129,* 229(45), *246*
Truce, W. E., 338, 352(104), 353(109), 354, *364, 365*
Trumbull, E. R., 277, 278(89), *283,* 297 (32), *303*
Trumbull, P. A., 297(32), *303*
Tsuno, Y., 220(27), 221(27), *245*
Tung, C. C., 315(72), 338(72), *364*

U

Ullmann, H., 219(19), *245*
Ulrich, H., 232, *246*
Ulrichs, K., 24(244), *131*

V

Valega, T. M., 97, *130*
Vandenbelt, J. M., 307, *362*
VanderWerf, C. A., 83(164), 84(164), 89, 90, 92(164), 93(174), 111 (212), 112(212), 115(217), *129, 130,* 173(89), 174(89), *191,* 331, *364*
Vandi, A., 232, *246*
Van Dormael, A., 70(145), *128*
Van Wazer, J. R., 82, *129*
Vaver, V. A., 23(28), 25(123), 32(28), 47(28), 58(123), 64(135), 103 (28), *126, 128,* 136(23), 185(93), 187(95), *190, 191*
Vetter, H. J., 219(22), *245*
Viehe, H. G., 24(243), *131*
Vill, J. J., 25(210), 111(210), *130,* 202 (19), 210(19), *216*
Virkhaus, R., 55, *128*
Vogt, F., 224(35), 225, *245*
von Eicken, S., 268(57), *282*

W

Wadsworth, D. H., 207, *216*
Wadsworth, W. S., 205, 206, 210, *216,* 235, *246*
Wagner, D., 299, 300, *303*
Walborsky, H. M., 112(214), *130,* 210 (41), *216*
Walker, D. M., 18, 21, 25(8, 12, 86), 26 (8, 12, 21), 27(21), 28(8, 21, 85, 86), 42, 46(8), 47, 48, 55, 76, 94, 103(8, 86), 104, 115, *125, 126, 127,* 137, 159, 188, *190, 191, 192,* 206, *216,* 230, *246*
Wall, M. E., 361(147), *366*
Walling, C., 309, 349, *363*
Wang, T. W., 287(12), 296(12), *303*
Wawzonek, S., 266, 274, *282*
Weaver, C., 357, *365*
Webster, B. C., 13, *15,* 306, *362*
Webster, O. W., 265(46), *282,* 360 (141), *366*
Wedekind, E., 352, *365*
Wegner, E., 81, *129*
Wehr, R., 360, *365*
Weigmann, H. D., 18(11), 24(11), *125,* 139(35), 144(35), 145(35), 154 (35), 156(35), 159(35), 160(35), 164(35), 167(35), *190*
Weil, T., 239(74), *246*
Weinstock, J., 277, *283*
Weis, W., 262(28), *282*
Welch, D. E., 194, *215*
Welcher, R. P., 57, *128*
Wells, A. F., 9(17), *14,* 284, *303*
Wepster, B. M., 75, *129*
Wessely, F., 360, *365*
West, R., 219(14), 220, *245*
Westheimer, F. H., 84, *129,* 173, *191*
Wetroff, G., 9(15), *14*
Wetterling, M. H., 3(6), *4,* 252, 253, 254(6), 262(6), *281*
Weygand, F., 255, 256, 274, 280, *281, 283,* 340, *364*
Wheatley, P. J., 10, *14,* 286, *303*
White, F. G., 343, *364*
Whitehead, J. K., 359, *365*
Whitehead, M. A., 87(172), *129*

Whiting, M. C., 20(17), *125*
Whitlock, H. H., Jr., 206, *216*
Wiberg, N., 219(13), *245*
Wiegrasse, W., 219(17), 221, 231, 232, *245*, *246*
Wieland, J. H. S., 18(9), 21(9), 30(9), 76(9), *125*
Wilhelm, K., 125, *131*
Wilkins, R. G., 287, 296(14), *303*
Williams, A. A., 10, *14*
Williams, J. L., 271(61), *282*, 313(35), 315(35), 321(35), 325(35), 333(35), 334(35), *363*
Wilson, C. V., 21, 30(20), *125*
Wineman, R. J., 63(132), *128*
Winkler, H., 124, *131*, 173, *191*, 196, 198, *216*, 223, 224, *245*
Winstein, S., 309, 345, *363*
Wippel, H. G., 32(98), 51(98), 69(98), *127*, 136(20), *190*, 193(1), 194(1), 195(7), 202(7), 205(1), *215*
Witschard, G., 23, 24(29, 236), 25(236), *126*, *131*, 137, 144, 149, 151(69), *190*, *191*
Wittig, G., 2(2), 3(7, 8), *4*, 8, 9, 10, *14*, 16, 17, 18, 20, 22, 24(1, 3, 4, 11, 48), 26(1, 23, 33, 242), 28(23), 30(15), 31(26), 33, 34, 36(47), 46, 53, 57(3, 15, 46), 59, 60, 62, 64, 71, 74, 88, 89(1), 98, 99, 103, 108(47), 110, *125*, *126*, *128*, *131*, 132(7), 133, 134(12), 135, 136, 137, 138, 140(12, 38), 143, 144, 145(1), 147, 148, 149, 152, 153, 154(35), 155(12), 156, 158(12), 159, 160, 161, 164, 165, 167, 168, 171(1), *189*, *190*, *191*, 195, *215*, 239(72), 240(72), *246*, *247*, 252, 253, 254(6), 255, 256, 258, 262(6, 12), 266, 272, 274(18, 19), 275, 278, 279, *281*, *283*, 285, 286, 288, 289, 290, 291, 294, 295(25), 296, 297, 298(17), 300, 301, 302, *303*, 311, 315(27), 316(40), 317(40), 324, 325(55, 56), 327, 350, *363*, *364*, *365*
Wittwer, C., 309(23, 24), *363*
Wolf, A. P., 26(211), 111(211), 112(211), *130*, 167, *191*, 202, 210, *216*
Wolff, H. A., 220, 227(25), *245*
Worrall, D. E., 8, *14*, 25(229), *130*
Wotiz, J., 287(12), 296(12), *303*

Y

Yanovskaya, L. A., 132, 142, 145(5), 171(5), *190*, 205(29), *216*
Yao, A., 315(72), 338(72), *364*
Yates, K., 169, *191*
Yeakey, E., 266, 274, *282*
Yudina, K. S., 11(25), *14*

Z

Zanger, M., 83(164), 84(164), 89(174), 90(174), 92(164), 93(174), *129*, 173(89), 174(89), *191*, 331(63), 360(142), *364*, *366*
Zbiral, E., 111, 112(213), 113, 122, *130*
Zecher, W., 262(28), *282*
Zeller, P., 24(238), *131*
Zhdanov, S. I., 265(44), *282*
Zhmurova, I. N., 231(54), 233(54), 234(63), 238(54), *246*
Zhuzhlikova, S. T., 27(201), 28(205), 107(201), *130*
Zienty, F. B., 40(59), *126*
Zimmer, H., 206, *216*, 219, 224(36), 225(10), 241, *245*, *247*, 362, *366*
Zincke, T., 352, *365*
Zollinger, H., 309, *363*

SUBJECT INDEX

A

Acenaphthylene, with sulfonium ylids, 326
Acylation
  of iminoarsenanes, 300
  of iminophosphoranes, 225
  of methylenedimethyloxysulfurane, 348, 361
  of methylsulfinyl carbanion, 351
  of phosphine oxides, 202
  of phosphiteimines, 234
  of phosphonium ylids, 45, 102
    intramolecular, 45, 102
  of pyridinium ylids, 262
  of sulfonium ylids, 361
  of sulfonyl carbanions, 352
  of trimethylammoniummethylide, 256
C-Acylation, of phosphonium ylids, 104
O-Acylation, of phosphonium ylids, 104
N-Acyliminophosphoranes, pyrolysis, 230
N-Acyliminotrihalophosphoranes, pyrolysis, 233
Alkylation
  of iminophosphoranes, 225
  of phenyl sulfide, 312
  of phosphinazines, 241
  of phosphiteimines, 234
  of phosphonium ylids, 43, 98
    intramolecular, 23
  of pyridinium ylids, 262
  of sulfonyl carbanions, 352
  of tertiary phosphines, 52
  of trimethylammoniumfluorenylide, 258
  of trimethylammoniummethylide, 256
  of triphenylarsine, 289
  of triphenylphosphine, 52
O-Alkylation, of ketoammonium imines, 267
  of ketophosphonium ylids, 69, 99
  of phosphinite carbanions, 213
  of phosphitemethylenes, 214
Alkynes
  from ketophosphonium ylids, 105
  with phosphinazines, 244
Allylic halides, alkylation of phosphines with, 53
Amides, dehydration of, 233
Ammonium imines, 266
  thermal rearrangement, 267
Ammonium ylids, 253–260
  with benzophenone, 255
  bis-ylids, 256
  decomposition to carbenes, 254
  intermediates in eliminations, 278
  rearrangement of, Sommelet, 275
    Steven's, 259
Antimony ylids, *see* Stibonium ylids
Arsonium imines, *see* Iminoarsenanes
Arsonium ylids, 288–299
  hydrolysis, 292
  properties, 291
  reactivity, 296
  synthesis, 288
  with carbonyl compounds, 294
  with electrophiles, 293
  with nitrosobenzene, 299
Autoxidation, of phosphonium ylids, 94
Azides, with tertiary phosphines, 219

B

Basicity
  phosphonium ylids, 65
  sulfonium ylids, 316
Benzhydrylidenetriphenylphosphorane, 8
  with carbonyls, 132
Benzonitrile, with phosphonium ylids, 115
p-Benzoquinone, addition of triphenylphosphine, 39
Benzylidenediphenylsulfurane, with carbonyl compounds, 333
Benzylidenetriphenylphosphorane, with benzylideneaniline, 221
Benzyne
  with methyl sulfide, 313
  with phosphines, 38
  with phosphonium ylids, 122
Betaine
  decomposition, 166, 296

direct interconversion, 172, 189
dissociation, 164, 189
interconversion, 172, 173, 189
isolation, 160
rate of formation, 163
rearrangement of, 188
reversible formation, 164, 189
stereochemistry, 172, 335
structure, 162
Bismuth ylids, 288
Bis-ylids, 20, 145
Bond dissociation energy, in iminophosphoranes, 237
Bond order, in phosphonium ylids, 125
Boranes, with phosphonium ylids, 106
Butylidenediphenylsulfurane, with carbonyl compounds, 333

C

Carbanion
hybridization, 80
stabilization, 9, 199, 285, 306, 308, 317, 319
electrostatic, 251
Carbenes
from ammonium ylids, 254, 258
from diazo compounds, 272
from phosphonium ylids, 21
from sulfonium ylids, 322, 324
with diazo compounds, 271
with phosphonium ylids, 107
with triphenylphosphine, 33
Carbenoids, 34, *see also* Carbenes
Carbodiimides, from isocyanates, 227, 269, 300, 302
Carbon disulfide, with phosphonium ylids, 115
Carbonyl compounds
with ammonium ylids, 253, 256
with arsonium ylids, 294
with diazo compounds, 270
with iminophosphoranes, 226
with methylenedimethyloxysulfurane, 346
with phosphinazines, 242
with phosphinite carbanions, 212
with phosphinoxy carbanions, 194
with phosphonate carbanions, 204
with phosphonium ylids, 88, 132–192
with phosphoramidate anions, 235
with stibonium ylids, 301
with sulfinyl carbanions, 349
with sulfonium ylids, 329
with sulfonyl carbanions, 351
Chloramine, with triphenylphosphine, 218
Chloramine-T
conversion to nitrene, 356, 359
with *N,N*-dimethylaniline, 266
with sulfides, 356
with triphenylarsine, 299
with triphenylphosphine, 217
with triphenylstibine, 302
Cinnamylidenetriphenylphosphorane, 71
Cuprous chloride, catalyst for phosphinazine decomposition, 36, 240
Cyclic phosphonium ylids, 84
Cyclic sulfonium ylids, 354
Cyclopentadienylidenedimethylsulfurane, 361
Cyclopentadienylidenetriphenylphosphorane, 70
Cyclopropanes
from diazo compounds, 272
from oxiranes, 111, 202, 209
from phosphonium ylids, 111

D

Deuterium exchange
ammonium salts, 12, 251
arsonium salts, 287
phosphonium salts, 11
stibonium salts, 287
sulfonium salts, 307
Diazo carbonyl compounds, with phosphonium ylids, 110
Diazo compounds, 270–273
decomposition to carbenes, 272
with carbenes, 271
with carbonyl compounds, 270
with nitrosobenzene, 271
with triphenylphosphine, 35
Diazocyclopentadiene, 35
Diazofluorene, 35, 270
Diazonium salts, with phosphonium ylids, 109
Dichlorocarbene, with fluorenylidenedimethylsulfurane, 323, 329
Dicyanomethylenedimethylsulfurane, 360
Diphenylcarbene, 35

Diphenyldiazomethane, 35
1,1-Diphenyl-l-phosphabenzene, 85
9,9-Diphenyl-9-phosphaphenanthrene, 86
Dipole moment
  fluorenylidenetriphenylarsenane, 292
  iminophosphoranes, 236
  phosphonium ylids, 63
    effect of phosphorus substituent, 74
  pyridinium ylids, 264
  sulfonium ylids, 316, 318, 360
  thiabenzenes, 355
d-Orbitals
  contraction of, 13, 305
  in multiple bonding, 10
  in sigma bonding, 9
d-Orbital overlap
  experimental evidence, 9, 306
  theoretical arguments, 13, 304
Double bond character
  of phosphonium ylids, 125
  of sulfonium ylids, 318
  of ylid bond, 76

E

α-Elimination, of sulfonium ylids, 324
α',β-Elimination
  of ammonium salts, 278
  of sulfonium salts, 340
Epoxides
  from arsonium ylids, 296, 297
  from diazo compounds, 270
  from sulfonium ylids, stereochemistry, 334
  from sulfur ylids, 329, 332, 346, 361
  with phosphinoxy carbanions, 202
  with phosphonate carbanions, 209
  with phosphonium ylids, 111
  with tri-*n*-butylphosphine, 183
  with triphenylphosphine, 178
Ethyl diphenylphosphinite, 40, 212

F

Fluorenylidenedimethylsulfurane, 310, 315
  with carbonyl compounds, 329
Fluorenylidenetri-*n*-butylphosphorane, 74
  basicity, 18
Fluorenylidenetrimethylarsenane, 289
Fluorenylidenetrimethylphosphorane, 74
  Wittig reaction of, 18
Fluorenylidenetriphenylarsenane, 290
Fluorenylidenetriphenylphosphorane, 8, 74
  basicity, 18
Fragmentation
  of phosphorus ylids, 20
  of sulfur ylids, 321

H

Halogenation
  of iminophosphoranes, 225
  of phosphonium salts, 49
  of phosphonium ylids, 49, 98
Halophosphinimines, 230
Hartree-Fock SCF Method, sulfur orbitals, 13, 306
Hauser rearrangement, *see* Sommelet rearrangement
Hofmann elimination
  of ammonium salts, 277
  of phosphonium salts, 100
  of sulfonium salts, 340
Hybridization
  in iminophosphoranes, 236
  in phosphonium ylids, 80–84
Hydrolysis
  phosphonium salts, 90
  phosphonium ylids, 88
Hydroxylamine-*O*-sulfonic acid, with triphenylphosphine, 218

I

Imidooxysulfuranes, 359
Imidosulfuranes, 356
Imines
  from iminophosphoranes, 226
  from phosphonium ylids, 114
  from phosphoramidate anions, 235
Iminoarsenanes, 299
Iminooxysulfuranes, 359
Iminophosphoranes, 217–247
  acylation, 225
  alkylation, 225
  basicity, 222
  bond dissociation energy, 237
  dimerization, 231, 238
  dipole moments, 236
  halogenation, 225
  hybridization, 236
  hydrolysis, 223

molecular structure, 236
preparation, 217
with alkynes, 229
with carbonyl compounds, 226
with Lewis acids, 224
with nitrosyl chloride, 224
Iminosulfuranes, 356, 358
optically active, 362
Iminotrialkoxyphosphoranes, 233
Iminotrihalophosphoranes, 230
Iminotriphenylarsenane, 299
Iminotriphenylphosphorane, 218
Isocyanates, to carbodiimides, 227, 269, 300, 302
Isoquinolinium ylids, 260

K

Ketoammonium imines
o-alkylation, 267
pyrolysis, 267
Keto-oxysulfuranes, photocleavage, 348
Ketophosphinimines, pyrolysis, 230
Ketophosphonium ylids
acylation, 104
alkylation, 99
o-alkylation, 69
enolate structure, 69
preparation, 45
pyrolysis to alkynes, 105
Ketosulfonium ylids, acylation, 361
Knoevenagel reaction, of pyridinium salts, 262

L

Ligand exchange, in phosphonium salts, 60

M

Metal complexes, *d*-orbital π bonding, 10
Metal halides, with phosphonium ylids, 107
Methylation, with methylenedimethyloxysulfurane, 348
Methylene, *see* Carbene
Methylenedimethyloxysulfurane, 345
acylation, 348, 361
with carbonyl compounds, 346
with conjugated carbonyls, 346
Methylenedimethylsulfurane, 308
with carbonyl compounds, 332
Methylenediphenyloxyphosphorane, 193
Methylene transfer by sulfonium ylids, 339
Methylenetrimethylphosphorane, 74
with carbonyls, 133
Methylenetriphenylarsenane, 289
with benzophenone, 296
Methylenetriphenylphosphorane, 8, 74
with carbonyls, 133
Methylsulfinyl carbanion
acylation, 351
with carbonyl compounds, 349

N

Nitrenes
with sulfides, 356
with sulfoxides, 359
Nitrogen imines, 266
Nitrogen ylids, 251–283
as rearrangement intermediates, 273
Nitrones
from arsonium ylids, 299
from diazo compounds, 271
from pyridinium ylids, 263
from sulfonium ylids, 338
from sulfonyl ylids, 352
Nitrosobenzene
with arsonium ylids, 299
with diazo compounds, 271
with phosphinazines, 244
with phosphonium ylids, 114
with pyridinium ylids, 263
with sulfonium ylids, 338
with sulfonyl ylids, 352
Non-stabilized ylids, 154

O

Olefins
from arsonium ylids, 294, 295
from phosphinoxy carbanions, 194
from phosphonate carbanions, 205
from phosphonium ylids, 94, 132
synthesis via Wittig reaction, 145
Ortho-substitution rearrangement, *see* Sommelet rearrangement
Oxidation
iminosulfuranes, 359
phosphinoxy carbanions, 202
phosphonium ylids, 94
Oxiranes, *see* Epoxides

Oxysulfidimides, *see* Imidooxysulfuranes
Oxysulfidimines, *see* Iminooxysulfuranes
Oxysulfonium ylids, *see* Oxysulfuranes
Oxysulfuranes, 345–359

P

Pentaphenylantimony, 285
Pentaphenylarsenane, 285
Pentaphenylphosphorane, 10
Pentavalent antimony, 285
Pentavalent arsenic, 285
Pentavalent nitrogen, 252
Pentavalent phosphorus, 9
Perkow reaction, 55
Phenacylidenedimethylsulfurane, 334, 361
Phenacylidenemethylphenylsulfurane, 315, 317, 333
Phenacylidenetriphenylarsenane, 289
Phenacylidenetriphenylphosphorane, 68, 69
Phenyl azide, with phosphonium ylids, 109
Phenyldiazomethane, 36
Phenyl isocyanate
  with amine oxides, 219
  with arsine oxides, 300
  with phosphine oxides, 227
  with phosphonium ylids, 48, 115
  with stibine oxides, 302
  with sulfoxides, 357
Phosphabenzenes, 85
Phosphinate carbanions, 212
Phosphinazines, 238–244
  alkylation, 241
  catalyzed decomposition, 240
  hydrolysis, 242
  preparation, 239
  pyrolytic decomposition, 35, 239
  with alkynes, 244
  with carbonyl compounds, 242
  with nitrosobenzene, 244
Phosphine dihalides
  with active methylenes, 50
  with amines, 218
Phosphine oxides
  acylation, 202
  conversion to phosphinoxy carbanions, 195
Phosphinimines, *see* Iminophosphoranes
Phosphinite carbanions, 212
Phosphinoxy carbanions, 193–203
  acylation, 202
  preparation, 193
  reactivity, 199
  with carbonyls, 194
  with oxiranes, 202
  with oxygen, 202
Phosphiteimines, *see* Iminotrialkoxyphosphoranes
Phosphitemethylenes, 213
Phosphonate carbanions, 203–212
  nucleophilicity, 206
  with carbonyl compounds, 204
  with oxiranes, 209
Phosphonium carbanions, *see* Phosphonium ylids
Phosphonium group
  conformation in ylids, 82
  hybridization, 82
Phosphonium salts
  conversion to phosphonium ylids, 56
  from tertiary phosphines, 52–56
  halogenation, 49
  hydrolysis, 89–92
  ligand exchange in, 60
  preparation, 52
  reduction, 93
Phosphonium ylids, 16–131
  acylation, 45–48, 102–104
  alkylation, 43–45, 98–102
  aromatic, 84
  chemical structure, 62
  cyclic, 84–87
  definition of, 16
  dipole moments, 63
  effect of carbanion substituents, 19, 63
    of phosphorus substituents, 17, 73
  from phosphonium salts, 56
  halogenation, 49, 97
  hybridization about carbon, 80
    about phosphorus, 82
  hydrolysis, 88
  methods of synthesis, 29–61
  molecular structure, 80, 124
  nomenclature, 3
  optical activity, 83
  oxidation, 94
  physical properties, 61–87

p-d π-overlap, 82
preparation of, 17–61
reactions, 88–123
reduction, 92
tables of, 24–32
with acyl halides, 46
with benzyne, 122
with conjugated alkenes, 41, 116
with conjugated alkynes, 42, 120
with conjugated ketones, 117
with electrophiles, 106
with epoxides, 111
with esters, 47
with phenylisocyanate, 48, 115
with thioesters, 46
Wittig reaction, 132
Phosphoramidate anions, 235
with carbonyl compounds, 235
Phosphorane, 4
Photochromic effect, 63
Photocleavage, of keto-oxysulfuranes, 348
pK_a
arsonium salts, 291
fluorenylidenedimethylsulfurane, 316
phosphonium salts, 9, 66, 74
pyridiniumcyclopentadienylide, 264
P^{31} NMR chemical shifts
in iminophosphoranes, 237
in phosphonium salts, 76
in phosphonium ylids, 76
"PO-activated" olefin formation, 200
Polymethylene
from ammonium ylids, 254
from sulfonium ylids, 327
$p\pi$-$d\pi$ Overlap
in arsonium and stibonium ylids, 287
in cyclic phosphonium ylids, 84
in phosphonium ylids, 82
in sulfonium ylids, 305, 308
theoretical treatment, 13
Pyridiniumcyclopentadienylide, 264
Pyridiniumdicyanomethylide, 265
9-Pyridiniumfluorenylide, 263
Pyridinium imines, 268
Pyridinium ylids, 260–266
acylation, 262
alkylation, 262
with carbonyl compounds, 262
with nitrosobenzene, 263

Q

Quinolinium ylids, 260
Quino-ylids, 50

R

Reduction
phosphonium salts, 93
phosphonium ylids, 92

S

Salt method of ylid synthesis, 52, 289, 310
Schiff bases, *see* Imines
Sommelet rearrangement
of ammonium ylids, 274
of sulfonium ylids, 320, 342
Stabilized ylid, 154
Steven's rearrangement, 273
of ammonium imines, 267
of ammonium ylids, 259, 273
of arsonium ylids, 290
of imidosulfuranes, 357
of stibonium ylids, 301
of sulfonium ylids, 342
Stibonium imines, 302
Stibonium ylids, 301
Sulfenes, 352
Sulfidimides, *see* Imidosulfuranes
Sulfidimines, *see* Iminosulfuranes
N-Sulfinylaniline, *see* Thionylaniline
Sulfinyl carbanions, 349
Sulfones, 309, 351
Sulfonium ylids, 310–344
acylation, 361
aromatic, 354
as reaction intermediates, 339
chemical stability, 319
cyclic, 354
decomposition, 322, 327
fragmentation, 321
hydrolysis, 319
physical properties, 314
reactions, 328–344
Sommelet rearrangement of, 320, 342
Steven's rearrangement of, 342
synthesis, 310
tables of, 315
thermal instability, 321
with carbonyl compounds, 329, 361

with conjugated carbonyls, 336
with nitrosobenzene, 338
Sulfonyl carbanions, 351
with nitrosobenzene, 352
Sulfoxides, 309, 349
Sulfur, with phosphonium ylids, 109
Sulfurdiimines, 361

T

Tertiary phosphines, *see also* Triphenylphosphine
alkylation of, 52
with azides, 219
with epoxides, 161, 178
Tetracyanoethylene oxide
with methyl sulfide, 360
with pyridine, 265
Tetramethylammonium salts
acidity, 253
deuterium exchange, 12
Tetramethylarsonium salts, deuterium exchange, 287
Tetramethylphosphonium salts, deuterium exchange, 11
Tetravalent sulfur, 305
Thiabenzenes, 354
Thiamine, 343
Thiazolium carbanion, 344
Thionylaniline, 358, 362
Trans effect, 287
Transylidation, 44, 65
Trialkoxymethylenephosphoranes, 213
Tribenzoylcyclopropane, 100, 361
Tri-*n*-butylphosphine
with dibenzoylethylene, 40
with epoxides, 183
Trimethylammoniumcyclopentadienylide, 259
Trimethylammoniumdicyanomethylide, 260
Trimethylammoniumfluorenylide, 258
Trimethylammoniummethylide
acylation, 256
alkylation, 256
complexing with LiBr, 254
preparation, 252
Trimethylphosphite, with dibenzoylethylene, 40, 214
Trimethylsulfonium salts, deuterium exchange, 307
Triphenylarsine, alkylation, 289
Triphenylarsine dichloride, with active methylene compounds, 290
Triphenylphosphine
addition to α,β-unsaturated carbonyls, 39–41, 50
alkylation of, 52
with azides, 219
with carbenes, 33
with chloramine, 218
with chloramine-T, 217
with diazo compounds, 35, 239
with epoxides, 161, 178
with *N*-haloiminotriphenylphosphorane, 221
with hydroxylamine-*O*-sulfonic acid, 218
Triphenylphosphine dibromide, with arylamines, 218
Triphenylstiboniummethylide, 301
with benzophenone, 301

U

Ultraviolet spectra, phosphonium ylids, 63

V

Valence shell expansion
antimony, 286
arsenic, 286, 291, 296
group VA atoms, 287
phosphinoxy group, 212
phosphorus, 11
sulfur, 319
Vinyldimethylsulfonium bromide, nucleophilic addition to, 251, 307
Vinylphosphonium salts, nucleophilic addition to, 37
Vinylphosphorus compounds, nucleophilic addition to, 11, 37
Vinyltrimethylammonium bromide, nucleophilic addition to, 10, 251

W

Wittig reaction, 8, 132–192
abnormal, 142
alternate mechanisms, 168
applications, 145
Bergelson mechanism, 168

betaine, decomposition, 166
  dissociation, 164, 189
  formation, 160
catalysis, 134
effect of carbonyl structure, 140, 156
  of ylid structure, 138, 157
experimental conditions, 134, 184
intramolecular, 151
kinetics, 155, 179
mechanism, 152–171, 189
rearrangement during, 188
scope, 138
solvent effect, 155, 185
stereochemistry, 171–187
  effect of additives, 135, 186
    of carbanion substituents, 174
    of phosphorus substituents, 175, 188
    of reaction conditions, 184
  with non-stabilized ylids, 181
  with stabilized ylids, 177

X

X-Ray analysis
  phosphonium ylids, 80, 124
  pyridiniumdicyanomethylide, 265

Y

Ylene, 2
Ylid, *see also* individual ylids
  definition of, 1
  derivation of, 2
  nomenclature of, 3
Ylide, *see also* Ylid, 2
Ylids
  ammonium, 253–260
  antimony, 301–302
  arsenic, 284–301
  arsonium, 288–299
  nitrogen, 251–283
  phosphonium, 16–131
  phosphorus, 7–216
  sulfonium, 310–344
  sulfur, 304–366